Molecular Liquids

Dynamics and Interactions

NATO ASI Series

Advanced Science Institutes Series

A series presenting the results of activities sponsored by the NATO Science Committee, which aims at the dissemination of advanced scientific and technological knowledge, with a view to strengthening links between scientific communities.

The series is published by an international board of publishers in conjunction with the NATO Scientific Affairs Division

A	Life Sciences	Plenum Publishing Corporation
B	Physics	London and New York
C	Mathematical and Physical Sciences	D. Reidel Publishing Company Dordrecht, Boston and Lancaster
D	Behavioural and Social Sciences	Martinus Nijhoff Publishers
E	Engineering and Materials Sciences	The Hague, Boston and Lancaster
F	Computer and Systems Sciences	Springer-Verlag
G	Ecological Sciences	Berlin, Heidelberg, New York and Tokyo

Series C: Mathematical and Physical Sciences Vol. 135

Molecular Liquids

Dynamics and Interactions

edited by

A. J. Barnes and W. J. Orville-Thomas

Department of Chemistry and Applied Chemistry,
University of Salford, Salford, U.K.

and

J. Yarwood

Department of Chemistry,
University of Durham, Durham, U.K.

Springer-Science+Business Media, B.V.

Proceedings of the NATO Advanced Study Institute on
Molecular Liquids – Dynamics and Interactions
Florence, Italy
June 26-July 8, 1983

Library of Congress Cataloging in Publication Data

Main entry under title:

Molecular liquids. Dynamics and interactions.

(NATO ASI series. Series C, Mathematical and physical sciences; v. 135)
"Proceedings of the NATO Advanced Study Institute on Molecular Liquids - Dynamics and Interactions, Florence, Italy, June 26–July 8, 1983"–P.
"Published in cooperation with NATO Scientific Affairs Division."
Bibliography: p.
Includes index.
1. Liquids–Congresses. 2. Molecular theory–Congresses. 3. Molecular dynamics–Congresses. I. Barnes, A. J. (Austin J.) II. Orville-Thomas, W. J., 1921- III. Yarwood, J., 1939- . IV. NATO Advanced Study Institute on Molecular Liquids–Dynamics and Interactions (1983: Florence, Italy) V. North Atlantic Treaty Organization. Scientific Affairs Division. VI. Series.
QC138.M65 1984 530.4'2 84–15039
ISBN 978-94-009-6465-5 ISBN 978-94-009-6463-1 (eBook)
DOI10.1007/ 978-94-009-6463-1

Published by D. Reidel Publishing Company
P.O. Box 17, 3300 AA Dordrecht, Holland

Sold and distributed in the U.S.A. and Canada
by Kluwer Academic Publishers,
190 Old Derby Street, Hingham, MA 02043, U.S.A.

In all other countries, sold and distributed
by Kluwer Academic Publishers Group,
P.O. Box 322, 3300 AH Dordrecht, Holland

D. Reidel Publishing Company is a member of the Kluwer Academic Publishers Group

CONTENTS

PREFACE

This ASI was planned to make a major contribution to the teaching of the principles and methods used in liquid phase research and to encourage the setting up of collaborative projects, as advocated by the European Molecular Liquids Group (secretary: Dr J. Yarwood, University of Durham, U.K.).

During the past five years considerable progress has been made in studying molecular liquids. The undoubted advantages of international collaboration led to the formation of the European Molecular Liquids Group (EMLG) in July 1981. The activities of the EMLG were widely disseminated in a special session of the European Congress on Molecular Spectroscopy (EUCMOS) held in September 1981 (for details, see J. Mol. Structure, 80 (1982) 375 - 421). Following the success of this meeting, it was thought that the aims and objectives of the EMLG would be best served by the organisation of a broader-based gathering designed to attract those interested in the study of the structure, dynamics and interactions in the liquid state. Thanks to the generous support by the Scientific Affairs Division of NATO, it was possible to hold a NATO ASI on Molecular Liquids at the Italian Centre of Stanford University, Florence, Italy during June-July 1983. This book is based on the lectures presented at that meeting.

The contents of this volume cover the three broad areas of current liquid phase research:

(a) Analytical theory. A description is presented of the methods used to produce a theory of molecular diffusion, which describes the ever increasing range of experimental information available. A great deal of emphasis is placed on the development of expressions for dynamic and interaction processes in terms of various observed quantities.

(b) Experimental investigations. These chapters describe how the dynamics and interactions of molecules in liquids are measured using a wide range of spectroscopic and diffraction techniques. These include recently developed non-linear laser spectroscopy and neutron diffraction techniques.

(c) Computer (Molecular Dynamics) simulation. In this section, a description is given of how a numerical solution of the equations of motion of up to 1000 interacting molecules is obtained, and of how the resulting molecular positions (in space and time) may be used to calculate observable properties.

The editors would like to thank the authors for the great care which they took with their manuscripts. In addition, we thank Dr Craig Sinclair, Director of the ASI programme, for constant help before and during the meeting.

A.J. Barnes, W.J. Orville-Thomas and J. Yarwood

INTERMOLECULAR FORCES

A. J. STONE

University Chemical Laboratory, Lensfield Road, Cambridge CB2 1EW, England.

1. MOLECULES AND FORCES

Intermolecular forces are fundamental to the study of molecular liquids. They are responsible not only for the very existence of condensed matter, but for all the ways in which the properties of a condensed fluid differ from those of an ideal gas of non-interacting particles. The fundamental interactions are the same as those responsible for the ordinary chemical forces which hold the atoms together in a molecule, but whereas the bonds in a molecule are strong enough for it to retain its identity in normal circumstances, the 'bonds' between molecules are weak enough to be constantly broken and reformed at ordinary thermal energies.

It is usually more helpful to think in terms of potentials rather than forces, and we can define a potential in various ways. We can write down the potential energy of an assembly of electrons and nuclei, but in practice we always need the energy of a nuclear configuration averaged over electron positions. The procedure for obtaining this is a familiar one: it involves the Born-Oppenheimer approximation and is very accurate. For many purposes we may also want to average over the internal coordinates (vibrations) of the molecules to obtain a rigid-molecule picture. If we are studying the effects of the interaction between a particular pair of molecules, we shall not wish to average over their rotational or translational motion, since these motions are of particular interest to us; but we might want to average over the positions of all other molecules. This would give us a potential of mean force, which is a Helmholtz free energy $A(\mathbf{R}) = U(\mathbf{R}) - TS(\mathbf{R})$, where both the internal energy U and the entropy S are functions of the temperature as well as the relative positions $\mathbf{R}$ of the molecules of

A. J. Barnes et al. (eds.), Molecular Liquids - Dynamics and Interactions, 1–34.

interest. Such an approach is essential for an understanding of phenomena such as the hydrophobic effect, where hydrocarbon molecules or residues attract each other in aqueous solution because the entropy term associated with the formation of a single solvent cage around a pair of molecules is more favourable than that for two separate cages.

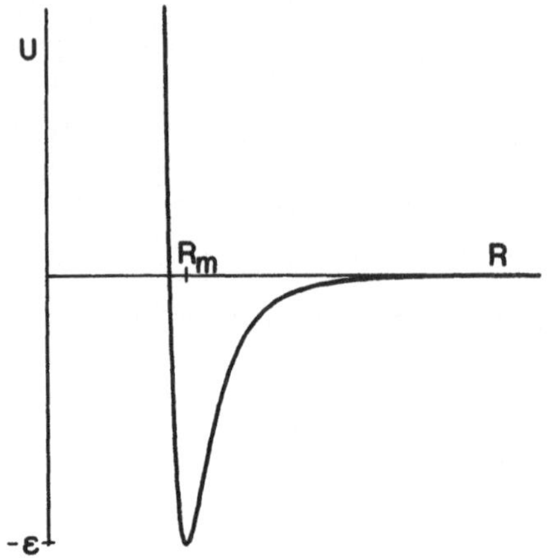

Figure 1. The energy U of interaction of two molecules as a function of the distance R between them.

However such averaged descriptions must be derived from a microscopic description. Consider then the intermolecular potential U of an isolated pair of molecules—that is, the amount by which the energy of the interacting pair exceeds the sum of the energies of the individual molecules. Regarded as a function of the distance R between the molecules (Figure 1), the intermolecular potential usually has a minimum, whose depth is conventionally denoted by ε, at a value of R denoted by R_m. At larger values of R the potential approaches zero asymptotically, while for smaller values of R it rises through zero and then rapidly to very high values. The attractive nature of the potential at long range is evident from the fact that condensed phases occur at all, while the strong repulsion at close range is apparent from the low compressibility of most liquids and solids. The depth of the well varies considerably, but for small closed-shell molecules it is typically of the order of a few kJ mol^{-1}, corresponding to a temperature of a few hundred kelvin, in accordance with the observed fact that solids liquefy and vapourize at such temperatures. An example is argon, which boils at 87.5 K and for which $\varepsilon = 1.18$ kJ mol^{-1}, equivalent to a temperature of 142 K.

The repulsion at short range is a consequence of the overlap of the electron clouds (it can be thought of as the energy needed to redistribute the electrons to avoid a breach of the Pauli principle) and it therefore varies with distance roughly like $\exp(-R/\rho)$, where ρ is a distance of the order of R_m. There may also be attractive effects, such as charge-transfer interactions, which depend on overlap between wavefunctions, but these are always swamped by repulsive effects at very short range. This energy of interaction is not pairwise additive: that is, the energy of interaction of three molecules A, B and C cannot be written as the sum of three pair interactions U_{AB}, U_{AC} and U_{BC}, because the presence of C modifies the interaction between A and B. These points are summarized in Table 1.

Term	Functional form	Sign	Pairwise additive?
Short range			
Overlap	$\exp(-R/\rho)$, $\rho \sim R_m$ (approx.)	either, but $\gg 0$ for small R	No
Long range			
Electrostatic	R^{-n}, $n \geq l_1+l_2+1$	either	Yes
Induction	R^{-n}, $n \geq \min(2l_1, 2l_2)+4$	<0 (for ground states)	No
Dispersion	R^{-n}, $n \geq 6$	<0 (for ground states)	Yes, approx.
Resonance	R^{-n}, $n \geq 3$	either	No
Magnetic	R^{-n}, $n \geq 3$	either, weak	Yes

Table 1. Contributions to the interaction energy of two molecules. l_1 and l_2 are the ranks of the lowest non-vanishing multipoles of the two molecules; that is, 0 for a charge (monopole), 1 for a dipole, 2 for a quadrupole, and so on.

The interaction energy at long range can be separated into a number of distinct contributions, with characteristics shown in Table 1. The electrostatic term is a straightforward classical interaction between the charge distributions of the two molecules, while the induction terms arise from the perturbation of each molecule by the electrostatic field of the other. The dispersion energy arises from correlated fluctuations in the charge distributions. The resonance term arises only when at least one of the molecules is in a degenerate or excited state, and will be ignored here, though it is a large effect if present. The magnetic term contributes only when both molecules are in electronically degenerate states, and is even then very weak; there is a further term arising from the magnetic interaction of any nuclear spins which may be present, but this is several orders of magnitude weaker still. These terms all have an R^{-n} dependence on the distance between the molecules; the value of n depends on the nature of the molecules—for example, on whether they are charged, or polar—in a way which we shall explore in Section 2.

The intermolecular distance is very far from being the only variable in the problem. Unless both molecules are actually atoms, we must specify their relative orientation. For an atom and a linear molecule, only one more coordinate is needed, and it can conveniently be taken as the angle between the intermolecular vector and the axis of the linear molecule. For two linear molecules, there are two such angles, and a third is needed to specify the angle between the two planes defined by the axis of each molecule and the intermolecular vector. When the molecules are non-linear, a further angle is needed to describe the angle of rotation of each molecule about its axis. Thus in the general case there are five variables describing the relative orientation of the molecules, and it is important to find an efficient way of describing the dependence of the potential on these variables.

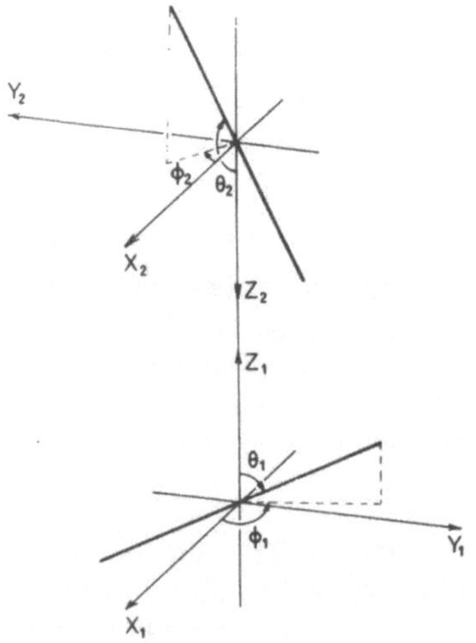

Figure 2. Definition of angular variables for two linear molecules (Pople)

Pople[1] described the potential between two linear molecules in terms of the coordinates shown in Figure 2. Using these coordinates the potential can be expanded in the form

$$U = \sum_{l_1 l_2 m} u_{l_1 l_2 : m}(R)\, C_{l_1,m}(\theta_1,\phi_1)\, C_{l_2,m}(\theta_2,\phi_2), \tag{1.1}$$

if the expansion converges. $C_{lm}(\theta,\phi)$ is a spherical harmonic normalized so that $C_{l0}(0,0) = 1$, this being the most convenient normalization for discussion of multipole moments; it is related to the conventional Y_{lm} by

$$C_{lm} = [4\pi/(2l+1)]^{\frac{1}{2}} Y_{lm}. \tag{1.2}$$

For (1.1) to be real it is necessary that $u_{l_1 l_2 : m}{}^* = u_{l_1 l_2 : -m}$. Although this function contains four angular variables it is independent of $(\phi_1 - \phi_2)$, and it is manifestly invariant with respect to a change of laboratory axes, as a scalar quantity like the energy must be, since only $(\phi_1 - \phi_2)$ depends on the choice of axes.

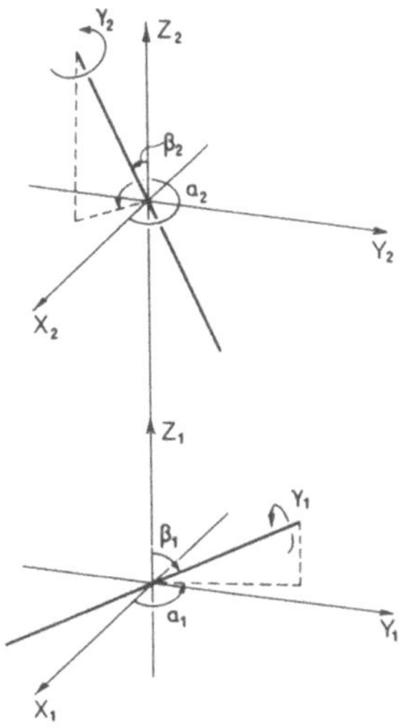

Figure 3. Definition of angular variables for two non-linear molecules

Steele[2] generalized this type of expansion to deal with non-linear molecules. Here the orientation is described by Euler angles (α,β,γ), β and α describing the direction of an arbitrary axis within the molecule (usually the highest-order symmetry axis) like θ and ϕ respectively in Figure 2, and γ describing the angle of rotation of the molecule about that axis as shown in Figure 3. Note that in this definition the two Z axes are parallel, whereas Pople took them to be antiparallel. In terms of these coordinates the expansion of the energy takes the form

$$U = \sum_{l_1 l_2 m k_1 k_2} u_{l_1 l_2 : k_1 k_2 m}(R)\, D^{l_1}_{m,k_1}(\Omega_1)\, D^{l_2}_{-m,k_2}(\Omega_2). \tag{1.3}$$

The functions D^l_{mk} are Wigner rotation matrices, which except for a normalization factor are the same as the wavefunctions for the symmetric top. The argument Ω_1 represents a set of Euler angles $(\alpha_1,\beta_1,\gamma_1)$, and Ω_2 similarly, and the Wigner matrices provide a complete set for expanding functions of these variables. In (1.3) there are six angular variables, but the energy is independent of $(\alpha_1 + \alpha_2)$. Again the variables are all internal and the expression is manifestly invariant with respect to change of laboratory axes.

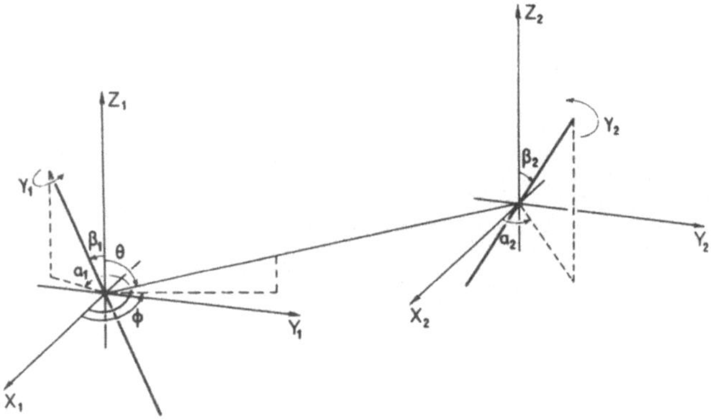

Figure 4. General definition of angular variables for two non-linear molecules.

It is not always convenient to refer the angles to a local frame of reference. It may well be more natural to use laboratory axes, defined perhaps in relation to an external field, to the sides of the box in a computer simulation, or to the geometry of a scattering experiment. In this case the angles can be defined as in Figure 4, and the functions

$$D^{l_1}_{m_1 k_1}(\Omega_1)\, D^{l_2}_{m_2 k_2}(\Omega_2)\, C_{jm}(\theta,\phi) \tag{1.4}$$

form a complete set of expansion functions in this coordinate system. Here however the coordinates do refer explicitly to an external frame of reference, and (1.4) is not as it stands invariant under a change of frame. However it is not difficult to show[3,4] that the following linear combinations of the functions in (1.4) form a complete set of scalar expansion functions:

$$S^{k_1k_2}_{l_1l_2j} = i^{l_1-l_2-j} \sum_{m_1m_2m} \begin{pmatrix} l_1 & l_2 & j \\ m_1 & m_2 & m \end{pmatrix} D^{l_1}_{m_1k_1}(\Omega_1)^* D^{l_2}_{m_2k_2}(\Omega_2)^* C_{jm}(\theta,\phi). \tag{1.5}$$

Here the factor in parentheses is a Wigner 3j symbol[3]. In terms of these 'S functions' the energy can be expressed in the form

$$U = \sum u^{k_1k_2}_{l_1l_2j}(R)\, S^{k_1k_2}_{l_1l_2j}(\Omega), \tag{1.6}$$

(always provided the expansion converges) where Ω is an abbreviation for the full set of angular variables.

$$S(000;00) = 1$$
$$S(101;00) = -\surd\tfrac{1}{3}\,\hat{z}_1\cdot\hat{R}$$
$$S(101;\pm10) = \pm\surd\tfrac{1}{6}\,(\hat{x}_1\cdot\hat{R} \mp \hat{y}_1\cdot\hat{R})$$
$$S(011;00) = \surd\tfrac{1}{3}\,\hat{z}_2\cdot\hat{R}$$
$$S(110;00) = -\surd\tfrac{1}{3}\,\hat{z}_1\cdot\hat{z}_2$$
$$S(112;00) = (30)^{-\frac{1}{2}}\left(\hat{z}_1\cdot\hat{z}_2 - 3(\hat{z}_1\cdot\hat{R})(\hat{z}_2\cdot\hat{R})\right)$$
$$S(224;00) = \tfrac{1}{4}(70)^{-\frac{1}{2}}\left(1 + 2(\hat{z}_1\cdot\hat{z}_2)^2 - 5(\hat{z}_1\cdot\hat{R})^2 - 5(\hat{z}_2\cdot\hat{R})^2 - 20(\hat{z}_1\cdot\hat{z}_2)(\hat{z}_1\cdot\hat{R})(\hat{z}_2\cdot\hat{R}) + 35(\hat{z}_1\cdot\hat{R})^2(\hat{z}_2\cdot\hat{R})^2\right)$$
$$S(kk0;00) = (-)^k\,(2k+1)^{-\frac{1}{2}}\,P_k(\hat{z}_1\cdot\hat{z}_2)$$

Table 2. Some of the simpler S functions expressed in terms of unit vectors $\hat{x}_i$, $\hat{y}_i$ and $\hat{z}_i$ along the molecular axes, and the unit vector $\hat{R}$ in the direction from molecule 1 to molecule 2. The notation $S(l_1l_2j;k_1k_2)$ is used for $S^{k_1k_2}_{l_1l_2j}$.

The S functions are suitable for expanding any scalar function of the relative orientation of a pair of molecules. They are much easier to evaluate than eq. (1.5) might suggest, since they depend only on scalar products involving unit vectors $\hat{x}_i$, $\hat{y}_i$ and $\hat{z}_i$ along cartesian axes fixed in the molecules, and the unit vector $\hat{R}$ in the direction from molecule 1 to molecule 2. Some of the simpler functions are given in Table 2.

Finally, we must take account of the internal structure of the molecules. The formulae given so far assume that the molecules are rigid, but there are vibrational coordinates which are likely to affect the intermolecular potential. Indeed, if the intermolecular potential were independent of vibrational coordinates there would be no way to transfer energy between translational and vibrational motion. There are various ways in which the internal coordinates may be brought into the intermolecular potential; for the present we note merely that the coefficients $u^{k_1 k_2}_{l_1 l_2 j}(R)$ may be allowed to depend on these coordinates as well as on the intermolecular separation.

2. LONG-RANGE FORCES

If the two molecules are far enough apart, we can regard the electrons as being definitely assigned to one molecule or the other. More formally, we can say that the matrix element vanishes for any observable $\hat{Q}$ between a wavefunction $\Psi_A\Psi_B$ describing the pair of molecules and a function $P_{AB}\Psi_A\Psi_B$ obtained by exchanging a pair of electrons between the molecules:

$$\langle\Psi_A\Psi_B|\hat{Q}|P_{AB}\Psi_A\Psi_B\rangle \approx 0. \tag{2.1}$$

Then although the wavefunction properly consists of a sum of many terms, comprising all possible assignments of the electrons to the two molecules, there are no cross matrix elements of any operator between terms arising from different assignments, and consequently the indistinguishability of the electrons on the two molecules has no observable consequences—there are no 'exchange effects'. At shorter distances, (2.1) does not hold and the theory becomes more difficult, as we shall see in section 3.

However, if we can assign each electron to a particular molecule, then we can write down unperturbed Hamiltonians for the individual molecules:

$$H_A = -\tfrac{1}{2}\sum_{i\,A}\nabla_i{}^2 - \sum_{i,k\in A}\frac{Z_k}{r_{ik}} + \sum_{k,k'\in A}\frac{Z_k Z_{k'}}{r_{kk'}} + \sum_{i,i'\in A}\frac{1}{r_{ii'}}, \tag{2.2}$$

and similarly for molecule B, while the perturbation is conveniently written as

$$V = \sum_{ab} e_a e_b r_{ab}{}^{-1}, \tag{2.3}$$

in which the sum is taken over all particles (electrons and nuclei) a in molecule A and b in molecule B.

For the isolated molecule we then have

$$H_A \Psi_m^A = W_m^A \Psi_m^A,$$
$$H_B \Psi_n^B = W_n^B \Psi_n^B. \qquad (2.4)$$

For the joint system, in the absence of any interaction, the unperturbed Hamiltonian is $H_0 = H_A + H_B$ and its eigenfunctions are the simple products $\psi_m^A \psi_n^B$, with eigenvalues $W_m^A + W_n^B$. Ψ_m^A is antisymmetric with respect to permutations of electrons within molecule A, as is ψ_n^B with respect to permutations within molecule B, but we take no account of exchange between the molecules. Because of (2.1) it would make no difference if we did.

Now standard Rayleigh-Schrödinger perturbation theory gives the perturbed energy:

$$W = W_A + W_B + U, \qquad (2.5)$$

where to second order

$$U = \langle 0_A 0_B | V | 0_A 0_B \rangle - \sum_{m_A n_B}{}' \frac{|\langle 0_A 0_B | V | m_A n_B \rangle|^2}{\Delta W_m^A + \Delta W_n^B}. \qquad (2.6)$$

Here we have used the Dirac ket notation $|m_A n_B\rangle$ for $\Psi_m^A \Psi_n^B$, and $\Delta W_m^A \equiv W_m^A - W_n^A$, while $|0_A 0_B\rangle$ denotes the ground state of the combined system which, as indicated by the prime, is excluded from the sum over states.

This is conventionally separated into three parts. The first-order energy is

$$U_{es} = \langle 0_A 0_B | V | 0_A 0_B \rangle, \qquad (2.7)$$

and is the classical electrostatic energy of interaction of the charge distributions of the unperturbed molecules. The second-order energy is conventionally separated into those terms in the sum over states in which only one molecule is excited, which together comprise the induction energy U_{ind}, and those in which both are excited, which comprise the dispersion energy U_{disp}:

$$U_{ind} = -\sum_{m\neq 0} \frac{|\langle 0_A 0_B|V|m_A 0_B\rangle|^2}{\Delta W_m^A} - \sum_{n\neq 0} \frac{|\langle 0_A 0_B|V|0_A n_B\rangle|^2}{\Delta W_n^B}, \tag{2.8}$$

$$U_{disp} = -\sum_{m\neq 0}\sum_{n\neq 0} \frac{|\langle 0_A 0_B|V|m_A n_B\rangle|^2}{\Delta W_m^A + \Delta W_n^B}. \tag{2.9}$$

The induction energy describes the stabilization which occurs when the electron distribution of molecule B relaxes under the influence of the unperturbed electrostatic field from molecule A, and vice versa. The dispersion energy describes the stabilization which results when the motion of the electrons in the two molecules is correlated, rather than uncorrelated as it is in the unperturbed description. This interpretation is more readily apparent if we express the perturbation in a different form.

2.1 The Multipole Expansion

The perturbation consists of the electrostatic interaction between the particles of A and those of B:

$$V = (4\pi\varepsilon_0)^{-1} \sum_{a\in A}\sum_{b\in B} e_a e_b / r_{ab}. \tag{2.10}$$

For completeness we include the factor $(4\pi\varepsilon_0)^{-1}$ in this section; it is equal to 1 in the atomic units used elsewhere in the chapter. Let $\mathbf{R}$ be the vector from the centre of molecule A to the centre of molecule B. The 'centre' is arbitrary, though it will often be determined by symmetry. In a molecule like HD, however, we have the choice between the centre of mass or the geometrical centre, and which is appropriate may depend on the application. We shall see, however, that for optimal convergence of the multipole expansion the centre is best chosen so as to minimise the greatest distance from it of significant charge density. We take $\mathbf{r}_a$ ($\mathbf{r}_b$) to be the position of particle a (b) relative to the centre of molecule A (B). Then we require an expression for $r_{ab}^{-1} = |\mathbf{R} + \mathbf{r}_b - \mathbf{r}_a|^{-1}$. The standard expansion of $|\mathbf{r}_1 - \mathbf{r}_2|^{-1}$ gives[3]

$$r_{ab}^{-1} = \sum_{lm} (-1)^m R_{lm}(\mathbf{r}_a)\, I_{l,-m}(\mathbf{R}+\mathbf{r}_b), \tag{2.11}$$

where $R_{lm}(\mathbf{r})$ and $I_{lm}(\mathbf{r})$ are respectively the regular and irregular spherical harmonics:

$$R_{lm}(\mathbf{r}) = r^{l}C_{lm}(\theta,\phi),$$
$$I_{lm}(\mathbf{r}) = r^{-l-1}C_{lm}(\theta,\phi). \quad (2.12)$$

Equation (2.11) is valid only when $|\mathbf{r}_a| < |\mathbf{R}+\mathbf{r}_b|$.

A less familiar addition theorem, a generalization of (2.11), states that if $|\mathbf{r}_b| < |\mathbf{R}|$ then[5]

$$I_{l_1,-m_1}(\mathbf{R}+\mathbf{r}_b) = (-1)^{l_1-m_1} \sum_{l_2=0}^{\infty} \sum_{m_2} \left[\frac{(2l_1+2l_2+1)!}{(2l_1)!(2l_2)!}\right]^{\frac{1}{2}}$$
$$\times R_{l_2m_2}(\mathbf{r}_b)\, I_{l_1+l_2,m}(\mathbf{R}) \begin{pmatrix} l_1 & l_2 & l_1+l_2 \\ m_1 & m_2 & m \end{pmatrix} \quad (2.13)$$

(where the last factor is a Wigner 3j symbol) and if we assemble these results the perturbation takes the form of the <u>multipole expansion</u>:

$$V = (4\pi\varepsilon_0)^{-1} \sum_{l_1,l_2=0} \sum_{m_1m_2m} (-1)^{l_1} \left[\frac{(2l_1+2l_2+1)!}{(2l_1)!(2l_2)!}\right]^{\frac{1}{2}}$$
$$\times \sum_{a\in A} e_a R_{l_1m_1}(\mathbf{r}_a) \sum_{b\in B} e_b R_{l_2m_2}(\mathbf{r}_b)$$
$$\times I_{l_1+l_2,m}(\mathbf{R}) \begin{pmatrix} l_1 & l_2 & l_1+l_2 \\ m_1 & m_2 & m \end{pmatrix}, \quad (2.14)$$

which is valid provided that $|\mathbf{r}_a|+|\mathbf{r}_b|<|\mathbf{R}|$ for all pairs of particles a and b. We see now that for rapid convergence we need both $|\mathbf{r}_a|/|\mathbf{R}|$ and $|\mathbf{r}_b|/|\mathbf{R}|$ to be small. Since the molecular charge distribution extends formally to infinity, the multipole expansion is not strictly convergent at <u>any</u> separation; it is an asymptotic series in R^{-1}[6]. Nevertheless it is very useful in practice if the molecules are not too close.

An expression equivalent to (2.14) can be obtained working in Cartesian tensors throughout[7]. The 'interaction tensors' $T_{\alpha\beta\ldots\nu}$ (n subscripts) which appear in that description correspond to the irregular spherical harmonics $I_{nm}(\mathbf{R})$ of (2.14)[5]. The cartesian form has been widely used but the spherical tensor form is more convenient when n>2, since $T_{\alpha\beta\ldots\nu}$ has 3^n components of which only 2n+1 are independent, while I_{nm} has only these 2n+1 components and their angular behaviour is at once apparent.

2.2 Multipole moments

The expression

$$\hat{Q}^A_{lm} \equiv \sum_{a\in A} e_a R_{lm}(\mathbf{r}_a) \tag{2.15}$$

which occurs in (2.14) is a multipole moment operator of rank l, and its expectation value is the multipole moment. For l=0 we have $R_{00}(\mathbf{r}_a) = 1$ and

$$Q^A_{00} = \langle\sum_a e_a\rangle = Q, \tag{2.16}$$

the total charge, or monopole moment. (Here Q is the conventional notation, not to be confused with the general symbol for a multipole moment). For higher values of l it is often convenient to work with the real and imaginary parts of the complex multipole moments and spherical harmonics, and we define, for m>0,

$$\begin{aligned} Q_{lmc} &= \sqrt{\tfrac{1}{2}}[(-)^m Q_{lm} + Q_{l,-m}], \\ iQ_{lms} &= \sqrt{\tfrac{1}{2}}[(-)^m Q_{lm} - Q_{l,-m}]. \end{aligned} \tag{2.17}$$

Then for example

$$\begin{aligned} Q^A_{11c} &= \sqrt{\tfrac{1}{2}}[-Q_{11} + Q_{1,-1}] \\ &= \langle \sum_{a\in A} e_a r_a \sin\theta_a \cos\phi_a\rangle = \langle \sum_{a\in A} e_a x_a\rangle = \mu^A_x, \end{aligned} \tag{2.18}$$

and similarly

$$Q_{11s} = \mu_y,$$

$$Q_{10} = \mu_z.$$

In the same way, $l = 2$ gives the components of the quadrupole moment:

$$\begin{aligned} Q^A_{20} &= \langle\sum_a e_a \tfrac{1}{2}(3z_a{}^2 - r_a{}^2)\rangle, \\ Q^A_{21c} &= \langle\sum_a e_a \sqrt{3}\, x_a z_a\rangle, \qquad Q^A_{21s} = \langle\sum_a e_a \sqrt{3}\, y_a z_a\rangle, \\ Q^A_{22c} &= \langle\sum_a e_a \tfrac{1}{2}\sqrt{3}\,(x_a{}^2 - y_a{}^2)\rangle, \quad Q^A_{22s} = \langle\sum_a e_a \sqrt{3}\, x_a y_a\rangle. \end{aligned} \tag{2.19}$$

It should be realised that the multipole moments are not generally invariant with respect to change of origin. The moments $Q_{lk}(S)$ evaluated with respect to an origin at S are related to the moments $Q_{jm}(S+a)$ with respect to an origin at S+a by[5,8]

$$Q_{lk}(S) = \sum_{j=0}^{l} \sum_{m=-j}^{j} \left[\binom{l+k}{j+m}\binom{l-k}{j-m}\right]^{\frac{1}{2}} Q_{jm}(S+a)\, R_{l-j,k-m}(a). \quad (2.20)$$

However eq. (2.20) shows that the moments of rank l are all independent of origin if all the moments of lower rank are zero.

The perturbation is now readily expressed in terms of the multipole moment operators:

$$V = (4\pi\varepsilon_0)^{-1} \sum_{l_1 l_2} \sum_{m_1 m_2 m} (-)^{l_1} [l_1;l_2]\, \hat{Q}^A_{l_1 m_1} \hat{Q}^B_{l_2 m_2}$$

$$\times\, I_{l_1+l_2,m}(R) \begin{pmatrix} l_1 & l_2 & l_1+l_2 \\ m_1 & m_2 & m \end{pmatrix}. \quad (2.21)$$

where $[l_1;l_2]$ is the numerical factor $[(2l_1+2l_2+1)!/(2l_1)!(2l_2)!]^{\frac{1}{2}}$. However this is still not quite in the form we need, because the multipole moments are referred to a laboratory-fixed frame of reference. It is more helpful to deal with multipole moments in a molecule-fixed axis system, where their values are independent of the molecular orientation. This is easily done; using a tilde to denote components in the molecule-fixed frame we have[3]

$$\hat{Q}^A_{lm} = \sum_k \tilde{Q}^A_{lk} D^l_{mk}(\Omega_1)^*, \quad (2.22)$$

where $\Omega_1 \equiv (\alpha_1,\beta_1,\gamma_1)$ is the set of Euler angles defining the orientation of the molecule.

The number of non-zero multipole moments in molecular axes is substantially reduced if the molecule has any symmetry. In particular,

(i) For linear molecules, $\tilde{Q}_{lk}$ vanishes when $k \neq 0$, provided that the molecular axis is chosen to be the z axis;

(ii) for centrosymmetric molecules, $\tilde{Q}_{lk}$ vanishes when l is odd, provided that the origin of coordinates for the molecule is taken at the centre of symmetry.

Thus the only multipole components which can appear for a homonuclear diatomic are Q_{00} (the charge or monopole moment), Q_{20} (the quadrupole moment), Q_{40} (the hexadecapole moment), and so on.

More generally, since the multipole moments Q_{lm} transform under proper and improper rotations in the same way as the spherical harmonics C_{lm}, we can use standard group theoretical techniques to

	0_g	1_g	2_g	3_g	4_g
$D_{\infty h}$	Σ_g^+	$\Sigma_g^-+\Pi_g$	$\Sigma_g^++\Pi_g+\Delta_g$	$\Sigma_g^-+\Pi_g+\Delta_g+\Phi_g$	$\Sigma_g^++\Pi_g+\Delta_g+\Phi_g+\Gamma_g$
$C_{\infty v}$	Σ^+	$\Sigma^-+\Pi$	$\Sigma^++\Pi+\Delta$	$\Sigma^-+\Pi+\Delta+\Phi$	$\Sigma^++\Pi+\Delta+\Phi+\Gamma$
O_h	A_{1g}	T_{1g}	E_g+T_{2g}	$A_{2g}+T_{1g}+T_{2g}$	$A_{1g}+E_g+T_{1g}+T_{2g}$
T_d	A_1	T_1	$E+T_2$	$A_2+T_1+T_2$	$A_1+E+T_1+T_2$
D_{3h}	A_1'	$A_2'+E''$	$A_1'+E'+E''$	$A_2'+E'+A_1''$ $+A_2''+E''$	$A_1'+2E'+A_1''$ $+A_2''+E''$
C_{2v}	A_1	$A_2+B_1+B_2$	$2A_1+A_2+B_1+B_2$	$A_1+2A_2+2B_1+2B_2$	$3A_1+2A_2+2B_1+2B_2$
	0_u	1_u	2_u	3_u	4_u
$D_{\infty h}$	Σ_u^-	$\Sigma_u^++\Pi_u$	$\Sigma_u^-+\Pi_u+\Delta_u$	$\Sigma_u^++\Pi_u+\Delta_u+\Phi_u$	$\Sigma_u^-+\Pi_u+\Delta_u+\Phi_u+\Gamma_u$
$C_{\infty v}$	Σ^-	$\Sigma^++\Pi$	$\Sigma^-+\Pi+\Delta$	$\Sigma^++\Pi+\Delta+\Phi$	$\Sigma^-+\Pi+\Delta+\Phi+\Gamma$
O_h	A_{1u}	T_{1u}	E_u+T_{2u}	$A_{2u}+T_{1u}+T_{2u}$	$A_{1u}+E_u+T_{1u}+T_{2u}$
T_d	A_2	T_2	$E+T_1$	$A_1+T_1+T_2$	$A_2+E+T_1+T_2$
D_{3h}	A_1''	$A_2''+E'$	$A_1''+E'+E''$	$A_1'+A_2'+E'$ $+A_2''+E''$	$A_1'+A_2'+E'$ $+A_1''+2E''$
C_{2v}	A_2	$A_1+B_1+B_2$	$A_1+2A_2+B_1+B_2$	$2A_1+A_2+2B_1+2B_2$	$2A_1+3A_2+2B_1+2B_2$

Table 3. Symmetry species in some common point groups of functions transforming like the spherical harmonics of ranks 0 to 4 under proper rotations and having even (g) or odd (u) parity under inversion.

find out how the components transform in any symmetry group. The results for a number of important symmetry groups are given in Table 3, which also shows the behaviour of entities transforming like spherical harmonics under proper rotations but having the opposite parity under inversion, since some polarizabilities have components behaving in this way. The number of non-vanishing components of the static multipole moment can now be found by counting the number of times the totally symmetric representation of the molecular symmetry group appears. For example, the dipole moment transforms like the spherical harmonics of rank 1 (the p functions) and is odd under inversion, so we find its properties in the 1_u column of Table 3. The dipole components of a C_{2v} molecule such as H_2O have symmetry species A_1, B_1 and B_2, so just one of them, the A_1 component, can be non-zero.

If we use eq. (2.22) in eq. (2.21), we can identify the S functions defined in eq. (1.5) and the perturbation becomes

$$V = (4\pi\varepsilon_0)^{-1} \sum_{l_1 l_2 k_1 k_2} (-)^{l_1+l_2} [l_1;l_2] \times R^{-l_1-l_2-1} \hat{Q}^A_{l_1k_1} \hat{Q}^B_{l_2k_2} S^{k_1k_2}_{l_1l_2,l_1+l_2}, \qquad (2.23)$$

The expectation value of this operator now gives the electrostatic energy:

$$U_{es} = (4\pi\varepsilon_0)^{-1} \sum_{l_1 l_2 k_1 k_2} (-)^{l_1+l_2} [l_1;l_2] \times R^{-l_1-l_2-1} Q^A_{l_1k_1} Q^B_{l_2k_2} S^{k_1k_2}_{l_1l_2,l_1+l_2}, \qquad (2.24)$$

where the Q are multipole moments in molecule-fixed axes. So the functions $S^{k_1k_2}_{l_1l_2j}$ defined previously emerge as natural functions for the expansion of the electrostatic energy, provided that the molecules are far enough apart for the series to converge. Even at shorter distances they are useful, however, because they form a complete set of scalar functions in the space of relative orientations. A convenient feature is that not all of them are needed if the molecules have any symmetry; the functions with $k_1 \neq 0$ are not required when molecule A is linear, and those with l_1 odd are not required when it is centrosymmetric. It is convenient to use the abbreviated notation $S_{l_1l_2j}$ when $k_1 = k_2 = 0$.

2.3 Molecules in Electric Fields

Consider molecule A in an electrostatic potential due to molecule B or some other external source. We expand the potential ϕ^B due to B in a Taylor series about the origin of A:

$$\phi^B(r) = \phi^B(A) + r_\alpha[\nabla_\alpha\phi^B]_A + \tfrac{1}{2}r_\alpha r_\beta[\nabla_\alpha\nabla_\beta\phi^B]_A + \ldots \qquad (2.25)$$

In terms of spherical tensors this is[5,9,10]

$$\phi^B(\mathbf{r}) = \sum_{lk} (l!)^{-1} R_{lk}(\mathbf{r}) \left[R_{lk}(\nabla)^* \phi^B\right]_A$$

$$= - \sum_{lk} R_{lk}(\mathbf{r}) F_{lk}{}^*. \tag{2.26}$$

The constants F_{lk} in this expression describe the electric field (for $l=1$), the electric field gradient (for $l=2$) and so on. Now the interaction of the charges in molecule A with this potential is described by the perturbation

$$\sum_a e_a \phi(\mathbf{r}_a) = - \sum_{lk} \tilde{Q}^A_{lk} F_{lk}{}^*. \tag{2.27}$$

In the case where the potential arises from another molecule, the F_{lk} can be obtained from the multipole expansion. Whatever the source of the potential, however, we can use perturbation theory to obtain the energy of molecule A in the external field:

$$W^{(1)} = - \sum_{lk} Q^A_{lk} F_{lk}{}^*, \tag{2.28}$$

$$W^{(2)} = - \sum_{ll'kk'} F_{lk}{}^* F_{l'k'}{}^* \sum_n{}' \frac{|\langle 0_A|\tilde{Q}^A_{lk}|n_A\rangle\langle n_A|\tilde{Q}^A_{l'k'}|0\rangle|}{\Delta W^A_n}, \tag{2.29}$$

and so on. The first-order energy describes the interaction of the static moments with the external field. In cartesian tensor form it is

$$W^{(1)} = Q\phi - \mu_\alpha F_\alpha - \tfrac{1}{3}\Theta_{\alpha\beta}F_{\alpha\beta} - \ldots \tag{2.30}$$

The second-order energy can be written in the form

$$W^{(2)} = -\tfrac{1}{2} \sum_{ll'kk'} F_{lk}{}^* F_{l'k'}{}^* \alpha(ll'kk'), \tag{2.31}$$

where $\alpha(ll'kk')$ is a polarizability:

$$\alpha(ll'kk') = \sum_n{}' \frac{[\langle 0|Q_{lk}|n\rangle\langle n|Q_{l'k'}|0\rangle + \langle 0|Q_{l'k'}|n\rangle\langle n|Q_{lk}|0\rangle]}{\Delta W_n}. \tag{2.32}$$

parameters $-F_{lk}^*$ describing the external field is the multipole moment operator $\tilde{Q}_{lk}$. It then follows from the Hellmann-Feynman theorem[11] that the derivative of the energy with respect to $-F_{lk}^*$ is the expectation value of the multipole moment Q_{lk}. Accordingly we see that $\alpha(ll'kk')$ describes the $l'k'$ multipole induced by the lk derivative of the potential, and also the lk multipole induced by the $l'k'$ derivative of the potential[12]. When $l = l' = 1$ we have the ordinary dipole polarizability, equivalent to the Cartesian form

$$\alpha_{\alpha\beta} = 2\sum_n{}' \frac{\langle 0|\mu_\alpha|n\rangle\langle n|\mu_\beta|0\rangle + \langle 0|\mu_\beta|n\rangle\langle n|\mu_\alpha|0\rangle}{\Delta W_n}. \tag{2.33}$$

Eq. (2.32) shows that the components of $\alpha(ll'kk')$ transform under rotations like the components of the direct product $l \times l'$. We can construct the irreducible tensor components by recoupling in the usual way[10,12]:

$$\alpha_{LK}(ll') = \sum_{kk'} \langle ll'kk'|LK\rangle\, \alpha(ll'kk'), \tag{2.34}$$

where $\langle ll'kk'|LK\rangle$ is a Clebsch-Gordan coefficient. The values of L which occur run from $l+l'$ to $|l-l'|$, except that when $l=l'$ odd values of L do not occur because of the symmetrization in (2.32).

For example the dipole polarizability given in (2.33) has spherical tensor components $\alpha_{00}(11)$ and $\alpha_{2K}(11)$. The dipole-quadrupole polarizability ($A_{\alpha;\beta\gamma}$ in Cartesian notation), which describes the quadrupole moment induced by an electric field or the dipole moment induced by an electric field gradient, has components $\alpha_{1K}(12)$, $\alpha_{2K}(12)$ and $\alpha_{3K}(12)$. The polarizabilities are even (g) or odd (u) under inversion according as $l+l'$ is even or odd. This information is then sufficient, with the help of Table 3, to determine the transformation properties in the molecular symmetry group. Any component which transforms according to the totally symmetric representation may have a non-zero value.

For some kinds of spectroscopy, it is necessary to know about the behaviour of derivatives of the polarizabities or multipole moments with respect to normal coordinates. Such a derivative transforms according to the direct product $\Gamma_\alpha \times \Gamma_Q$ of the representation Γ_α for the polarizabilities and the representation Γ_Q for the vibration, and can be non-zero if $\Gamma_\alpha = \Gamma_Q$. In other words, the derivative can be non-zero, so that the transition is allowed, if the normal mode transforms in the same way as some component of the polarizability.

The perturbation series may be continued to higher order, in which case we derive hyperpolarizabilities $\beta(ll'l''kk'k'')$:

$$\beta(ll'l''kk'k'') = S\Big\{ \sum_{mn} \frac{\langle 0|\tilde{Q}_{lk}|m\rangle\langle m|\tilde{Q}_{l'k'}|n\rangle\langle n|\tilde{Q}_{l''k''}|0\rangle}{\Delta W_m\, \Delta W_n} - \langle 0|\tilde{Q}_{lk}|0\rangle \sum_n \frac{\langle 0|\tilde{Q}_{l'k'}|n\rangle\langle n|\tilde{Q}_{l''k''}|0\rangle}{(\Delta W_n)^2} \Big\}. \quad (2.35)$$

where the S denotes that we are to take the sum of all six expressions obtained by permuting lk, l'k' and l"k". Higher hyperpolarizabilities are defined similarly[12].

If we contemplate the response of the molecule to oscillating fields rather than static ones, we can define the polarizability $\alpha(ll'kk';\omega)$ which describes the oscillating multipole Q_{lk} induced by the Fourier component of the parameter $-F_{l'k'}(\omega)$ of the electric field. Time-dependent perturbation theory[13] shows this to be

$$\alpha(ll'kk';\omega) = \sum_n{}' \frac{\Delta W_n \left[\langle 0|Q_{lk}|n\rangle\langle n|Q_{l'k'}|0\rangle + \langle 0|Q_{l'k'}|n\rangle\langle n|Q_{lk}|0\rangle\right]}{(\Delta W_n)^2 - \hbar^2\omega^2}. \quad (2.36)$$

2.4 The Induction Energy

If the multipole expansion of the perturbation is substituted into the expression for the induction energy we obtain terms of the form

$$R^{-l_1-l_2-l_1'-l_2'-2}\, S^{k_1k_2}_{l_1l_2l_1+l_2}\, S^{k_1'k_2'}_{l_1'l_2'l_1'+l_2'} \times \sum_{n\neq 0} \frac{\langle 0_A0_B|\tilde{Q}^A_{l_1k_1}\tilde{Q}^B_{l_2k_2}|n_A0_B\rangle\langle n_A0_B|\hat{Q}^A_{l_1'k_1'}\hat{Q}^B_{l_2'k_2'}|0_A0_B\rangle}{\Delta W^A_n}$$

$$= R^{l_1-l_2-l_1'-l_2'-2}\, S^{k_1k_2}_{l_1l_2l_1+l_2}\, S^{k_1'k_2'}_{l_1'l_2'l_1'+l_2'} \times Q^B_{l_2m_2}\, Q^B_{l_2'm_2'} \sum_{n\neq 0} \frac{\langle 0_A|\tilde{Q}^A_{l_1m_1}|m_A\rangle\langle n_A|\tilde{Q}^A_{l_1'm_1'}|0_A\rangle}{\Delta W^A_n}. \quad (2.37)$$

(Numerical coefficients have been omitted.) Here the sum-over-states is a polarizability. When $l_2 = l_2' = 1$ we obtain a component of the ordinary dipole polarizability α, but (2.37) includes higher polarizabilities describing the magnitude of the moment of rank l_1' induced in A by the l_1th gradient of the electric potential. Thus we can understand (2.37) physically as describing the induction of a moment of rank l_1' in A by the l_1th derivative of the potential due to the rank l_2 moment of B, and the interaction of this induced moment with the rank l_2' moment of B.

The dipole-dipole polarizability is an accurately known property for many simple molecules, so evaluation of the leading term in the induction energy is relatively easy for isolated pairs of molecules.

2.5 The Dispersion Energy

In the dispersion energy we find terms like

$$R^{-l_1-l_2-l_1'-l_2'-2}\, S^{k_1k_2}_{l_1l_2l_1+l_2}\, S^{k_1'k_2'}_{l_1'l_2'l_1'+l_2'}$$

$$\times \sum_m{}' \sum_n{}' \frac{\langle 0_A0_B|\hat{Q}^A_{l_1k_1}\hat{Q}^B_{l_2k_2}|m_An_B\rangle\langle m_An_B|\hat{Q}^A_{l_1'k_1'}\hat{Q}^B_{l_2'k_2'}|0_A0_B\rangle}{\Delta W^A_m + \Delta W^B_n}$$

$$= R^{l_1-l_2-l_1'-l_2'-2}\, S^{k_1k_2}_{l_1l_2l_1+l_2}\, S^{k_1'k_2'}_{l_1'l_2'l_1'+l_2'}$$

$$\times \sum_m{}' \sum_n{}' \frac{\langle 0_A|\hat{Q}^A_{l_1k_1}|m_A\rangle\langle m_A|\hat{Q}^A_{l_1'k_1'}|0_A\rangle\langle 0_B|\hat{Q}^B_{l_2k_2}|n_B\rangle\langle n_B|\hat{Q}^B_{l_2'k_2'}|0_B\rangle}{\Delta W^A_m + \Delta W^B_n}. \tag{2.38}$$

There are two ways to handle this expression. The first is to replace the energy denominator by the following approximation:

$$(\Delta W^A_m + \Delta W^B_m)^{-1} \approx \frac{U_A U_B}{U_A+U_B} (\Delta W^A_m)^{-1}(\Delta W^B_n)^{-1}, \tag{2.39}$$

where U_A and U_B are average excitation energies, commonly assumed to be roughly equal to the ionization potential. Then the sum over states in (2.38) factorizes into a product of polarizabilities of the form

$$\frac{1}{4}\frac{U_A U_B}{U_A + U_B}\, \alpha^A(l_1l_1'k_1k_1')\, \alpha^B(l_2l_2'k_2k_2'). \tag{2.40}$$

The alternative is to use the Casimir-Polder identity

$$\frac{1}{A+B} = \frac{2}{\pi} \int_0^\infty \frac{A\,B}{(A^2+u^2)(B^2+u^2)}\,du. \tag{2.41}$$

Then the sum over states in (2.38) factorizes in a different way to give

$$(1/2\pi)\int_0^\infty \alpha^A(l_1 l_1' k_1 k_1'; iu)\, \alpha^B(l_2 l_2' k_2 k_2'; iu)\, du, \tag{2.42}$$

where $\alpha(ll'kk';iu)$ is a polarizability at the imaginary frequency iu:

$$\alpha(ll'kk';iu) = 2\sum_n{}' \frac{\Delta W_n\big(\langle 0|\hat{Q}_{lm}|n\rangle\langle n|\hat{Q}_{l'k'}|0\rangle + \langle 0|\hat{Q}_{l'k'}|n\rangle\langle n|\hat{Q}_{lm}|0\rangle\big)}{(\Delta W_n)^2 + u^2}. \tag{2.43}$$

The dipole polarizability at imaginary frequency is not usefully regarded as describing a physical property. Rather it is a formal analytic continuation of the ordinary time-dependent polarizability, eq. (2.36), into the complex freqency domain. Nevertheless it is a well-defined quantity, and is well-behaved, as a function of u, since it goes monotonically to zero as u tends to infinity, whereas $\alpha(ll'kk';\omega)$ has singularities at the transition frequencies when ω is real. It can be estimated quite well in favourable cases by the use of sum rules for oscillator strengths and related properties[14-17], and in such cases the leading coefficient C_6 in the R^{-1} expansion of the dispersion energy can be estimated reasonably accurately.

The angular dependence of the induction and dispersion energies involves a product of S functions. This can be expressed as a sum of single S functions using the following product formula:

$$S^{k_1'k_2'}_{l_1'l_2'j'}\, S^{k_1''k_2''}_{l_1''l_2''j''} = \sum_{l_1 l_2 j} i^{l_1'-l_2'-j'+l_1''-l_2''-j''+l_1-l_2-j}$$

$$\times\ (-1)^{k_1'+k_2'+k_1''+k_2''}(2l_1+1)(2l_2+1)(2j+1)S^{k_1k_2}_{l_1l_2j}$$

$$\times \begin{pmatrix} l_1' & l_1'' & l_1 \\ k_1' & k_2' & -k_1 \end{pmatrix}\begin{pmatrix} l_2' & l_2'' & l_2 \\ k_2' & k_2'' & -k_2 \end{pmatrix}\begin{pmatrix} j' & j'' & j \\ 0 & 0 & 0 \end{pmatrix}\begin{Bmatrix} l_1' & l_1'' & l_1 \\ l_2' & l_2'' & l_2 \\ j' & j'' & j \end{Bmatrix}. \tag{2.44}$$

where the last factor is a Wigner 9j symbol[3]. For example, the leading R^{-6} term in the dispersion energy arises from the dipole-dipole term in the perturbation, and gives terms proportional to S_{000}, S_{220}, S_{202}, S_{022}, S_{222} and S_{224}. The first of these is isotropic, and has the form

$$-\frac{3\hbar}{\pi}(4\pi\varepsilon_0)^{-2}R^{-6}\int_0^\infty \bar{\alpha}^A(iu)\,\bar{\alpha}^B(iu)\,du. \tag{2.45}$$

This is the only term which arises for atoms, where the polarizability is isotropic, but all the other terms are present when there is anisotropy in the polarizability.

3. SHORT-RANGE FORCES

A number of complications arise when the molecules are close enough for overlap between their wavefunctions to be significant. Because (2.1) is no longer valid, the antisymmetrization of the overall wavefunction has to be dealt with explicitly. If now we envisage a separation of the Hamiltonian of the form $H = H_0 + \lambda V$, as used previously, with λ being a formal expansion parameter, then we can see from eqs. (2.2) and (2.3) that although H is symmetric with respect to all electron permutations, H_0 and V separately are not. That is, if P_{AB} exchanges electrons between the molecules, then

$$[H,P_{AB}] = [H_0+\lambda V,P_{AB}] = 0, \tag{3.1}$$

but

$$[H_0,P_{AB}] = -[\lambda V,P_{AB}] \neq 0. \tag{3.2}$$

Equation (3.2) states that the 'zeroth-order' quantity $[H_0,P_{AB}]$ is equal to the 'first-order' quantity $-[\lambda V,P_{AB}]$. This means that we can no longer separate the Rayleigh-Schrödinger expansion into orders of perturbation in the conventional way, by collecting together terms with the same power of λ. It does not mean that no such separation can be made, but rather that there is no unique way of doing it.

A further difficulty arises with the 'zeroth-order' functions used for the perturbation expansion. The natural choice is the set of product functions $\psi_m^A\psi_n^B$, but these are not antisymmetric. If antisymmetrized they become an overcomplete set, though this is not a problem with the finite basis sets used in practical calculations. Whether antisymmetrized or not, they are not orthogonal in the short-range region. Consequently they cannot be eigenfunctions of any Hermitian operator, so that a zeroth-order

Hamiltonian cannot be defined with these functions as eigenfunctions.

Most perturbation theory methods seek to overcome these difficulties by treating the overlap itself as part of the perturbation, and performing an expansion in powers of overlap. This is unsatisfactory because the expansion converges rather slowly at distances around R_m. An alternative method is to perform an orthogonalization, but this also has serious disadvantages; in practice the orthogonalization is applied to the molecular orbitals rather than the overall wavefunctions, and has the effect of obscuring physical effects by mixing virtual orbitals with occupied ones and orbitals of one molecule with orbitals of the other.

It now seems, however, that these difficulties can be avoided[18]. There is in fact no need to define a zeroth-order Hamiltonian; by using a Brillouin-Wigner type of expansion only the complete Hamiltonian for the combined system is needed, and this is properly symmetric. The most suitable expansion functions are not the product functions $\Psi_m^A\Psi_b^B$ but determinants comprising the occupied orbitals of both molecules, and these are properly antisymmetric. The non-orthogonality of the expansion functions is a nuisance, but they can nevertheless be used without orthogonalization at the cost of a certain amount of additional labour. Early results suggest that the method has considerable promise[18,19].

New contributions to the interaction energy arise at short range. Because of the arbitrariness inherent in the perturbation expansion, terms which are given the same name by different workers may in fact describe somewhat different quantities, so it is difficult to compare methods. The sum of the zeroth-order and first-order energies is the expectation value of the energy for the unperturbed wavefunction, but even this may not be what it seems because in some methods the orthogonalization used may change the form of the 'zeroth-order' wave function. However it is usually possible to identify the electrostatic energy, which is the classical electrostatic interaction between the unperturbed charge distributions and correlates directly with the same quantity in the long-range theory. It cannot be represented by a multipole expansion, both because the expansion fails to converge and also because it takes no account of the overlap of the charge distributions. A further reasonably well-defined contribution is the exchange energy:

$$E_{ex} = -\tfrac{1}{2} \sum_{i\in A}^{occ} \sum_{j\in B}^{occ} (ij|ji), \qquad (3.3)$$

in which $(ij|ji)$ is the exchange integral between orbital i on A and orbital j on B. The exchange energy is attractive and is pairwise additive when more than two molecules are involved. The remainder when these two terms and the zeroth-order energy have been accounted for is a strongly repulsive, non-additive contribution which can be called the repulsion energy. However the term 'exchange energy' is also used for the sum of (3.3) and the repulsion energy[20].

At second order we can, as in the long-range theory, distinguish between one-electron and two-electron excitations. The one-electron excitations contributing to the induction energy can be categorized as [A→A] and [B→B], the notation indicating that the excitation is from an occupied orbital to a virtual orbital on the same molecule. The same contributions are present in the short-range case, but the form of the energy contribution looks very different because the multipole expansion cannot be used and overlap effects must be taken into account. In addition, there are contributions of the form [A→B] and [B→A], called charge-transfer terms, in which the excitation is from an occupied orbital on one molecule to a virtual orbital on the other. They make a significant contribution in donor-acceptor interactions, for example when hydrogen bonding is involved.

The two-electron excitations include

(i) the dispersion terms [AB→AB], which like the induction terms are modified from the long-range form when overlap becomes important.

(ii) [AB→AA] and [AB→BB]. These may be called charge-transfer correlation terms.

(iii) [AA→AA] and [BB→BB]. These terms represent intramolecular correlation, and since they are present in the separated molecules they are not properly part of the interaction. However their effects are modified by the presence of overlap, and so they may have some effect on the interaction energy.

(iv) [AA→AB] and [BB→AB]. These are called extension correlation terms, for a reason which we discuss later.

(v) [AA→BB] and [BB→AA]. These describe a double charge transfer.

A natural alternative to perturbation theory which is widely used is the so-called supermolecule approach, in which a calculation is carried out on the combined system and compared with the energies of the separated molecules to obtain an interaction energy[21]. This is straightforward in principle but is also not without its difficulties. The interaction energy is usually a small fraction of the total energy — of the order of 10^{-3} hartree, compared with a total energy of hundreds of hartrees. Consequently accurate calculation is essential. The energies themselves are variationally bounded, but their difference is not, and may be in error in either direction.

A very important source of error is the 'basis set superposition error': if the basis used for molecule A is inadequate, the virtual orbitals of molecule B may be able to improve the description of A in a way which has nothing to do with the interaction, and this leads to a spurious stabilization. It is conventional to correct for it by means of the 'functional counterpoise method'[22] in which reference calculations are performed for each molecule in the presence of the basis functions, but not the electrons or nuclei, of the other. This procedure overcorrects for the effect, since it makes available to molecule A the occupied space of molecule B as well as the virtual space. It is possible to carry out a reference calculation for A in which the occupied orbitals of B are projected out of the basis[23], but although this gives better results than the normal procedure it is probably too cumbersome for routine use. Note that the function counterpoise method demands a separate reference calculation for each of the interacting molecules at every relative position, since the basis extension error varies with the position of the orbitals of the other molecule, and the procedure is therefore very time-consuming.

The only satisfactory solution at present appears to be the use of basis sets which are large enough to describe both molecules accurately. This is unfortunately often impracticable, since such a large basis can only be used for small molecules. Even in such cases difficulties remain, since electron correlation must be taken into account if dispersion effects are to be described. It is becoming clear that basis extension error is a much more serious problem in correlated calculations than in SCF ones[24], both because it is larger at a given level of basis and because it is difficult to correct for it in a consistent way[25]. A recent accurate calculation of the interaction between pairs of Be atoms[26] required s, p, d and f basis functions on each atom for a reasonably accurate result; omission of the f functions led to a 40% error in the well depth.

Perturbation theory offers some advantages in this respect. It is also subject to basis superposition error, but the contributions in which such errors may occur can be identified, and it is possible to make an estimate of the error and to correct for it to some extent. The charge-transfer energy is subject to basis superposition error, but it is possible to estimate the contribution of this error to the result. The extension correlation and double charge transfer terms are wholly due to basis superposition effects, to lowest order in overlap at least[18], and can be discarded. The charge-transfer correlation and dispersion terms, on the other hand, can have no basis superposition error at all because they can only arise when occupied orbitals of both molecules are present[19].

However it has to be concluded that at present, although progress is being made in both supermolecule and perturbation methods, no method currently available allows accurate calculation of intermolecular interactions without a great deal of computation using very high quality basis sets. Basis superposition error is a serious limitation in calculations on large systems, and the 'corrections' which can be obtained for it are best regarded as a qualitative guide to the unreliability of the calculation.

4. NON-ADDITIVE EFFECTS

The treatment so far given of both the long-range and the short-range interactions has assumed that only two molecules are considered at a time. The first approximation to the total interaction energy in a system of many molecules is the sum of all these pair-wise interactions. However, as noted in the Introduction, this will not usually be exact.

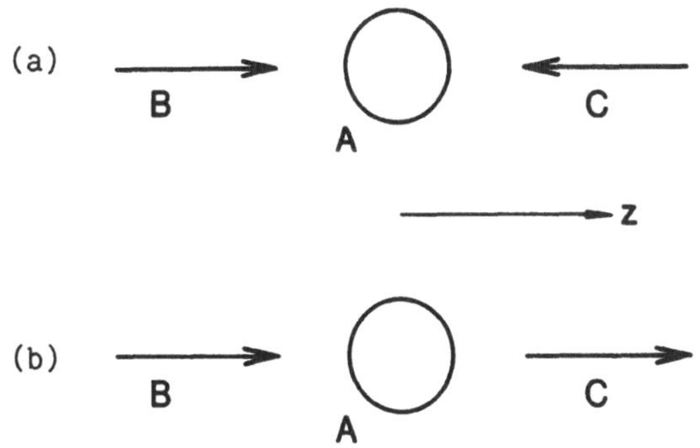

Figure 5. Induction energy of a polarizable molecule A in the field of two dipoles B and C.

For example, non-additivity is an important feature of the induction energy. This is readily seen from (2.31) which shows that the energy is quadratic in the field of the rest of the system. Because of this non-linearity, the induction energy is strongly non-additive. If a polarizable molecule A is placed symmetrically between two dipolar molecules B and C, as shown in Figure 5a, the total field F_{10} at A is zero, so the leading term in the induction energy, namely $-\frac{1}{2}(F_{10}^*)^2\,\alpha^A(1100)$, vanishes. (The field gradient is non-zero, so there is a small contribution $-\frac{1}{2}(F_{20}^*)^2\,\alpha^A(2200)$.) If the dipoles are arranged as in Figure 5b, however, the total field F_{10} is twice as large as that of a single molecule, and the induction energy four times as big.

A further important source of non-additivity arises as follows. If in Figure 5 molecules B and C are polarizable, then their dipole moments depend on the field they experience. This will in turn depend on the magnitude of the dipole they jointly induce in A, and will be different for the two arrangements shown. The determination of the induction energy for an assembly of polarizable molecules involves the solution of a set of linear equations for all the molecules in the assembly[27] and is a time-consuming procedure. Nevertheless Monte Carlo simulations have been carried out for polarizable dipolar hard spheres[28] and for a polarizable model of water[29], and it is concluded that the effect of the polarizability is substantial.

The other non-additive effect which has attracted most attention is the Axilrod-Teller triple-dipole dispersion energy, which for three atoms or rotationally-averaged molecules A, B and C is[30]

$$W^{(3)} = (3 \cos \theta_A \cos \theta_B \cos \theta_C + 1)\, C_{ABC}\, R_{AB}^{-3} R_{BC}^{-3} R_{CA}^{-3}, \quad (4.1)$$

where R_{AB} etc. and θ_A etc. are the sides and angles of the triangle formed by the three atoms, and C_{ABC} is a positive coefficient. It can be obtained by methods similar to those used for the ordinary second-order dispersion energy, and values have been calculated for many atoms and small molecules[31]. The triple-dipole energy is believed to contribute adversely to the binding energy in the inert gas solids, to the extent of about 10% of the total binding energy. Its effect in liquids is harder to estimate. In any case there is some doubt as to whether the Axilrod-Teller term provides an adequate description of the non-additive effect at short range[32]; a belief that there is a convenient cancellation between other significant contributions has recently been called into question.

In most studies there is no attempt made to describe three-body and higher order effects explicitly. This means that if reasonable agreement is to be obtained with experiment for condensed-phase properties, the pair potential must incorporate in some average way these many-body effects. The extent to which this can be done satisfactorily is not well understood at present. However it is clear that such effective condensed-phase pair potentials cannot be expected to give a good account of gas-phase properties such as second virial coefficients where many-body effects are rigorously absent.

5. MODEL POTENTIALS

For some applications, notably the Monte Carlo and molecular dynamics simulations described later in this book, the potential must be computed for many different relative positions of the molecules. Even when ab initio calculations become reliable enough, they are likely to remain orders of magnitude too slow for such applications. Consequently it is necessary to find some kind of simple model which incorporates whatever information can be assembled about the potential. It is usually necessary to draw this information from both theoretical and experimental sources.

One very old model potential for atoms which is still used is the Lennard-Jones or 6-12 potential:

$$u_{LJ} = \varepsilon[(R_m/R)^{12} - 2(R_m/R)^6]. \tag{5.1}$$

This has a minimum of depth ε at $R = R_m$. It has a convenient functional form, and can be used when experimental data are not good enough to parametrize better functions, but it has few other virtues. The R^{-6} term gives a crude representation of the dispersion energy, but the value of $2\varepsilon R_m^6$ which it gives for the C_6 coefficient is usually incorrect, while the repulsion is not well represented by the R^{-12} term.

At the other end of the scale are the potentials obtained for pairs of argon atoms by Barker and his colleagues[33]. These potentials are of the general form

$$u(R) = f_n(\bar{r}-1)\exp[\alpha(1-\bar{r})] - C_6/(\delta+\bar{r}^6) - C_8/(\delta+\bar{r}^8) - C_{10}/(\delta+\bar{r}^{10}), \tag{5.2}$$

where f_n is a polynomial of degree n (usually about 5), $\bar{r} = R/R_m$, and δ is a small positive constant. Variants on this form give a very good account of the properties of argon, and are generally agreed to be very accurate. These are only two examples of the multitude of empirical potential functions which have been proposed. A comprehensive survey of such inert gas potentials is given by Maitland, Rigby, Smith and Wakeham[34].

The description of molecule-molecule potentials is greatly complicated by the orientation dependence. One approach is to expand the potential using the type of function described in Section 1, but this is unsatisfactory for all except nearly spherical molecules like H_2 and perhaps a few others, because the expansion diverges at physically accessible distances. Another is to treat a molecule as a semi-rigid collection of spherical atoms, and this usually gives a better account of the dispersion and the repulsive part of the potential. To this may be added the multipole expansion of the electrostatic energy.

5.1 Distributed Multipole Models

It is now becoming clear, however, that conventional multipole expansions are quite inadequate for describing the electrostatic behaviour. A much better account is given by using a distributed multipole expansion[8,35], in which the charge distribution is represented by a number of charges, dipoles, quadrupoles etc. at various sites in the molecule, usually on the nuclei. An alternative preferred by some is to use point charges only, distributing them so as to reproduce the observed or calculated multipole moments[36] or the calculated electrostatic potential[37]. Some examples of the success of this approach are:

(i) Calculations of the delectric virial coefficient of HCl, using a hard-sphere model with embedded point multipoles, show that a one-centre multipole description totally fails to account for the experimental result. A distributed multipole model, on the other hand, gives quite good qualitative agreement[38].

(ii) The structures of a number of hydrogen-bonded systems were quite accurately predicted by a very similar model in which the repulsive potential was again of hard-sphere type, using standard Van der Waals radii for atoms other than hydrogen and ignoring hydrogen atoms (which fall inside the van der Waals spheres of the heavier atoms)[39]. The electrostatic interactions were modelled using distributed multipoles on the heavy atoms. The success of this simple model cannot be taken to mean that other contributions to the hydrogen bond energy can be ignored, but it is clear that a good description of the charge distribution is essential. Moreover it is clear from this work that a one-centre multipole model cannot be relied on to give the correct geometry.

(iii) The librational frequencies calculated for N_2 and CO_2 were in much better agreement with experiment when the molecular charge distribution was modelled using 5 charges than when central multipoles were used[40].

A reliable way to obtain such distributed multipole models is the distributed multipole analysis (DMA) method[8]. This takes advantage of the special properties of the Gaussian basis functions which are used in almost all modern wavefunction calculations. The electron density is a sum of products of such functions, and any product of Gaussian functions can be represented as a Gaussian centred at some intermediate point[41]:

$$\exp[-\alpha(\mathbf{r}-\mathbf{A})^2]\exp[-\beta(\mathbf{r}-\mathbf{B})^2] = \exp[-\alpha\beta(\mathbf{A}-\mathbf{B})^2/(\alpha+\beta)]\exp[-(\alpha+\beta)(\mathbf{r}-\mathbf{P})^2], \quad (5.3)$$

where

$$\mathbf{P} = (\alpha\mathbf{A}+\beta\mathbf{B})/(\alpha+\beta). \tag{5.4}$$

The Gaussian at **P** can then be represented by a point charge if charge overlap effects can be ignored. If the basis functions have angular momenta l_1 and l_2, then the angular dependence of these is represented by factors $R_{l_1m_1}(\mathbf{r-A})$ and $R_{l_2m_2}(\mathbf{r-B})$ respectively, and the product function has to be represented by a collection of point multipoles of ranks up to l_1+l_2.

This procedure generates a very large (though finite) number of point multipoles, since the point **P** is different for each pair of gaussian exponents α and β. However eq. (2.20) is invoked to represent each of these point multipoles in terms of a multipole expansion about the nearest of a small number of sites within the molecule. In this way the multipole expansion about each site has to represent only the charge distribution in its immediate neighbourhood, and the convergence properties are greatly improved.

The method gives results which depend to some extent on the basis set used in the wavefunction calculation, but it is still much more direct and unambiguous than fitting procedures. For example, the quadrupole moment of nitrogen, which is $-1.16\,ea_0^2$, might be represented by point charges (which requires charges of $-0.54\,e$ at each nucleus and $+1.08\,e$ at the centre of the bond) or by dipoles (which requires a dipole of $0.28\,ea_0$ at each nucleus) or by a quadrupole of $-0.58\,ea_0^2$ at each nucleus. The third of these gives much the best account of the overall hexadecapole, which comes out at $-7.48\,ea_0^4$, in comparison with the best theoretical estimate[42] of $-7.45\,ea_0^4$. However a distributed multipole analysis of N_2 shows that a much better description involves
(i) charges of $+0.46\,e$ at each nucleus and $-0.92\,e$ at the centre of the bond (note that the signs are now in agreement with chemical intuition);
(ii) dipoles of $0.88\,ea_0$ on each atom (considerably larger than required to fit the quadrupole moment alone);
(iii) a quadrupole of $2.11\,ea_0^2$ at the centre of the bond (larger in magnitude than and opposite in sign to the total quadrupole, but of the correct sign to describe the shape of the charge density in the bond).

Thus it is unwise to attempt to obtain distributed multipole models solely by reference to experimental data, especially as experimental data for the multipole moments are usually very unreliable for all but the first one or two non-vanishing moments. There is no difficulty in obtaining the information needed for the distributed multipole description from ab initio calculations; the distributed multipole analysis takes a fraction of the time required for the wavefunction calculation, and can be applied to CI functions as well as to SCF ones. Modern wavefunction calculations can give very good values for such properties, but it is of course possible to adjust the values in a calculated model to bring it into agreement with any reliable experimental data that may be available. The higher-order multipoles required for the equivalent one-centre description can also be obtained from ab. initio calculations, but because they involve operators containing high powers of r they are very sensitive to wavefunction errors in regions distant from the expansion centre. In any case the one-centre expansion would not perform as well even if the multipole moments were known accurately because it does not converge satisfactorily.

An obvious failing of multipole expansions of any kind is that they represent an extended charge distribution by means of point multipoles. Even if the expansion of the electrostatic energy converges, the result may still be wrong if the charge distributions overlap significantly. Such charge overlap effects have been shown to be quite important at short range[43], and indeed it is evident that the electrostatic interaction must tend to a finite value, or at worst diverge like R^{-1}, as the molecules are brought together, whereas the multipole expansion diverges like R^{-n} for some arbitrarily large n. Consequently some sort of damping ought to be incorporated at short range. In practice the repulsive part of the potential becomes more important at such distances, and a suitable model of the repulsion can absorb errors in the electrostatic description if they are not too large[44,45]. The same applies to the dispersion interaction.

5.2 Models of the repulsion

Accurate calculations of the repulsive part of the potential are difficult to achieve, for reasons explained above, so the choice of model is often made on grounds of computational convenience, and excused on the basis that nothing better is available. However we have found[46] that a repulsive interaction potential for pairs of H_2 molecules of the form

$$U_{rep} = \left[1 + b(\Omega)R\right] \exp\left[-\alpha\left(R - \rho(\Omega)\right)\right] \tag{5.5}$$

was able, in conjunction with suitable descriptions of the electrostatic and dispersion terms, to reproduce extremely accurately an SCF-CI potential surface calculated at a wide variety of distances and orientations. The parameters b and ρ were described by S function expansions requiring only one or two terms for convergence; α could have been allowed to depend on orientation too, but satisfactory results were obtained with a constant value.

H_2 is nearly spherical, unlike most molecules, but the success of (5.5) suggests that for larger molecules an anisotropic atom-atom potential should be used. Since atoms are nearly spherical, even in a molecule, it seems reasonable to attempt to describe the small departure from sphericity by an expansion technique. A suitable general form for a non-spherical atom-atom potential, by analogy with (5.5), would be the following:

$$u(R,\Omega) = f_n(R,\Omega)\exp\left[-\alpha(\Omega)[R-\rho(\Omega)]\right]. \tag{5.6}$$

In this function, $\rho(\Omega)$ is a sort of collision diameter, dependent on the full set of orientation variables denoted by Ω. It gives a description of the shape of the atoms. $\alpha(\Omega)$ describes the 'hardness' of the repulsive wall, and also depends on orientation. f_n is a polynomial in R, whose coefficients may depend on Ω. Normally the intermolecular potential would contain a term like (5.6) for each pair of interacting atoms, but extra centres may be added to obtain a better account of the repulsion in small molecules[47], or several atoms may be lumped together in a single interaction centre to obtain a more economical description of large molecules[48,49].

The virtue of this type of function is that it is fairly general in form, though very similar to the repulsive part of the potential (5.2) which has proved so successful for inert-gas atoms, and is expressed in terms of quantities ρ, α and f_n which are likely to be only weakly dependent on orientation. Consequently their angular dependence can be described using an S function expansion, which will converge very rapidly. The particular functional form given in eq. (5.6) is only one of a number of possibilities; for example several workers[50-52] have described a Lennard-Jones potential in which the well-depth and the position of the minimum were allowed to be functions of orientation in the same way. The idea of incorporating angular dependence in the potential parameters, rather than expanding the potential itself in terms of a set of angular expansion functions, seems to be due to Corner[53].

Most workers have preferred to use spherical atom-atom functions in simulations, because they are easier to handle computationally. However the formulae for handling anisotropic atom-atom functions are available[54], and explicit expressions have been given in terms of scalar and vector products of the vectors describing relative positions and orientations[55]. These vectors are already used in most simulation techniques. Representations of charge distributions in terms of arrays of point charges require more interaction centres (and an increase in computational time proportional to the square of the number of centres) and do not address the description of anisotropy in the repulsion. The necessity for using anisotropic atom-atom repulsion potentials is illustrated by the success of a recent model of this kind used to account for the crystal structure of Cl_2[56]. In this case the form of the function $\rho(\Omega)$ in the expression (5.6) used for the atom-atom repulsive potential is such as to describe a slight increase in the effective atomic radius in the directions of maximum lone-pair density, so that it accounts for the structure in a chemically sensible as well as mathematically efficient way.

Note that an atom-atom potential depends implicitly on the internal coordinates, since these determine the position of the atoms relative to the molecular coordinate frame. The potential parameters may be allowed to depend explicitly on the internal coordinates, but it is unlikely that experimental data will be good enough to characterize such dependence for some time. Nevertheless it is clear that anisotropic atom-atom potential functions of the type described are capable not only of summarizing our present knowledge but of incorporating future developments.

References

[1] J. A. Pople, 1954, Proc. Roy. Soc. A, **221**, 498.
[2] W. A. Steele, 1963, J. Chem. Phys, **39**, 3197.
[3] D. M. Brink & G. R. Satchler, 'Angular Momentum', 1968, Oxford University Press.
[4] L. Blum & A. J. Torruella, 1972, J. Chem. Phys., **51**, 303.
[5] A. J. Stone & R. J. A. Tough, 1977, J. Phys. A, **10**, 1261.
[6] R. Ahlrichs, 1976, Theor. Chim. Acta, **41**, 7.
[7] A. D. Buckingham, 1967, Adv. Chem. Phys., **12**, 107.
[8] A. J. Stone, 1981, Chem. Phys. Letters, **83**, 233.
[9] A. J. Stone, 1976, J. Phys. A, **9**, 485.
[10] R. J. A. Tough, Ph.D. Thesis, University of Cambridge, 1977.
[11] J. O. Hirschfelder, W. Byers Brown & S. T. Epstein, 1964, Adv. Quantum Chem., **1**, 255.
[12] C. G. Gray & B. W. N. Lo, 1976, Chem. Phys., **14**, 73.
[13] P. W. Atkins, 'Molecular Quantum Mechanics' (Oxford, 1983).
[14] G. Starkschall & R. G. Gordon, 1971, J. Chem. Phys., **54**, 663; 1972, ibid., **56**, 2801.
[15] P. W. Langhoff & M. Karplus, 1970, J. Chem. Phys., **53**, 233; P. W. Langhoff, 1972, ibid., **57**, 2604.
[16] G. D. Zeiss & W. J. Meath, 1977, Molec. Phys., **33**, 1155.
[17] R. T. Pack, 1982, J. Phys. Chem., **86**, 2794.
[18] I. C. Hayes & A. J. Stone, Mol. Phys., submitted for publication.
[19] A. J. Stone & I. C. Hayes, 1982, Faraday Discussions Chem. Soc., **73**, 19, 109, 113.
[20] J. N. Murrell, M. Randic & D. R. Williams, 1965, Proc. Roy. Soc. A, **284**, 566; J. N. Murrell & G. Shaw, 1967, Molec. Phys., **12**, 475.
[21] J. A. Pople, 1982, Faraday Discussions Chem. Soc., **73**, 7.
[22] S. F. Boys & F. Bernardi, 1970, Molec. Phys., **19**, 553.

[23] P. A. Madden & P. W. Fowler, J. Chem. Phys., submitted for publication.
[24] P. G. Burton, P. D. Gray & U. E. Senff, 1982, Molec. Phys., **47**, 785; P. G. Burton, 1982, Faraday Discussions Chem. Soc., **73**, 111.
[25] S. L. Price & A. J. Stone, 1979, Chem. Phys. Letters, **65**, 127.
[26] R. J. Harrison & N. C. Handy, 1983, Chem. Phys. Letters, **98**, 97.
[27] J. Applequist, 1977, Acc. Chem. Res., **10**, 79.
[28] G. N. Patey, G. M. Torrie & J. P. Valleau, 1979, J. Chem. Phys., **71**, 96.
[29] P. Barnes, J. L. Finney, J. D. Nicholas & J. E. Quinn, 1979, Nature, **282**, 5738.
[30] B. M. Axilrod & E. Teller, 1943, J. Chem. Phys., **11**, 299.
[31] D. J. Margoliash, T. R. Proctor, G. D. Zeiss & W. J. Meath, 1978, Molec. Phys., **35**, 747.
[32] S. F. O'Shea & W. J. Meath, 1976, Molec. Phys., **31**, 515.
[33] J. A. Barker & A. Pompe, 1968, Aust. J. Chem., **21**, 1683; J. A. Barker, R. A. Fisher and R. O. Watts, 1971, Molec. Phys., **21**, 657; J. A. Barker, R. O. Watts, J. K. Lee, T. P. Schafer & Y. T. Lee, 1974, J. chem. Phys., **61**, 3081.
[34] G. G. Maitland, M. Rigby, E. B. Smith & W. A. Wakeham, 'Intermolecular Forces', Oxford University Press (1981).
[35] J. Bentley, in 'Chemical Applications of Atomic and Molecular Electrostatic Potentials', eds. P. Politzer & D. G. Truhlar (Plenum Press, New York, 1981).
[36] J. T. Brobjer & J. N. Murrell, 1982, Faraday Trans. Chem. Soc. II, **78**, 1853.
[37] S. R. Cox & D. E. Williams, 1981, J. Comput. Chem., **2**, 304.
[38] C. G. Joslin, private communication.
[39] A. D. Buckingham & P. W. Fowler, J. Chem. Phys., submitted for publication.
[40] C. S. Murthy, S. F. O'Shea & I. R. McDonald, Molec. Phys., submitted for publication.
[41] S. F. Boys, 1950, Proc. Roy. Soc. A, **200**, 542.
[42] R. D. Amos, 1980, Molec. Phys., **39**, 1.

[43] K. Ng, W. J. Meath & A. R. Allnatt, 1976, Molec. Phys., **32**, 177; 1977, ibid., **33**, 699; 1979, ibid., **38**, 449.
[44] J. N. Murrell & J. J. C. Teixeira-Dias, 1970, Molec. Phys., **19**, 521.
[45] K.-C. Ng, W. J. Meath & A. R. Allnatt, 1979, Molec. Phys., **37**, 237.
[46] S. L. Price & A. J. Stone, 1980, Molec. Phys., **40**, 805.
[47] J. C. Raich & N. S. Gillis, 1977, J. chem. Phys., **66**, 846.
[48] D. J. Evans & R. O. Watts, 1976, Molec. Phys., **31**, 83; 1976, ibid., **32**, 93.
[49] S. L. Price & A. J. Stone, 1983, Chem. Phys. Letters, **98**, 419; Molec. Phys., submitted for publication.
[50] B. J. Berne & P. Pechukas, 1972, J. chem. Phys., **56**, 4213.
[51] R. T. Pack, 1978, Chem. Phys. Letters, **55**, 197.
[52] S. H. Walmsley, 1977, Chem. Phys. Letters, **49**, 390.

[53] J. Corner, 1948, Proc. Roy. Soc. A, **192**, 275.
[54] A. J. Stone, 1978, Molec. Phys., **36**, 241.
[55] S. L. Price & A. J. Stone, Molec. Phys., submitted for publication.
[56] S. L. Price & A. J. Stone, 1982, Molec. Phys., **47**, 1457.

DIFFUSION IN LIQUIDS*)

F.P. Ricci, D. Rocca

Dipartimento di Fisica, Università degli Studi
"La Sapienza" - P.le Aldo Moro, 2 - 00185 Roma - Italy

*) Partially supported by the G.N.S.M. - C.N.R.

INTRODUCTION

Transport phenomena are the irreversible decay processes that occur spontaneously when a system, pertubed to a non equilibrium state by some external or internal disturbance, advances through a series of non equilibrium states until equilibrium is reached. A system in absence of external forces is in equilibrium if the intensive state variable (for example density, local velocity, temperature, composition etc.) are constant at all times and have the same magnitude at all position in the system. Therefore at equilibrium there are no gradients of the state parameters in the system and there is no net flux of matter or momentum or energy through the system. If the system is perturbed to a non equilibrium state at least one state parameter becomes function of position and spontaneously fluxes of the appropriate thermodynamic quantities are generated by these gradients. In these lectures we will be restricted to the transport process called isothermal isobaric diffusion i.e. the transport process we have in a multicomponent system, in absence of external forces at mechanical equilibrium and at uniform temperature, when a composition gradient is present. These lectures are organized as follows: in the first chapter we will derive the Fick's law in the framework of the non equilibrium thermodynamics; in such a way it can be generalized and we will discuss the problem of the frame of reference, the physical meaning of the various diffusion coefficients generally used and finally the range of validity of this generalized Fick's law.
In chapter two we will discuss two general feature of the self diffusion process in liquids from triple to critical point: the dependence on the peculiarities of the intermolecular potential and the meaning of the diffusion as an "activated process".

A. J. Barnes et al. (eds.), Molecular Liquids - Dynamics and Interactions, 35–58.

In chapter three we will present what we think to be the most promising theoretical model for the diffusion process in liquids and will show the agreement between its theoretical predictions and the experimental data. We want to stress that until now no satisfactory theory for the transport processes in liquids is available in the literature; the theory we present here, altough it has still some open problems, contains the most important physical insights and reproduces very well the experimental self diffusion behaviour.

In chapter four we will discuss the self diffusion in water showing quantitatively the influence of the H-bond in this process.

1. PHENOMENOLOGICAL FRAMEWORK(1)

For any particular transport process there is an empirical relationship between the fluxes and the gradients. For example in the case of binary mixture, in mechanical equilibrium and without external forces and at uniform temperature, if we have composition gradient, $\frac{\partial c_i}{\partial x}$, we will have also flux of component i, J_i, and the following empirical relationship has been found to hold:

$$J_i = -D \frac{\partial c_i}{\partial x} \tag{1.1}$$

where D is the diffusion coefficient; eq. (1.1) is called the Fick's first law for diffusion.

The empirical laws can be organized in the framework of the non equilibrium thermodynamics. To accomplish this goal, we start assuming three postulates:

I) Local equilibrium: "for a system in which irreversible processes are taking place, all thermodynamic functions of state exist for each small element of the system. These thermodynamic quantities for the non equilibrium system are the same functions of the local state variables as the corresponding thermodynamic quantities in an equilibrium system".

II) Validity of the linear approximation: "the fluxes J_i (expressed in number of grams of component i crossing a plane of unit area in unit time) are linear homogeneous functions of the forces X_i". Following this hypothesis we can write

$$\underline{J}_i = \sum_k L_{ik} \underline{X}_k \tag{1.2}$$

the L_{ik} are the phenomenological coefficients. Eq. (1.2) can be implemented by the so - called "Curie's theorem" wich states that: "in an isotropic system there are no interactions among entities whose tensorial characters differ by an odd integer".

III) The Onsager linear relationships: "if we write the linear relationships (1.2) using fluxes and forces defined in an "appropriate" way (2), the matrix L_{ik} is symmetric i.e. $L_{ik} = L_{ki}$. This postulate does not hold in the cases in which the space is not isotropic as in presence of an external magnetic field or for a rotating system. "Appropriate" way means that if J_i is a time derivative of the thermodynamic variable a_i, i.e. $J_i = \frac{da_i}{dt}$ the conjugate force X_i is defined by the relationship $X_i = T \frac{\partial \sigma}{\partial a_i}$ where σ is the entropy of the isolated system. With these definitions we can write the rate of internal entropy production for unit volume, Φ_s, as

$$\Phi_s = \frac{1}{T} \sum_i \underline{J}_i \ X_i \tag{1.3}$$

and if we define the quantity Φ as $\Phi = \Phi_s . T$ we have

$$\Phi = \sum_i \underline{J}_i \underline{X}_i \tag{1.4}$$

Eq. (1.2) are the general expressions for the laws which govern the various transport processes. To write down them explicitly for each transport process, we must identify the fluxes and the forces. This can be done looking at the expression for the rate of internal entropy production written in terms of fluxes and forces [see eq. (1.3)].

The rate per unit volume of the internal entropy production, using the assumption of local equilibrium, is given by

$$\rho \frac{ds}{dt} = \Phi_s - \underline{\nabla} \cdot \underline{J}_s \tag{1.5}$$

where ρ is the mass density of the system, $\frac{d}{dt}$ is the substantial derivative, s is the specific entropy and $\underline{J}_s$ is the total entropy flux (diffusion and heat flow) measured in respect to the local center of mass frame of reference (i.e. the frame of reference for which the local center of mass velocity $\underline{u}$ vanishes: $\sum_{i=1}^{r} \rho_i \underline{u}_i = 0$

where r is the number of the components ρ_i the partial mass density of component i and $\underline{u}_i$ its local mean velocity; in the following we will call this frame of reference the mass fixed frame of reference and the fluxes measured in this frame of reference are denoted by small J). We recall that for a fluid

$$Tds = de + pd \left(\frac{1}{\rho}\right) - \sum_{i=1}^{r} \mu_i \, dx_i \tag{1.6}$$

where e is the specific internal energy, μ_i is the chemical potential of component i and x_i its weight fraction; moreover we use the conservation of energy and matter, taken from hydrodynamics, which in the case of isobaric (i.e. a system in mechanical equilibrium and in absence of external forces) isothermal mixture of r components are

$$\frac{de}{dt} = \frac{1}{\rho}\{ -\underline{\nabla}\cdot\underline{q} - T\,\underline{\nabla}\cdot\sum_{i=1}^{r} \underline{j}_i\, s_i - \underline{\nabla}\cdot\sum_{i=1}^{r} \underline{j}_i\, \mu_i\};\ \frac{d\rho}{dt} = 0\ ;\rho\,\frac{dx_i}{dt} = -\underline{\nabla}\cdot\underline{j}_i$$

Introducing these conservation laws in (1.6) and noting that $\underline{j}_s = \frac{q}{T} + \sum_{i=1}^{r} \underline{j}_i\, s_i$, we obtain for the quantity Φ, in the case of an isothermal isobaric mixture of r components,

$$\Phi = -\sum_{i=1}^{r} \underline{j}_i\cdot \underline{\nabla}_{T,p}\mu_i \tag{1.7}$$

where $\nabla_{T,p}$ means the gradient taken at constant temperature and pressure. We can compare eq. (1.7) with eq. (1.4).

Therefore from eq.(1.7) we derive that in the case of a multicomponent system in isothermal isobaric conditions and with gradients of chemical potentials, the forces which drive the diffusion process are the gradients of chemical potentials of the various components.

Using eq.(1.2), the phenomenological relationships for isothermal isobaric diffusion process in an r - components system is

$$\underline{j}_i = -\sum_{l=1}^{r} \Omega_{il}\, \underline{\nabla}_{T,p}\, \mu_l \qquad (i=1,2,\dots r) \tag{1.8}$$

where Ω_{il} are the phenomenological coefficients related to the L_{il} of eq.(1.2) simply as $\Omega_{il} = \frac{1}{T} L_{il}$

There are some important relationships:

$$\sum_{i=1}^{r} \Omega_{il} = 0 \qquad (l = 1, 2 \dots r) \tag{1.9}$$

$$\sum_{i=1}^{r} \rho_i\, \underline{\nabla}_{T,p}\mu_i = 0 \tag{1.10}$$

Eq. (1.9) follows from eq. (1.8) and from the fact that the fluxes $\underline{j}_i$ are measured in the local center of mass frame of reference in which, by definition, $\sum_{i=1}^{r} \underline{j}_i = 0$; eq. (1.10) is the Gibbs-Duhem equation written for a system in isothermal and isobaric conditions.

Since the chemical potentials are functions of the temperature T, the pressure p and of r-1 partial mass densities, ρ_r, i.e.

$\mu_l = \mu_l\,(T, p, \rho_1, \rho_2, \dots, \rho_{r-1})$ we can write $\nabla_{T,p}\,\mu_l = \sum \frac{\partial \mu_l}{\partial \rho_k} \nabla_{T,p}\rho_k$

Defining the diffusion coefficient D_{ik} as

$$D_{ik} = \sum_{l=1}^{r} \frac{\partial \mu_l}{\partial \rho_k} \Omega_{il} \qquad (1.11)$$

eq. (1.8) becomes:

$$-\underline{j}_i = \sum_{k=1}^{r-1} D_{ik} \underline{\nabla} \rho_k \qquad (i=1, 2, \dots r) \qquad (1.12)$$

Eq. (1.12) is the generalization of eq. (1.1) i.e. of the Fick's empirical relationship. It is clear that in a multicomponent system, $(r > 2)$, the diffusion flux of component i is driven not only by its concentrations gradients but also by the concentration gradients of the other components.

Since it is easier to measure the concentration gradients instead of the chemical potential ones, in real experiments eq. (1.12) is used and therefore the D_{ik}, practical diffusion coefficients, are determined instead of using eq. (1.8) from which the theoretically more meaningful Ω_{ik} could be obtained.

It is obvious that, using eq. (1.11) we can go from the D_{ik} to the Ω_{ik} if we know the dependence of the chemical potentials on the partial mass densities. We think important to stress that in our derivation of the Fick's law, we measured the diffusion flux relative to the velocity of the local center of mass, therefore also the D_{ik} are referred to the mass fixed frame of reference. However other frames of reference are of some interest (3): the solvent fixed one, the volume fixed one and the cell frame of reference. Since if we change frame of reference the fluxes $\underline{j}_i$ can change, we must rewrite eq. (1.12), or eq. (1.8), in the new reference system. The starting point is that if we measure the same diffusion flux first relative to the frame of reference R, $(\underline{j}_i)_R$, and secondly to frame S, $(j_i)_S$, we have:

$$(\underline{j}_i)_R = (\underline{j}_i)_S + \rho_i \underline{u}_{SR} \qquad (1.13)$$

where $\underline{u}_{SR}$ is the velocity of the frame S relative to the frame R; of course u_{SR} is function of the position and the time. Each frame of reference is defined by a condition on the fluxes j_i:

1. For the mass fixed frame of reference, which is indicated in the following, by $(\ )_M$,

$$\sum_{i=1}^{r} (j_i)_M = 0 \qquad (1.14)$$

2. For the solvent fixed frame of reference, which is indicated by $(\ )_r$ since r is the solvent component,

$$(\underline{j}_r)_r = 0 \tag{1.14b}$$

3. For the volume fixed frame of reference, which is indicated by $(\ )_v$,

$$\sum_{i=1}^{r} v_i (\underline{j}_i)_v = 0 \tag{1.14c}$$

where v_i is the partial specific volume of component i. This relationship is a consequence that in this frame of reference there is no net flow of volume.
From eq. (1.13) and (1.14) we can obtain the various u_{sr} and rewrite eq. (1.8) and (1.12) and consequently define the D_{ik} and Ω_{ik} in the new frame of reference in terms of the same quantities defined in the old frame of reference. For example in the case of the mass fixed, taken as S, and solvent fixed, taken as R, frames of the reference we have from eq. (1.13) written in the case of i = r

$$\underline{u}_{Mr} = - \frac{(\underline{j}_r)_M}{\rho_r}$$

Therefore $(\underline{j}_i)_r = (\underline{j}_i)_M - \frac{\rho_i}{\rho_r} (\underline{j}_r)_M$ and using eq. (1.8)

$$(\underline{j}_i)_r = - \sum_{l=1}^{r} (\Omega_{il})_M \underline{\nabla}_{T,p}\, \mu_l + \frac{\rho_i}{\rho_r} \sum_{l=1}^{r} (\Omega_{rl})_M \underline{\nabla}_{T,p}\, \mu_l \tag{1.15}$$

Writing $\nabla_{T,P}\, \mu_r$ as a function of $\nabla_{T,P}\mu_i$ (with i= 1,2 r-1) by means of the Gibbs -Duhem eq., (see eq. 1.10), and defining

$$(\Omega_{il})_r = (\Omega_{il})_M - \frac{\rho_i}{\rho_r} (\Omega_{rl})_M - \frac{\rho_l}{\rho_r} (\Omega_{ir})_M + \frac{\rho_i \rho_l}{\rho_r^2} (\Omega_{rr})_M \tag{1.16}$$

we have that in the solvent fixed frame of reference the phenomenological eq.(1.15) becomes

$$(\underline{j}_i)_r = - \sum_{l=1}^{r-1} (\Omega_{il})_r \underline{\nabla}_{T,p}\, \mu_l \qquad (i=1,2,\ldots r-1) \tag{1.17}$$

or

$$(\underline{j}_i)_r = - \sum_{k=1}^{r-1} (D_{ik})_r \underline{\nabla}\rho_k \qquad (i=1,2,\ldots r-1) \tag{1.18}$$

with

$$(D_{ik})_r = \sum_{l=1}^{r-1} \frac{\partial \mu_l}{\partial \rho_k} (\Omega_{il})_r \tag{1.19}$$

It is easy to show that $(\Omega_{ij})_r = (\Omega_{ji})_r$ and $(D_{ik})_r = (D_{ik})_M - \frac{\rho_i}{\rho_r}(D_{rk})_M$
In similar way we can derive that $(D_{ik})_v = (D_{ik})_r - \rho_i \sum_{l=1}^{r} v_l (D_{lk})_r$.
It is different the case of the cell frame of reference. In fact, as it is clearly shown in Ref. 3 for the case in which the gradients are only in the x direction, eq. (1.13), written between the cell frame of reference, $(\ )_c$, and another frame of reference R, becomes

$$(j_i)_c = (j_i)_R - \rho_i \sum_{l=1}^{r} v_i (j_i)_R + \rho_i \int_{+\infty}^{x} \sum_{l=1}^{r} (j_i)_r \sum_{k=1}^{r} \left(\frac{\partial v_l}{\partial c_k}\right)_{m\neq k} \frac{\partial c_k}{\partial x}\, dx \tag{1.20}$$

where $(\ )_{m\neq k}$ means that all the concentrations are held constant except the k and r concentrations.
If we specialize the frame R with the solvent fixed frame of reference, the diffusion equation written in the cell frame of reference is:

$$(j_i)_c = - \sum_{k=1}^{r-1} \left[(D_{ik})_r - \rho_i \sum_{l=1}^{r-1} v_l\ (D_{lk})_r\right] \frac{\partial \rho_k}{\partial x} - \rho_i \int_{+\infty}^{x} \sum_{l=1}^{r-1} \sum_{k=1}^{r-1} \sum_{m=1}^{r-1} \left(\frac{\partial v_l}{\partial \rho_m}\right) (D_{lk})_r \frac{\partial \rho_k}{\partial x} \left(\frac{\partial \rho_m}{\partial x}\right) dx \tag{1.21}$$

analogous relationship we can derive in the case in which R is identified with the volume fixed frame of reference.
Therefore one must be very careful in the case in which the fluxes are measured in respect to a frame of reference fixed with the cell; in fact in this case one cannot apply the Fick's law to derive the diffusion coefficient unless the partial specific volumes are nearly independent from the composition or the concentration gradients are sufficiently small. It is worthwhile to notice that also in the case in which we can neglect the integral in eq. (1.21)the Fick's law will not give the diffusion coefficient by a linear combination among them with a weight given by the concentrations and the partial molar volumes. We now go back to eq. (1.12), i.e. we suppose to measure the fluxes in the mass fixed frame of reference, and we want to discuss the case of a binary mixture (i.e. r=2). In such a situation we have from eq. (1.9) and the Onsager reciprocal relationships:

$$\Omega_{11} = -\Omega_{12} = -\Omega_{21} = \Omega_{22} = \Omega \tag{1.22}$$

Inserting eq. (1.22) in eq. (1.9) we have

$$\begin{cases} \underline{j}_1 = -D_{11}\, \underline{\nabla}\, \rho_1 \\ \underline{j}_2 = -D_{21}\, \underline{\nabla}\, \rho_1 \end{cases} \quad \text{or} \quad \begin{cases} \underline{j}_1 = -D_{12}\, \underline{\nabla}\, \rho_2 \\ \underline{j}_2 = -D_{22}\, \underline{\nabla}\, \rho_2 \end{cases} \tag{1.23}$$

and using eq. (1.10), (1.11) and (1.22)

$$D_{11} = \Omega \left(\frac{\partial \mu_1}{\partial \rho_1} - \frac{\partial \mu_2}{\partial \rho_2}\right) = \frac{\Omega}{x_2} \frac{\partial \mu_1}{\partial \rho_1} \qquad (1.24)$$

$$D_{21} = - D_{11}$$

The same for D_{12} and D_{22}; and $D_{22} = \frac{V_2}{V_1} D_{11}$. This last relationschip is a consequence that D_{ii} is defined in respect to $\underline{\nabla}\rho_i$.

Therefore in the case of a binary mixture we measure only one diffusion coefficient, D_M, which is called the mutual diffusion coefficient and its inverse is a measure of the relaxation time, i.e. the inertia, with which the binary mixture recover its equilibrium state following a local disturbance in the concentration.
If we use the solvent fixed frame of reference, eq. (1.23) and (1.24) are substituted by

$$\begin{cases} (\underline{j}_1)_r = - (D_{11})_r \underline{\nabla}\rho_1 \\ (\underline{j}_2)_r = 0 \end{cases} \qquad (1.25)$$

and

$$(D_{11})_r = (1 + \frac{\rho_1}{\rho_2}) D_{11} \qquad (1.26)$$

If we use the volume fixed frame of reference eq. (1.23) and (1.24) are substituted by

$$\begin{cases} (\underline{j}_1)_v = -(D_{11})_v \underline{\nabla}\rho_1 \\ (\underline{j}_2)_v = (D_{21})_v \underline{\nabla}\rho_1 \end{cases} \qquad (1.27)$$

and

$$\begin{cases} (D_{11})_v = \rho v_2 D_{11} \\ (D_{21})_v = -\rho v_1 D_{11} \end{cases} \qquad (1.28)$$

It is interesting to discuss two particular cases:

a) - a binary mixture in which the two components have the same molecular properties, i.e. are isotopes, and one component (i.e. component 1) is at infinite dilution. In this case there is no difference among the various frames of reference and we call the mutual diffusion coefficient as the "self-diffusion coefficient, D_s".

b) - a binary mixture in which the component 1 is at infinite dilution (i.e. tracer). In this case the mutual diffusion coeffi-

cient is practically equal to the diffusion coefficient of the tracer in respect to the solvent (see eq. (1.26); the tracer diffusion coefficient, D_i, in a binary mixture is practically the self diffusion coefficient of the tracer itself in the binary mixture (to be correct to measure the self diffusion coefficient of species 1 in a two component system we should have a three component system $1 + 1^* + 2$).
Finally few words about the range of validity of the Fick's law. It is clear, from our derivation of the Fick's law in the framework of the non equilibrium thermodynamics, that we must satisfy the following conditions:

1) - the coarse graining in time must be such that the longest memory time present in the microscopic dynamics of our system can be considered negligible.

2) - the coarse graining in space must be such that we can neglect distance of the order of the average intermolecular distance so that the system can be approximated with a continuum medium.

3) - the gradients of concentrations must be small enough so that the linear approximation holds. There are no clear rules to say when "the gradients are small enough" but there are many suggestions (4) that gradients of the order generally found in nature are well ithin the linear approximation.

2. SOME GENERAL FEATURES OF THE SELF DIFFUSION PROCESS IN LIQUIDS.

If we want to build up a microscopic theory of the diffusion process in liquids, it is worth while to clarify, looking at the experimental data available in the literature, the following general questions:
a) - How strongly the behaviour of the diffusion coefficient is dependent on the exact shape of the intermolecular potential?

b) - Can the diffusion process in liquids be considered as an "activated process"?

The first question is relevant if we want to use an effective hamiltonian having the same functional form for a large number of liquids.
The second question is motivated by the strong similarity that many authors (5) claimed between the solid and the liquid state; moreover the analysis of the experimental data has been frequently given in terms of an activated model with some conclusion about the "activation energies" which are strongly model dependent.
We try to answer to question a) using the corresponding state approach (6).
The corresponding state principle says that if the various systems

can be treated classically, as far as the translational degrees of freedom and can be described by an effective intermolecular potential of universal shape $\left[U = \varepsilon f\left(\frac{r_1}{\sigma}, \frac{r_2}{\sigma} \ldots \frac{r_N}{\sigma}\right)\right.$ where f is the same function for all the molecular species, ε and σ are the molecural parameters], the diffusion coefficients, measured in reduced units and expressed in function of reduced state parameters, must follow an universal curve.

In fact, as it is clearly deduced in Ref. 6c, we can define reduced forms for all the quantities like the hamiltonian, the streaming operator, the momentum, the temperature, the time, the volume etc. Therefore the Kubo expression (6c) for the selfdiffusion coefficient

$$D_s = \frac{1}{m^2} \int_0^\infty < p_x(0)\ p_x(t) > dt \qquad (2.1)$$

can be written as

$$D_s = \sigma\sqrt{\frac{\varepsilon}{m}} \int_0^\infty < p_x^*(0)\ p_x^*(t^*) >^* dt^* \qquad (2.2)$$

where $p_x^*(t^*) = \frac{p_x\left(\frac{t}{\sigma}\sqrt{\frac{\varepsilon}{m}}\right)}{\sqrt{m\varepsilon}}$, $t^* = \frac{t}{\sigma}\sqrt{\frac{\varepsilon}{m}}$ and $\langle \ldots \rangle^*$ means the thermodynamic average for a canonical ensemble described by the reduced hamiltonian $H^* = \sum_{i=1}^{N} \frac{p_i^{*2}}{2} + f\,(r_1^*, r_2^* \ldots r_N^*)$. As a consequence the quantity $D_s^* = \frac{D_s}{\sigma}\sqrt{\frac{m}{\varepsilon}}$ must be independent of the molecular species; it is only function of $\rho^* = \frac{\rho\sigma^3}{m}$ and $T^* = \frac{K_B T}{\varepsilon}$.

Therefore, for the same values of the reduced state parameters, the D_s^* value must be the same for all the systems. We can also use another version of the reduced quantities. In fact since the critical point is, by definition, a "correspondent point" it follows that the critical temperature T_c is proportional to ε and the critical molar volume v_c is proportional to σ^3 with constant of proportionality independent of the particular substance. Therefore, to define the reduced quantities, we can use T_c and v_c instead of ε and σ as it has been done already in the case of equilibrium properties (7).

Coming back to question a) we consider the experimental selfdiffusion data available in the literature for various liquids having different intermolecular interactions. First of all we confine ourselves to the coexistence curve where we have only one thermodynamic state parameter; we choose the temperature. Secondly, since we are mainly interested in the behaviour of D_s as a function of T, instead of considering D_s^* (T^*) we use the values normalized with the selfdiffusion value at a fixed point T_k^* on the coexisting curve; we choose T_k=0.667T_c (i.e. T_k^*=0.667) so that this normalization point is just in the middle of the liquid range. Recalling that

$$D^*(T^*) = \frac{1}{V_c^{4/3}} \sqrt{\frac{m}{K_B T_C}} \; D_s\,(T)$$

we have

$$\Gamma = \frac{D^*(T^*)}{D^*(0.667)} = \frac{D_s\,(T)}{D_s(0.667T_c)} \qquad (2.3)$$

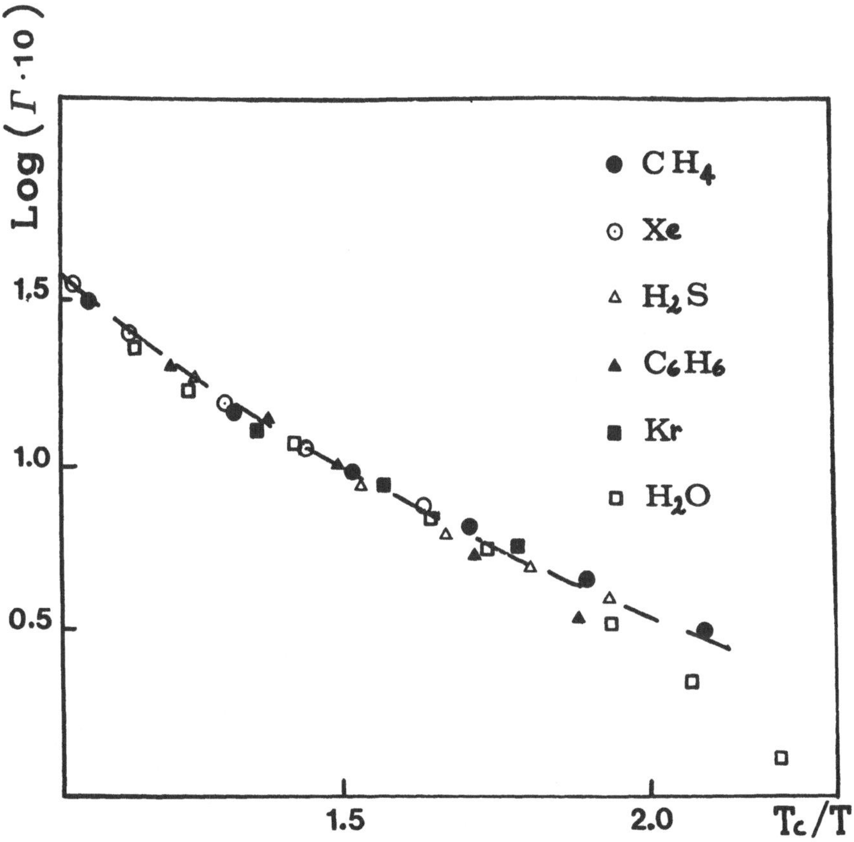

Fig. 1 Reduced self diffusion coefficient (see the text, eq. (2.3), for the definition of Γ) as a function of the reduced temperature along the liquid side of the coexistence curve. To focus the attention of the temperature behaviour the data have been normalized at their values at T=0.667T_c. The experimental data are taken: for Kr from Ref. 8a; for Xe from Ref. 8b; for CH_4 from Ref. 9; for H_2S from Ref. 11 and for C_6H_6 and H_2O from Ref. 10.

In fig. 1 we report these normalized reduced selfdiffusion experimental data for various systems having different intermolecular potential:

- monoatomic liquid as Kr and Xe
- a liquid of simple globular molecules like CH_4
- a liquid of planar molecules like C_6H_6
- a dipolar liquid like H_2S
- an H-bonded liquid like H_2O

From fig. 1 it is impressive to note that, except H_2O at low temperature, the reduced selfdiffusion data of all these different systems follow the same curve showing unambigously that "the behaviour of selfdiffusion coefficient is independent of the exact shape of the intermolecular potential". This is just the reply to question a.

This conclusion enables us to simplify the problem of the theoretical model for the diffusion process in liquids; in fact, if we do not consider the associated liquids, the orientational effects or the coupling with internal degrees of freedom does not seems to be important in the liquid state as far as the behaviour of D_s. This means that they are either negligible or constant as the temperature and density vary from the triple point to near the critical one.

In our opinion the physical reason why the behaviour at D_s is insensitive to the exact shape of the intermolecular potential, must be found in the fact that the diffusion coefficient is an integral property over the microscopic dynamics of the molecule (see eq. 2.1). The time integration washes out the structures of the velocity autocorrelation function; in some sense it is similar to what happens for the specific heat in respect to the pseudo "density of states". It is worthwhile to note that for the same reason one must be careful in deriving information on the microscopic dynamics of the molecules just from self diffusion measurements.

Coming to question b) we want to stress that many selfdiffusion data, along the liquid side of the coexistence curve, can be fitted by an expression like $D = D_0 \exp - (\frac{\Delta E}{RT})$;

this fact has been the main reason why many authors describe the diffusion process in liquids as an activated one and ΔE has been interpreted as an activation energy. We do not agree with this point of view because it is not supported by the experimental data. The "activated process" in the usual sense is a process in which a molecule jumps from one "quasi-equilibrium" position to another one and during this step it overcomes an energy barrier of height ΔE;if the model makes sense it is necessary that $\Delta E \gg KT$ to explain the low values of D_S in liquids ($10^{-4} < D_S < 10^{-5}$ cm^2/s). Of course if we want to interpret ΔE as the energy barrier between the two equilibrium position we must follow the diffusion process at constant volume. The rate process theory (12) is a typical theoretical approach to describe the diffusion process as an activated one. We now consider the experimental results. In fig. 2 we report the experimental data of the selfdiffusion coefficient in CH_4 as a function of temperature along the isochore ρ= 0.38 gr /cm^3; this value of the density is about 2.4 times the critical density so that at 0.38 gr /cm^3 we have a quite dense liquid. As we can see from fig. 2 the experimental data can be fitted with an expression $D= D_0 \exp - \frac{\Delta E}{RT}$ however the value of $\frac{\Delta E}{R}$ which comes out from the fit,is 220°K so that ΔE is roughly equal to $K_B T$. The same result we find in the case of diffusions in CCl_4 (13) although in this case the temperature range is much narrower. Therefore we think to be correct to conclude that the diffusion process in liquids is not an activated process in the usual sense. The large temperature dependence we find in the behaviour D_S along the coexistence curve is essentially due to the work done in reducing the density going from the triple point up to the critical one. This is the reason why it has been found (14) that the activation energy for the diffusion process along the coexisting curve is proportional to the vaporization energy. The main limitation to the diffusion process is the excluded volume effect and not the existence of an energy barrier.

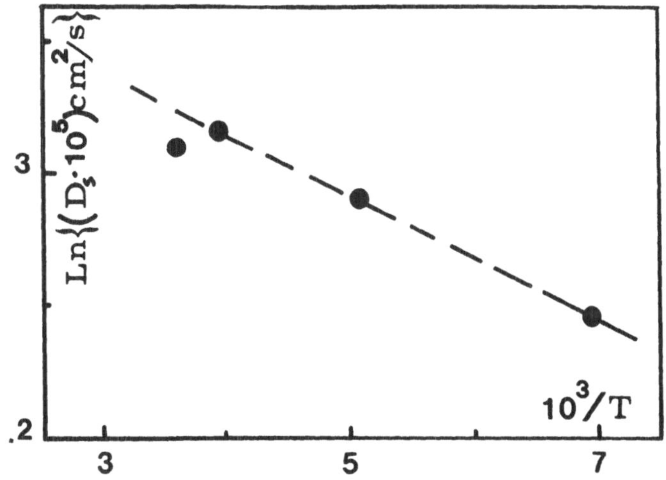

Fig. 2. Self diffusion coefficient of CH_4 along the isochore ρ =0.38 gr/cm^3. The experimental data are taken from ref. 9

3. A MODEL FOR THE DIFFUSION PROCESS IN LIQUIDS.

We already saw that the behaviour of the selfdiffusion coefficient in liquids is practically insensitive to the exact shape of the intermolecular potential unless we are concerned with H-bonded liquids. Therefore it makes sense to study the diffusion process in a simple model system where exact calculations can be perfomed. The system we want to study is the one composed of hard-spheres. For this system a very interesting theory has been proposed long time ago by Cohen and Turnbull (15). The point of wiew of these authors, exposed in a pictorial way, is the following: in a dense hard-spheres system the dynamics of the molecules is symilar to what happens in a crowded bus during rush hours. In this bus each person is confined in very small region of space, something like a cage, due to the continuous and random repulsive impacts with its immediate neighbours. In this "free volume" everyone is moving with small and random displacements around the "quasi-equilibrium" position, i.e. the center of this "free volume". He is rattling around this "quasi-equilibrium" position since there is no space enough so that he can migrate away. However sometime the bus has unexpected stop so that all the people is pushed one against the other and the "free volumes" are squeezed practically everywhere and, since there is total volume conservation, this fact implies that at some point large voids become available. Therefore for the people on the border of these voids it is possible to run in this available space. But the people do not want to be strongly compressed, therefore all the passangers, with their continuous rattling motions, push away their neighbours so to recover the "free volume" available for each one before the unexpected stop. In such a way the "equilibrium" is restored and the large voids disappear and the passenger, who ran in one of this void, remain trapped in a new "quasi-equilibrium" position far from the previous one he had before the stop. In such a way we have the diffusion of the passengers along the bus. What it is important to realize is that such diffusive walk is driven essentially by local density fluctuations which must be larger than the excluded volume of each passenger so that, during the density fluctuation, he can change, at least in one direction, his rattling motion in a "quasi-linear" trajectory.
In the case of an hard spheres system this model can be worked out in a quantitative way; we will refer to the Cohen Turnbull paper (15). These authors, according to the general scheme of their model, write the diffusion coefficient as

$$D(V,T) = \int_{v_f^*}^{\infty} D(v_f)\ p(v_f)\ dv_f \tag{3.1}$$

where $D(v_f)$ is the diffusion coefficient in a fluid of local free volume v_f and clearly $D(v_f)=0$ unless v_f is greater than a threshold value v_f^*; $p(v_f)$ is the probability that the free volume per molecule is between v_f and $v_f + dv_f$; $v_f = v - \frac{\pi}{6}\sigma^3$ where v is the local volume per molecule and σ the hard sphere diameter; V is the total molar volume. First of all we calculate the average distribution $p(v_f)$. Due to the shape of the hard-sphere intermolecular potential, in such a system no energy change is associated with a redistribution of the free volume; therefore the evaluation of $p(v_f)$ is a combinatorial problem. We define the average free volume per molecule $v_f^\circ = \frac{V_f}{N}$ where V_f is the total free volume and N the total number of molecule of the system; moreover we divide the total range of the values of the local free volume per molecule into small regions i having the average value v_{fi} and we call N_i the number of molecule having their free volume in the i-th region. We define also the function W which is the number of ways of redistributing the free volume without changing the N_i, i.e. $W = \frac{N!}{\prod_i N_i!}$. Next we require W be a maximum for a given N and V_f (i.e. we search the maximun of W as a function of the various N_i under the conditions $\Sigma_i N_i = N$ and $\gamma \Sigma_i N_i v_{fi} = V_f$ where γ is a numerical factor of the order of unity to correct for overlap of free volumes). This maximum condition determines the various N_i and passing to the continuum limit we obtain

$$p(v_f) = \frac{\gamma}{v_f^\circ} \exp\left(-\frac{\gamma v_f}{v_f^\circ}\right) \tag{3.2}$$

The relationship (3.2) is exact. Next Cohen and Turnbull, propose for $D(v_f)$ the relationship

$$D(v_f) = g(v_f^*)^{1/3}\, u \tag{3.3}$$

where g is a geometric factor and u is the thermal velocity; the expression (3.3) is a very rough approximation since it implies a gas like diffusion with mean free-path proportional to $(v_f^*)^{1/3}$. Clearly what in our pictorial description was the rearrangement of the molecules, which implies the decay of the free-volume fluctuation, it is not present in eq. (3.3). In fact this rearrangement takes place through cooperative motions of the neighbouring molecules; in (3.3) this cooperative effect has not been taken into account. However the temperature dependence of eq. (3.3) is certainly correct and also it is plausible that the right order of magnitude of the diffusive step is $(v_f^*)^{1/3}$.

The effect of the cooperativity must be mainly reflected in the dependence of D from the trace and solvent molecular masses and certainly would not enter in the dependence of D from V and T. In the case of self diffusion it would only mean a different value for g.
Putting eq. (3.2) and (3.3) in (3.1) we obtain (16).

$$D(V,T) = gV_0^{1/3}\sqrt{\frac{3RT}{M_i}}\left(\frac{\frac{V}{V_o}-1}{\gamma N}\right)^{1/3}\int_{\bar{z}^*}^{\infty} z^{1/3}\, e^{-z} dz \tag{3.4}$$

where $V_o = \frac{N \sigma^3}{\sqrt{2}}$, $z = \gamma\, v_f^* \frac{N}{V - V_o}$ and M_i is the tracer's mass.

With reasonable value of the constant γ, in the range $1.5 \leq \frac{V}{V_o} \leq 3$, which is about the range from triple to near the critical point, eq. (3.4) can be very well approximated with

$$D(V,T) = g'\left(\frac{V_o}{N}\right)^{1/3}\sqrt{\frac{3R\,T}{M_i}}\ \exp\left(-\frac{\gamma' v_f^* N}{V - V_o}\right) \tag{3.5}$$

where g' and γ' are new constants obviously related to g and γ. Eq. (3.5) is similar to the one originally proposed in ref. 15. To compare the theoretical prediction with the experimental results we limit ourselves to the case of selfdiffusion since we are mainly interested in the dependence of D from V and T. We refer to Ref. 16 for the tracer's diffusion when the tracer's mass and/or diameter differ from those of the solvent. The experimental results we consider are the molecular dynamics "experiments" by Alder and co-workers (17). We think that one of the reason why the Cohen and Turnbull theory was not complementely appreciated it is that in 1959 extensive hard-sphere computer experiments were not available so the theory was tested against real fluid experiments and in such a way is obvious the reason why the theory was not able to reproduce the experimental results.
In fig. 3 we report the comparison between the theory and the hard-spheres "experiments". To eliminate the trivial temperature dependence and to use a pseudo reduced quantity, in fig. 3 we plot the quantity $y = \frac{D_s}{\sqrt{\frac{T}{M}} \cdot V_o^{1/3}}$ vs. $\frac{V_o}{V - V_o}$; from eq. (3.5) we can see that in the dense fluid region the Cohen-Turnbull theory predicts $y = C \exp -\left(\frac{\gamma' V^*}{V - V_o}\right)$. where C is a constant and $V^* = N\, v_f^*$.

Looking at fig. 3 it is impressive the agreement between theory and experiments in the range $1.5 \leq \frac{V}{V_o} \leq 3$; if we take $C = 8.04 \times 10^{-5}$

and $\gamma' V^* = 1.69\ V_o$ which are values well in agreement with the hypothesis of the theory. The reason why for $V > 4V_o$ the theory gives values of D_s significantly lower than the experimental results and that the discrepancy increases lowering the density, it is that at low densities the picture of a molecule confined in "cage" has not more meaning. Fig. 3 is the experimental evidence that for an hard spheres fluid, in the dense region ($V < 4V_o$), it is correct the idea that the diffusion process is driven by density fluctuations and therefore we must have, as it really happens, very different temperature dependence of D_s if we move along an isochore or along the coexistence curve. The temperature dependence along the coexistence curve it is essentially the strong dependence of D_s from $(V-V_o)$; therefore there is no physical meaning in deriving numbers for the activation energy just by the temperature dependence of D along the coexistence curve.

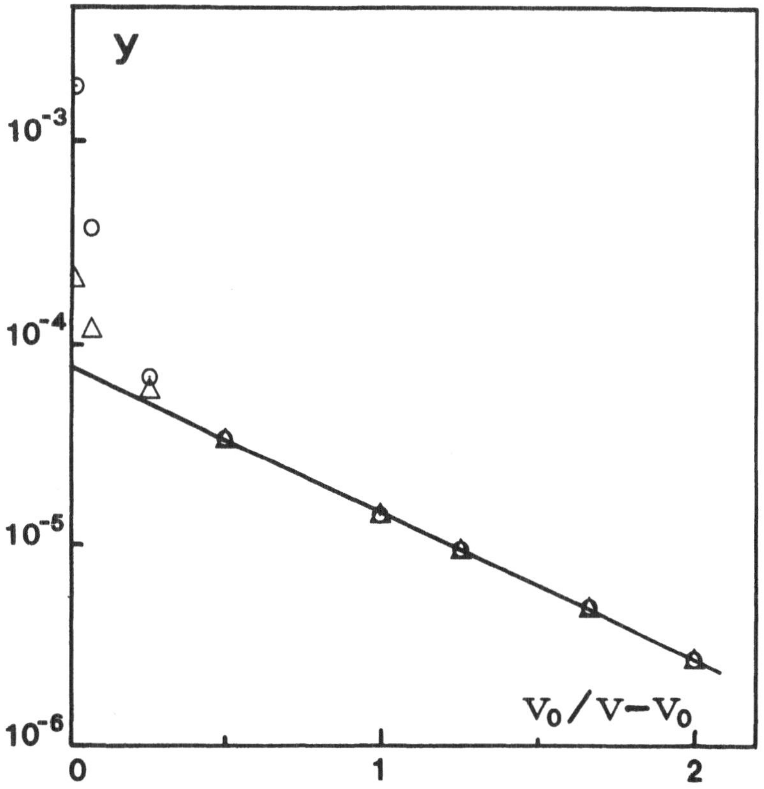

Fig. 3. y vs $\frac{V_o}{V-V_o}$ for an hard spheres fluid; △ refer to eq. (3.4) with $\gamma V^* = 1.69\ V_o$ and $g = 0.28\gamma^{1/3}$. O are the "experimental" results of Alder and coworkers quoted in Ref. 17. The full line represents the expression

$$D_s = 8.04 \times 10^{-5}\ V_o^{1/3}\sqrt{\frac{T}{M}}\ \exp\left(\frac{-1.69\ V_o}{V - V_o}\right)$$

Since the Cohen-Turnbull theory works so well in the case of self-diffusion in an hard-sphere fluid, it is worth while to generalize it to real fluids. As we said previously, there is no sense in trying to apply this theory to real fluids without some minor change; in fact going from hard-spheres to real molecules we introduce an attractive part in the intermolecular potential. However we can handle this change as a pertubation as it has been very successfully done in the case of equilibrium properties. This point of view was first proposed by Macedo and Litovitz (18); they suggest that in real fluids the diffusion process is not only driven by density fluctuations, but also by energy fluctuations. However, since the attractive part of the potential is only a perturbation, we can make the approximation that the energy and the density fluctuations do not interact. The dependence from the energy fluctuations gives in D an explicit temperature dependence which can be easily evaluated (19) following the same procedure as for the free volume fluctuation in the case of the hard sphere fluid. For density fluctuations, Macedo and Litovitz take for D the expression found in the case of hard spheres. So we have, see Ref. (16),

$$D_s(V,T) = M \cdot D_{HS}(V,T) \cdot \exp - \left(\frac{\Delta E}{RT}\right) \qquad (3.6)$$

and using eq. 3.5 and the results of Fig. 3 (see eq. of the full line)

$$D_s(V,T) = M \cdot 8.04 \cdot 10^{-5} \cdot V_o^{1/3} \sqrt{\frac{T}{M}} \; \exp - \left[\frac{1.69 V_o}{V - V_o} + \frac{\Delta E}{RT}\right] \qquad (3.7)$$

where M is a numerical constant and ΔE the energy threshold to allow the diffusion (ΔE has a physical meaning analogous to Nv_f^*). Eq. (3.6) is called the generalized Macedo-Litovitz expression. We can compare eq. (3.6) with the experimental results and the agreement is quite satisfactory; we report in fig. 4 the comparison in the case of CH_4. One could object that the comparison of eq. (3.6) with experimental data is only a three parameters fitting. We disagree with such a point of view because in building up eq. (3.6) there is a clear physical model and moreover the values for the three parameters turn out to be of the right order of magnitude according to the hypothesis of the model. It is true that more work must be done to deduce theoretical expressions for the values of M, V_o and ΔE starting from microscopic dynamics instead to derive their values by the fitting.
Finally we want to note that, using the orthobaric densities from PVT data and the D_s values from eq. (3.6), we find that the following relationship holds for thermodynamic states along the coexistence curve

$$\rho D_s(T) = AT^{\alpha} \qquad (3.8)$$

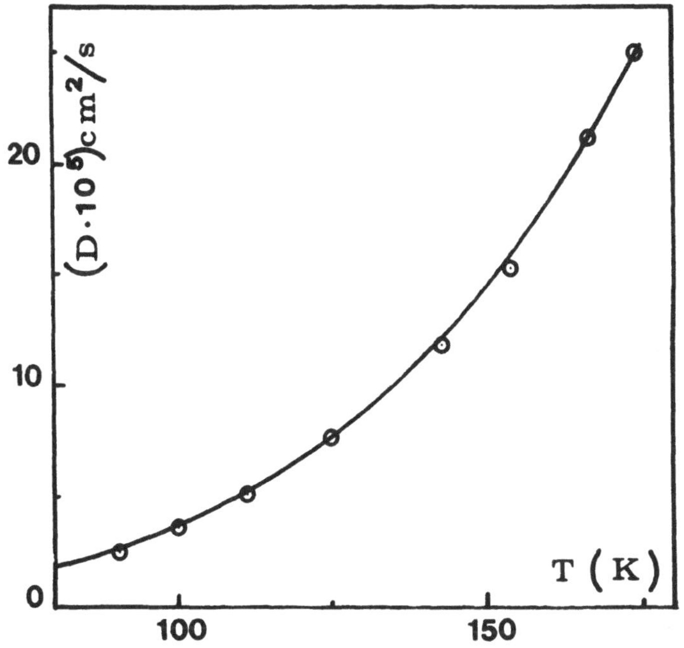

Fig. 4. Selfdiffusion of CH_4 along the coexistence curve. o are the experimental data taken from Ref. 9. The full line is the generalized Macedo-Litovitz expression

$D_s = 10.1 \cdot 10^{-5} \cdot \sqrt{T}\ \exp - \{\frac{1.69 \cdot 20.2}{V - 20.2} + \frac{105}{T}\}$; with T in K and the molar volume V in cm^3 we obtain D_s in cm^2/s

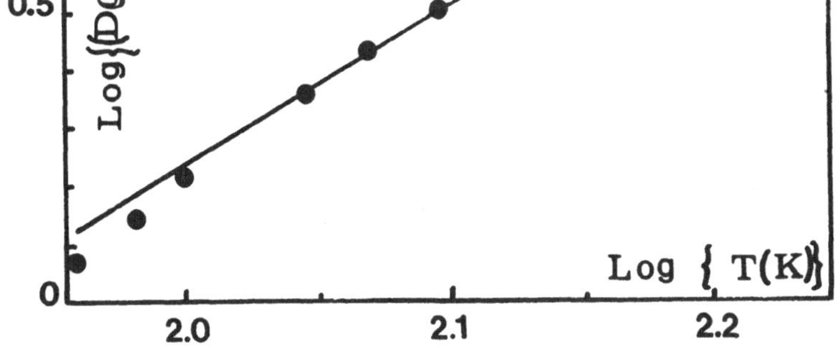

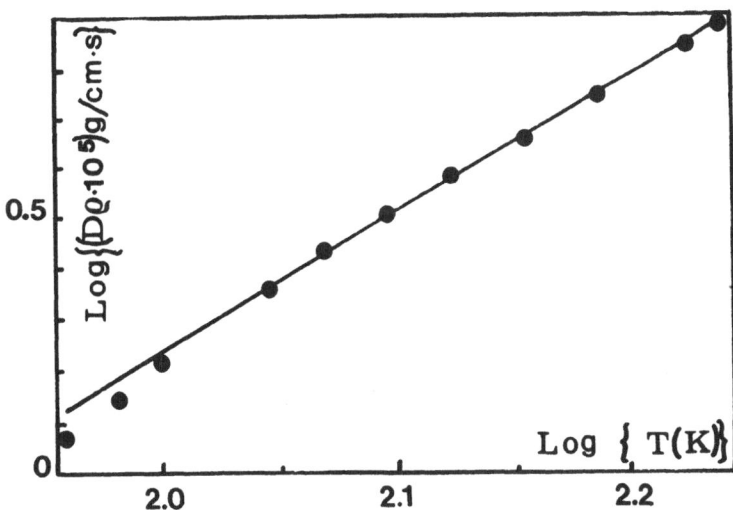

Fig. 5. Lg $D_s\rho$ vs. lg T for CH_4 along the coexistence curve; • are the experimental data from Ref. 9; the full line is the expression $\rho D_s = 4.9 \cdot 10^{-11} \cdot T^{2.76}$ where ρ is in gr/cm^3, T in K and D_s in cm^{2}/s

where A and α are constants. This relationship was already suggested on pure empirical basis (20). In fig. (5) we show how eq. (3.8) is very well verified in the case of CH_4. Moreover using eq. (3.8) we find that in many liquids the α values are very near to the number 2.7. Eq.(3.8) is a very useful representation since it implies that at orthobaric densities in ordinary liquids $\ln D_s$ vs. $\ln T$ is a straight line of slope around 2.7.

4. SELF DIFFUSION IN WATER.

Already some years ago (21) it has been pointed out that at low temperature the self diffusion coefficient in water is lower than we can expect in an ordinary liquid and also that its temperature behaviour is anomalous. These considerations are also qualitatively evident in fig. 1 where we can see that for $T < 100°C$ the water's data are significantly lower than the universal curve followed by all other liquids; moreover this discrepancy increases as the temperature decreases. The water behaviour has been connected with the presence of H-bonds in the intermolecular interaction. In this chapter we want to show quantitatively this influence of the H-bonds in the selfdiffusion in water for thermodynamic states along the coexistence curve.
Since the behaviour of water molecules is driven not only by the H-bonds but also by an average intermolecular field as in "normal" liquid, the first step to do it is to identify the region were D_s behaves normally. We will do by means of eq. (3.8); using for D_s the experimental values of Ref.(10). The result is shown in fig.6. From fig. 6 it is clear that while for $T > 400°K$ the curve is a good straight line, for $T < 400°K$ it has a curvature which becomes greater as we go to lower temperatures. Therefore this fact is a strong indication that for $T > 400°K$ the D_s in water has a normal behaviour. We use the experimental D_s values in the range $T > 400°K$, to fit them with a generalized Macedo-Litovitz expression, i.e. eq. (3.8), and we will call it $D_{s,n}$ to mend the selfdiffusion in "normal" H_2O.
We obtain

$$D_{s,n}(V,T)=9.3\cdot 10^{-5}\sqrt{T}\exp -\left(\frac{16.9}{V-10}+\frac{358}{T}\right) \qquad (4.1)$$

It is important to know that the V_o and ΔE values are in agreement with the general relationships found in ordinary liquids:

$V_o \simeq \dfrac{V_c}{4.77}$ and $\dfrac{\Delta E}{R} \simeq 0.55T_c$ where V_c and T_c are the critical point parameters.
We propose to use eq. (4.1) to evaluate $D_{s,n}$ also for $T < 400°K$ and therefore the influence of H-bond on the selfdiffusion process in water is measured by the quantity $\dfrac{D_{s,n}}{D_s}$ reported in fig. 7. How we can see from fig. 7 the H-

bond effect on the diffusion process is a decrease in D_s of about 16% at 100°C, a factor 3 at 20°C and a factor 20 at -29°C. To explain theoretically the curve in fig. 8, one could try an analysis following the Cohen-Turnbull kind of approach. This means to suppose that the diffusion process is driven not only by density and energy fluctuations but also by H-bond fluctuation. However the main problem it is that it is impossible to make the hypothesis that these fluctuations, mainly density and H-bond fluctuation, are not coupled. Another attempt to explain the anomalous water behaviour has been done using the percolation theory (22) however this approach fails dramatically for $T > 50°C$. The reason why it is instead successfully in the supercooled region, it is obvious from fig. 7 which shows that in this region the H-bond fluctuations are the most important one; therefore in this case we can treat both the energy and density fluctuations as perturbations. As we increase the temperature this approximation becomes worse and for temperatures higher than room temperature the relative weights of the various fluctuations are reversed.
Very valuable information could be obtained from the measurement of diffusion of impurities,not able to form H-bonds with water molecules, along the water coexistence curve in a quite wide temperature range. In fact the percolation theory by Stanley and Texeira, predicts for D_i a behaviour similar to $D_{s,n}$ while the H-bonds fluctuation approach in the Cohen-Turnbull's scheme would inply for D_i a behaviour similar to D_s since the local dynamics is a result of a large region cooperative behaviour. Unfortunately such experimental data are not available in the literature.

5. CONCLUSIONS

In these two lectures we gave a short review on the isothermal isobaric diffusion in liquids. We want now to stress what are the most interesting problems still open in this field. In the case of ordinary liquids we think that, as far as the experimental approach is concerned, there is need of more data on impurity diffusions varying the impurity's molecular parameters to obtain information about the dependence of D from the masses and diameters of the solute and the solvent. So we can have insights on the cooperative nature of the fluctuaions' relaxation; also computer "experiments" would be valuable to clarify this point. As far the theory it would be necessary to have a more satisfactory formulation of the generalized Macedo-Litovitz equation particularly there is need of theoretical expressions for V_o and ΔE in terms of the molecular parameters. As far as associated liquids we already pointed out that measurements of D_i for non H-bonded impurity will be extremely useful to discriminate among the various theoretical points of view.

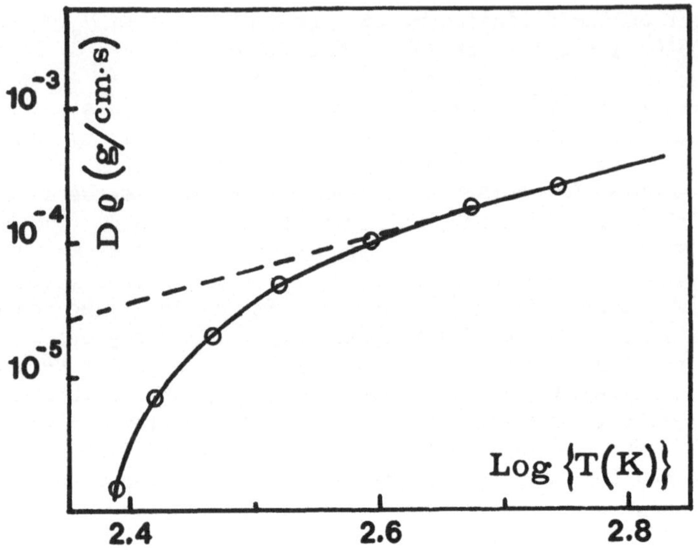

Fig. 6. $D\rho$ vs. T for selfdiffusion of H_2O along the coexistence curve. The dotted line is the relationship $D\rho = 3.46 \cdot 10^{-11}\ T^{2.51}$ with the same units as in Fig. 5; o are the experimental data of Ref. (10); the full line is the behaviour of the experimental data. The dotted line represents what it would be the "normal" behaviour of H_2O.

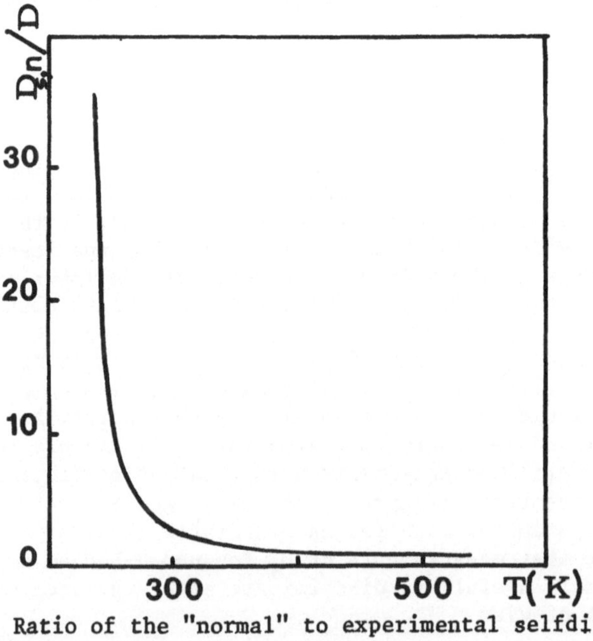

Fig. 7. Ratio of the "normal" to experimental selfdiffusion coefficient in H_2O along the coexistence curve. $D_{s,n}$ is obtained from eq. (4.1); D_s are the experimental data of Ref. (10).

REFERENCES

(1) Fitts D.D. (1962) "Non equilibrium thermodynamics" Mc Graw Hill, New York.

(2) See Appendix B of ref. 1.

(3) Kirkwood J.G., Boldwin R.L., Dunlop P.J., Gosting L.J., Kegeles G. (1966) J. Chem. Phys. 33: 1505.

(4) See Ch. 4 of Ref. 1; also Ciccotti G. and Jacucci G. (1975) Phys. Rev. Letters 35: 789.

(5) Baker A.J. (1963) in"The International Encyclopedia of Physical hemistry and Chemical Physics: Topic 10, the fluid state", J.S. Rowlinson ed. Vol. I, Lattice Theories of Liquid State, Macmillan, New York.

(6) a) Cini-Castagnoli G., Pizzella G., Ricci F.P. (1959), Nuovo Cimento 11: 466;
b) Helfand E., Rice S.A. (1960) J. Chem. Phys. 32: 1642.
c) Gubbins K.E. (1973) in "Statistical Mechanics" the Chemical Society, Burlington House,London W1VOBN Vol. 1, ch. 4, pag. 241.

(7) Reed T.M., Gubbins K.E. (1973) "Applied Statistical Mechanics" Mc Graw-Hill New York. See ch. 11.

(8) a) Nagazadeh J. and Rice S.A. (1962) J. Chem. Phys. 36: 2710;
b) Ehrlich R.S., Can H.Y. (1970) Phys. Rev. Letters 25: 314;

(9) Oosting P.H. and Trappeniers N.J. (1971) Physical 51: 418.

(10) Hausser F., Maier G., Noack F. (1966) Z. Naturforsch 21a: 1410.

(11) Duprè F., Piaggesi D., Ricci F.P. (1980) Phys. Letters 80A: 178.

(12) Glaistone S., Laidler K.J., Egring H.(1941) "The theory of rate processes" Mc Graw-Hill New York, see Ch. IX.

(13) Watts H., Alder B.J., Hildebrand J.H. (1955) J. Chem. Phys. 23: 659.

(14) See Ref. 12 pag. 491.

(15) Cohen M.H. and Turnbull D. (1959) J. Chem. Phys. 31: 1164.

(16) Ricci F.P., Ricci M.A., Rocca D. (1977) J. Phys. Chem. 81:171.

(17) Alder B.J., Gars D.M., Wainwright T.E. (1970) J. Chem Phys. 53: 3813;
Herman P.T. and Alder B.J. (1972) J. Chem. Phys. 56:987;
Alder B.J., Alley W.E., Dymond J.H. (1974) J. Chem. Phys. 61: 1415.

(18) Macedo P.B. and Litovitz T.A. (1965) J. Chem. Phys. 42: 245.

(19) Chung H.S. (1966) J. Chem. Phys. 44: 1362.

(20) Noble J.D. and Bloom M. (1965) Phys. Rev. Letters 14: 250; see also Ref. 9.

(21) Ricci F.P., Ricci M.A., Rocca D. (1974) Phys. Letters A48: 289.

(22) Stanley H.E. and Texeira J. (1980) J. Chem. Phys. 73: 3404.

DIELECTRIC POLARIZATION AND RELAXATION

Robert H. Cole

Chemistry Department, Brown University,
Providence, Rhode Island, U.S.A.

A review of field quantities and formal results is followed by discussion of equilibrium theories and simulations for polar liquids. Relaxation processes are considered primarily in terms of correlation function treatments, starting with models of behavior in simple systems and going on to cooperative behavior in more complex systems as a function of temperature, with some reference to polymers and molecular crystals. Other aspects discussed include correlation functions from transient birefringence, ion-solvent interaction effects in electrolytes and relaxation in mixtures of polar liquids.

1. BASIC CONCEPTS AND FORMAL THEORIES

The province of conventional dielectric measurements is here taken to be the determination of the relations of the polarization $\mathbf{P}$ and current density $\mathbf{J}$ to the electric field $\mathbf{E}$ in the macroscopic Maxwell equations. Proper theory should account for these relations in condensed phases, as a function of state variables, time dependence of applied fields, and molecular parameters, by appropriate statistical averaging over molecular displacements determined by the equations of motion in terms of molecular forces and fields. Simplifying assumptions and approximations are of course necessary. One kind often made and debated is use of an effective or mean local field at a molecule rather than the sum of microscopic

A. J. Barnes et al. (eds.), Molecular Liquids - Dynamics and Interactions, 59–110.

fields (which when suitably averaged give the macroscopic $\mathbf{E}$).

In this first section, general aspects of these three subjects are considered briefly, to provide background for more specific developments, and hopefully, to clarify some questions raised in discussing them.

1.1 Maxwell Equations

Following Jackson (1), we assume the microscopic Maxwell equations

$$\begin{aligned} &\nabla\cdot\underline{e} = 4\pi n = 4\pi\sum_j e_j\,\delta(\underline{r}-\underline{r}_j) \\ &\nabla\times\underline{b} - \frac{1}{c}\frac{\partial\underline{e}}{\partial t} = \frac{4\pi}{c}\underline{j} = \frac{4\pi}{c}\sum_j e_j\underline{v}_j\,\delta(\underline{r}-\underline{r}_j) \\ &\nabla\cdot\underline{b} = 0 \\ &\nabla\times\underline{e} + \frac{1}{c}\frac{\partial\underline{b}}{\partial t} = 0 \end{aligned} \qquad (1)$$

where $\mathbf{e}$ and $\mathbf{b}$ are the microscopic electric and magnetic field vectors at position $\mathbf{r}$, and n and $\mathbf{j}$ are the microscopic charge and current densities of all point charges q_j at positions $\mathbf{r}_j$ with velocities $\mathbf{v}_j$. The fields so defined fluctuate rapidly in both space and time by virtue of the motions of the charges q_j, with unnecessary complexity to describe macroscopic phenomena.

The average of $\mathbf{e}$ and $\mathbf{b}$ over a suitable volume large compared to atomic dimensions yield the macroscopic fields $\mathbf{E}$ and $\mathbf{B}$ adequate to describe electromagnetic phenomena down to distances of order 10^{-5}cm. This figure is cited by Jackson as 'the absolute lower limit' on the grounds that reflections and refraction in the visible and near ultraviolet are adequately described by the macroscopic Maxwell equations, but diffraction of X-rays is not. Applying the averaging denoted by angle brackets <..> to the last two of equations (1) gives trivially both $\nabla\cdot\langle\underline{b}\rangle = \nabla\cdot\mathbf{B} = 0$ and $\nabla\times\langle\underline{e}\rangle - (1/c)\partial\langle b\rangle/\partial t = \nabla\times\mathbf{E} + (1/c)\partial\mathbf{B}/\partial T = 0$.

The spatial averages of n and $\mathbf{j}$ in the first two of equations (1) result in the macroscopic charge, electric moment, and current densities of the macroscopic Maxwell equations after classifying charges q_j according to molecules n in which they are found and expanding the delta functions in powers of small displacements $\mathbf{r}_{jn} = \mathbf{r}_j - \mathbf{r}_n$ from molecular reference points, usually the centers of mass. The first of equations (1) becomes, through terms of second order,

$$\nabla\cdot\langle\underline{e}\rangle = \nabla\cdot\underline{E} = 4\pi\left(\rho - \nabla\cdot\underline{P} + \tfrac{1}{6}\nabla\cdot\nabla\underline{Q} + \cdots\right) \qquad (2)$$

where the macroscopic charge, dipole, and quadrupole moments densities are given by

$$\begin{aligned}\rho &= \langle\textstyle\sum_n(\sum_{j(n)} e_j)\delta(\underline{r}-\underline{r}_n)\rangle\\ \underline{P} &= \langle\textstyle\sum_n(\sum_{j(n)} e_j \underline{r}_{jn}\,\delta(\underline{r}-\underline{r}_n)\rangle\\ \underline{Q} &= \langle\textstyle\sum_n(3\sum_{j(n)} e_j \underline{r}_{jn}\underline{r}_{jn})\delta(\underline{r}-\underline{r}_n)\rangle\end{aligned} \qquad (3)$$

The first two terms on the right of equation (2) give the usual Maxwell relation, with the third and higher terms omitted as usually negligible. The requirement for spatial variation of the Q tensor to contribute significantly to E or $\underline{D} = \underline{E} + 4\pi\underline{P}$ is that there be appreciable changes in electric field over a molecular dimension, as the energy of interaction of a molecular quadrupole with an external field is proportional to the product of the appropriate moment tensor element $Q_{\alpha\beta} = 3\sum_{j(n)} e_j(\underline{r}_{jn})_\alpha(\underline{r}_{jn})_\beta$ with the field gradient. The Buckingham-Disch experiment (2) is an example of extreme measures to produce measurable molecular quadrupole orientation effects by strong gradients of fields between long wires, and one can envisage other possibilities with laser fields. The usual grounds for dismissing quadrupolar and higher order effects do not apply, however, to interactions of molecules with multipole fields of nearby molecules, as these can be both large and rapidly varying with distance. Some examples of the effects are discussed in 2.

The averaging of the second of equations (1) gives, by similar but more complicated manipulations

$$\begin{aligned}\nabla\times\langle\underline{b}\rangle - \frac{1}{c}\left\langle\frac{\partial\underline{e}}{\partial t}\right\rangle &= \nabla\times\underline{B} - \frac{1}{c}\frac{\partial\underline{E}}{\partial t}\\ &= \frac{4\pi}{c}\left(\underline{J} + \frac{\partial\underline{P}}{\partial t} + c\nabla\times\underline{M} + \cdots\right)\end{aligned} \qquad (4)$$

where the macroscopic current and magnetic moment densities are given by

$$\begin{aligned}\underline{J} &= \langle\textstyle\sum_n(\sum_{j(n)} e_j\underline{v}_{jn})\delta(\underline{r}-\underline{r}_n)\rangle\\ \underline{M} &= \langle\textstyle\sum_n(\sum_{j(n)}\frac{1}{2c}e_j(\underline{r}_{jn}\times\underline{v}_{jn})\delta(\underline{r}-\underline{r}_n)\rangle\end{aligned} \qquad (5)$$

the omitted higher order terms involve products of molecular electric multipole moments and velocities which can be significant in moving media, but are not of concern for present purposes.

With these developments, the macroscopic Maxwell equations relate to fields to macroscopic charge, current, and moment densities, with further statistical averaging over appropriate ensembles giving expectation values. Before going on to these operations, two points which sometimes cause difficulty should be mentioned. The first is that only 'total' current density $\underline{J}_t = \underline{J} + \partial \underline{P}/\partial t$ is related to fields and hence that only $\underline{J}_t$ can be determined by purely electromagnetic measurements, but not conduction current $\underline{J}$, if significant, separately from polarization current $\partial \underline{P}/\partial t$. This is evident in equation (4). It may not seem so for $\nabla \cdot \underline{E} = 4\pi(\rho - \nabla \cdot \underline{P})$, but combining this with $\partial \rho/\partial t + \nabla \cdot \underline{J} = 0$ for continuity gives $\nabla \cdot (\partial \underline{E}/\partial t) = -4\pi(\underline{J} + \partial \underline{P}/\partial t)$.

A second point as a corollary is that strictly speaking, a true static or equilibrium relation of $\underline{P}$, or $\underline{D}$ to field $\underline{E}$ cannot be measured in the presence of conduction current $\underline{J}$. For example, a sinusoidal field with time dependence exp $(i\omega t)$ and linear response of $\underline{P}$ gives $\underline{J}_t = \underline{J} + i\omega \underline{P}$, with vanishing contribution from $\underline{P}$ in the limit $\omega = 0$. Practically, one can hope to distinguish a limiting low frequency behavior of a frequency independent real component of $\underline{J}_t$ and an imaginary component proportional to ω. The first is then reasonably assigned to conductivity and the second to a 'static' permittivity $\epsilon_s = \lim_{\omega \to 0} \mathrm{Im}(\underline{J}/\underline{E})/\omega$. (Electrode space charge effects of course complicate the problem considerably for highly conducting solutions at low frequencies.)

1.2 Response Theory and Dipole Correlation Functions.

In order to use statistical mechanics to calculate the macroscopic properties introduced only by spatial averaging, one needs to perform ensemble averages over states. For present purposes, classical mechanics will suffice, using the phase space of conjugate coordinates and momenta q_j and p_j with a time dependent probablity density function f (q_j, p_j, t). The mean expectation value of a function A (q_j, p_j, t) is denoted by $A(t) = \langle A(q_j, p_j, t)\, f(q_j, p_j, t)\rangle$, where the angle brackets now refer to the ensemble average over the q_j and p_j.

The macroscopic polarization $\underline{P}(t)$ is ordinarily taken to be for a sum of classical permanent and induced electric dipole moments of molecules or constituent segments, as in polymers for example. This approximation is justified quantum mechanically for spacings of energy states involved which are either small or large relative to thermal energy k_BT (3). In terms of frequency, k_BT at 300°k corresponds to 6.25 THz or 210 cm^{-1}, which ordinarily lies between most rotational and vibrational regines and is far above relaxation ranges.

Denoting permanent moments by $\underline{\mu}_n$ and induced moments of electronic or vibrational oscillator displacements by $e\underline{r}_n$, the moments to be averaged are sums of terms of the form $\underline{M} = \sum_n (\underline{\mu}_n + e\underline{r}_n)$. Often, the effects of induced moments are taken to be equivalent to using $e\underline{r}_n = \alpha_n \underline{E}$, where α_n is a polarizability and $\underline{E}$ an appropriate field at the oscillator. This is discussed in 1.3.

The time evolution of a function A (p_j, q_j, t) is given classically by

$$\frac{dA}{dt} = \frac{\partial A}{\partial t} + \sum_j \left[\left(\frac{\partial H}{\partial p_j}\right)\left(\frac{\partial A}{\partial q_j}\right) - \left(\frac{\partial H}{\partial q_j}\right)\left(\frac{\partial A}{\partial p_j}\right)\right] = \left(\frac{\partial A}{\partial t}\right) + LA \quad (6)$$

where L is the classical Liouville operator for the Hamiltonian H(p_j, q_j, t). For polarization of moments $\underline{M}$ not depending explicitly on time, one has d $\underline{M}$/dt = L$\underline{M}$, and if further one has a time independent Hamiltonian H_o and Liouville operator L_o the formal solution is

$$\underline{M}(t) = \exp(tL_o)\underline{M}(o) \quad (7)$$

where $\underline{M}$(o) is the value at t=o in terms of the p_j and q_j.

Finally in these preliminaries to writing formal response theory expressions for P(t), equation (6) applied to the distribution function f gives Liouville's equation

$$\frac{\partial f}{\partial t} = -Lf \quad (8)$$

as df/dt = 0 for particles moving in phase space according to the equations of motion. For equilibrium, one has f = N exp ($-\beta$H) and in the absence of applied fields F_o = N exp ($-\beta H_o$) where $\beta = 1/k_BT$ and N is the normalization factor.

In the case of polarization P(t) = $\langle M(p,q)f(t)\rangle$, response theory is usually developed by writing $H = H_o - M_z E_{o(t)}$, where $M_z = M\cos\theta$ is the component of $\underline{M}$ in the direction z of applied field E_o (see 1.3). The Liouville operator L is then schematically $L = Lo - (\partial M_z/\partial\theta)E_o(t)(\partial/\partial p_\theta)$ and the time dependent f(t) satisfies

$$\frac{\partial f}{\partial t} = -L_o f + E_o(t)\left(\frac{\partial M_z}{\partial \theta}\right)(\partial f/\partial p_\theta) \tag{9}$$

To first order in field E_o, the operator $\partial/\partial p_\theta$ acts on the equillibrium distribution f_o. As $(\partial M_z/\partial\theta)(\partial f_o/\partial p_\theta) = -\beta \dot{M}_z f_o$, the first order perturbation f_1 of f from f_o satisfies

$$\frac{\partial f_1}{\partial t} + L_o f_1 = -\beta \dot{M}_z f_o$$

with formal solution (4)

$$f_1(t) = -\beta \int^t dt_1 E_o(t_1)\exp[-(t-t_1)L_o]\dot{M}_z \tag{10}$$

The response of polarization $P_z(t)$ to a constant field E_o applied at t = o is then given by

$$P_z(t) = \beta\left[\langle M_z(o)^2 f_o\rangle - \langle M_z(o)M_z(t)f_o\rangle\right]E_o \tag{11}$$

while the relaxation after removal of E_o at t = 0 is

$$P_z(t) = \beta\langle M_z(o)M_z(t)f_o\rangle E_o \tag{12}$$

and the steady state response to $E_{o(t)} = E_o \exp(i\omega t)$ is

$$P_z(i\omega) = \beta\mathcal{L}\langle -M_z(o)\dot{M}_z(t)f_o\rangle E_o \tag{13}$$

where $\mathcal{L}$ denotes the Laplace transform to frequency ω and $M_z(t) = \exp(tLo)M_z(0)$ is the evolution in time of M_z by the 'natural motion' of the unperturbed system described by H_o and L_o Thus the response formalism relates observable transient or A.C. responses to equilibrium ensemble averages for these motions, with the equilibrium polarization $P_z = \langle M_z(0)^2 fo\rangle\ Eo$ as a limiting case.

The normalized response $\Phi(t) = \langle M_z(0)M_z(t)fo\rangle/\langle M_z(0)^2 fo\rangle$ is usually called the macroscopic correlation function, as it is for a sample containing many molecules with both self and joint (pairwise) correlations. As for other macroscopic transport properties, one obtains only limited information, about 'dipole active' molecular dynamics in the present case. This makes it important to be able to compare with response theory and measurements of other kinds of correlations, in nuclear magnetic resonance, light scattering, and electric birefringence, for example.

1.3 Local and Macroscopic Fields

In the preceding sketch of first order response theory, various problems in details of the development have not been exhibited explicitly or have been glossed over, particularly in dealing with the roles of both induced and permanent dipoles and in relating local molecular fields to the Maxwell $\mathbf{E}$ and to external fields. The need for distinguishing between these fields has long been recognized: a molecule sees the fields of other external charges from its own special position, but the macroscopic field $\mathbf{E}$ to which one relates the polarization experimentally is determined by the averaging over a physically small volume of the fields of all charges in the volume.

Lorentz was the first to consider such problems for a reasonably defensible model of induced dipoles, derive the local Lorentz field $\mathbf{E}_L$, and from this obtain the venerable Clausius-Mossotti (or perhaps more properly Lorentz-Lorenz) formula. As shown schematically in Figure 1 (a), the molecules are assumed to be at sites on a cubic lattice with uniform macroscopic $\mathbf{E}$ along the z axis.

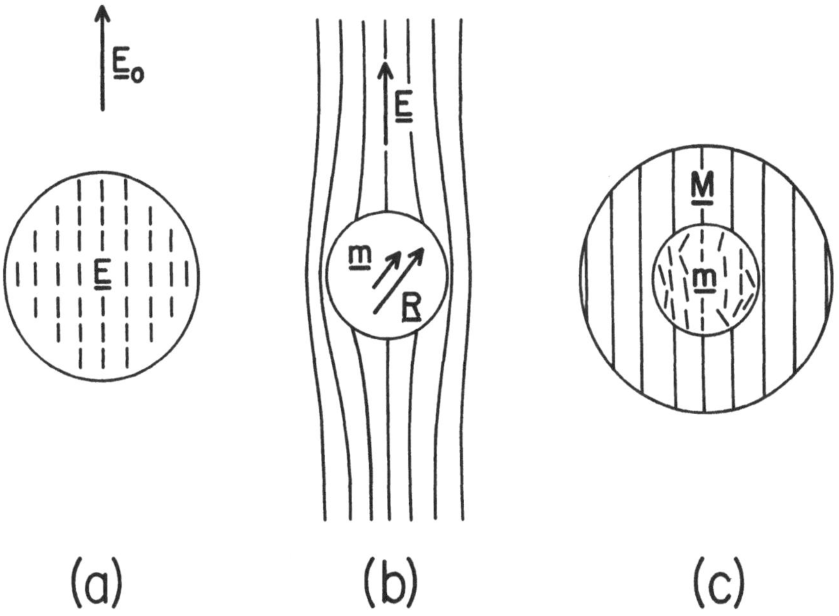

Figure 1. (a) The Lorentz local field model, (b) the Onsager

cavity and fields, (c) the macroscopic and inner spheres of Kirkwood - Fröhlich theory.

The field E_L is then the same for all molecules and parallel to E. as are the induced moments m, taken to be given by $m = \alpha E_L$, where α is a simple scalar polarizability. The gist of the Lorentz argument (7) is that the resultant field at any one molecule i from all the other dipoles j in a sphere surrounding the one vanishes, as it is given by the sum $\Sigma R_{ij}^{-3}[1 - 3\cos^2(R_{ij}, z)]\alpha E_L$ over lattice distances, which is zero for cubic symmetry (or an isotropic continuum), leaving $E_L = E_o$, the field of charges external to the sphere if it is in a vacuum. The macroscopic E in the sphere from Eo and the macroscopic P is by electrostatics $E = E_o - (4\pi/3)P$, and the Lorentz field is

$$E_L = E + (4\pi/3)P = (\epsilon_\infty + 2)E/3 \tag{14}$$

where ϵ_∞ is used to denote a permittivity of induced moments. As P for the model is given by $P = (\epsilon_\infty - 1)E/4 = (N/V)\alpha E_L$, where N/V is the dipole number density, the Clausius-Mossotti (CM) formula for ϵ_∞ in terms of α follows immediately

$$\frac{\epsilon_\infty - 1}{\epsilon_\infty + 2} = \frac{4\pi N}{3V}\alpha \tag{15}$$

The original Lorentz argument is a static one developed long before molecular theory of polarizability as related to charge displacements. Van Vleck (8), however, showed that the CM formula remains valid if the induced moments are of harmonic oscillators on a cubic lattice when the full range of their displacements r_a is considered and not just the equilibrium value from $k\, r_{ieq} = e\, E_o$, where k is the force constant and e the effective charge. This follows from the harmonic oscillator Hamiltonian for the field Eo and fields of other oscillators at distances Rij (16)

$$H(p_i, r_i) = \sum_i [p_i^2/2m + kr_i^2/2 - er_i\cdot(E_o - \nabla_R \sum' e r_j \cdot R_{ij}/R_{ij}^3)]$$

In terms of displacements $r^1{}_i = r_i - r_{ieq}$ from the equilibrium values satisfying

$$kr_{ieq} = e[E_o - \nabla_R \sum_j e r_{jeq}\cdot R_{ij}/R_{ij}^3] \tag{17}$$

the Hamiltonian becomes

$$H(\underline{p}_i, \underline{r}_i) = H(\underline{p}_i, \underline{r}_i', E_o = 0) - \frac{1}{2}\sum_i e\underline{r}_{ieq} \cdot \underline{E}_o \tag{18}$$

The macroscopic energy and polarization in the field E_o are then for equilibrium displacements calculable by the Lorentz argument, with the Clausius-Massotti formula as the result for cubic lattices or a uniform continuum, with $\alpha = er_{ieq}/E_o = e^2/k$.

The early work of Van Vleck on induced dipoles and fields is still timely for several reasons. Although the harmonic oscillator model has its limitations, it is even so physically more satisfying than simply defining a polarizability by $\underline{m} = \alpha \underline{E}$, as is so often done, and provides a basis for doing so , and it also makes both the considerable success and the limitations of the CM formula for induced polarization more understandable. The device of using equilibrium displacements, as in equations (17) and (18), made possible by the quadratic form for harmonic oscillator energy has proved useful also in treating induced moments of real polarizable polar molecules (9), (10), (11).

The use of the Lorentz field for polar molecules with permanent moments μ and polarizabilities α leads to the familiar Langevin-Debye formula

$$\frac{\epsilon_s - 1}{\epsilon_s + 2} = \frac{4\pi N}{3V}\left(\alpha + \frac{1}{3}\beta\mu^2\right) \tag{19}$$

for the static permittivity, denoted by ϵ_s. As Debye recognized at the beginning, this formula is in gross disagreement with experiment even for rather weakly polar liquids and fails spectacularly for strongly polar ones, e.g., by predicting the '4 /3 catastrophe' when the right side of equation (19) is unity, requiring $\epsilon_s = \infty$ which of course does not happen for liquids.

The first notable success in identifying the principal difficulty was of course provided in Onsager's classic 1936 paper (12). Onsager pointed out that the Lorentz treatment for molecular dipoles lying parallel to $\underline{E}$ should not be used for variable orientations of permanent moments, as part of the field at a given molecule, the reaction field $\underline{R}$, results from polarization of surrounding molecules produced by its own dipole moment field and cannot play a role in reorienting it. In order to derive a potential of mean torque for calculating orientational averages, Onsager used the model of a polarizable point dipole in a spherical cavity of radius a, arbitrarily assigned a volume $4\pi a^3/3 = V/N$ for liquids, with the surrounding molecules represented by a continium with macroscopic permittivity ϵ_s, as shown in Figure 1 (b).

Electrostatic boundary conditions applied at the cavity surface (which is thus a 'real' cavity in this sense, rather than an arbitrary mathematical one, as used in some Lorentz field arguments) then gave the cavity field $\underline{G}$ for field $\underline{E}$ in the surrounding dielectric as $\underline{G} = 3\epsilon_s E/(2\epsilon_s + 1)$, the reaction field $\underline{R}$ of a point dipole $\underline{m}$ at the center as $\underline{R} = 2(\epsilon_s - 1)\,\underline{m}/(2\epsilon_s + 1)\,a^3$ and the resultant field $\underline{F} = \underline{G} + \underline{R}$ inducing a moment $\underline{M}_{ind} = \alpha\,\underline{R}$. Solution for the mean polarization of the N molecules in volume V then gives the famous Onsager equation, often written in the form

$$\frac{(\epsilon_s - \epsilon_\infty)(2\epsilon_s + \epsilon_\infty)}{3\epsilon_s} = \frac{4\pi N\beta}{3V}\left(\frac{\epsilon_\infty + 2}{3}\mu\right)^2 \qquad (20)$$

where ϵ_∞ is the permittivity of induced moments satisfying $(\epsilon_\infty - 1)/(\epsilon_\infty + 2) = 4\pi N\alpha/3V$. (For details of derivation of this and related formulas with extensive discussion, see Reference (13)).

Onsager's equation has been found over the years to be astonishingly successful in predicting permittivities of polar liquids. The agreement between the left and right sides of equation (20) is usually to better than 20 percent, except for liquids ordinarily considered to be hydrogen bonding or associated by other criteria. This is surely uncommon, if not unique, for such a simple model of a polar liquid. This degree of success has stimulated both numerous elaborations of the original model, as by using ellipsoidal cavities and off center dipoles in the cavity, and extension of it, to solutions for example. At the same time, there have been numerous criticisms of the model, ranging from 'correction' of the electrostatic formulas (all refuted, except for Scholte's (14) ellipsoidal cavity formulas (15)) to more serious questions of the propriety of using the cavity model and macroscopic electrostatics to describe interactions of a molecule with even its nearest neighbors.

1.4 The Kirkwood-Fröhlich Formulation

In the last 1930's, Kirkwood (16) and Fröhlich (17) independently developed statistical mechanical formulations which avoided introduction of a molecular cavity, and instead introduced the resultant moment of representative molecules and discrete neighbors in a sphere of radius R_1 large enough for it and an exterior spherical shell of outer radius R_M to be treated macroscopically as a large sphere in vacuum with external field E_0 prior to its insertion, as shown in Figure 1 (c).

The result of electrostatic boundary value conditions at R_1 to relate moments and polarizations of the two regions with statistical averaging is the Kirkwood - Fröhlich formula (differing from Kirkwood's original result because of the dubious expedient he used to include induced moment effects):

$$\frac{(\epsilon_s - \epsilon_\infty)(2\epsilon_s + \epsilon_\infty)}{3\epsilon_s} = \frac{4\pi N\beta}{3V}\left(\frac{\epsilon_\infty + 2}{3}\right)^2 \langle \underline{\mu} \cdot \textstyle\sum_{R_1} \underline{\mu}_i \rangle \quad (21)$$

where the sum is over molecules in the inner sphere and the statistical average is strictly to be evaluated in the double limit $R_1 \to \infty$, $R_M/R_1 \to \infty$.

The practical advantage of introducing the inner sphere average (for Eo = 0) is that it is roughly proportional to $\epsilon_s - 1$, whereas calculation of the average over all molecules in the macroscopic sphere R_M is proportional to $(\epsilon_s - 1)(\epsilon_s + 2)$ (as seen for example, from equation (14) with E_0 from charges external to the sphere in vacuum). The consequence is that large errors in large values of ϵ_s result from small inaccuracies in practicable evaluations of the average. One also can hope, and there are grounds discussed in 2 for belief, that practically a region containing only a few shells of neighbors is necessary. Even so, such a calculation is, as Kirkwood remarked, 'a difficult task in statistical mechanics', as later developments have clearly shown.

1.5 Fulton's Use of Charge-Current Densities as Sources.

Before going on to discuss recent developments in theory of static or equilibrium permittivity for specific models in the next section, mention should be made of different kinds of formalism developed by Fulton (18) and by Felderhof and Titulacr (19). The reader may have noticed that in the theories described so far, the macroscopic polarization is evaluated as a statistical mechanical average consistent with the basic definition for the Maxwell equations discussed in 1.1, while the macroscopic $\mathbf{E}$, and thence ϵ_s, is not so computed, but is instead introduced by electrostatic or cavity arguments. Both Fulton and Felderhof-Titulaer have dispensed with cavities in their treatments relating the permittivity to polarization fluctuations (as expressed by $\langle(\sum m_i)^2\rangle$ for example).

In his most recent papers (20), Fulton has put his case in the statement 'we take the view that if nature does not endow herself with cavities, we should not have to introduce them . . .' He avoids such introduction by taking an 'external' charge and current density as sources of electromagnetic field in the medium, and by using methods of quantum electrodynamics obtains solutions of microscopic and macroscopic field equations for the polarizations and fields, with susceptibility and permittivity obtained as functional derivatives of polarization with respect to source field _ and macroscopic E.

The formal developments have been used for several purposes, including derivation of the Kirkwood-Fröhlich equation (21) by averages over a spherical region in the medium and derivation of Onsager type limiting formulas, without introducing boundaries or cavities. The theory is not limited to equilibrium, and another application was to derivation of an Onsager form for the complex permittivity essentially in agreement with the Fatuzzo-Mason (21) and Titulaer-Deutch (22) results. In the latest papers (20), the methods have been applied to non-linear permittivities at high field strengths, for polar but non-polarizable molecules (rigid dipoles) and for nonpolar molecules. In such problems, the use of cavities is particularly suspect because such superpositions as adding cavity and reaction fields are no longer valid. The results are quite different from those of earlier work (cited in (20)), and may prove useful for increasing interests in non-linear effects.

This necessarily cursory account of Fulton's work is perhaps adequate to suggest its importance and potential value for many applicatons. Unfortunately, the formalism is formidable for the initiated, including the writer, and one can express the pedestrian hope that the methods could be set out in more elementary, if less elegant, terms.

2. EQUILIBRIUM THEORIES AND SIMULATIONS

The theories described in 1 were all considered in general terms, either for very simple models of molecules and intermolecular forces or without introducing more realistic descriptions at the molecular level. In this section, we describe for equilibrium behavior results obtained by statistical mechanical evaluations and computer simulation for various models of molecular structure and pair interactions. For background, the nature and usefulness of information obtained from measurements on imperfect gases are considered first; this is followed by a brief review of observed patterns of behavior in the static permittivities of polar liquids as a function of dipole strength $y = 4\pi N\mu^2/3k_bTV$.

2.1 Polarization in Gases.

At even quite moderate pressures of a few atmospheres, the static permittivites of gases show significant deviations from simple linear proportionality to density expected for ideal gases without significant effects of molecular interactions. As with the equation of state for pressure as a function of density, these deviations can be described by a virial series in powers of density with second and higher order dielectric virial coefficients. To introduce these in convenient form for theoretical analysis, a macroscopic spherical sample in vacuum is assumed, with a uniform field E_0 (before insertion of the sample) from external charges. As the macroscopic E in the sample is then given by $E = 3E_0/(\epsilon_s + 2)$, this leads to a Clausius-Mossotti function for permittivity $\epsilon_s - 1 = 4\pi P/E$.

$$\frac{\epsilon_s - 1}{\epsilon_s + 2} \cdot \frac{V}{N} = \frac{4\pi}{3}\langle M_z f \rangle \tag{22}$$

where linear response is assumed and the equilibrium distribution function f is for energies of the molecules on to external field E_0 and other molecules in the sample. Dielectric virial coefficients are conveniently defined by the expansion

$$\frac{\epsilon_s - 1}{\epsilon_s + 2} \cdot \frac{V}{N} = A_\epsilon + B_\epsilon \frac{n}{V} + C_\epsilon \left(\frac{n}{V}\right)^2 + \ldots \tag{23}$$

where n/V is the molar density. The ideal gas value A_ϵ is then just the classic Debye function or equivalent for single molecules

$$A_\epsilon = (4\pi N_A)(\alpha + \tfrac{1}{3}\beta\mu^2)$$

where N_A is Avogadro's number.

The second dielectric virial coefficient B_ϵ is due to interactions of pairs of molecules and as formulated by Buckingham and Pople (21) is given by the statistical average

$$B_\epsilon = \frac{2\pi N_A^2}{3}\int dV_N\left[(\alpha_p - 2\alpha) + \frac{1}{3}\beta(\mu_p^2 - 2\mu^2)\right]\exp(-\beta\mathcal{U}_p) \quad (24)$$

where α_p and μ_p are the polarizability and dipole moments (for no external field of a pair at variable separations. The difference $\alpha_p - 2\alpha$ is thus the fluctuation by pair at interaction of moments induced by E_0, such as those from electrostatic induction and by overlap. They are significant for atoms, but for molecules are almost always small compared to 'permanent' dipole fluctuation effects in $\mu_p^2 - 2\mu^2$ which include moments induced in one by the fields of the other's permanent multipole moments. Finally $\mathcal{U}_p$ is the potential energy of pair interactions for $E_0 = 0$, including all multipole and induced dipole-multipole terms, and dV_N is a normalized volume element in the configuration space.

Because of orientation-dependent terms in both the moments and the Boltzmann factor, values of B_ϵ are much more sensitive to molecular anisotropics than the pressure virial coefficient or the gas shear viscosity as a function of temperature. For nonpolar molecules, quadrupole moment effects are large in the case of CO_2, for example, demonstrating the importance of quadrupole moments $\theta = 4.2 \times 10^{-26}$ esu cm^2 inferred from B_ϵ, while octopole and even hexadecapole effects can be recognized for more symmetrical molecules, e.g., CH_4 and SF_6. For polar molecules, permanent dipole interactions also come into play and anisotropy of repulsive forces (shape) is also important. The result is a very wide range in magnitudes and sign of B_ϵ even for relatively simple molecules, and comparison of calculated values with experiment is a sensitive test of multipole moments and anisotropies of U_p used in the calculation. All these matters are discussed in detail by Sutter (21).

Of the prototype molecules selected for collaborative efforts, only CH_3F has been studied to obtain reliable B_ϵ data. The value at 323°K is -607 in units cm6 mole^{-2}, while CHF_3 with a slightly smaller dipole moment (1.65 vs 1.85 debye) has a $B_\epsilon = +1125 cm^6$ mole^{-2} at 323° K with a much stronger inverse temperature dependence. The obvious interpretation is that because of prolate and oblate 'shapes' in the two cases, CH_3F pairs preferentially orient antiparallel while CHF_3 pairs minimize their dipole energy at closest approach by parallel alignment. Detailed model calculations (22) with Lennard-Jones 12-6 parameters from gas viscosities, known μ and α, and rather large axial quadrupole moments (4.5 and 2.5 X 10^{-26} esu) estimated from B_p temperature dependence gave agreement to ca 5 percent of calculated and experimental B_ϵ and B_p with moderate anisotropy of the R^{-12} repulsion.

The evidence just cited illustrates potential usefulness of dielectric virial data when coupled with pressure virial and viscosity data. Similar information for CH_3I and CH_2Cl_2, among others, should provide much useful information. The case of CH_3I seems particularly interesting, because of the anomalously low static permittivities of it and other iodo compounds in the liquid state. Unfortunately, measurements of the necessary precision (a few parts per million in ϵ_s and 1 part in 10^4 for relative density, with the former easier than the latter) are not a popular activity.

The formulas and calculations for third and higher dielectric virial coefficients become increasingly complicated, to the extent that detailed results have been obtained only for C_ϵ and by using radial pair potentials for the potential energy. The limited experimental data indicate the expected negative contributions to decrease the CM function, with permittivities better described by Onsager or Kirkwood-Fröhlich types of formulas at liquid densities. This is also seen in the integrated intensities of the associated far infrared absorption, suggesting obviously that fluctuation effects are largely suppressed, but detailed discussions are best left to specialists in theory of collision induced processes.

2.2 The Onsager Equation: Usefulness and Difficulties

In this section, we sketch briefly the nature and extent of agreement of Onsager's theory with experimental values of ϵ_s for a large number of polar liquids with known dipole moments from gas or dilute solution data, with a few remarks at the end about early theoretical indications of difficulties in understanding why the Onsager formula should work as well as it does for many non-associated liquids.

Comparisons can be made in several ways. We mention here results compiled in Reference (13) for the dipole moments μ_{ons} calculated from ϵ_s and estimates of ϵ_∞ for more than thirty non-associated liquids, using the Onsager equation (20). These are to be compared with the experimental μ_{gas} from dilute gas (or solution) measurements, with values of

μ_{ons}/μ_{gas} less than one signifying that the predicted ϵ_s from (20) is less then than the observed value. The ratios found for reasonably simple molecules differ from unity in many cases by ± 10 percent.

There is a rather striking correlation between μ_{ons}/μ_{gas} and axial ratios of ellipsoids approximating molecular shapes from interatomic distances and atomic radii, scale models, and the like. Almost without exception, ratios μ_{ons}/μ_{gas} are less than or greater than one for the ratio b/a of transverse to dipole axis less than or greater than one. This can obviously be pictured as preferred anti-parallel alignment of neighboring dipoles for 'cigar shaped' molecules (with respect to the dipole axis) and preferred parallel alignment for 'disk shaped' or flattened molecules. Some modest success has been achieved in reproducing this behavior by using ellipsoidal rather than spherical molecular cavities in the Onsager model, but there are some difficulties with the method used by Scholte (see Ref (13)), and results of revised calculations are in prospect (15).

The major deviations from Onsager's equation are for hydrogen bonding liquids, with water, several aliphatic alcohols, sulfuric, formic, and acetic acids, HF and HCN, and several amides as prime examples. The extent of the deviations is usually expressed by values of the Kirkwood g-factor, defined on the basis of equation (21) by

$$g = \langle \mu \cdot \Sigma_{R_1} \mu_i \rangle / \mu^2 \tag{25}$$

as for g = 1, one has the Onsager formula (20). Values of g derived from measured ϵ_s of the liquids just cited range from 1.5 to 4 or more, with the record seemingly held by N-methylpropionamide for which ϵ_s = 348 at -40°C gives g = 8. (The exception to g > 1 in the group is acetic acid, with g ≈ 0.6 at 10°C plausibly explained by non-polar closed dimers.)

Numerous 'chemical' models of intermolecular bonding have been developed to account for such g values and their temperature dependence, as discussed in (13) for example. Kirkwood's first estimate of g for water (16) with four-coordinated nearest neighbors that $g = 1 + 4\cos^2(105°/2) = 2.48$ gives ϵ_s about 20 percent less than experiment; various elaborations can be made to do better. Models of chainwise association in alcohols, HF, HCN, and amides also produce reasonable agreement for bonding of neighbors consistent with X-ray structures of the solids.

Treatments of the kind just sketched led some ten years ago to the rather comfortable situation that Onsager's equation gave a quite decent first approximation to static permittivities of non-associated polar liquids, with modest deviations understandable in terms of anisotropies in molecular 'shape' and polarizability, while major differences often could be accounted for quite well by simple models of specific short range interactions.

The plausible inference from the foregoing that the model of a point dipole in an electrostatic continuum was a good way of describing short range interactions of real molecules has not stood up in view of more recent results of molecular theory. Before going on to these in the next section, we first remark on already disquieting indications from earlier work. Even the simplest conceivable molecular model, of point dipoles on regular lattice sites with electrostatic dipole -dipole coupling, leads to problems in evaluation, for example, of the partition function or mean moment for a macroscopic sample. This is of course because of the long range character of dipole interaction energy $\mathcal{U}(\underline{m})$ in the Boltzmann factor

$$\mathcal{U}(\underline{m}) = -\sum_i \sum_{j>i} R_{ij}^{-3} \left[\underline{m}_i \cdot \underline{m}_j - 3R_{ij}^{-2} (\underline{m}_i \cdot \underline{R}_{ij})(\underline{m}_j \cdot \underline{R}_{ij})\right] \quad (26)$$

This energy, supplemented by a field interaction energy - $-\sum_i \underline{m}_i \underline{E}_0$ and relevant molecular energies (as of kinetic energies of oscillator induced moments) must be used in exp ($-\beta$H) for averaging. There is no exact solution of the problem, but a variety of approximate treatments have been produced. One is by high temperature expansion of exp ($-\beta$H) in powers of βH, i.e. of 1/T. Van Vleck's second order result (24) for rigid dipoles with $m_i = \mu_i$ and $\epsilon_\infty = 1$ gave agreement with the Onsager result in this limit provided the lattice sums were evaluated for a continuum with $4\pi a^3/3 = V/N$, but smaller reductions of the Lorentz result for discrete sums over more realistic cubic lattice points. Extensions to third order dipole interactions, for rigid dipoles by Rosenberg and Lax (25) and with harmonic oscillator induced dipoles included by Cole (26), showed further differences from the Onsager result even for a continuum, with the conclusion that for the model the true result lies somewhere between the Lorentz and Onsager field expressions.

The high temperature expansion approach is of little use for even moderately polar conditions because of convergence problems. A quite different approximation by Toupin and Lax (27) to obtain results at reasonable temperatures for the dipole lattice model also gave the same sort of indications, but went largely unnoticed. Their very interesting approach and results, with possibilities for further studies of the dipole coupling problem, are described briefly in the next section.

2.3 Molecular Theories.

The beginning of serious molecular theory of dipolar fluids can be set at about 1970 with the papers of Nienhuis and Deutch (28), dealing with the short and long range aspects of dipole correlations in formal terms, and of Wertheim (29), giving an analytic solution for rigid (i.e., non-polarizable permanent dipoles) of the mean spherical model (MSM or MSA). Both invoked the Ornstein-Zernike equation relating the pair distribution function $g(\underline{R}_{12})$ to the direct correlation function $C(\underline{R}_{12})$

$$g(R_{12}) - 1 = C(\underline{R}_{12}) + \int d\underline{R}_3 C(\underline{R}_{13}) h(\underline{R}_{32}) = h(\underline{R}_{12}) \quad (27)$$

in order to work with C which is expected to have simpler properties than h. In the mean spherical model, $C(\underline{R}_{12})$ is taken to satisfy $g(\underline{R}_{12}) = 0$ for $R_{12}<R_o$, a hard core radius, and is approximated by $C(\underline{R}_{12}) = -\beta U(\underline{R}_{12})$ for $R_{12}>R_o$, where $U(\underline{R}_{12})$ as given by (27) for dipoles μ_1 and μ_2. Wertheim's solution by generalizing the Percus-Yevick equation in this way relates $(\epsilon_s - 1)/(\epsilon_s + 2)$ to the dipole strength $y = (4\pi N/3V(\beta\mu^2/3)$ by equations for each in terms of $\eta = \pi R_o^3 N/6V$.

The result of the MSM calculation for rigid dipoles is sketched in Figure 2, together with the Lorentz and Onsager results as a function of y. One sees that the Wertheim

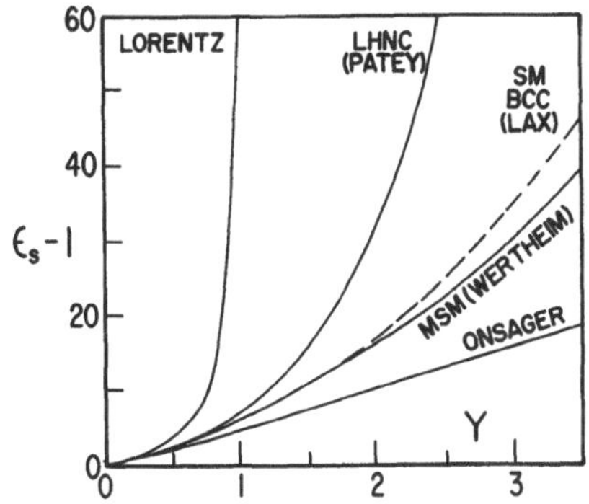

Figure 2. Results for $\epsilon_s - 1$ as a function of $y = (4\pi N/3V)(\beta\mu_s^2/3)$ of rigid dipoles for the models (from left to right): Lorentz field, linear HNC (Patsy), mean spherical model (Wertheim), face center cubic lattice (spherical model of Tonpur and Lax), and the Onsager field.

MSM result lies between the two for the Lorentz and Onsager local fields, and for a given y is much larger than the latter. For y = 4.2 of a 'rigid water molecule' at 0°C, ϵ_s is about 70 from MSM and 20 from the Onsager model. Neither is of course to be compared directly with the experimental ϵ_s = 86 for water at 0°C or with values of ϵ_s for other polar molecules as a function of y, as all real molecules are polarizable. Unfortunately, the inclusion of molecular polarizabilities leads to far more complicated problems and methods of attack, for example 'a formidable superstructure of new graph theory' developed by Wertheim (30). A somewhat ad hoc short cut in lieu of these approaches is suggested by the lattice theory developed by Toupin and Lax, which we sketch before going on to some of the more recent developments for non-polarizable dipoles.

Toupin and Lax consider the problem of permanent and induced dipoles on cubic lattice sites (or continuum) with the latter represented by harmonic oscillators as in Van Vleck's early work described in 1.3. The device of introducing fluctuations from equilibrium displacements works for harmonic oscillators because the integrations over the formula to evaluate averages are for all values from $-\infty$ to $+\infty$ and unchanged by the shifts in origin. A similar device is not possible for proper averages over possible permanent dipole moments $\underline{\mu}_i$, as the ranges are restricted by the N constraints $\underline{\mu}_i \cdot \underline{\mu}_i = \mu^2$, but becomes so if these are replaced by the single 'spherical' constraint $\sum_N \mu_i \cdot \mu_i = N\mu^2$. This can be expressed by the Fourier transform representation of a delta function, and the Hamiltonian for the permanent dipole energies can also be spatially transformed to yield tractable algebraic functions. Evaluation of the partition function integral over the μ-space by the saddle point method then gives the desired expression for polarization P in external field E_0 and the permittivity ϵ_s.

Denoting the k-transforms of the permanent dipole interaction energies (equation (26) with $m = \mu$)by

$$\lambda(\underline{k}) = -\frac{1}{\mu^2}\sum_i \sum_{j>i} u_{ij} \exp(i\underline{k}\cdot\underline{R}_{ij}) \tag{28}$$

which are sometimes called dipole wave sums, the spherical model solution in the limit $E_0 = 0$ can be written in the form

$$F(\epsilon_s) + (\epsilon_\infty - 1)\left(\frac{\epsilon_s - 1}{\epsilon_s - \epsilon_\infty}\right) = \left(\frac{\epsilon_s - 1}{\epsilon_s - \epsilon_\infty}\right)^2 \left(\frac{\epsilon_\infty + 2}{3}\right)^2 y \tag{29}$$

where y is the permanent dipole strength and ϵ_∞ for induced polarization is given by the Clausius-Mossotti expression (15). The function F (ϵ_s) is

$$F(\epsilon_s) = \frac{\epsilon_s - 1}{\epsilon_s + 2}\left[\frac{2}{3}\int dk' \frac{1}{1 - \left(\frac{\epsilon_s - 1}{\epsilon_s + 2}\right)\left(\frac{3\lambda_t}{4\pi}\right)} + \frac{1}{3}\int dk' \frac{1}{1 - \left(\frac{\epsilon_s - 1}{\epsilon_s + 2}\right)\left(\frac{3\lambda_\ell}{4\pi}\right)}\right] \tag{30}$$

where the (normalized) integrals are over the primitive cell in k-space for the cubic lattice and the λ_t and λ_ℓ are the two transverse and one lonagitudinal components of the dipole wave sums for $\mathbf{k} \cdot \mathbf{R} = 0$ and $\mathbf{k} \times \mathbf{R} = 0$.

There are several interesting features of the results (26) and (27). In the absence of dipole interactions, and hence $\lambda(\mathbf{k}) = 0$, equation (26) reduces to the Langevin-Debye formula for the Lorentz field, while use of the k = 0 (long wavelength) limiting values $\lambda_t(0) = 4\pi/3$, $\lambda_\ell(0) = -8\pi/3$ for all k gives the Onsager equation. Detailed evaluations (31) of the k-integrals in equations (27) from numerical value of the $\lambda(\mathbf{k})$ by direct and Ewald sums (32) depend on the lattice type, but give very similar results for FCC and BCC lattices and equivalent continuum (requiring the proper number and density of states). The results are indicated schematically by the curve labelled FCC (LAX), which is seen to lie near the MSM result of Wertheim, but extrapolates to $\epsilon_s = \infty$, indicating a phase transition, at $y \sim 5$ (for rigid dipoles).

A little study of equation (30) and the $\lambda(k)$ shows that except for ϵ_s not much greater than ϵ_∞, the longitudinal term is of order $1/\epsilon_s$ relative to the two transverse ones. If it is neglected, the effect of including induced movements is then to increase the permanent dipole strength y by the factor $[(\epsilon_\infty+2)/3]^2$, or equivalently to replace μ by $[(\epsilon_\infty+2)/3]\mu$. This suggests that other results for non-polarizable dipoles can in reasonable approximation be extended to polarizable ones simply by using $(\epsilon_\infty+2)/3$ as a scaling factor, a conclusion also reached for the Onsager and Lorentz models and seemingly indicated by Wertheim's developments.

An approximate extension of MSM by Wertheim (30) to include polarizability and the Toupin-Lax result give $\epsilon_s \sim$ 120 for y = 4.2 of water at 0°C, which is considerably larger than the experimental 86, as well as several times the value 33 from the Onsager equation. The situation is further divergent because neither of these theories is very reliable, as a result of the very different 'spherical' approximations used, and it appears that use of better approximations for the pair distribution function can only increase calculated values for the point dipole model. Space does not permit discussion of the numerous efforts (ones to 1979 have been comprehensively reviewd by Wertheim (30)), but the linear hypernitted chain (HNC) or single superchain approximation should be mentioned as an example. Calculations by Patey (88) for this model with reduced density $p^* = 0.8$ give for rigid dipoles the curve labelled HNC in Figure 2, which is well above those for the spherical models.

With these and other results, it seems abundantly clear that calculations, however refined, for point dipole models lead to values of ϵ_s considerably larger than that of water and a fortiori of 'normal' polar liquids, which as discussed in 2.2 are moderately well described by Onsager's equation. This might be called an experimentalist's assessment. That of theorists seems to be embodied in such statements ad 'We now know that the Onsager model seriously underestimates ϵ_s'. (30) and 'breakdown of Onsager's theory' (31). To the writer, these are true in the sense of calculations for point dipoles, but beg the question of why Onsager's equation, for a point dipole in continuum model, works as well as it does.

Some progress toward answering the question just raised is provided by a series of calculations by Patey and associates which incorporate point quadrupole as well as point dipole interactions. Patey and Valleau (32) showed by Monte Carlo simulation that addition of axial quadrupoles $\theta_1 = \sum e_i (z_i^2 - x_i^2)$, where z′ is the dipole axis, of moderate strength reduced the calculated ϵ_s for water-like parameters by 30 percent or more, and linear HNC integral equation solutions (33) gave similar results. This sort of quadrupole is, however, not appropriate for the water molecule, as for tetrahedral symmetry θ_1 vanishes, but a tetrahedral quadrupole $\theta_2 = \sum e_i (y_i^2 - x_i^2)$ is. In the most recent calculations, Carnie and Patey (33) included the interactions of such a quadrupole $\theta_T = 2.5 \cdot 10^{-26}$ esucm^2 and with a hard sphere diameter of 2.8 A reproduced the ϵ_s of water from 25° to 200°C to 3 percent.

Patey's results thus lend much credence to the belief that short range quadrupole interactions greatly reduce the positive correlations produced by dipole interactions alone, and to the extent that good quantitive agreement of ϵ_s with experiment is obtained for water. Questions still reamin, such as the long standing one of whether point dipoles plus quadrupoles are really adequate models of real molecules generally, and of whether the success in an average sense of Onsager's equation is to be understood in such a simple way, rather than as some sort of compromise among these and a variety of other contributing interaction effects.

2.4 Computer Simulations.

As the preceding sketches of developments in molecular theories of polar liquids are intended to suggest, the molecular models so far used are in most cases too simplified and idealized for the results to bear very close comparison with experimentally accessible properties of real fluids. One of the important uses of computer simulations is of course not only to do 'experiments' giving such results for the assumed model, but also to evaluate for comparison integrals, correlation functions, and the like treated approximately in the theories and not otherwise subject to direct experimental test.

In the case of polar liquids, computer experiments, whether by molecular dynamics or Monte Carlo methods, present difficulties arising from the long range character of the dipole interaction energy, and resolution of the problems for the modest sizes of systems in simulations is a task for statistical mechanics rather than ingenious computation methods. Wertheim (35) has summarized the results of the two main kinds of effort to improve on truncation, whether by cutoff of the potential or by periodic boundary conditions, namely use of mean field approximations for longer distances and use of Ewald or other summation methods. There is much of interest and instruction value in comparison of results so far obtained; the reader is referred to the recent review by Alder and Pollock (34) for discussions.

A rather different approach for polar molecules has recently been explored by Bossis and Brot. This is to calculate correlation functions for dipole, in an inner spherical region R_1 which is part of a larger spherical region R_M, the motivation being to gain some idea of how well such simulations can approximate the inner sphere function in the Kirkwood-Fröhlich (KF) equation (21), which strictly should be evaluated in the double limits R_1, $R_M/R_1 \rightarrow \infty$. Molecular dynamics computer simulations have so far been done (35) for a two-dimensional system, i.e. disks rather than spheres, for 324 molecules in most cases with a Lennard-Jones 12-6 radial plus rigid dipole pair potential, with the system confined within an outer radius $R_M = 13.2\sigma$ by a 'soft dish' potential barrier.

From a variety of computer runs for two different dipole strengths y, values of small sphere correlation functions for values of R_1/R_M from 0.2 to 0.6 were found to be consistent with disk versions of the KF equation (21) for R_1/R_M finite after averaging over center positions of the disk and excluding contributions of molecules with centers too close to R_1 and R_M. The lower limit is remarkably small, indicating that only a few near neighbors determine the inner sphere correlation function for the system studied. The derived values of ϵ_s are also considerably greater than the Onsager 2-D values, particularly for the larger y, much as for the results of 3-D calculations, while a note without details reports that inclusion of quadrupole moments reduces the values considerably, which is again in qualitative agreement with the 3-D results of Patey et al. This work with what might be called KF simulation has thus given very interesting results in the two dimensions, with obvious possibilities for further studies.

Very recently in work not yet published (15), Hassis-Bezot, Bossis, and Brot have done 3-D molecular dynamics simulations of a system of dipolar molecules interacting with a Stockmayer pair potential (Lennard-Jones plus point dipole-dipole coupling), with 913 and 1472 molecules in a spherical dish potential of the sort just described. Computer runs were done at reduced density $\rho^* = \sigma^3\rho = 0.8$, temperature $T^* = k_BT/\epsilon_{LJ} = 1.35$, and dipole moment $\mu^{*2} = \mu^2/\sigma^3\epsilon_{LJ} = 2.7$, these values being nearly the same as in previous simulations by Pollock and Alder (36) and by Patey, Levesque, and Weis (37) and roughly corresponding to liquid CH_3F at 206 K.

The derived values of Lennard Jones and dipole

interaction energies for the spherical sample agree closely with those from the simulations with periodic boundary conditions. The static permittivity from the 1472 molecule run was found to be $\epsilon_s = 28 \pm 1$. This is in good agreement with the value 29 ± 1 of Pollock and Alder using the Ewald sum method and with the value 31.6 ± 4 obtained by Patey et al (38) using a spherical cutoff with reaction field correction, but for hard spheres rather than the L-J potential. Although the agreement is encouraging, there are unresolved questions as to why the periodic boundary conditions and correction methods should work as well as they seem to.

The simulation values of ϵ_S correspond to a Kirkwood g factor equal to 2.7, again consistent with the general conclusion that centered point dipoles are not a good model for normal polar liquids. Brot et al have also examined the simulation consequences of assuming an 'off-center' dipole at a distance $r = 0.3\sigma$ from the Lennard-Jones center, and so approximately equivalent to a central dipole plus a central axial quadrupole. This results in a greatly reduced permittivity $\epsilon_s = 11 \pm 1$. With an approximate polarizability correction, the permittivity of this model for CH_3F is about ten percent less than the experimental $\epsilon_s = 22.5$ at 206 K.

Finally, in this section, it seems appropriate to mention some early Monte Carlo calculations for crystalline 1,2,3 trichloro 4,5,6 tri methyl benzene by Brot and Darmon (39), because these quite clearly show the inadequacy of point dipole models for predicting the stable partially ordered phase which develops on cooling the disordered high temperature solid below the melting point or the electrostatic energy involved. Calculations for a twelve point charge model of the molecular charge distribution, however, both gave the correct minimum energy configuration and reasonable decrease in energy; calculated static permittivities ϵ_s also showed a decrease with ordering at decreasing temperature in fair agreement with that observed experimentally. The very interesting results obtained for complex permittivity are discussed in 3.

3. DIELECTRIC RELAXATION.

Studies of the time or frequency dependence of charge displacements in varying electric fields obviously can provide much more information about the underlying molecular dynamical processes than results for equilibrium or 'quasi-static' limiting behavior, but the theoretical interpretation is correspondingly more difficult, as one must find tractable methods for approximate time-dependent solutions of relevant equations of motion rather than just energy integrals. The result is that relaxation theory is almost always at a cruder level of description than the more refined equilibrium treatments, and there are more or less continuous catching-up processes.

This section begins with a rough indication of the time scales in study of relaxation processes to which discussion is limited, with brief surveys of experimental methods and of the extent and quality of results available or in prospect. A review of primitive early relaxation models and circumstances for simple Debye relaxation is followed by sketches of observed patterns of relaxation behavior, models proposed to account for them, and developments in more sophisticated formal methods.

3.1 Relaxation Domains and Experimental Methods.

There is no sharp or single division between time scales for damped resonant response, as usually studied by optical spectroscopy, on the one hand, and the slower responses characteristic of further time evolution, as usually studied by guided wave and lumped circuit methods, on the other. Somewhat arbitrarily, then, the dividing line for present purposes is taken to be of the order of a picosecond, or for frequencies somewhere between tens of gigahertz (10^9Hz) and one terahertz (10^{12}Hz), with the larger values of the latter usually appropriate for small molecules in condensed phases. A very rough justification for numbers of such size is that times between collisions of molecules at liquid densities with loss in correlations of momentum are typically in the picosecond range, followed by much slower loss of correlations of position. Attention here will be focussed on the latter, as the former are considered by other contributors to these Proceedings. This is not to say that the two are mutually exclusive or that features of resonance processes, such as spectral line shapes, do not give information about slower relaxation processes, for of course they do.

The theoretical division between time scales is for

roughly the frequencies at which one shifts from optical to line or circuit kinds of measurements, with the latter being for frequencies of order 20 to 40 GHz. Dielectrically active relaxations in liquids can occur over the gamut from these values for small molecules and ordinary temperatures to ones well below 1Hz if sufficiently supercooled toward a glass transition. The most accessible range in the past has been from, say, 10Hz to 1MHz because of availability of precise, relatively simple transformer bridges and the like, with the result that there is a comparative wealth of information about relaxation processes in the range. A variety of ultralow frequency and transient methods have also long been available which are satisfactory for dielectrics if they are not too highly conducting, but are not as widely useful and have not helped much to reduce the notorious problems with electrode polarization in electrolyte systems.

Measurements above 1MHz have long been possible, but increasingly tedious, difficult, and inaccurate, to 10GHz or so, with some success using various waveguide techniques to 100GHz or more. This is unfortunately just the range of primary interest for studying dipolar liquids with small molecules, and there is a comparative dearth of adequate information about them as a result. Fortunately, this situation is improving, thanks to several developments in the last ten years and others underway or possible. The key elements can be identified as development of detection devices capable of picosecond time resolution or of phase as well as amplitude measurements in the frequency domain, both thanks to sampling techniques for repetitive wave forms.

The time domain alternatives are particularly attractive in principle, because a single time record acquired by signal averaging for a minute or less can by numerical Laplace transformation give data over two or more decades in frequency. Fast tunnel diode switching to produce step like pulses with rise times of 30ps or less makes the procedure feasible to 10GHz. In early stages, the implementation was rather crude, with simplified models of system performance and neglect, or inadequate account for precision taken, of various artifacts in the system response and sample cell designs. Fortunately, most of these can now be handled by surprisingly simple overall calibrations (40). This can be illustrated by the basic working equations for a precision bridge method developed by the writer and associates (40). If reflected signals $R_x(t)$ and $R_s(t)$ from the admittance y_x of a sample cell containing the unknown dielectric and from the admittance y_s of a reference standard e.g., a second sample cell with known dielectric, their difference is given by the bilinear expression

$$y_x(\omega) - y_s(\omega) = \frac{A(\omega)\rho(\omega)}{1 - B(\omega)\rho(\omega)}$$
$$\rho(\omega) = \frac{\mathcal{L}\,[R_s(t) - R_x(t)]}{\mathcal{L}\,[R_s(t) + R_x(t)]} \quad (31)$$

where $\rho(\omega)$ is the ratio of (numerical) Laplace transforms of the difference and sum of reflected signals. The two complex parameters $A(\omega)$ and $B(\omega)$ take account of all linear system response characteristics and are readily determined by calibration using two reference standards for y_x. There are of course some matters of detail, but nothing really more complicated, and so the way seems to be open for many measurements with comparative ease.

There are also alternatives in the frequency domain, by use of programmable network or admittance analyzers to sweep frequencies from 1MHz to 1GHz, or more, for example. Unfortunately, commercially available instruments are designed for comparisons to 50 ohms rather than a reference capacitance. This largely nullifies the advantage of automation, as several cells must be used to cover an extended range. There seems to be no problems in principle of adapting such instrumentations, intended for electrical engineering purposes, to dielectric measurements, but no such developments are known to the writer.

Finally, there remains the problem of dielectric measurements above 10GHz, until recently only possible with highly developed, and expensive, instrumentation in a few laboratories, with the result that there is an even greater dearth of experimental results in this range than at low gigahertz frequencies. Fortunately, this also is changing, as by the work of Gerschel reported in this Proceedings volume. Alternative routes to measurements of correlation functions from other effects in this time range are rapidly being developed, notably using linear and non-linear laser techniques, as discussed by Kenney-Wallace. Some comments are made in Section 4 about the possibilities for dynamic Kerr effect studies and information obtainable in relation to that from dielectric measurements.

3.2. Simple Relaxation Models.

Perhaps the simplest defensible model of the statistical

outcome of molecular reorientations is to assume that the probability of a molecular dipole turning from an initial angle θ_1, with respect to applied field direction or other fixed frame, to a new value θ_2 is described by a transition probability $k(\vartheta)$ which is some function of the angle ϑ between the two orientations, but not of θ_1 or θ_2 separately. Using detailed balances of such jumps to and from all other orientations and the resultant rate equation, the dipole correlation function is

$$\langle \cos\theta(0)\cos\theta(t)\rangle = \langle \cos\theta(0)^2\rangle \exp(-k_1 t) \tag{32}$$

$$k_1 = \int d\omega(\vartheta)k(\vartheta)(1-\cos\vartheta) \tag{33}$$

This result, originally given by Van Vleck and Weisskopf (41), is obtained very easily in correlation function terms (39) by using $\cos\theta_2 = \cos\theta_1\cos\vartheta + \sin\theta_1\sin\vartheta\cos\varphi$ of spherical trigonometry. A simple exponential decay is thus predicted without any specific assumptions about the kinds of 'jumps', i.e., the functional dependence of $k(\vartheta)$ on ϑ. If large jumps are infrequent so that $k(\vartheta)$ is small unless $\vartheta \ll 1$, the rate constant k_1 is

$$k_1 = \int d\omega(\vartheta)\,k(\vartheta)\left(\frac{\vartheta^2}{2}\right) \tag{34}$$

which is thus a mean square angular displacement $\langle\vartheta^2/2\rangle$ of the sort in theories of Brownian motion. A model at the other extreme of only discrete large jumps clearly gives a quite different interpretation of k_1, but the same form of time dependence. Hence there is no basis for deciding between these or other possibilities from the observable time history of dipole correlations alone, and evidence favoring one or another must come from other sources.

One kind of evidence, about which a bit more will be said in 4.1, can be from experiments which measure correlations of $P_2(\cos\theta) = (1/2)(3\cos^2\theta-1)$. When the preceding analysis is repeated, one obtains an expression of the same form as (32) except for the expression for k, which becomes

$$k_2 = (3/2)\int d\omega(\vartheta)k(\vartheta)(1-\cos^2\vartheta) \tag{35}$$

on using the cosine formula twice, or equivalently, the addition theorum for spherical harmonics. In the Brownian motion limit of $k(\vartheta)$ small unless $\vartheta \ll 1$, one obtains instead of (34)

$$k_2 = (3/2)\int d\omega(\vartheta)k(\vartheta)\vartheta^2 = 3k_1 \tag{36}$$

and a rate of decay three times faster than for $P_1(\cos\vartheta) = \cos\vartheta$ correlations. In the other extreme of all or none transitions to effect loss of correlations, k_2 and k_1 are the same. Thus <u>unambiguous</u> comparison of experiments measuring different aspects of the same molecular motions <u>can</u> give information to help distinguish between different possibilities. (The reasons for underlining will soon appear.)

Before considering diffusion models for k_1 and k_2, we touch briefly on the time dependence $\exp(-kt)$. As is well known, this cannot be a proper form, as correlations such as $\langle P_1P_1(t)\rangle$ and $\langle P_2P_2(t)\rangle$ must by time reversibility of equations of motion be even functions of time, to which $\exp(-kt)$ can only be an approximation introduced by the model. The difficulty is sometimes patched up by writing $\exp(-k|t|)$, but there is the further difficulty at short times that the loss of correlation must be as t^2, rather than as t, because of inertial delay neglected in the model, with effects of intermolecular torques first showing up in coefficients of t^4 in a short time expansion. Proper treatment of these effects, conspicuous in the short time or high frequency limit prior to onset of relaxation dominated time evolution, is the subject of other contributions to these proceedings.

Debye's original rotational diffusion model corresponds to $k_1 = 2D_w$, where D_w is the rotational diffusion coefficient, which for his model of the molecule as a sphere of volume V reorienting in a medium of shear viscosity η is $D_w = k_bT/6\eta V$, giving the 'Debye' relaxation time τ_{1D}.

$$\tau_{1D} = \frac{1}{k_1} = \frac{3\eta V}{k_B T} \tag{37}$$

Values of τ_{1D} from this equation for small molecules of polar liquids at room temperature are typically several picoseconds and in rough agreement with macroscopic dielectric relaxation times τ_M, usually obtained by fitting measured complex permittivities $\epsilon(\omega)$ to a Debye relaxation equation.

The adequacy of this representation is considered in 3.3; here the proposition is accepted that, very high frequencies excepted, equation (35) does fit data for small molecules of simple structure at ordinary liquid temperatures quite well for suitable choices of ϵ_s, ϵ_∞, and τ_M.

Unfortunately, for simplicity, the relaxation times τ_1 and τ_M are not the same, as τ_1 is for molecular reorientations and τ_M for macroscopic polarization, with the relation between the two dictated by molecular theory. More generally put in terms of response theory, one needs the relation between microscopic time-dependent dipole correlations $\langle \mu \cdot \Sigma_j \mu_j(t) \rangle$, for a molecule and correlated near neighbors, and the observed function $\langle M \Sigma M_j(t) \rangle$ for a macroscopic sample. A second digression is thus in order to discuss this question before returning to diffusion models.

Clear cut, simple relations between τ_1 and τ_M have been established only for the primitive Lorentz and (with more difficulty) Onsager local field models. Simple generalization of the Langevin-Debye formula (19) to relaxation gives a formula of the form (35) with

$$\tau_1 = \frac{\epsilon_\infty + 2}{\epsilon_s + 2} \tau_M \qquad (39)$$

Generalization of the Onsager model to time-dependent processes has presented problems and generated considerable controversy about proper treatment of cavity boundary value problems. There is now general agreement that the correct result, thanks primarily to the work of Titulaer and Deutch (42) and of Fulton (43), is the Fatuzzo and Mason (44) formula

$$\left[\frac{(\epsilon - \epsilon_\infty)(2\epsilon + \epsilon_\infty)}{3\epsilon}\right] \Big/ \left[\frac{(\epsilon_s - \epsilon_\infty)(2\epsilon_s + \epsilon_\infty)}{3\epsilon_s}\right] = \mathcal{L}(-\dot{\varphi}) \qquad (40)$$

where $\varphi(t)$ is the normalized microscopic correlation function either of a single molecule (Onsager limit) or inner spherical region (parenthetically, Nee and Zwanzig have referred to a simpler version of this, proposed long ago by the writer (45) on a quite ad hoc basis, as 'the Onsager-Cole Formula'. Since then, Titulaer and Deutch have described the writer's later change of heart, because of Glarum's treatment of the Kirkwood-Frohlich model, as 'a mild act of apostacy'.) For further discussion, see the review by Vaughan (46).

Because equation (37) is non-linear in $\epsilon(\omega) = \epsilon$, it does not lead to Debye relaxation for an exponential $\varphi(t)$ =

$\exp(-t/\tau_1)$, but to a complex locus lying a little outside the 'Debye semicircle'. This objectionable feature is probably to be overcome by more detailed considerations of intermolecular forces and torques, as in Berne's forced diffusion model (47) with coupling of rotation and translation leading to transverse and longitudinal components of orientational fluctuations. Unless $\epsilon_s-\epsilon_\infty$ is not larger than ϵ_∞, however, the result very nearly corresponds to Debye relaxation with

$$\tau_1 = \frac{2\epsilon_s^2+1}{\epsilon_s(2\epsilon_s+1)}\tau_M \qquad (41)$$

for the case of non-polarizable rigid dipoles and the limiting low frequency behavior. (A commonly used relation, proposed semiempirically by Powles and obtained from Glarum's treatment, is $\tau_1=[(2\epsilon_s+\epsilon_\infty)/3\epsilon_s]\tau_M$).

It seems clear that, particularly for strongly polar liquids, use of the Lorentz field formula (36) gives too small values of τ_1. The much smaller reductions using (38), by at most 10 per cent for non-polarizable molecules, seem more reasonable, at least for non-associating polar liquids with $\epsilon_s-\epsilon_\infty$ not far from the Onsager prediction. In the cases of hydrogen bonding liquids, the situation is less clear: the static behavior is understood in at best crude terms, except possibly for water, and there is even less progress in really understanding the molecular dynamics, diffusionlike or otherwise. In such cases, τ_1 can at best be taken as a correlation time for a molecule and its near neighbors, with the self-correlation time for a single molecule unknown; but presumably shorter.

After all these complicating factors are considered, one can at best conclude from the limited evidence that a relation $\tau_1 = 3\tau_2$ indicating diffusionlike behavior is not inconsistent with experimental results for simple polar liquids at ordinary liquid temperatures. As discussed by Williams in his contribution, the situation can change considerably at low temperatures.

The problems for hydrogen bonding liquids can be suggested in bald and simple terms. For water, the quite precise macroscopic relaxation time τ_M is 8ps at 25 C, with no certainty about τ_1 and some uncertainty about a value of τ_2 from other measurements, is <u>larger</u> than τ_M = 4ps for the much larger acetone molecule, for example. Similarly, the macroscopic times τ_M = 50, 170, and 320 ps for methanol, ethanol, and 1-propanol at 25 C are all much larger than for non-hydrogen bonding molecules of the same size. Satisfying explanations in molecular terms for these special, but important, liquids are challenges for future work.

3.3 Time Dependence, Forms of Relaxation Functions.

There are numerous examples of polar liquids, at room temperature at least, which exhibit Debye, or nearly so, behavior (in the sense of conformity to equation (35)) over most of the frequency range of dispersion. Only a few of these are mentioned here; others can be found, for example, in Böttcher-Bordewijk (48) and in the tables compiled by Buckley and Maryott (49). This is true of substituted methanes (50) (including methanol!), such as chloroform, $CHCl_3$, with τ_M = 5.7ps at 25 C. Even for these and other simple molecules, there are small residual effects, as indicated schematically in Figure 3(a). These, seen as millimeter-far infrared absorption, are a resultant of inertial effects and Poley absorption variously attributed to librations, coupling to translational modes, and the subject of ot\er contributions to these proceedings.

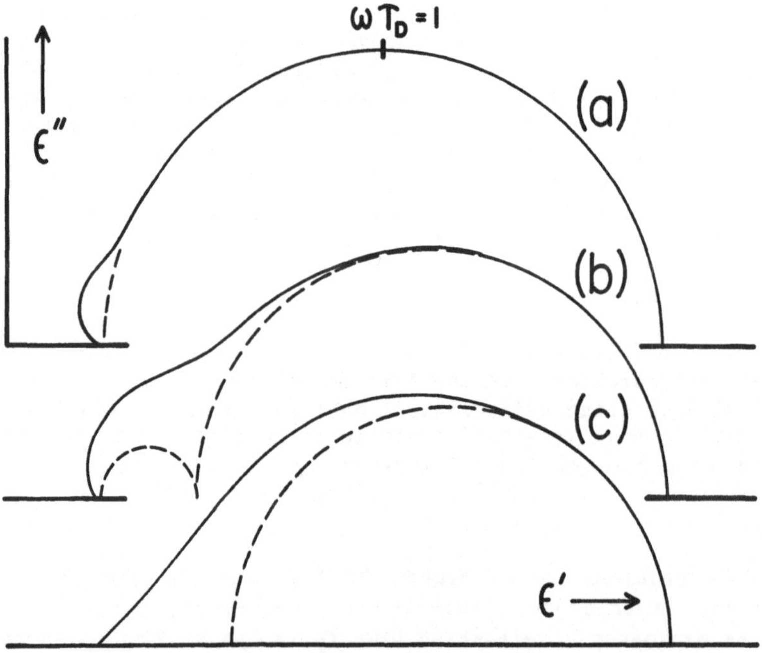

Fig. 3. Complex plane loci of permittivity. (a) Simple molecules, (b) Molecules with secondary relaxations, (c) Extended range of high frequency deviations from the Debye semicircle.

For more complex molecules; larger departures are often seen, which can sometimes by represented by a sum of two or more Debye relaxations, as indicated in Figure 3(b), and plausibly attributed to relative reorientations of polar groups for example (26). Similar behavior has been found in some solids, where the discrete set of relaxations has been explained (51) in terms of jumps between non-equivalent orientation sites, thus violating the condition for deriving equation (29) for simple exponential decay.

The intention here is not to dwell on these and other examples of complications resulting from specifics of molecular constitution, but rather to distinguish them from a more general kind of deviation from Debye behavior in liquids, which usually becomes increasingly apparent at lower temperatures. As shown by the complex plane locus in Figure 3(c), the low frequency region resembles Debye behavior, but relaxation at higher frequencies is over a broader range, often with an asymptotic dependence $\epsilon \sim (i\omega)^{-\beta}$, with β less than the value unity for Debye relaxation. Most of the examples for liquids in the literature are for molecules of at least moderate complexity, but this is at least partly due to the fact that until quite recently there have been few definitive measurements for simpler molecules because of the high frequencies needed.

This kind of behavior found in an increasing number and variety of systems can often be described remarkably well by either of two empirical relaxation functions: the 'skewed arc' function of Davidson and Cole (52) and the Williams-Watts function (53). The former is simpler in the frequency domain, with the form

$$\epsilon = \epsilon_\infty + \frac{\epsilon_s - \epsilon_\infty}{(1 + i\omega\tau_0)^{\beta_c}}, \quad \beta_c \leqslant 1 \tag{42}$$

This expression was first conjured up by the writer after seeing unexpected results for, of all things, supercooled glycerol and being reminded of a similar function, with β_c = 0.5, for the admittance of a semi-infinite transmission line with series resistance and shunt capasitance. The initial supposition that it somehow resulted from complexities of hydrogen bonding in glycerol and other glycols, as contrasted with aliphatic alcohols exhibiting simple Debye behavior, was quickly destroyed by results for several 'supercoolable' alkyl halides (54) notably iso-amyl bromide, which also could be so represented with β_c in the range 0.6 to 0.7 for the low temperatures at which the relaxation was then measurable.

Somewhat later, Williams and Watts found that data for a number of solute molecules in supercooled o-terphenyl and for some amorphous solid polymers could be very well represented by a seemingly rather different function. This is simpler in the time domain, the normalized relaxation function being given by

$$\varphi(t) = \exp\left[-(t/\tau_0)^{\beta_W}\right], \quad \beta_W \leqslant 1 \tag{43}$$

while the counterpart from the Laplace transform of $(1+i\omega\tau_0)^{-\beta_c}$ is in terms of the incomplete beta-function related to the gamma function.

With suitable choices of β_c and β_W, the two relaxation functions can be matched quite closely, with β_c = 0.50 and β_W = 0.63 giving very nearly the same results, for example, so a choice between the two, or preference for neither, purely on experimental grounds requires data of good precision. Opinions are divided, with no clear majority for either. The writer is of the opinion that the question is far less important than the one of understanding the behavior in terms of resonable models which may give one or the other, but can perfectly well result in still other functions or series expansions with very similar time and frequency dependence.

Before going on to discuss progress in understanding origins of conspicuous deviations from exponential decay over extended ranges of time and frequency, two other empirical functions should be mentioned for purposes of reference in the discussion. These are the circular arc (Cole-Cole, (55)) and Havriliak-Negami (56) relaxation functions

$$\epsilon = \epsilon_\infty + \frac{(\epsilon_s - \epsilon_\infty)}{1 + (i\omega\tau_0)^{1-\alpha}}, \quad \alpha \geqslant 0 \tag{44}$$

$$\epsilon = \epsilon_\infty + \frac{(\epsilon_s - \epsilon_\infty)}{[1 + (i\omega\tau_0)^{1-\alpha}]^{\beta}}, \quad \alpha \geqslant 0, \beta \leqslant 1 \tag{45}$$

The deviations from exponential relaxation at long, as well as short, times ('long time tails') represented by these equations are not often found for simple molecules in liquids, but are common in molecular solids, amorphous polymers and polymer solutions, raising questions as to why this should be.

3.4 Models of Cooperative Relaxation.

There seems to be a growing consensus that the answers to the questions posed by non-exponential behavior are to be found as cooperative processes of motions coupled together by intra and intermolecular forces. The solutions of a series of coupled linear dynamical equations can in principle be expressed in terms of series of normal, in our case relaxation, modes and a resultant set of exponential decays with associated ratio or relaxation times. As mentioned in 3.3, this can work well for discrete site models, but can generate enormous numbers of active modes for many body problems and these indeed are what would be required to predict the deviations from non-exponentiality expressed by β_c or β_W significantly different from unity. The formal device of solving for the necessary broad distribution of relaxation times to fit observed data, as developed by Colonomos and Gordon (57) for example, can be very helpful in visualizing the outcome in these terms, but understanding the reasons is another matter. Thus attributing different molecular relaxation times to a distribution of environments, for example, begs such questions as what they are, how produced, how long lasting, and how apt the description is. We here take the view that approaches in which solutions are otherwise obtained are more likely to be fruitful.

An early attempt to describe effects of coupled motions in relaxing dipole orientations was the defect diffusion model of Glarum (58), in which it is assumed that a dipole relaxes to a randomorientation either by exponential decay or by arrival of a nearest 'defect' which completely destroys the correlation existing just before it arrives, assumed to be determined by a random walk or diffusion process in one dimension, but otherwise unspecified. With τ_0 as the rotational relaxation time and τ_d as the diffusional time, use of the diffusion solution for the absorbing wall problem (59) gives the frequency domain relaxation function

$$\mathcal{L}(-\dot{\varphi}) = \frac{1}{1+i\omega\tau_0}\left[1 + \frac{i\omega\tau_0}{1+(\tau_d/\tau_0)^{1/2}(1+i\omega\tau_0)^{1/2}}\right] \qquad (46)$$

The introduction of one parameter τ_d results in an astonishing variety of simulations of the empirical relaxation functions discussed in 3.3. for $\tau_d \gg \tau_0$ and no defect mechanism, the Debye function is obtained; at the opposite extreme, $\tau_d \ll \tau_0$ and no rotational 'local' relaxation, the circular arc function for $\alpha = 0.5$.; and for $\tau_d = \tau_0$, the skewed arc with $\beta = 0.5$. For $\tau_d > \tau_0$, a reasonable simulation of the skewed arc with $0.5 < \beta_c < 1$ can be obtained for different ratios τ_d/τ_0, and for $\tau_d < \tau_0$, a resemblance to the Haveriliak-Nagami function. An intriguing small difference, however, is the prediction of a high-frequency (short time) tail over a limited region in which $\mathcal{L}(-\dot{\phi})$ varies a $(i\omega\tau_0)^{-1}$, as indications of just this behavior have been found in supercooled alkyl halides (60). The further indication that this square root behavior over extended ranges may be approached as a limiting one, when extended cooperative diffusion aspects are dominant, has also been seen in a collection of results for isoamyl bromide by several methods over a very wide range of relaxation time τ_0 and temperature. In this synthesis, the values of β_c in equation (39) to fit the data do decrease from $\beta_c = 0.75$ to a limiting $\beta_c = 0.5$ at the lowest temperatures (60).

While it should be noted also that these changes occur in the supercooled state of the liquid and are accompanied by grossly non-Arrhenius behavior or 'critical slowing down' of the relaxation as set of glassy behavior is approached, values of β_c or β_W appreciably less than one are encountered at temperatures above the normal melting point and when there are no obvious indications of more than exponential dependence of relaxation times τ_0 on reciprocal temperature (i.e., on $\beta = 1/k_BT$).

The very simple and somewhat ad hoc form of Glarum's assumptions, coupled with the one-dimensional diffusion model used, have led to several extensions of the original treatment; these include relaxation by next nearest and other neighboring defects, again by diffusion or random walk models, and to the three dimensioanl case by such models. These go well beyond the motivation of the original idea, to see whether a simple cooperative mechanism could account for marked short time deviations from simple models of rotational relaxation of a single dipole in a mean field approximation. For a review of much of this, see Böttcher-Bordewijk (48). Particularly in the three dimensional case, however, there is an increasing question as to how far one can, or should, invoke isotropic diffusion processes to relax the component of a molecular electric moment parallel to that at an earlier time.

The Glarum one-dimensional treatment of correlations affected by coupling to neighbors is obviously particularly appropriate to the problem of correlations in extended unbranched polymer chains, for which dipole correlations are often found to give rise to Williams-Watts or Havriliak-Negemi relaxation functions. A pioneer effort at a dynamical treatment of couplings between segments in such polymer chains was made by Shore and Zwanzig (61). They considered a model in which relative orientations of chain segments could be described by a system of N 'spins' or vectors perpendicular to the chain of segments, but not to be confused with dipoles so oriented, because any number of thse could be assigned bond or group dipole moments. The coupling energy between 'spins' or segments was assumed to be a quadratic function of relative angular displacements $\theta_i - \theta_{i+1}$ of angles θ_i for rotations around the chain axis (i.e., a simulation, by harmonic oscillator displacements, of small fluctuations around potential minima) and interactions with the external medium were represented by rotational segment diffusion with diffusion constant D. With increasing coupling energy relative to k_BT, the resulting relaxation spectrum for a single dipole in the chain over a range of times from 1/D to N/D showed increasing deviations from Debye behavior described quite well in some cases by Williams-Watts or Havriliak-Negami relaxation functions. An interesting feature of the results is that the deviations became much smaller when increasing number of 'spins' were assigned dipole moments, but with the interaction energy <u>not</u> of electric dipole couplings.

The Zwanzig and Shore calculations are instructive in showing effects of next enighbor interactions on the response function of a diple, but the model is clearly not realistic for actual molecular constraints of representative linear polymers for example. Two models to account for observed fast relaxation by local conformational changes have recently been proposed which also lead to deviations from Debye behavior. Helfand (62) has shown analytically and by Brownian dynamics simulations that conformational changes take place involving correlated conformational transitions of neighboring bonds between trans (t) and gauche (g^+, g^-) orientations, as symbolized by $g^{\pm}tt \rightleftharpoons ttg^{\pm}$ and $ttt \rightleftharpoons g^{\pm}tt$, and obtains orientational correlation functions involving products of exponentials with modified Bessel Functions. Skinner (63) has worked out an Ising model of 'spin' flips resulting in a diffusion-like motion of defects, or misaligned spins, separating sequences of aligned spins. These produce a correlation function of Williams-Watts form with β_W in the range 0.74 to 0.50 for increasing defect densities and there is a resemblance of the model both to the original Glarum defect diffusion and to development of soliton modes by non-linear couplings.

Another form of chain dynamics calculations which may be relevant to dielectric relaxation in alkyl halides has recently been developed by G. Evans (64). In his model for alkanes such as nonane, realistic forms for intramolecular torques and reorientations of chain segments are used while interactions with solvent are modelled by stochastic random forces in computer simulations with no adjustable parameters. Correlation functions for the terminal bond then obtained are roughly of skewed arc form with $\beta_c = 0.6$ for nonane, which is at least qualitatively like the behavior of alkyl halides such as n-octyl iodide (as neat liquids). Solution measurements are planned for closer comparison with the model.

The linear chain models evidently have no immediate relevance for relaxation of simple polar molecules in liquids, but to the writer at least the results suggest strongly the importance, particularly at low temperatures, of cooperative interactions of molecules with their neighbors which may, but need not, involve electric dipoles as the source of intermolecular torques. When the energies involved are appreciable relative to k_BT, it seems important to develop and study more realistic but still tractable models for analytic and simulation calculations. There have not yet been many serious attempts or promising results in analytical two and three dimensional theories except for diffusion-like models discussed by other contributors to this volume, but a few developments which have some relevance can be mentioned.

Zwanzig (65) considered a lattice model of dipoles reorienting by rotational diffusion and coupled by electrostatic dipole-dipole interaction energy. He was able to obtain results by a high temperature expansion of the Boltzmann factor in the linear response theory correlation function expression and evaluation for the corresponding Langevin form of the Liouville operator through third order terms in powers of dipole interaction. This is just enough to produce a secondary higher frequency Debye relaxation with relaxation time τ_2 one-fifth that for the principal diffusion generated relaxation. It thus seems possible that higher order terms needed for realistic temperatures would generate further relaxations, but the algebra rapidly becomes intractable and the convergence is at best too slow for such a brute force approach to be attractive, so alternative approximate methods would be preferable.

Finally, it is appropriate to mention the Scher-Montroll treatment (66) of anomalous charge transport in doped amorphous solids and polymers as an example of stochastic non-Markoffian rate expressions which result in the empirical Cole-Cole circular arc function. The memory function used is not, contrary to some opinions, purely empirical. as Scher and Lax (67) have derived a similar function by consideration of transitions between randomly distributed trapping sites and evaluation of overlap integrals for the transition probability. (Some further discussion of this approach and its possible relevance to dielectric relaxation in solids is in a review by the writer (68)).

3.5 Computer Simulations.

A pioneer effort to the account for electrostatic interaction effects in dipole reorientations and correlation functions was made by Brot and Darmon (39) in their Monte Carlo simulations for the partially ordered solid phase of 1,2,3 trichloro 4,5,6 trimethyl benzene (TCTMB) using the point charge model already mentioned in 2.4. Calculations of transition rates between 6 fold rotational wells of fluctuating depth as a result of changing neighbor orientations resulted in essentially Debye relaxation at 300°K, but a second simulation at 186°K, for which considerable rotational ordering is present, produced very nearly a circular arc with $\alpha = 0.28$ as compared to the experimental $\alpha = 0.39$.

The Brot-Darmon calculation is for solid rather than liquid like behavior on the plastic crystal phase of TCTMB, as shown by the positive value of $\partial\epsilon_s/\partial T$, corresponding to an increased entropy with applied field ($\partial S/\partial E^2 > 0$) from the Maxwell thermodynamic relation. Other chloromethyl benzenes with weaker dipole interactions have liquid like polarization in the sense that $\partial\epsilon_s/\partial T$ is negative, and also exhibit skewed arc behavior experimentally at low temperatures. It would be of considerable interest to do similar Monte Carlo calculations with appropriate point charge or point multiple models for these systems to see if the observed form of relaxation is obtained.

The molecular dynamics simulations by Brot, et al. of dipoles interacting with a Stockmayer or 'off center Stockmayer potential' in a two dimensional disk and three dimensional sphere as described in 2.4 provide P_1 and P_2 rotational correlation functions as well as equilibrium properties. The more extensive 2D calculations (at less cost) produced single molecule, inner disk, and macroscopic disk in vacuo correlations (as defined in 1.4 and 3.2) together with the experimental macrosopic relaxation function $\varphi_D(t)$ for $\epsilon(\omega)$. A distinction needs to be made here between the single particle function, denoted by $f^{(1)}(t)$ of the form $\langle\underline{\mu}_1(0)\cdot\underline{\mu}_1(t)\rangle$, the inner disk or sphere function $\varphi_m(t) = \langle\underline{\mu}_1\cdot\sum'\underline{\mu}_i(t)\rangle$ for the sum over dipoles in the inner space, the macroscopic disk or sphere function $\varphi_M(t)$ and the macroscopic $\varphi_D(t)$. It was found for values 1.0 and 2.0 of dipole strength $\gamma = N\mu^2/3k_BTV$ that $f^{(1)}(t)$ relaxed most rapidly with damped oscillatory behavior, somewhat as for free rotation but on a somewhat longer time scale indicative of 'dielectric friction' effects. The macroscopic disk function $\varphi_M(t)$ is rather similar, but the inner disk function $\varphi_m(t)$ after an initial inertial period soon changes to a considerably slower nearly exponential decay, particularly for the stronger dipole interaction. The relation of $\varphi_m(t)$ to $\varphi_M(t)$ was found to be very nearly that predicted by the Fatuzzo-Mason treatment, and the derived $\varphi_P(t)$ most closely related to $\varphi_m(t)$ gives a relaxation for $\epsilon(\omega)$ which is very nearly a semicircle except for a small 'hump' near the high frequency limit as a result of inertia with an indication of an itinerant oscillator type of effect for the higher dipole strength.

All the features just listed are eminently reasonable, and together with other specific conclusions too numerous to be given here, provide a substantial basis for the assertion (69) that 'MD calculations on an isolated system can yield reliably the complex permittivity of the system' - for, of course, the model used. The necessarily more elaborate 3D computations for a single dipole strength give qualitatively similar results. The derived value of dipole relaxation time τ_D is found to be about 0.35 of the experimental value for CH_3F at 206°K, leading to the authors' conclusion that anisotropic steric forces (torques) not included in the interaction model are as important as the dipole torques in slowing the relaxation process.

Very recently Levesque, Weis, and Oxtoby (70) have reported MD simulations for a model of liquid HCl, using 256 particles with perrodic boundary conditions and Ewald sums to take account of long range electrostatic interactions. The complete interaction potential used was a sum of atom-atom potentials for short range forces (the C^* potential of Klein and McDonald (71)), plus a three point charge model of electrostatic forces. The results give a P_2 correlation function time of 0.2 ps at 195°K from the long time exponential decay, whereas experimental values from Raman measurements indicate 0.35 ps (but there are some questions about the latter because of seeming inconsistency with NMR values). The dipole P_1 relaxation time is found to be 0.47 ps with roughly Debye behavior. The value is not too far from the factor 3 larger than for P_2 from simple diffusion theory, but there seem to be no dielectric measurements for comparison.

A disturbing feature of the results is that a Kirkwood g = 2.6 ± 0.5 is obtained from the static permittivity, whereas the experimental value for (polarizable) liquid HCl is close to unity (72). Whether this and other discrepancies are to be attributed to the point charge model, too small anisotropy in the short range potential, uncertainty in the vibrational line shape analysis, or a combination of these, remains to be determined. Altogether, however, computer simulations of the kinds just described are yielding encouraging results on the whole, with considerable promise for future development.

4. RELATED TOPICS.

In this section, three aspects of dielectric behavior in liquids and solutions with long histories are reviewed in the light of recent developments which make them subjects of wider or renewed interest.

4.1. Kerr Effect Relaxation.

The Kerr effect consists in measurement of the optical birefringence (difference of refractive indices) produced when anisotropically polarizable nonpolar or polar molecules are oriented in a strong electric field applied at right angles to the direction of a polarized light beam. If this bias field is changing in time, the rates of orientational response and hence changes in birefringence are governed by the rotational dynamics of the molecules, and hence can provide a probe of the dynamics.

Transient Kerr effect measurements have been used primarily to study large molecules (polymers and biopolymers) or smaller molecules at sufficiently low temperatures, as discussed by G. Williams. The limitations are because the relaxation time scale must not be shorter than the time resolutions for available transient field generators and optical detectors. Until recently, this has set short time limits of 10^{-6} to 10^{-9} seconds at best. These are too large for useful studies of small molecules in liquids at ordinary temperatures, but developments in pump and probe laser techniques have already made some measurements possible with picosecond resolution (as discussed by Kenney-Wallace in this volume) and it should be possible to do much more with availability of fast solid state switching and optical coupling devices.

The kinds of information obtainable, as well as complications in analysis of the non-linear effects, can be shown by considering for simplicity an axially symmetric polar molecule, with dipole moment μ along the axis and anisotropy $\Delta\alpha = \alpha_p - \alpha_s$ of polarizabilities parallel and perpendicular to the axis is in the second rank polarizability tensor. The relevant theoretical quantity is the average $\langle \Delta\alpha P_2(\cos\theta) f(t) \rangle$

where $P_2(\cos\theta) = \frac{1}{2}(3\cos^2 -1)$ is for the angle θ between the molecular axis and the applied field $E_0(t)$ and $f(t)$ is the time dependent distribution function in the field. The Hamiltonian and Liouville operator for the time evolution of $f(t)$ are

$$H = H_0 - \mu E_0(t)\cos\theta - \Delta\alpha E_0^2(t)P_2(\cos\theta) \tag{47}$$

$$L = L_0 + \mu E_0(t)\sin\theta\frac{\partial}{\partial p_\theta} + \Delta\alpha E_0^2 3\sin\theta\cos\theta\frac{\partial}{\partial p_\theta} \tag{48}$$

So by symmetry the response must be an even function of $E_0(t)$, one looks first for leading terms in E_0^2 in the response theory perturbation treatment. These will evidently result from first order effects of $\Delta\alpha$ polarizability torques involving $P_2(\cos\theta)$ only, but for polar molecules there can also be second order effects of torques on the permanent dipoles involving $P_1(\cos\theta) = \cos\theta$ as well as P_2. It is here that complications of non-linearity develop, as the effect of the operator $\partial/\partial p_\theta$ on the first order perturbation $f_1(t)$ and hence, from equation (10), on $P_1(t)$ must be considered unless $P_1(t) = P_1(0)$ which is independent of p_θ.

In a classic paper, Benoit (73) extended Debye's original diffusion model for dipole orientational relaxation to include the effects of both torques by solutions to order E_0^2 of the forced diffusion equation. The simplest solutions are for the responses to removal and application of a constant field E_0 at $t = 0$. These are

$$\langle\Delta\alpha P_2 f(t)\rangle = \frac{1}{5}\beta\Delta\alpha\left(\Delta\alpha + \frac{1}{3}\beta\mu^2\right)E_0^2\exp(-6Dt) \tag{49}$$

$$\langle\Delta\alpha P_2 f(t)\rangle = \frac{1}{5}\beta\Delta\alpha\left(\Delta\alpha + \frac{1}{3}\beta\mu^2\right)E_0^2[1-\exp(-6Dt)] \tag{50}$$

$$+\frac{1}{10}\beta^2\Delta\alpha\mu^2E_0^2[\exp(-6Dt) - \exp(-2Dt)]$$

The field off solution is simply P_2 relaxation, for which the correlation function is $\exp(-6Dt)$, as the second order effect of the dipole $\partial/\partial p_\theta$ operator is on $P_1(0)$ and hence zero for field free conditions. For field applied at $t = 0$, however, thee is both the P_2 response function and the second term involving $\exp(-2Dt)$ of P_1 response as well.

Kerr effect studies thus offer the perhaps unique possibility of determining both P_2 and P_1 correlations from basically the same experiment, at least for diffusional dynamics. The model seems reasonably good for solutions of biopolymers, such as globular proteins, and of other polymers at ordinary temperatures, but definitely is not for a variety of other molecules, as discussed by Williams. He has found, for example, Williams-Watts relaxation functions rather than exponentials, and little or no difference in form of field off and field on response. There is thus a need for, and usefulness of, a more general response theory treatment to obtain expressions in terms of appropriate correlation functions with a minimum of restrictive assumptions about the underlying dynamics. This prompted the writer, at Williams' instigation, to look into the problem.

The chief problem in extending response theory to second order effects of dipole torques proves to be that of dealing with such derivatives as $\partial P_1(t)/\partial p_\theta$ appearing in expressions for response in an applied field. The crucial role of 'collisions' becomes apparent, as evaluation of this derivative for free rotation gives a divergence proportional to t. In condensed phases with molecules in virtually continuous states of collision, persistence of angular momentum is very short as compared to the time scale for appreciable changes in angular correlations. From such considerations, the writer adopted the ansatz that the effect of an increment Δp_θ in momentum was to add an increment to the orientational displacements at all later times of interest. With this and some further manipulations (74), it was then possible to obtain simple counterparts of Benoit's results

$$\langle \Delta\alpha P_2 f(t)\rangle = \Delta\alpha^2 \beta E_0^2 \langle P_2 P_2(t)\rangle + \tfrac{1}{2}\Delta\alpha\beta^2\mu^2 E_0^2 \langle P_2 P_1^2(t)\rangle \tag{51}$$

$$\begin{aligned}\langle \Delta\alpha P_2 f(t)\rangle = {}& \Delta\alpha^2 \beta E_0^2 \langle P_2[P_2 - P_2(t)]\rangle + \tfrac{1}{2}\Delta\alpha\beta^2\mu^2 E_0^2 \langle P_2[P_1^2 - P_1^2(t)]\rangle \\ & - \tfrac{3}{4}\Delta\alpha\beta^2\mu^2 E_0^2 \langle \dot{P}_2 P_1(t)[P_1 - P_1(t)]\rangle \end{aligned} \tag{52}$$

These results reduce to Benoit's if diffusional motion is assumed. Introducing joint correlations leads to replacing $P_1(t)$ by $\sum_i P_{1i}(t)$, etc. (74).

Without going into further details, the conclusion from the analysis is that one can indeed obtain both P_2 and P_1 correlation functions quite directly from the two responses, at least for the simple molecular anisotropy and simple field pulses assumed. Further generalization to molecules of arbitrary symmetry presents problems discussed by Rosato and Williams (75), while other pulse shapes require more complicated treatments than for linear effects with simple superposition principles (from unpublished calculations of the writer). There is also the question of 'local field' effects.

One can hope that these will not greatly affect comparisons of macroscopic relaxation functions rather than microscopic functions and that better treatments, as from Fulton's methods for example, will clarify these questions. Even so, it seems fair to claim that a better basis now exists for extracting useful information from Kerr effect measurements and to explore questions of whether rotational reorientations in time are diffusion, or Brownian motion, like at one extreme, infrequently by large jumps at the other, or something in between. With developments in instrumentation of the sort suggested above, there appear to be real possibilities for studies of dynamics of simpler molecules to complement those by other methods.

4.2. Ion-Solvent Interactions.

The decreases in permittivity when ions are added to a polar solvent were traditionally interpreted in terms of saturation or solvation of local ionic environments (76) until Hubbard and Onsager (77),(78) worked out a continuum theory of the kinetic depolarization effect. This arises from the fact that part of the electric field solvent dipoles near an ion is from the moving ion, and similarly for the ion in the field of reorienting dipoles, with the consequences that both responses are delayed in proportion to the relaxation time of the solvent polarization. The remarkably simple Hubbard-Onsager expression for the resulting decrement of 'static' (or better, limiting low frequency) permittivity can be written

$$\Delta\epsilon_s = -0.057\left(\frac{\epsilon_s - \epsilon_\infty}{\epsilon_s}\right)(p + p')\tau_D K_s \qquad (53)$$

Here τ_D is the assumed Debye relaxation time of the solvent in ps and K_s the low frequency specific conductance of the solution in mho/m., while p and p' are either 2/3 or 1 for slip or stick boundary conditions of continuum solvent at the ion surface. This prediction qualitatively explained a large fraction of the observed decrements in water and methanol (76), (79) and the relatively enormous decrements (relative to total ion concentrations) in sulfuric acid solutions (80) as a result of the large τ_D = 400 ps at 25°C (as compared to 8 ps for water and 52 ps for methanol).

In more recent theoretical work, Hubbard, Colonomos, and Wolynes (81) developed a molecular theory of the dynamics of the two depolarization processes which predicts specific effects of ion size, reducing to the Hubbard-Onsager result for slip boundary conditions in the limit of large ion radii. Numerical results for methanol were later tested by measurements of Winsor and Cole (82) for five salts, which gave decrements in semi-quantitative agreement except for unresolved ambiguities in such comparisons discussed below. Other studies for NaI in six other polar solvents gave decrements larger by factors 1.5 to 6 than predicted by equation (50), with the difference roughly in proportion to solvent molecule dipole moments (rather than saturation permittivity).

The results cited show that kinetic effects play an important role in determining measured permittivities of electrolyte solutions, and must be considered as well as static solvation effects. Properly, a self consistent unified theory of both is needed (83), but as yet does not exist. In the case of aqueous salt solutions, Patey and Carnie (84) have recently applied LHNC theory to calculations of the equilibrium permittivity and found reasonably good agreement with experimental values at low concentrations (79) after taking account of an assumed additive kinetic contribution calculated by HO theory. For further discussion of the equilibrium problem, reference should be made to the original paper and to work of Friedman (85).

One further problem which should be mentioned is of the effect of ion-ion interactions and ion atmosphere relaxation on the permittivity, particularly at moderate or high concentrations. The classic Debye-Falkenhagen theory (83) for high dilutions predicting an increment in permittivity proportional to $C^{1/2}$ relaxing at frequencies of order f_{DF}(GHz) $= 1800\kappa_S(\text{mho/m})/\epsilon_s$ has received some support from low frequency measurements at low concentrations (79), but to the best of the writer's knowledge the predicted frequency dependence has not been directly observed nor is there unambiguous evidence about the magnitude of the effect at other than low concentrations. The necessary measurements appear to be feasible, although not easy, with current instrumentation, and should be useful for interpretation of dielectric behavior and for electrolyte theory.

4.3. Mixed Solvents.

Schallamach (86) many years ago advanced the proposition that mixtures of two polar liquids exhibit one relaxation region at intermediate frequencies, rather than two resolved ones, when both or neither of the components form an associated liquid, but that two are to be found if one component associates and the other does not. His original supporting evidence was somewhat inconclusive because he made measurements at a single frequency as a function of temperature and for large molecules with relaxation in the millisecond to microsecond range. Measurements of slow relaxation as a function of frequency at low temperatures which supported the thesis were made by Denney and Cole for methanol /1-propanol (87), by Denney for i-butyl bromide/i-butyl chloride and 1-propanol/i-amyl bromide (88), and by Bos (89) for bromobenzene with benzyl alcohol and with cyclohexanol.

One naturally wonders whether the 'merging' of relaxations is characteristic of slow processes, often under supercooled conditions, but is more or less lost at ordinary temperatures. The available information (90) is meagre, and inconclusive in most cases because the natural width of even a Debye relaxation is 1.1 decades in frequency making resolution of overlapping processes difficult if these are much less than a decade apart, particularly from information for a limited number and range of frequencies. Results of measurements at room temperatures do seem to indicate in a few cases, however, that merging remains in mixtures of alcohols, but that separation occurs for non-associated mixtures.

The available results present something of a problem for explanation in terms of rotational diffusion models, even after recognizing that two distinct species with similar static properties and molecular sizes but different relaxation rates can relax at nearly the same rate in an environment with a common (i.e., mutual) viscosity, as Kivelson has pointed out to the writer (91). Such conditions are not met in some of the available examples. Ones from very recent work are for mixtures of methanol, ethanol, and 2-propanol with added water up to 0.5 mole fraction for the first two and at 0.15 for the last by Bertolini, Cassettori, and Salvetti (92) at frequencies from 470 MHz to 3.75 GHz. They found essentially single Debye relaxations in all cases, with indications of slight broadening at -43°C.

Perl (93) has made measurements of 1-propanol solutions with water added to mole fraction 0.75 at 25°C at frequencies from 10 MHz to 5 or 10GHz. These showed a principal Debye relaxation with times τ_1 from 320 ps for pure alcohol to 48 ps for the water rich solution, roughly as expected, and also a smaller but increasing higher frequency relaxation with time τ_2 of approximately 20 ps at all concentrations. This is very nearly the value found by Garg and Smyth for the secondary relaxation of neat 1-propanol. This result suggests some sort of conformational motions in hydrogen bonded networks of the two species, as will be discussed elsewhere (93).

The foregoing discussion suggests a variety of

possibilities for further work aimed at adequate definition of relaxation effects in solutions with more than one kind of polar molecules to help elucidate the nature of the molecular reorientations, particularly as a function of temperature. With present availability of wide band time and frequency domain digital methods, such work is a less formidable undertaking than in the past.

Acknowledgements.

The writer is indebted to Professors Claude Brot and Daniel Kivelson for preprints of unpublished work and for permission to quote some of their results. Most of the writer's contributions to the subjects discussed were made possible by support from the National Science Foundation (USA).

REFERENCES

1. Jackson, J.D. 1975, 'Classical Electrodynamics,'2nd Ed. (Wiley, New York), pp. 226-235.

2. Buckingham, A.D., Disch, R. 1963, Proc. Roy. Soc. (London) A273, pp. 275-280.

3. Van Vleck, J.H. 1932, 'Theory of Electric and Magnetic Susceptibilities' (Oxford)

4. Kubo, R. 1957, J. Phys. Soc. Japan 12, pp. 570-586.

5. Glarum, S.H. 1960, J. Chem. Phys. 33, pp. 1371-1375.

6. Cole, R.H. 1965, J. Chem. Phys. 42, pp. 637-643.

7. Lorentz, H.A. 1909, 'The Theory of Electrons' (Teubner, Leipzig).

8. Van Vleck, J.H. 1937, J. Chem. Phys. 5, pp. 556-568.

9. Cole, R.H. 1957, J. Chem. Phys. 27, pp. 33-35.

10. Toupin, R.A., Lax, M. 1957, J. Chem. Phys. 27, pp. 458-464.

11. Cole, R.H. 1963, J. Chem. Phys. 39, pp. 2602-2609.

12. Onsager, L. 1936, J. Am. Chem. Soc. 58, pp. 1486-1493.

13. Böttcher, C.J.F. 1973, 'Theory of Electric Polarization',

2nd Ed., Vol. I, 'Dielectrics in Static Fields', completely revised by van Belle, O.C., Bordewijk, P., Rip, A. (Elsevier, Amsterdam).

14. See Ref. (13) for details.

15. C. Brot, personal communication.

16. Kirkwood, J.G. 1939, J. Chem. Phys. 7, pp. 911-919.

17. Frohlich, H. 1956, Physics 22, pp. 898-904.

18. Fulton, R.L. 1975, Mol. Phys. 29, pp. 405-413.

19. Felderhof, B.U. 1977, J. Chem. Phys. 67, pp. 493-500; Titulaer, U.M., Felderhof, B.U. 1977, J. Chem. Phys. 67, pp. 3126-3132.

20. Fulton, R.L. 1983, J. Chem. Phys. 78, pp. 6865-6876, 6877-6884.

21. Buckingham, A.D., Pople, J.A. 1956, Discuss. Far. Soc. 22, p. 17.

22. Sutter, H. 1972, 'Dielectric and Related Molecular Processes' Vol. I, M. Davies, Senior Reporter (Chem. Soc. London) pp. 64-99.

23. Copeland, T.G., Cole, R.H. 1976, J. Chem. Phys. 64, pp. 1741-1746.

24. Van Vleck, J.H. 1937, J. Chem. Phys. 5, pp. 556-568.

25. Rosenberg, R., Lax, M. 1953, J. Chem. Phys. 21, pp. 424-428.

26. Cole, R.H. 1963, J. Chem. Phys. 39, pp. 2603-2609.

27. Toupin, R.A., Lax, M. 1957, J. Chem. Phys. 27, pp. 458-464.

28. Nienhuis, G., Deutch, J.M. 1972, J. Chem. Phys. 56, pp. 235-247, pp.1819-1834.

29. Wertheim, M.S. 1971, J. Chem. Phys. 55, pp. 4291-4298.

30. Wertheim, M.S. 1979, Ann. Rev. Phys. Chem. 30, pp. 471-501.

31. Berkowitz, M., Adelman, S.A. 1980, J. Chem. Phys. 72,

pp. 4796-4798.

32. Patey, G.N., Valleau, J.P. 1976, J. Chem. Phys. 64, pp. 170-184.

33. Carnie, S.L., Patey, G.N. 1982, Mol. Phys. 47, pp. 1129-1154.

34. Alder, B.J., Pollock, E.L. 1981, Ann. Rev. Phys. Chem. 32, pp. 311-330.

35. Bossis, G., Quentrec, B., Brot, C. 1980, Mol. Phys. 39, pp. 1233-1249.

36. Pollock, E.L., Alder, B.J. 1980, Physica 102A, pp. 1-21.

37. Patey, G.N., Levesque, D., Weis, J.J. 1979, Mol. Phys. 38, pp. 1635-1654.

38. Patey, G.N., Levesque, D., Weis, J.J. 1982, Mol. Phys. 45, pp.733-746.

39. Brot, C., Darmon, I. 1970, J. Chem. Phys. 53, pp. 2271-2280.

40. Cole, R.H. 1983, IEEE Trans. Instrum. Meas. IM-32, pp.42-47.

41. Van Vleck, J.H., Weisskopf, V.F. 1945, Rev. Mod. Phys. 17, pp. 227-236.

42. Titulaer, U.M., Deutch, J.M. 1974, J. Chem. Phys. 60, pp. 1502-1513.

43. Fulton, R.L. 1975, J. Chem. Phys. 62, pp. 4355-4359.

44. Fatuzzo, E., Mason, P.R. 1967, Proc. Phys. Soc. 90, pp. 729-741.

45. Cole, R.H. 1938, J. Chem. Phys. 6, pp.385-391.

46. Vaughan. W. 1979, Ann. Rev. Phys. Chem. 30, pp. 103-124.

47. Berne, B.J. 1975, J. Chem. Phys. 62, pp. 1154-1160.

48. Böttcher, C.J.F., Bordewijk, P. 1978, 'Theory of Electric Polarization', 2nd Ed., Vol. II (Elsevier, Amsterdam).

49. Buckley, F., Maryott, A.A. 1954, J. Res. Nat'l. Bur. Std. 53, pp. 229-234.

50. Gerschel, A. Brot, C., Dimicoli, I., Rion, A. 1977, Mol. Phys. 33, pp. 527-541.

51. Hoffman, J.D. 1955, J. Chem. Phys. 23, pp. 1331-1339.

52. Davidson, D.W., Cole, R.H. 1951, J. Chem. Phys. 19, pp. 1484-1490.

53. Williams, G., Watts, D.C. 1970, Trans. Far. Soc. 66, pp. 80-85.

54. Denney, D.J. 1957, J. Chem. Phys. 27, pp. 259-264.

55. Cole, K.S., Cole, R.H. 1941, J. Chem. Phys. 9, pp. 341-351.

56. Havriliak, S., Negami, S. 1966, J. Polymer Sci. C14, p. 99.

57. Colonomos, P., Gordon, R.G. 1979, J. Chem. Phys. 91, pp. 1159-1166.

58. Glarum, S.H. 1960, J. Chem. Phys. 33, pp. 639-643.

59. Chandresekhar, S. 1943, Rev. Mod. Phys. 15, pp. 1-89.

60. Berberian, J.G., Cole, R.H. 1968, J. Am. Chem. Soc. 90, pp. 3100-3104.

61. Shore, J.E., Zwanzig, R.W. 1975, J. Chem. Phys. 63, pp. 5445-5458.

62. Helfand, E. 1983, to appear in J. Polymer Sci., Polymer Physics Ed.

63. Skinner, J.L. 1983, to appear in J. Chem. Phys.

64. Evans, G.J., personal communication.

65. Zwanzig, R.W. 1963, J. Chem. Phys. 38, pp. 2766-2772.

66. Scher, H., Montroll, E.W. 1975, Phys. Rev. B12, pp. 2455-2463.

67. Scher, H., Lax, M. 1973, Phys. Rev. B7, pp. 4491-4501.

68. Cole, R.H. 1980, (British) Inst. of Phys. Conf. Series No. 58, pp. 1-21.

69. Brot, C., Bossis, G., Hesse-Bezot, C. 1980, Mol. Phys.

40, pp. 1053-1072.

70. Levesque, D., Weis, J.J., Oxtoby, D.W. 1983, J. Chem. Phys. 79, pp. 917-925.

71. Klein, M.L., McDonald, I.R. 1981, Mol. Phys. 42, pp. 243-247.

72. Swenson, R.W., Cole, R.H. 1954, J. Chem. Phys. 22, pp. 284-288.

73. Benoit, H. 1951, Ann. Phys. (Paris) 6, p. 561.

74. Cole, R.H. 1982, J. Phys. Chem. 86, pp. 4700-4704.

75. Rosato, V., Williams, G. 1981, J. Chem. Soc. Far. Trans 2, 77, pp. 1767-1778.

76. For a review, see Hasted, J.B. 1973, 'Aqueous Dielectrics' (Chapman and Hall, London).

77. Hubbard, J.B., Onsager, L. 1977, J. Chem. Phys. 67, pp. 4850-4857.

78. Hubbard, J.B., Onsager, L. 1978, J. Chem. Phys. 68, pp. 1649-1664.

79. Hubbard, J.B., Onsager, L., van Beek, W.M., Mandel, M. 1977, Proc. Nat. Acad. Sci. 7, pp. 401-404.

80. Hall, D.G., Cole, R.H. 1981, J. Phys. Chem. 85, pp. 1065-1069.

81. Hubbard, J.B., Colonomos, P., Wolynes, P.G. 1979, J. Chem. Phys. 71, pp. 2652-2661.

82. Winsor, P., Cole, R.H. 1982, J. Phys. Chem. 86, pp. 2486-2490, pp. 2491-2496.

83. Wolynes, P.G. 1980, Ann. Rev. Phys. Chem. 31, pp. 345-376.

84. Patey, G.N., Carnie, S.L. 1983, J. Chem. Phys. 78, pp. 5183-5190.

85. Friedman, H.L. 1983, J. Chem. Phys. 76, pp. 1092-1105.

86. Schallamach, A. 1946, Trans. Far. Soc. 42A, pp. 180-186.

87. Denney, D.J., Cole, R.H. 1955, J. Chem. Phys. 23, pp.

1767-1772.

88. Denney, D.J. 1959, J. Chem. Phys. 30, pp. 1019-1024.

89. Bos, F.F. 1958, doctoral thesis, Leiden. See also Ref.(48).

90. Kivelson, D., personal communication.

91. Bertolini, D., Cassettari, M., Salvetti, G. 1982, J. Chem. Phys. 76, pp. 3285-3290.

92. Perl, J., Wasan, D.T., Winsor, P., Cole, R.H. To be submitted to J. Mol. Liquids.

THE STATISTICAL MECHANICS OF VIBRATION-ROTATION SPECTRA IN DENSE PHASES

William A. Steele

Department of Chemistry, The Pennsylvania State University, University Park, Pennsylvania 16802

The time-dependent classical statistical mechanics of systems of simple molecules is reviewed. The Liouville equation is derived; the relationship between the generalized susceptibility and time-correlation function of molecular variables is obtained; and a derivation of the generalized Langevin equation from the Liouville equation is given. The G.L.E. is then simplified and/or approximated by introducing physical assumptions that are appropriate to the problem of rotational motion in a dense fluid. Finally, the well-known expressions for spectral intensity of infrared and Raman vibration-rotation bands are reformulated in terms of time correlation functions. As an illustration, a brief discussion of the application of these results to the analysis of spectral data for liquid benzene is presented.

1. TIME-DEPENDENT STATISTICAL MECHANICS

We will first show how one can obtain the time-correlation function expression for the susceptibility in a classical statistical ensemble of particles which exhibits a linear response to an externally applied perturbation. This will be followed by an outline of the argument that leads to the generalized Langevin equation for the time-dependence of an arbitrary function of the molecular canonical coordinates. In both cases, derivations with minor modifications have been presented previously in numerous reviews, monographs, etc. However, the results are employed in a large fraction of current descriptions of dynamical processes in dense phases and thus, it seems worthwhile to again show the basic ideas underlying the formalism.

A. J. Barnes et al. (eds.), Molecular Liquids - Dynamics and Interactions, 111–150.

Consider an isolated system of N interacting particles with coordinates $\underset{\sim}{q}_1 \ldots \underset{\sim}{q}_N$ and momenta $\underset{\sim}{p}_1, \ldots \underset{\sim}{p}_N$. The equations of motion for this system can be derived from the Hamiltonian H. Thus

$$\dot{\underset{\sim}{q}}_i = \frac{\partial H}{\partial \underset{\sim}{p}_i} \tag{1.1}$$

$$\dot{\underset{\sim}{p}}_i = - \frac{\partial H}{\partial \underset{\sim}{q}_i} \tag{1.2}$$

These equations of motion define the time evolution of both microscopic and macroscopic variables. Coordinates and momenta can be written for each nucleus or, by well-known transformations, for the translation, rotation and vibration of each molecule. In addition to Hamilton's equations, the laws of continuum mechanics place constraints on the time-dependence of macroscopic conserved quantity such as number, momentum and energy density. For example, if the macroscopic number density of particles at point $\underset{\sim}{r}$ is denoted by $\rho(\underset{\sim}{r},t)$, one has a microscopic definition

$$\rho(\underset{\sim}{r},t) = \sum_{i=1}^{N} \delta(\underset{\sim}{q}_i(t) - \underset{\sim}{r}) \tag{1.3}$$

The macroscopic conservation law merely states that the flow of particles through the surface S of a volume element V is the only reason for change (in the absence of chemical reaction); thus

$$\int_V \frac{\partial \rho(\underset{\sim}{r},t)}{\partial t} d\underset{\sim}{r} = - \int_S \rho(\underset{\sim}{r},t)\underset{\sim}{u}(\underset{\sim}{r},t) d\underset{\sim}{A} \tag{1.4}$$

where $\underset{\sim}{u}(\underset{\sim}{r},t)$ is the flow velocity. The surface integral can be converted to a volume integral by Gauss' theorem. Since V is arbitrary, one must have

$$\frac{\partial \rho}{\partial t} = - (\underset{\sim}{\nabla} \cdot \rho \underset{\sim}{u}) \tag{1.5}$$

The basic goal here is to derive equations for the time-dependence of those macroscopic (observable) variables which can be written as functions of the microscopic coordinates and momenta. This might include quantities such as the total dipole moment (of unit volume) of a fluid, or the various probability distributions for the positions and/or momenta of 1,2,... N molecules. To this end, we define a 6N-dimensional vector $\underset{\sim}{\Gamma}(t) = \underset{\sim}{q}_1(t), \ldots \underset{\sim}{q}_N(t), \underset{\sim}{p}_1(t) \ldots \underset{\sim}{p}_N(t)$.

From classical mechanics we can write down an equation of motion for the vector $\underset{\sim}{\Gamma}$:

$$\frac{d\underset{\sim}{\Gamma}}{dt} = \sum_{i=1}^{N} \left(\dot{\underset{\sim}{q}}_i \cdot \frac{\partial}{\partial \underset{\sim}{q}_i} + \dot{\underset{\sim}{p}}_i \cdot \frac{\partial}{\partial \underset{\sim}{p}_i} \right) \underset{\sim}{\Gamma}$$

$$= \Sigma \left(\frac{\partial H}{\partial \underset{\sim}{p}_i} \cdot \frac{\partial}{\partial \underset{\sim}{q}_i} - \frac{\partial H}{\partial \underset{\sim}{q}_i} \cdot \frac{\partial}{\partial \underset{\sim}{p}_i} \right) \underset{\sim}{\Gamma} \equiv iL\underset{\sim}{\Gamma} \ . \tag{1.6}$$

Eq. (1.6) serves to define the Liouville operator L; the factor $i = \sqrt{-1}$ is introduced to make this operator be Hermitian. The formal solution to (1.6) is

$$\underset{\sim}{\Gamma}(t) = e^{iLt}\underset{\sim}{\Gamma}(0) \tag{1.7}$$

(The exponential of an operator is defined only in terms of its power series expansion).

Consider the time dependence of some variable A that is a function of some or all of the variables in $\underset{\sim}{\Gamma}$. Evidently,

$$A(t) = A(e^{iLt}\underset{\sim}{\Gamma}(0)) \tag{1.8}$$

Alternative forms of this equation are

$$A(t) = e^{iLt}A(0) \tag{1.9}$$

or

$$\frac{\partial A(t)}{\partial t} = iL\ A(t). \tag{1.10}$$

The Liouville equation is the basic equation of non-equilibrium statistical mechanics. It gives the time dependence of the N-particle distribution function $f^{(N)}(\underset{\sim}{\Gamma},t)$, and can be derived quite simply if one realizes that $f^{(N)}(\underset{\sim}{\Gamma},t)$ gives the number of systems with a given $\underset{\sim}{\Gamma}$ in an ensemble; the main point that follows is that the total number of systems in the ensemble is constant in time, so that one has a conservation law for $f^{(N)}(\underset{\sim}{\Gamma},t)$ which is the many-variable generalization of that for number density $\rho(\underset{\sim}{r},t)$ given above. Thus, the total number of systems N inside a hyper-volume V changes only due to passage of systems through the walls. In this case, the flow velocity is $\dot{\underset{\sim}{\Gamma}}$ so that one has

$$\int \frac{dN}{dt} = \int_V \frac{\partial f^N}{\partial t}\, d\underset{\sim}{\Gamma} \tag{1.11}$$

$$= -\int_S \dot{\underset{\sim}{\Gamma}} f^N \cdot d\underset{\sim}{A} \tag{1.12}$$

We again use Gauss' divergence theorem to convert eq. (1.12) from a surface to a volume integral and, since V is arbitrary, we find

$$\frac{\partial f^N}{\partial t} + (\underset{\sim}{\nabla}_\Gamma \cdot \dot{\underset{\sim}{\Gamma}} f^N) = 0 \tag{1.13}$$

or

$$\frac{\partial f^N}{\partial t} + \dot{\underset{\sim}{\Gamma}} \cdot \underset{\sim}{\nabla}_\Gamma (f^N) + f^N (\underset{\sim}{\nabla}_\Gamma \cdot \dot{\underset{\sim}{\Gamma}}) = 0 \tag{1.14}$$

From Hamilton's equations, we have

$$\sum_i \left(\dot{\underset{\sim}{q}}_i \frac{\partial}{\partial \underset{\sim}{q}_i} + \dot{\underset{\sim}{p}}_i \frac{\partial}{\partial \underset{\sim}{p}_i} \right) \equiv \underset{\sim}{\nabla}_\Gamma \cdot \dot{\underset{\sim}{\Gamma}} = 0 \tag{1.15}$$

This leads immediately to Liouville's equation:

$$\frac{\partial f^N}{\partial t} + \sum_i \left(\dot{\underset{\sim}{q}}_i \cdot \frac{\partial f^N}{\partial \underset{\sim}{q}_i} + \dot{\underset{\sim}{p}}_i \cdot \frac{\partial f^N}{\partial \underset{\sim}{p}_i} \right) = 0 \tag{1.16}$$

or

$$\frac{\partial f^N}{\partial t} = -\, iLf^N \tag{1.17}$$

The ensemble average of a phase function $A(\underset{\sim}{\Gamma})$ at time t can now be written in dual forms,

$$\begin{aligned} < A(t) > &= \int d\underset{\sim}{\Gamma} f^N(\underset{\sim}{\Gamma}, t) A(\underset{\sim}{\Gamma}) \\ &= \int d\underset{\sim}{\Gamma} (e^{-iLt} f^N(\underset{\sim}{\Gamma})) A(\underset{\sim}{\Gamma}) \end{aligned} \tag{1.18}$$

or

$$= \int d\underset{\sim}{\Gamma} f^N(\underset{\sim}{\Gamma}) e^{iLt} A(\underset{\sim}{\Gamma}) \tag{1.19}$$

This result is particularly useful in calculating the linear response of a system which is under the influence of an external field (often, a time-dependent electromagnetic field). We assume that the interaction of such a field with the system can be represented by adding terms to the Hamiltonian of the form $A(\underset{\sim}{\Gamma})F(t)$, where $F(t)$ represents the explicit time-dependence of the field. The Liouville equation for such a system can be written as

$$\frac{\partial f^{(N)}}{\partial t} = - i(L_0 + \Delta L)(f_0^{\ N} + \Delta f^N) \tag{1.20}$$

where the subscript zero denotes quantities for the unperturbed system, Δf^N is the change in the N-particle distribution function due to the presence of the external field, and

$$i\Delta L = F(t)\Sigma\left(\frac{\partial A}{\partial p_i}\cdot\frac{\partial}{\partial q_i} - \frac{\partial A}{\partial q_i}\frac{\partial}{\partial p_i}\right) \tag{1.21}$$

If we keep only terms linear in the perturbation, eq. (1.20) becomes

$$\frac{\partial \Delta f^N}{\partial t} = - iL_o\Delta f^N - i\Delta L f_o^{\ N} \tag{1.22}$$

If the equilibrium distribution function is canonical,

$$f_o^{\ N} = \frac{1}{Q}\, e^{-H_o/kT} \tag{1.23}$$

The linearized equation of motion becomes

$$\frac{\partial \Delta f^N}{\partial t} = -iL_o\Delta f^N - \frac{1}{kT}\, f_o^{\ N}\dot{A}_o \,. \tag{1.24}$$

This equation has the solution

$$\Delta f^N(t) = \frac{-1}{kT}\int_0^t d\tau\; e^{-iL_o(t-\tau)}\; f_o^N\dot{A}_o F(\tau) \tag{1.25}$$

Now if the phase variable $B(\underset{\sim}{\Gamma})$ has an average value of zero at equilibrium, we can use (1.25) to calculate its non-equilibrium linear response to the perturbation:

$$< B(t) > = \int d\underset{\sim}{\Gamma}\; B(\underset{\sim}{\Gamma})\; \Delta f^N(t) \tag{1.26}$$

$$= -\frac{1}{kT}\int_0^t d\tau \int d\underset{\sim}{\Gamma}\, B(\underset{\sim}{\tau})e^{-iL_0(t-\tau)} A_o f_o^N(\tau)$$

$$= -\int_0^t d\tau\, \chi(t-\tau)F(\tau) \tag{1.27}$$

where the susceptibility $\chi(t) = \frac{1}{kT} < B(t)\dot{A}(0) >_o$ (1.28)

The pointed brackets denote an equilibrium ensemble average.

The function appearing in eq. (1.28) is of course a time-correlation function. These quantities play a central role in current theories of transport and spectral band shapes. More generally, they are defined as

$$C_{AB}(t) \equiv \int d\underset{\sim}{\Gamma}\, \rho_o(\Gamma)[e^{iLt}A(\underset{\sim}{\Gamma})]B^*(\underset{\sim}{\Gamma}) \tag{1.29}$$

where $\rho_o(\underset{\sim}{\Gamma})$ is the equilibrium distribution function previously denoted by $f_o^N(\underset{\sim}{\Gamma})$.

Of course, the problem of determining the time dependence of $B(\underset{\sim}{r}^N,\underset{\sim}{p}^N)$ remains. At present, the physically interesting variables can at best be approximately modelled, for dense phases. However, many of these models are (or can be) obtained by simplifying a rigorous, general equation of motion known as the generalized Langevin equation (GLE). We now show how this expression can be derived from the Liouville equation.

The scalar product of two functions A and B is defined as

$$(A,B^*) = \int d\underset{\sim}{\Gamma}\, \rho_o(\underset{\sim}{\Gamma})A(\underset{\sim}{\Gamma})B^*(\underset{\sim}{\Gamma}). \tag{1.30}$$

Then

$$C_{AB}(t) = (e^{iLt}A,B^*) \tag{1.31}$$

A set of functions A, B···· and this definition of scalar product defines a space in which the functions can be thought of as vectors and operators transform these vectors. The length of vector A is defined as $(A,A^*)^{1/2}$ and we note one important feature of this vector space: since (A(t), A*(t)) is independent of time, any time-displacement operator can rotate the vector but must leave its length unchanged. In particular, exp(iLt) is a generalized rotation operator. Also, $C_{AA}(t)$ is a measure of the component of A(t) parallel to A(0); i.e., it is the projection of A(t) onto A(0). This suggests that we define a projection operation P which, when it acts on an arbitrary vector B, projects B onto A. Thus,

$$PB = (B,A^*)(A,A^*)^{-1}A \tag{1.32}$$

Also, the complementary projection operator Q is defined by

$$QB = B - (B,A^*)(A,A^*)^{-1}A \tag{1.33}$$

Clearly

$$Q + P = 1, \tag{1.34}$$

the identity operator.

Several properties of projection operators are

$$\begin{aligned} &PA = A, \quad P^2 = P, \quad (PA,B^*)^* = (PB^*,A) \\ &QA = 0, \quad Q^2 = Q, \quad (QB,A^*) = 0 \end{aligned} \tag{1.35}$$

If we write $iL = i(P + Q)L$, it is possible to derive (by means of Laplace transforms) an important result which is

$$e^{iLt} = e^{iQLt} + \int_0^t d\tau \; e^{iL(t-\tau)} iPLe^{iQL\tau} \tag{1.36}$$

We can now derive an exact equation of motion for an arbitrary function $A(t)$ which has the form of the Langevin equation of Brownian motion theory. To this end, the rate of change of A is written as

$$\frac{dA(t)}{dt} = e^{iLt}(P+Q)iLA \tag{1.37}$$

Now

$$\begin{aligned} e^{iLt}\; PiLA &= e^{iLt}(iLA,A^*)(A,A^*)^{-1}A \\ &= (iLA,A^*)(A,A^*)^{-1}e^{iLt}A \\ &= i\Omega A(t) \end{aligned} \tag{1.38}$$

Ω is called the frequency.

Substitution of (1.36) and (1.38) into (1.37) yields

$$\begin{aligned} \dot{A}(t) &= i\Omega A(t) + e^{iLt}QiLA \\ &= i\Omega A(t) + \int_0^t d\tau \; e^{iL(t-\tau)} iPLe^{iQL\tau}QiLA \\ &\quad + e^{iQLt}QiLA \end{aligned} \tag{1.39}$$

We now define the random force F(t) as

$$F(\tau) = e^{iQL\tau} \cdot QiLA \tag{1.40}$$

This random force is so described because for all time it is uncorrelated with A. From $Q^n = Q$, $F(\tau) = QF(\tau)$. Therefore, $F(\tau)$ is a variable that is projected to be orthogonal to A so that

$$C_{FA}(t) \equiv (PF(t),A^*) \tag{1.41}$$

$$= 0$$

If we examine the convolution in eq. (1.39) in detail and make use of the Hermitian property of L, we see that

$$\begin{aligned} iPLF(\tau) &= iPLQF(\tau) \\ &= (iLQF(\tau),A^*)(A,A^*)^{-1}A \\ &= (F(\tau),\ (QiLA)^*)(A,A^*)^{-1}A \\ &= -(F(\tau),\ F^*(0))(A,A^*)^{-1}A. \end{aligned} \tag{1.42}$$

Defining the memory function by

$$K(\tau) = (F(\tau),\ F^*(0))(A,A^*)^{-1} \tag{1.43}$$

we now obtain the generalized Langevin equation

$$\dot{A}(t) = i\Omega A(t) - \int_0^t d\tau\ K(\tau)\ A(t-\tau) + F(t). \tag{1.44}$$

By taking the scalar product of (1.44) with A* and using (1.31) we obtain the memory function equation

$$\frac{d}{dt} C_{AA}(t) = i\Omega C_{AA}(t) - \int_0^t d\tau K(\tau) C_{AA}(t-\tau) \tag{1.45}$$

We note that the memory function K is proportional to the auto-correlation function of the random force. Also,

$$\Omega = (iLA,A^*)\ (A,A^*)^{-1} \tag{1.46}$$

Let $q \to q$ $p \to -p$ (time reversal);
then $\rho_o \to \rho_o$, $\underset{\sim}{\Gamma} \to \underset{\sim}{\Gamma}$ and

$$(\dot{A},A^*) \to -(\dot{A},A^*) \tag{1.47}$$

Since $iLA = \dot{A}$, (1.47) means that $\Omega = 0$ for the single variable case under consideration.

An extremely useful extension to the single-variable generalized Langevin equation can be obtained if one projects the time-dependence of A not only upon itself, but upon a set of other variables B, C,.... Thus, one defines a vector A = A,B,C...., a generalized projection operator $P = P_A, P_B, P_C \ldots$ and one complementary projection operator $Q = 1 - \Sigma P_i$. The individual operators now project A onto A,B,C····. It is straightforward to show that

$$\underset{\sim}{A}(t) = i\underset{\approx}{\Omega}\cdot\underset{\sim}{A}(t) - \int_0^t d\tau\, \underset{\approx}{K}(\tau)\cdot\underset{\sim}{A}(t-\tau)+\underset{\sim}{F}(t) \tag{1.48}$$

where the vector $\underset{\sim}{F}(t)$ is given by a set of equations of the form of (1.40) but containing A,B,C···· as operands; each of these components of $\underset{\sim}{F}$ is thus orthogonal to the corresponding variable A,B,C····. The frequency $\underset{\approx}{\Omega}$ is now a matrix whose elements are $(iLN,M^*)(N,N^*)^{-1}$ with N,M = A,B,C····. It is important to realize that only the diagonal elements of $\underset{\approx}{\Omega}$ are necessarily zero; the off-diagonal elements for a given variable A will thus account for much of its time dependence. The elements of the row memory matrix $\underset{\approx}{K}(\tau)$ corresponding to variable M are given by an equation of the form of (1.43) which is $(F_M(\tau),F_N^*(0))(M,M^*)^{-1}$.

The purpose of using such a multi-variable Langevin equation is basically to simplify the convolution integral involving the (unknown) memory function. It is hoped that one can include all the (relatively) slowly varying coupled variables in the set $\underset{\sim}{A}$. Since the matrix $\underset{\approx}{K}(\tau)$ involves variables that are projected to be orthogonal to this set of slowly varying ones, the memory functions should be rapidly varying in time. This then allows one to make the Markovian approximation that

$$\int_0^t K_{MN}(\tau)\; M(t-\tau)d\tau = \Gamma_{MN}M(t) \tag{1.49}$$

$$\Gamma_{MN} = \int_0^\infty K_{MN}(\tau)d\tau \tag{1.50}$$

For the single variable case, the result is essentially the Langevin equation:

$$\dot{A}(t) = \Gamma A(t) + F(t) \tag{1.51}$$

where Γ is now a friction constant.

In the actual calculation (or modelling) of the memory functions defined in these equations, it should be remembered the time dependence of these quantities is given by a modified propagator e^{iQLt}, not e^{iLt}. Of course, this means that $K(\tau)$ is not an ordinary time-correlation function (except for a few conserved variables) and thus, that no straightforward prescription can be given for precise modeling or computer simulation of these functions.

2. SOME APPLICATIONS OF THE G.L.E. TO REORIENTATION

In order to use the generalized Langevin equation in a particular physical situation, one must first choose the variable or variables of interest and then make appropriate approximations to obtain a soluble expression. We will illustrate some of the ways that one might use the G.L.E., first by discussing the reorientation of a linear molecule in an isotropic fluid. The basic quantities of interest are the orientational correlation functions $C_\ell^*(t) = \langle P_\ell(\cos\delta\theta) \rangle$, where $\delta\theta$ is the change in the orientation angle of the molecule in time t and $P_\ell(x)$ is a Legendre polynomial. In addition to this (infinite) set, one might also be interested in angular velocity autocorrelations; in torque autocorrelations; and in the analogous mutual correlations between a molecule and its neighbors. We will discuss some of these questions below; at present, we write

$$C_\ell(t) = \sum_m \langle P_{\ell,m}(\Omega(0))P_{\ell,m}^*(\Omega(t)) \rangle$$

where C_ℓ and C_ℓ^* differ by the normalization factor for the spherical harmonics $P_{\ell,m}(\theta,\phi)$. It can be shown that the time dependence of these functions is independent of m, so we need consider explicitly only $P_{\ell,0}(\theta)$ which we denote by $Q_\ell(t)/\sqrt{2\ell+1}$.

The simplest Langevin formalism is the single-variable expression; this now becomes

$$-\frac{d}{dt} C_\ell(t) = \int_0^t C_\ell(t-\tau)M_\ell(\tau)d\tau \tag{2.1}$$

where M_ℓ is the memory function associated with Q_ℓ. This equation can be solved for either function if the other is known, by direct step-by-step numerical integration (1) or by Laplace transforms. What is needed is a realistic ansatz (for $M_\ell(t)$, in most cases). Inasmuch as the formal expression for the time dependence of this function involves the projection operator, it is essentially impossible to generate exact or nearly exact $M_\ell(t)$ by computer simulation or explicit theory. The best one can do is to evaluate the first few coefficients in an expansion for M_ℓ in powers of t^2. If one writes

$$C_\ell(t) = \sum_{n=0}^{\infty} \frac{\gamma_\ell^{(2)} t^n}{n!} \tag{2.2}$$

$$M_\ell(t) = \sum_{n=0}^{\infty} \frac{\alpha_\ell^{(n)} t^n}{n!} \tag{2.3}$$

it is readily shown that for a linear molecule (1,2)

$$\gamma_\ell^{(0)} = 1$$

$$\gamma_\ell^{(2)} = \frac{kT}{I}\,\ell(\ell+1)$$

$$\gamma_\ell^{(4)} = \left(\frac{kT}{I}\right)^2 \ell(\ell+1)\left(3\ell(\ell+1) - 2 + \frac{\langle N_\perp^2\rangle}{2(kT)^2}\right) \tag{2.4}$$

$$\alpha_\ell^{(0)} = \gamma_\ell^{(2)}$$

$$\alpha_\ell^{(2)} = \gamma_\ell^{(4)} - (\gamma_\ell^{(2)})^2 \tag{2.5}$$

where $\langle N_\perp^2 \rangle$ is the mean square torque exerted on the molecular axis. In practice, this approach is of little use, since the long-time behavior of each function is the most interesting part of the problem. However, if one is modelling $M_\ell(t)$, knowledge of the correct short-time behavior can often be helpful. For example, one might write

$$M_\ell(t) = M_\ell(0)\exp(-t/\tau_r) \tag{2.6}$$

with τ_r equal to a relaxation time (that might be ℓ-dependent). Clearly, $M_\ell(0) = \gamma_\ell^{(2)}$; in addition, one sees that the exponential decay, while convenient for taking Laplace transforms, does not have the correct time dependence. A more suitable approximation might be

$$M_\ell(t) = M_\ell(0)\exp(-\alpha_\ell^{(2)} t^2/2) \tag{2.7}$$

An even better expression is

$$M_\ell(t) = (M_\ell)_{\substack{\text{free}\\ \text{rotor}}} \exp(-a_\ell t^2/2) \tag{2.8}$$

with $a_\ell = \langle N_\perp^2 \rangle / 2(kT)^2$. Of course the free rotor $M_\ell(t)$ can be obtained from eq. (2.1) if one inserts the known free $C_\ell(t)$ in it. Eq. (2.8) has the advantage that it becomes exact in the limit of small torque; it is indistinguishable from eq. (2.7) for large torque. Except for the fact that equations (2.6) - (2.8) do not display the negative dips generally observed in the $M_\ell(t)$ obtained from experimental or simulated $C_\ell(t)$ for large-torque systems, these simple expressions are reasonably successful, giving results that are in qualitative agreement with experiment and simulation for many systems with small to moderate torques (3,4). A somewhat better approximation than eq. (2.6) is

$$M_\ell(t) = (M_\ell)_{\substack{\text{free}\\ \text{rotor}}} \exp(-\beta t) \tag{2.9}$$

This expression gives rise to the J-diffusion model suggested by Gordon (5,6) who described it as free rotation interrupted by instantaneous collisions occurring with a frequency β. Eq. (2.9) is better than (2.6) in approaching the correct limit for vanishing collision frequencies, but computer simulations have revealed that the J-diffusion model is at best semi-quantitative for dense fluids. Furthermore, the physical significance of the parameter β is not easy to see. For liquid-like densities, the duration of the collisions is believed to be significant compared to the time interval between them, which does not agree with the model. The additional assumption that the angular velocity after collision is uncorrelated with its value before is questionable, and finally, the role of the intermolecular interactions is determining the dynamics tends to be obscured by this collisional picture. A re-interpretation of the parameter β has removed some of these difficulties (6), but the fact remains that the theoretical $C_\ell(t)$ often disagree with simulation and experiment.

Before progressing to more complex formulations of the G.L.E., it should be noted that one can obtain model-free results by assuming that one is in a regime where the variation in $C_\ell(t)$ is small in the time interval required for $M_\ell(t)$ to decay to zero. In that case, one can write

$$C_\ell(t-\tau) = C_\ell(t) - \tau \left(\frac{dC_\ell}{dt}\right)_{\tau=t} \tag{2.10}$$

Eq. (2.1) now becomes

$$-\frac{dC_\ell}{dt} = C_\ell(t)\tau_c^{-1} - \frac{dC_\ell}{dt} u \tag{2.11}$$

where the correlation time τ_c is

$$\tau_c = \left(\int_0^\infty M_\ell(\tau)\,d\tau\right)^{-1} \tag{2.12}$$

$$= \int_0^\infty C_\ell(\tau)d\tau \tag{2.13}$$

Eq. (2.13) follows immediately from the Laplace transform solution of (2.1):

$$- p \tilde{C}_\ell(p) + C_\ell(0) = \tilde{C}_\ell(p)\tilde{M}_\ell(p) \tag{2.14}$$

One takes the p=0 limit of (2.14) to obtain time integrals of C_ℓ and M_ℓ, and recollects that $C_\ell(0) \equiv 1$. Note that eq. (2.11) indicates that $C_\ell(t)$ will often show an exponential decay with

$$C_\ell(t) = C \exp(-t/\tau_r) \tag{2.15}$$

where C is an arbitrary constant. The relaxation time τ_r is related to the correlation time by

$$\tau_r = \frac{\tau_c}{1-u} \tag{2.16}$$

$$u = \int_0^\infty \tau \, M_\ell(\tau)d\tau \tag{2.17}$$

This simple argument leads one to deduce one other feature of the memory function $M_\ell(t)$: in those systems where $C_\ell(t)$ does not decay to zero at long times, the memory function will also be non-zero and will most likely show oscillation about zero as the time becomes very large. This is definitely the case for the best-known example of non-zero limit for $C_\ell(t)$, which is that of the free rotor (all ℓ for the spherical top and even ℓ for the linear molecule).

Since it appears that reorientation in large-torque systems is not handled very well in these single-variable approaches, let us consider an alternative use of the G.L.E. which might be better adapted to such systems.

The form to be chosen depends upon the intuitive feeling that some degree of libration might appear even in the liquid if the torques are large. Since generalized position and velocity are highly coupled in vibrating (or librating) systems, one approach would be to use the multi-variable form of the G.L.E. to describe the time dependence of Q_ℓ and its time-derivatives. The analogous form of the theory has been examined in detail for translational motion in liquids (7); it is not difficult to adapt these results to the rotational problem.

For simplicity, we consider the G.L.E. for two coupled variables which are Q_ℓ and $\dot{Q}_\ell$. The static correlation matrix for these quantities is

$$(A,A^*) = \begin{vmatrix} 1 & 0 \\ 0 & \omega_\ell^2 \end{vmatrix} \tag{2.18}$$

where $\omega_\ell^2 = \ell(\ell+1)(kT/I)$. The frequency matrix is

$$i\underset{\approx}{\Omega} = \begin{vmatrix} 0 & 1 \\ -\omega_\ell^2 & 0 \end{vmatrix} \tag{2.19}$$

and the memory function matrix contains only one non-zero element $M_\ell(t)$:

$$\underset{\approx}{M}(t) = \begin{vmatrix} 0 & 0 \\ 0 & M_\ell(t) \end{vmatrix} \tag{2.20}$$

(This is due to the fact that the projection operators for the elements of $\underset{\approx}{M}(t)$ leave only those parts of Q_ℓ, $\dot{Q}_\ell$ that are orthogonal to the space Q_ℓ, $\dot{Q}_\ell$.) We now find that the Laplace transform of the correlation matrix $\underset{\approx}{C}_\ell(t)$ is

$$\tilde{\underset{\approx}{C}}_\ell(p) = \begin{vmatrix} p & -1 \\ \omega_\ell^2 & p+\tilde{M}_\ell(p) \end{vmatrix}^{-1} \tag{2.21}$$

It follows that

$$\tilde{\underset{\approx}{C}}_\ell(p) = \frac{1}{\mathrm{Det}} \begin{vmatrix} p+\tilde{M}_\ell(p) & 1 \\ -\omega_\ell^2 & p \end{vmatrix} \tag{2.22}$$

with

$$\mathrm{Det} = p^2 + p\,\tilde{M}_\ell(p) + \omega_\ell^2 \tag{2.23}$$

Before proceeding further, it is worth noting that the (22) element of $\underset{\approx}{C}_\ell(t)$ is an angular velocity-orientation correlation function which is just

$$G_\ell(t) = \langle \dot{Q}_\ell(0)\dot{Q}_\ell(t) \rangle = -\frac{d^2}{dt^2} \langle Q_\ell(0)Q_\ell(t) \rangle \tag{2.24}$$

Remembering that the autocorrelation functions are normalized to unity at t=0, eq. (2.24) means that the Laplace Transforms obey

$$- \omega_\ell^2 \tilde{C}_{22}(p) = p^2\tilde{C}_{11}(p) - p \tag{2.25}$$

which can readily be confirmed by substituting the appropriate elements from eq. (2.22).

To complete the calculation, one needs an ansatz for the memory function. For lack of a better approximation, we begin with the simple exponential of eq. (2.6); thus,

$$\tilde{M}(p) = M_0 \; / \; (p+\tau_r^{-1}) \tag{2.26}$$

and

$$\tilde{C}_{11}(p) = \frac{p^2 + p\tau_r^{-1} + M_0}{p^3+p^2\tau_r^{-1}+p(M_0+u_\ell^2)+\omega_\ell^2\tau_r^{-1}} \tag{2.27}$$

$$\tilde{C}_{22}(p) = \frac{p^2 + p\tau_r^{-1}}{p^3+p^2\tau_r^{-1}+p(M_0+\omega_\ell^2)+\omega_\ell^2\tau_r^{-1}} \tag{2.28}$$

The inverses of eqs. (2.27) and (2.28) are given in standard tables of Laplace transforms; we here give only the result for $C_\ell(t) = C_{11}(t)$. If α,β,γ are roots of the cubic polynomial in the denominator of eq. (2.27), one finds that

$$C_\ell(t) = \frac{\alpha^2-\alpha\tau_r^{-1}+M_0}{(\alpha-\beta)(\alpha-\gamma)}\, e^{-\alpha t} + \frac{\beta^2-\beta\tau_r^{-1}+M_0}{(\beta-\alpha)(\beta-\gamma)}\, e^{-\beta t} + \frac{\gamma^2-\gamma\tau_r^{-1}+M_0}{(\gamma-\alpha)(\gamma-\beta)}\, e^{-\gamma t} \tag{2.29}$$

(The result for $G_\ell(t) = C_{22}(t)$ is the same as eq. (2.29), with the M_0's deleted). Any of the roots can be complex, so that the correlation functions can have either damped oscillatory behavior or a more-or-less exponential decay. Evaluation of the roots of a cubic polynomial is a straightforward but laborious exercise which will not be discussed here. It is perhaps of more interest to try other memory functions. One could go either in the direction of choosing a less convenient but more precise function such as a Gaussian with the correct decay constant, or one might try an even simpler representation than eq. (2.6). For example, suppose that the p-dependence of eq. (2.26) is neglected so that one has

$$\tilde{M}_{\ell}(p) = M_0\tau_r \tag{2.30}$$

(This implies a rapid decay of the memory, on the time-scale of the decay of the correlation functions.) It is easy to show that this gives

$$G_{\ell}(t) = \ell(\ell+1)\frac{kT}{I}[\cosh\delta t - \frac{1}{2\delta\tau_{\omega}}\sinh\delta t]e^{-t/2\tau_{\omega}} \tag{2.31}$$

$$C_{\ell}(t) = [\cosh\delta t + \frac{1}{2\delta\tau_{\omega}}\sinh\delta t]e^{-t/2\tau_{\omega}} \tag{2.32}$$

where $\tau_{\omega} = 1/M_0\tau_r$ and $\delta = [\tau_{\omega}^{-2} - \omega_{\ell}^{2}]^{1/2}$. Of course, if δ is complex, the hyperbolic functions will become sinusoidal, so that even this simplified model is capable of reproducing a range of behavior from oscillatory to exponential decays. The approximation made for $\tilde{M}_{\ell}(p)$ implies that the results may be in error at large p or thus, at short t where the decay of $M_{\ell}(t)$ is not yet complete. Inspection of the expansions of eqs. (2.31) and (2.32) reveals that $C_{\ell}(t)$ is exact to order t^2 but that errors appear at t^4 for C_{ℓ} and at t^2 for G_{ℓ}.

An alternative use of the G.L.E. which appears to be quite different from the multi-variable formalism is based on the realization that one can write an entire hierarchy of memory function equations. This arises from the fact that the memory function itself is a phase variable and thus obeys its own G.L.E. If the n'th memory function is denoted by $M^{(n)}(t)$, we can write

$$-\frac{d}{dt}M^{(1)}(t) = \int_0^t M^{(2)}(\tau)M^{(1)}(t-\tau)d\tau \tag{2.33}$$

and so on. Laplace transforming, one has

$$\tilde{C}(p) = \frac{1}{p+\tilde{M}^{(1)}(p)} \tag{2.34}$$

$$\tilde{M}^{(1)}(p) = \frac{M^{(1)}(0)}{p+\tilde{M}^{(2)}(p)} \tag{2.35}$$

$$\tilde{M}^{(n)}(p) = \frac{M^{(n)}(0)}{p+\tilde{M}^{(n+1)}(p)} \tag{2.36}$$

$$\int_0^t M(\tau)C(t-\tau)d\tau = \Gamma\, C(t) \tag{2.39}$$

$$\Gamma = \int_0^\infty M(\tau)d\tau \tag{2.40}$$

It emerges that asymmetric rotational diffusion can be obtained from this simplified G.L.E. only if one takes coupled sets of variables which are the $D_{km}^{\ j}(\Omega)$ for fixed j, k, but with m varying between j and -j. Since it can be shown that the results are independent of k (see below), we set k=0 for convenience.

We now construct the G.L.E. for these orientational variables. Since all the coupled variables are functions only of angle, the frequency matrix is identically zero. The memory function matrix for this problem will now be shown to have elements $< \frac{d}{dt} D_{km}^{\ j}(\Omega(0)) \frac{d}{dt} D_{km'}^{*\ j}(\Omega(t)) >^{\dagger}$ (where the dagger indicates that the time dependence is governed by the projection operator Q, as usual). We write

$$\frac{df(\Omega)}{dt} = \sum_i \omega_i \frac{df(\Omega)}{d\phi_i} \tag{2.41}$$

where ω_i is the angular velocity and ϕ_i is the angle of twist around the i'th principal axis of the molecule. In this way, one can show that

$$< \frac{d}{dt} D_{0,m}^{\ j}(\Omega)0)) \frac{d}{dt} D_{0,m'}^{*\ j}(\Omega(t)) >^{\dagger} =$$

$$- < D_{0,m}^{\ j}(\Omega(0)) \frac{d^2}{dt^2} D_{0,m'}^{\ j}(\Omega(t)) >^{\dagger} \tag{2.42}$$

$$= - < \sum_{i,i'} \omega_i(0)\omega_{i'}(t) D_{0,m}^{\ j}(\Omega(0)) \frac{\partial^2}{\partial\phi_i \partial\phi_{i'}} D_{0,m'}^{*\ j}(\Omega(t)) > \tag{2.43}$$

Since the elements of the memory function matrix are here assumed to decay rapidly compared to the rate of change of the orientation angles, the time dependence of eq. (2.43) must come from changes in the angular velocities and we can thus write

$$\Gamma_{m,m'}^{\ j} = \sum_i R_i H_i(mm') \tag{2.44}$$

This system of equations can be closed by approximating $M^{(n+1)}(t)$ - most often, by an exponentially decaying function. One then needs values of all of the $M^{(n)}(t)$ at t=0 to define the remaining parameters of the problem.

We now apply a simple version of this formalism to a calculation of the reorientational correlation functions for the linear molecule. Thus, we evaluate $\tilde{C}_\ell(p)$ by assuming that $M_\ell^{(2)}(p)$ is given by eq. (2.26). It is readily shown that

$$C_\ell(p) = \frac{p^2 + p\tau^{-1} + M_\ell^{(2)}(0)}{p^3+p^2\tau_r^{-1}+p(M_\ell^{(1)}(0)+M_\ell^{(2)}(0))+M_\ell^{(1)}(0)\tau_r^{-1}} \qquad (2.37)$$

This equation has an identical form to eq. (2.28), obtained previously from the approximate multi-variable approach! Indeed, it is easy to show here that $M_\ell^{(1)}(0) = \ell(\ell+1)(kT/I) = \omega_\ell^2$. However, the exact expression for $M_\ell^{(2)}(0)$ is $(kT/I)(2\ell(\ell+1)-2+ \langle N_\perp^2\rangle / 2(kT)^2)$; which is not quite the same as that in eq. (2.28), i.e., $M_0 = (kT/I) \langle N_\perp^2\rangle / 2(kT)^2$. Nevertheless, at small ℓ and large mean square torque $\langle N_\perp^2\rangle$, these two constants will be essentially equal.

The similarity exhibited here between the multi-variable and the memory hierarchy formulations of the orientational problem is obviously not a general result; in fact, it derives from the choice of a variable Q_ℓ and its time derivative $\dot{Q}_\ell$ in the multi-variable theory. Other variations on these themes can readily be devised; for example, one could couple Q_ℓ, $\dot{Q}_\ell$ and $\ddot{Q}_\ell$ (orthogonalized to Q_ℓ, however) (8); or one could combine higher-order memory functions with a multi-variable theory; or one could couple translational and rotational variables (9). The enormous flexibility of the G.L.E. means that the intuition of the user will play a particularly significant role in determining the success of the outcome. To illustrate this point, we now briefly recapitulate how the G.L.E. applies to a quite different orientational problem; namely, anisotropic rotational diffusion (10).

Consider a non-linear molecule with orientation described by Euler angles $(\alpha,\beta,\gamma=\Omega)$. Its angular time correlation functions will often have the form

$$C_{j,k,m}(t) = \langle D_{r,k}^{j}(\Omega(0))\, D_{r,m}^{*j}(\Omega(t)) \rangle \qquad (2.38)$$

where the D_{km}^{j} are Wigner functions here defined according to the convention of Rose (11). If we are interested only in the diffusional limit, we can assume that the memory functions for this system decay rapidly so that the convolution integrals can be written:

Here,

$$H_i(m,m') = \left< D_{0,m}^{\;j}(\Omega(0)) \frac{\partial^2}{\partial\phi_i^{\,2}} D_{0,m'}^{*\,j}(\Omega(0)) \right> \tag{2.45}$$

and it has been assumed that principal axes can be found such that $<\omega_i(0)\omega_{i'}(t)>^{\dagger}=0$ for $i \neq i'$. The rotational diffusion constants are given by

$$\mathcal{R}_i = \int_0^{\infty} <\omega_i(0)\omega_i(t)>^{\dagger} dt \tag{2.46}$$

Because the operators $\partial/\partial\phi_i$ are proportional to the quantum mechanical angular momentum operators, the matrix elements $H_i(m,m')$ are well known (they play an important role in determining the energy levels of a freely rotating asymmetric top). One finds that

$$\Gamma_{m,m}^{\;j} = \frac{1}{2}(\mathcal{R}_x + \mathcal{R}_y)(j(j+1)-m^2) + \mathcal{R}_z m^2$$

$$\Gamma_{m,m\pm2}^{\;j} = \frac{1}{4}(\mathcal{R}_x-\mathcal{R}_y)[(j\mp m)(j\pm m+1)(j\mp m-1)(j\pm m+2)]^{1/2} \tag{2.47}$$

Other elements are zero.

It is the presence of the non-zero elements $\Gamma_{m,m\pm2}^{\;j}$ which brings about a coupling of the memory function equations for the correlation functions with different m subscripts (but a fixed j superscript). It can be seen that the matrix of "memory constant" is diagonal when $\mathcal{R}_x = \mathcal{R}_y$; for these symmetric diffusers, the equations for each m decouple and one recovers the well-known result:

$$C_{j,m,m}(t) = \exp(-\Gamma_{m,m}^{\;j} t) \tag{2.48}$$

To solve the fully asymmetric problem, one must find eigenvalues and eigenfunctions for the non-diagonal matrices that characterize the motion; for example,

$$\underset{\approx}{\Gamma}^1 = \begin{vmatrix} \frac{1}{2}(\mathcal{R}_x+\mathcal{R}_y)+\mathcal{R}_z & 0 & \frac{1}{2}(\mathcal{R}_x-\mathcal{R}_y) \\ 0 & (\mathcal{R}_x+\mathcal{R}_y) & 0 \\ \frac{1}{2}(\mathcal{R}_x-\mathcal{R}_y) & 0 & \frac{1}{2}(\mathcal{R}_x+\mathcal{R}_y)+\mathcal{R}_z \end{vmatrix} \tag{2.49}$$

The solutions to this problem are well-known for j=1 (and for j=2) (12) and will not be repeated here.

One of the virtues of using the G.L.E. to derive such (relatively) simple results is that one can go back and improve on the original model by refining the approximations made. For example, "rotational diffusion with memory" has been proposed by Nee and Zwanzig (13), who noted that one need not assume that the memory functions decay instantaneously on the time scale of the reorientation. They retain the approximation that all the time dependence comes from changes in the angular velocity, but suggested that the constants $\Gamma_{m,m'}{}^{j}$ should be replaced by memory functions $M_{m,m'}{}^{j}(t)$ that are proportional to the (unprojected) angular velocity autocorrelation functions. We write the result of this approximation for the symmetric diffuser only:

$$M_{m,m}{}^{j}(t) = \frac{1}{2}[j(j+1)-m^2] < \omega_{\perp}(0)\omega_{\perp}(t) > + m^2 < \omega_{\parallel}(0)\omega_{\parallel}(t) > \quad (2.50)$$

where $\omega_{\perp}$ and $\omega_{\parallel}$ are the angular velocities perpendicular and parallel to the molecular symmetry axis. This model seems to work fairly well for high-torque linear molecules (m≡0) (4); it has not yet been tested against data for either symmetric or asymmetric diffusers.

Use of the G.L.E. can lead to other refinements to the rotational diffusion model; for example, Keyes (10) has noted that mutual rotational diffusion, where the relative reorientation of pairs of molecules is analyzed, can also be modelled in this way, with results obtainable not only for the simple linear diffusers treated earlier by Keyes and Kivelson (8), but for the asymmetric diffuser as well.

As a final example, the Keyes-Kivelson application of the G.L.E. to this important problem of relative reorientation will be briefly described. Because the reorientations of all the molecules in the fluid are now the object of attention, it is evident that the multi-variable G.L.E. is the applicable formulation. Limiting the problem to linear molecules (or to reorientations of the symmetry axes of non-linears), the variables of interest are $P_{\ell}(\cos\theta_i)$, where θ_i is the orientation angle of the axis of the i'th molecule. We denote these variables by $Q_{\ell,i}$ and, following Kivelson and Madden (8), take the entire set of N variables (for fixed ℓ) to be in $\underset{\sim}{A}$. In addition, these authors close the memory function equations by also including the $\dot{Q}_{\ell,i}$ and then assuming that the memory function matrix which characterize this 2N-variable set decay rapidly enought to be replaced by a matrix $\underset{\approx}{K}$ of constants. (Note that we have discussed similar formulations previously; the (2x2) example given is actually the limiting case of the present theory in which mutual correlations are negligible; also, the replacement of memory functions by constants was discussed in connection both with rotational diffusion and with a simple version of the (2x2) problem.)

We order the variables $Q_{\ell 1}$, $\dot{Q}_{\ell 1}$, $Q_{\ell 2}$, $\dot{Q}_{\ell 2}, \ldots$. Then the static correlation matrix has diagonal elements

$$< \underset{\sim}{A}, \underset{\sim}{A}^* >_{mm} = 1/(2\ell+1) \qquad \text{n odd} \tag{2.51a}$$

$$= \omega_\ell^{\,2}/(2\ell+1) \qquad \text{n even} \tag{2.51b}$$

where $\omega_\ell^{\,2} = \ell(\ell+1)(kT/I)$ as before, and the factor of $2\ell+1$ arises from the normalization of the ℓ'th Legendre polynomial. In addition, non-zero off-diagonal elements will be found when orientational correlations are present; we define

$$f_\ell = (2\ell+1) < P_\ell(\cos\theta_1(0))P_\ell(\cos\theta_2(0)) > \tag{2.52}$$

so that one can write another set of non-zero elements which is

$$< \underset{\sim}{A}, \underset{\sim}{A}^* >_{nm} = f_\ell/(2\ell+1) \qquad \text{n,m both odd, } n \neq m \tag{2.51c}$$

Since static mutual velocity and velocity-orientation correlations are all identically zero, this completes the specification of $< \underset{\sim}{A}, \underset{\sim}{A}^* >$. (These results should be compared with eq. (2.18)). The inverse of this (2Nx2N) matrix can now be calculated; it emerges that its non-zero elements are

$$< A, A^* >^{-1}_{mm} = 2\ell+1 \qquad \text{m odd}$$

$$= (2\ell+1)/\omega_\ell^{\,2} \qquad \text{m even}$$

and

$$< A, A^* >^{-1}_{nm} = (2\ell+1)f_\ell/(1+Nf_\ell), \quad \text{n,m both odd, } n \neq m. \tag{2.53}$$

Note that f_ℓ is a small quantity, but Nf_ℓ is of the order of unity, often amounting to 0.1 to 0.5 in fluids of small but non-spherical molecules. The frequency matrix has, as usual, diagonal elements equal to zero and non-zero off-diagonal elements given by:

$$(i\Omega)_{nm} = 1 \qquad \text{m even, } n = m-1$$

$$= -\omega_\ell^{\,2} \qquad \text{n even, } m = n-1$$

$$= \omega_\ell^{\,2} f_\ell/(1+Nf_\ell), \qquad \text{n even, m odd, } \neq n-1 \tag{2.54}$$

(Compare with eq. (2.19)). After replacing the memory functions with δ-functions, the non-zero elements of the time integral of the matrix can be written

$$K_{nn} = \tau_{\ell,\omega}^{-1} \qquad n \text{ even}$$

$$K_{nm} = \tau_{\ell,\omega}^{-1} h \qquad n,m \text{ even, } n \neq m \tag{2.55}$$

with

$$\tau_{\ell,\omega}^{-1} = \frac{2\ell+1}{\omega_\ell^2} \int_0^\infty < \dot{Q}_{\ell 1}(t)\dot{Q}_{\ell 1}(0) >^{\dagger} dt \tag{2.56}$$

where the dagger again indicates that projected operators govern the time dependence of the variables; also,

$$\tau_{\ell,\omega}^{-1} h_\ell = \frac{2\ell+1}{\omega_\ell^2} \int_0^\infty < \dot{Q}_{\ell 1}(t)\dot{Q}_{\ell 2}(0) >^{\dagger} dt \tag{2.57}$$

One might suppose that the problem is now to solve this system of 2N coupled equations; however, one is actually interested in only two aspects of the solution, which are the auto and the total (auto + mutual) orientational correlation functions. Consequently, one keeps the two equations for the autocorrelations $<Q_{\ell 1}(0)Q_{\ell 1}(t)>$ and $<\dot{Q}_{\ell 1}(0)\dot{Q}_{\ell 1}(t)>$, but sums the equations for the correlations $< Q_{\ell 1}(0)Q_{\ell k}(t) >$ and $< \dot{Q}_{\ell 1}(0)\dot{Q}_{\ell k}(t) >$ over all k (including k=1). The result is four coupled equations, of which only two are non-trivial. One finds

$$\frac{d}{dt} < Q_{\ell 1}(0)Q_{\ell 1}(t) > = < Q_{\ell 1}(0)\dot{Q}_{\ell 1}(t) > \tag{2.58}$$

$$\frac{d}{dt} < \sum_k Q_{\ell 1}(0)Q_{\ell k}(t) > = < \sum_k Q_{\ell 1}(0)\dot{Q}_{\ell k}(t) > \tag{2.59}$$

$$\frac{d}{dt} < Q_{\ell 1}(0)Q_{\ell 1}(t) > = -\left(\omega_\ell^2 + \frac{1}{\tau_{\ell,\omega}} \frac{d}{dt}\right) < Q_{\ell 1}(0)Q_{\ell 1}(t) > \tag{2.60}$$

where two terms of order 1/N proportional to f_ℓ and h_ℓ have been dropped; and

$$\frac{d}{dt} < \sum_k Q_{\ell 1}(0)\dot{Q}_{\ell k}(t) > = - \left(\frac{\omega_\ell^2}{1+Nf_\ell} + \frac{1}{\tau_{\ell,\tau}(1+Nh_\ell)} \frac{d}{dt}\right) < \sum_k Q_{\ell 1}(0)\dot{Q}_{\ell k}(t) > \qquad (2.61)$$

Thus, the calculation is decoupled and one can solve separately for the self $C_{s\ell}(t) = < Q_{\ell 1}(0)Q_{\ell 1}(t) >$ and the total $C_{M\ell}(t) = < \sum_k Q_{\ell 1}(0)Q_{\ell k}(t) >$. The actual equations to be solved are

$$\ddot{C}_{s\ell} + \frac{1}{\tau_{\ell,\omega}} \dot{C}_{s\ell} + \omega_\ell^2 C_{s\ell} = 0 \qquad (2.62)$$

and

$$C_{M\ell} + \frac{1}{\tau_{M\ell}} C_{M\ell} + \omega_M^2 C_{M\ell} = 0 \qquad (2.63)$$

with

$$\omega_M^2 = \omega_\ell^2 /(1 + Nf_\ell)$$

$$\tau_{M\ell} = \tau_{\ell,\omega} / (1 + Nh_\ell) \qquad (2.64)$$

It is interesting to note that the Laplace transforms of these equations have the same form as the (11) element of eq. (2.22) with $M_\ell(p)$ = constant. In fact, the solutions of these equations show the similarity quite clearly, since one has

$$C_{s\ell}(t) = \exp\left[-t/2\tau_{\ell,\omega}\right] \left(\cosh \delta_s t + \frac{1}{2\delta_s \tau_{\ell,\omega}} \sinh \delta_s t\right) \qquad (2.65)$$

where $\delta_s = [1/4\,\tau_{\ell,\omega}^2 - \omega_\ell^2]^{1/2}$. It is pleasing to see that this expression is identical to eq. (2.31). The total orientational correlation function $C_{M\ell}(t)$ is given by an expression of the form of eq. (2.65) but with $\tau_{\ell,\omega}$, δ_s replaced by $\tau_{M\ell}$, δ_M which are given by

$$\tau_{M\ell} = \tau_{\ell,\omega} / (1 + Nh_\ell)$$

$$\delta_M = \left(\frac{1}{4\tau_M^2} - \frac{\omega_\ell^2}{1+Nf_\ell}\right)^{1/2} \qquad (2.66)$$

Of course, the hyperbolic functions will be replaced by the appropriate sinusoidal functions when δ_s and/or δ_M is complex. Thus, the theory in this form is capable of representing a wide range of reorientational behavior. One sees that the time dependence of $C_{M\ell}(t)$ is affected by the presence of orientational correlations ($N\dot{f}_\ell$) and of correlations in second time derivatives of the angular functions (Nh_ℓ). The usual form seen in the literature for the Keyes-Kivelson theory is a simplified version of the present one which can be retrieved by assuming first that $1/4\tau_M^2$ is large compared to the second term in δ_M; then

$$\delta_M \simeq \frac{1}{2\tau_{M\ell}} - \frac{\tau_{M\ell}\omega_\ell^2}{1+N\dot{f}_\ell} \tag{2.67}$$

A similar approximation is made for δ_s. Secondly, one takes $\delta_M t$ large so that $\cosh \delta_M t \simeq \sinh \delta_M t \simeq \frac{1}{2}\exp(\delta_M t)$; this assumption is also made for $\cosh \delta_s t$ and $\sinh \delta_s t$. Consequently,

$$C_{s\ell}(t) \simeq \exp(-t/\tau_c) \tag{2.68}$$

with $1/\tau_c = \tau_{\ell,\omega}\omega_\ell^2$; and

$$C_{M\ell}(t) \simeq \exp(-t/\tau_t) \tag{2.69}$$

$$\tau_t = \tau_c(1+N\dot{f}_\ell)(1+Nh_\ell) \tag{2.70}$$

Finally, one rather arbitrarily sets

$$1 + Nh_\ell = 1/(1+N\dot{f}_\ell). \tag{2.71}$$

Since it is often argued that $N\dot{f}$ or Nh_ℓ is negligible, the weakness of eq. (2.71) is more apparent than real.

The various problems treated here serve to illustrate the power and flexibility of the G.L.E. while simultaneously presenting some of the better-known descriptions of molecular reorientation in dense fluids. It should be emphasized that great care must be exercised in using this theory, as it can lead to physically unrealistic expressions containing parameters that are difficult if not impossible to evaluate independently. Nevertheless, the G.L.E. has proved to give a common starting point for a wide variety of successful theories and it appears that it continues to offer the possibility for refinements and extensions of these theories as well as for the development of quite new descriptions of molecular motion.

3. TIME-CORRELATION FUNCTIONS FOR VIBRATION-ROTATION SPECTRA

The calculation of spectroscopic intensities can be formulated in a number of ways. The goal here is to show how spectral band shapes and intensities can be expressed in terms of time-correlation functions. A possible approach is to use the result obtained in Sec. 1 which relates the generalized susceptibility to an arbitrary externally applied perturbation (the electromagnetic field, in this case). Indeed, the polarization induced in a polar fluid is ordinarily treated just this way. However, at microwave frequencies and higher, it is ordinarily the absorption of energy which is measured rather than the dielectric response. These quantities are related via the Kramers-Kronig equation, but we choose not to follow this line of argument, for two reasons: we would like to obtain a theory for Raman spectra as well as for infrared and microwave absorption; and we would like to show how the well-known quantum mechanical expressions for spectroscopic intensity can be rigorously reformulated in terms of time-correlation functions.

The basic quantity of interest in absorption spectroscopy (and in a number of other spectroscopic quantities such as nuclear or electron magnetic relaxation times) is B_{mn}, the probability that a molecule will undergo a transition from state m to state n when it is perturbed by some time-dependent external field whose coupling to the molecule is expressible by a term in the Hamiltonian $H_p(t)$. The net rate of absorption of energy by a molecule when irradiated by light at some frequency ν_{mn} corresponding to the energy difference $\Delta E_{mn}/h$ is

$$\text{Rate} = h\nu_{mn}\,[-N_n B_{mn} + N_m B_{mn}]\,\rho(\nu_{mn}) \tag{3.1}$$

where N_n, N_m are the numbers of molecules in the final state n and in the initial state m, respectively, $B_{mn} = B_{nm}$, and $\rho(\nu_{mn})$ is the radiation density at frequency ν_{mn}; it is related to I_o, the incident light intensity at that frequency, by $\rho(\nu_{mn}) = cI_o/8\pi$ where c = velocity of light. Remembering that $N_n/N_m = \exp[-h\nu_{mn}/kT]$ if the absorption does not significantly perturb the system from thermal equilibrium, one can write

$$\text{Rate} = h\nu_{mn} N_m B_{mn}\left(1-e^{-h\nu_{mn}/kT}\right) c\, I_o/8\pi \tag{3.2}$$

The calculation of the transition probability B_{mn} is standard. The perturbation due to the interaction of radiation with matter can be written

$$-H_p(t) = \Sigma\, e_i \underset{\sim}{r}_i \cdot \underset{\sim}{E}_o \cos(2\pi\nu t) \tag{3.3}$$

where e_i is the i'th charge located at $\underset{\sim}{r}_i$, $\underset{\sim}{E}_o$ is the electric field of the light wave and ν is its frequency. The time-dependence of the wave function for the system is calculated using standard time-dependent perturbation theory. After introducing the relationship between $\rho(\nu_{mn})$ and $\underset{\sim}{E}_o$, the final result is

$$B_{mn} = \frac{8\pi^3}{3h^2} \underset{\sim}{M}_{mn} \quad \underset{\sim}{M}_{mn}^* \tag{3.4}$$

where the dipole transition matrix element is given by

$$\underset{\sim}{M}_{mn} = < \psi_m^{(0)} \quad \Sigma\, e_i \underset{\sim}{r}_i \; \psi_n^{*(0)} > \tag{3.5}$$

Up to this point, we have been following the usual argument given for spectral absorption in texts on quantum mechanics. The result for B_{mn} has its greatest utility for low density gases (or for electronic transitions) where one can identify individual lines corresponding to a specific $m \to n$ transition. However, in dense gases and liquids, individual rotational lines are broadened to the extent of merging into a continuous spectrum; in addition, the broadening of vibrational lines is a significant (and interesting) part of the problem of spectral interpretation. An alternative formulation which is well adapted to treat these problems can be obtained starting from the absorption coefficient $\alpha(\omega)$, which gives the fraction of incident radiation absorbed at angular frequency $\omega = 2\pi\nu$ per unit path length in the sample. The previous result yields

$$\alpha(\omega) = \frac{8\pi^3}{3c\hbar V} \sum_{m,n} \omega_{mn} \left(1 - e^{-\hbar\omega_{mn}/kT}\right) N_m \cdot$$
$$\underset{\sim}{M}_{mn} \cdot \underset{\sim}{M}_{mn}^* \; \delta(\omega - \omega_{mn}) \tag{3.6}$$

The Dirac delta function ensures that only those transitions with frequencies ω_{mn} equal to the incident light frequency takes part in the absorption, and N_m is the number of molecules in initial state m in the volume of fluid V. The delta function is now written in its Fourier representation

$$\delta(\omega - \omega_{mn}) = \frac{1}{2\pi} \int_{-\infty}^{\infty} e^{i(\omega - \omega_{mn})t}\, dt \tag{3.7}$$

Remember that $\omega_{mn} = (E_n - E_m)/\hbar$ and

$$e^{-iE_m t/\hbar} < \psi_m^{(0)} \underset{\sim}{M}(0) \psi_n^{(0)} > e^{iE_n t/\hbar} =$$

$$< \psi_m^{(0)} \underset{\sim}{M}(t) \psi_n^{*(0)} > \tag{3.8}$$

where the operator $\underset{\sim}{M}$ is now to be evaluated at time t or zero. The absorption coefficient $\alpha(\omega)$ can now be written as (12)

$$\frac{\hbar\alpha(\omega)}{\omega[1-e^{-\hbar\omega/kT}]} = \frac{4\pi^2}{3kTcV}\int_{-\infty}^{\infty} e^{i\omega t} C_M(t)dt \tag{3.9}$$

where

$$C_M(t) = \sum_{m,n} N_m < \psi_m^{(0)} \underset{\sim}{M}(t) \psi_n^{*(0)}> < \psi_n^{(0)} \underset{\sim}{M}(0) \psi_m^{*(0)} >$$

$$= < \underset{\sim}{M}(t) \cdot \underset{\sim}{M}(0) > \tag{3.10}$$

where the final angular brackets denote an ensemble average of the product of dipole moments for the fluid. This ensemble average can be done either by classical or quantum statistical mechanics, whichever is more appropriate. The function $C_M(t)$ is of course a time-correlation function for the total dipole moment of the molecules in the fluid. The total moment is ordinarily written as a sum of the individual dipole moments $\underset{\sim}{\mu}(\underset{\sim}{R}_i)$ for molecules with positions and orientations specified by $\underset{\sim}{R}_i \ldots \underset{\sim}{R}_n$, with $\underset{\sim}{R}_i = \underset{\sim}{r}_i, \Omega_i$, the center of mass vector and the Eulerian orientation angles of the molecular axes. There are a number of possible contributors to this time dependent molecular dipole moment, including:

(1) the permanent dipole moment, with time dependence due to the reorientation of the molecule-fixed vector.

(2) changes in the permanent moment due to vibrational motion, with time dependence due both to changes in the molecular electronic distribution as the molecule vibrates and to the reorientation of the vibration-induced dipole moment.

(3) induced dipole moments due to the interaction of an external applied electric field with a polarizable molecule.

(4) induced dipole moments due to the interactions of the electric fields produced by the electric multipoles of neighbouring molecules with a polarizable molecule. Here, the time dependence of the induced moment arises primarily from the translations and rotations of the neighbours, but also involves the reorientation of

the anisotropic part of the polarizability tensor for the molecule of interest as well as vibrational modulation of this tensor.

The second kind of experiment to be discussed here is Raman spectroscopy, which can be viewed as a consequence of effect #3. However, the theory of the Raman effect is made somewhat more complicated by the use of polarized incident and scattered radiation in modern laser Raman studies as a technique to extract more information. Suppose the incident radiation is polarized in the Z direction; the usual arrangement is to measure both the intensity of scattered light polarized in the Z direction and the scattered light polarized perpendicular to Z (call it X). We consider only these two geometries, since no additional information is ordinarily obtained from other polarizations, at least for fluids where the range of the correlations (or molecular size) is small compared to the wavelength of the light used.

The dipoles induced in a fluid by an incident light wave of frequency ω will oscillate at that frequency and consequently, will radiate to give Rayleigh scattering. In addition, the incident radiation can induce emission at frequencies other than the incident. A transition from state i to state f is associated with this emission. The Raman experiment consists of a measurement of R, the fraction of the incident light scattered into solid angle $d\Omega$ in frequency interval $d\omega$. The theory for this Raman scattering intensity was formulated in quantum terms by Placzek (13) who showed that

$$R_{IS} = \left(\frac{\omega}{c}\right)^4 \sum_{i,f} \rho_i \left| (\underset{\sim}{\varepsilon}_I \cdot \underset{\approx}{A} \cdot \underset{\sim}{\varepsilon}_S)_{if} \right|^2 \cdot \delta(\omega-\omega_{if}) \qquad (3.11)$$

where ρ_i is the number density of molecules in state i, $\underset{\sim}{\varepsilon}_I, \underset{\sim}{\varepsilon}_S$ are unit vectors parallel to the polarizations of the incident and scattered beams, respectively, and $\underset{\approx}{A}$ is the operator associated with the polarizability tensor of the molecules in the fluid. For the scattering geometries of interest, the relevant tensor elements are A_{ZZ} (for $R_{\parallel}$) and A_{XZ} (for $R_{\perp}$).

Just as in the infrared case, we wish to go from eq. (3.11), which is most useful when individual transitions i→f can be identified as discrete Raman lines, to the case where the spectrum is more or less continuous. If one again introduces the Fourier representation of the delta function, it can easily be shown that (12)

$$R = \left(\frac{\omega}{c}\right)^4 \frac{1}{2\pi} \int_{-\infty}^{\infty} e^{-i\omega t} \langle \underset{\sim}{\varepsilon}_I \cdot \underset{\approx}{A}(0) \cdot \underset{\sim}{\varepsilon}_S \underset{\sim}{\varepsilon}_I \cdot \underset{\approx}{A}(t) \cdot \underset{\sim}{\varepsilon}_S \rangle \, dt \qquad (3.12)$$

where $\omega = \omega_f - \omega_i = \omega_{inc} - \omega_{scatt}$, and the brackets now indicate an ensemble average of the time-dependent polarizability tensor element for the fluid. One generally writes this polarizability as a sum of molecular polarizabilities just as the total dipole moment was written as a sum of molecular contributions. The sources of the time dependence of this quantity are also similar to those for the dipole moment. One has:

(1) For molecules of symmetry lower than tetrahedral, the permanent polarizability tensor is anisotropic; consequently, the space-fixed molecular α_{zz} and α_{zx} vary as the molecule reorients.

(2) The tensor changes as vibrational motions distort the molecular frame; the vibrational changes in a space-fixed element also are affected by reorientation.

(3) Interactions with neighboring molecules can alter the polarizability in two ways: distortion of the electron clouds in an interacting pair or distortion of the nuclear framework in a colliding pair. Clearly, the time dependence of these effects is due primarily to a combination of translational and rotational motion.

To sum up the main features of the development so far, both infrared and Raman spectral intensities can be written in terms of the Fourier transform of the time-correlation of a molecular quantity. As often happens in Fourier transform spectroscopy, all the spectral information is formally contained in the time-correlation function. However, if one can isolate certain spectral regions which can be assigned to a particular motion, one can work with part of the time-correlation associated with that motion and thus gain considerable insight. To see this a bit differently, suppose one has an isolated spectral band (vibration-rotation, for instance). A Fourier transform of the appropriate intensity function (see eqs. 3.9 or 3.12) over the band will directly yield a time-correlation function ($C_M(t)$, for example), but one whose time-dependence reflects only the motions giving rise to the particular band of interest. Consequently, we can calculate (and compare with experiment!) time-correlation functions for pure rotational motion, for vibration-rotation motion for the ℓ'th normal mode or, in favorable cases, for the translation-rotation motions that give rise to an induced spectrum.

VIBRATION-ROTATION SPECTRA

In order to progress further, we need explicit expressions for the dipole moment and/or polarizability tensor of a molecule. (We will not consider the problem of induced spectra further in this discussion). Suppose that the normal coordinate associated with the ℓ'th vibrational mode of a molecule is denoted by Q_ℓ. For molecule i, one has

$$\mu(\underset{\sim}{R}_i(t)) = \mu_o \underset{\sim}{n}_i(t) + \sum_{\ell} Q_{\ell i}(t) \underset{\sim}{v}_{\ell i}(t) \tag{3.13}$$

where higher terms in the Taylor expansion are neglected; μ_o is the magnitude of the permanent moment, $\underset{\sim}{n}_i(t)$ is a unit vector parallel to the permanent dipole at time t and $\underset{\sim}{v}_{\ell i}(t) = (\partial\underset{\sim}{\mu}_i/\partial Q_\ell)_{Q_\ell=0}$. The normal coordinate $Q_{\ell i}(t)$ varies with time as $\exp(-2\pi i\nu_\ell t)$ for an unperturbed harmonic oscillator. In a dense fluid, there will be additional time dependence in Q_ℓ which will be denoted by a function $V_\ell(t)$ that reflects primarily modulations in the phase of Q_ℓ and, to a lesser extent, interruptions in the motion due to a change in vibrational state. Both $\underset{\sim}{n}$ and $\underset{\sim}{v}_\ell$ vary with time as the molecular frame reorients; perhaps the most useful way to represent this reorientation is to work with spherical harmonics or their generalization, the Wigner functions of the three Eulerian angles. To this end, it is convenient to use a spherical basis for these vectors rather than Cartesian coordinates. In this basis, an arbitrary vector $\underset{\sim}{v}$ is

$$\underset{\sim}{v} = \begin{vmatrix} v_{-1} \\ v_0 \\ v_{+1} \end{vmatrix} = \begin{vmatrix} -\frac{1}{\sqrt{2}}(v_x+iv_y) \\ v_z \\ \frac{1}{\sqrt{2}}(v_x-iv_y) \end{vmatrix} \tag{3.14}$$

Wigner D-functions are defined using the convention of Rose (11) thus,

$$D_{km}{}^j(\Omega) = e^{-i(k\alpha+m\gamma)} d_{km}{}^j(\beta) \tag{3.15}$$

These functions form a complete orthogonal set and are equal to the complex conjugate spherical harmonics for k or m equal to zero, except for a normalizing constant. They possess the useful property that use of the spherical basis yields

$$[v_m(t)]_{\text{space}} = \sum_{m'} [v_{m'}]_{\text{body}} D^{*}_{mm'}{}^1(\Omega(t)) \tag{3.16}$$

where $\underset{\sim}{v}]_{\text{body}}$ is the body-fixed vector. It will be convenient to use the addition theorem:

$$D_{mm'}{}^j(\Omega(t)) = \sum_{r=-j}^{j} D_{mr}{}^j(\Omega(0))D_{rm'}{}^j(\delta\Omega) \tag{3.17}$$

where $\delta\Omega$ is now the change in the Euler angles for the molecular frame in time t. This machinery will allow us to derive explicit expressions for the reorientational functions whose Fourier transform gives the band shape; the formalism is sufficiently general to include various types of vibration-rotation bands as well as pure rotation spectra.

Before deriving explicit expressions for absorption spectra, we will outline an analogous argument for the Raman problem. Here, the quantity of interest is a space-fixed (Cartesian) component of the polarizability tensor.

To do the body→space transformation, it is convenient to again use the spherical basis. For a symmetric tensor $\underset{\approx}{T}$, one defines

$$\bar{T} = \frac{1}{3}\,\mathrm{Tr}(\underset{\approx}{T})$$

$$T_0 = \sqrt{\frac{2}{3}}\,[T_{zz} - \frac{1}{2}(T_{xx}+T_{yy})]$$

$$T_{\pm 1} = \pm(T_{xy} \pm i\,T_{yz})$$

$$T_{\pm 2} = \frac{1}{2}(T_{xx}-T_{yy} \pm 2i\,T_{xy}) \qquad (3.18)$$

Of course, $\bar{T}$ is invariant to rotation of the molecular frame; the other components transform according to

$$[T_m(t)]_{\text{space}} = \sum_{m'} [T_{m'}]_{\text{body}}\, D^{*\,2}_{mm'}(\Omega(t)) \qquad (3.19)$$

The Raman intensities for parallel and perpendicular geometries are related to the spherical tensor components in a simple way; namely,

$$A_{ZZ} = \bar{A} + \sqrt{\frac{2}{3}}\,[A_0]_{\text{space}} \qquad (3.20)$$

$$A_{XZ} = \frac{1}{2}\,[A_{-1} - A_{+1}]_{\text{space}} \qquad (3.21)$$

Finally, the molecular polarizabilities (whose sum is equal to A) are written as a permanent polarizability $\underset{\approx}{\alpha}^{o}(R_i)$ plus vibrationally-modulated terms:

$$\underset{\approx}{\alpha}(R_i) = \underset{\approx}{\alpha}_i^{o} + \sum_{\ell} Q_{\ell i}(t)\, \underset{\approx}{t}_{\ell i}(t) \qquad (3.22)$$

where $\underset{\approx}{t}_{\ell i}(t) = (\partial\underset{\approx}{\alpha}(R_i)/\partial Q_\ell)_{Q_\ell=0}$. Thus, the Raman and the absorption spectra can be expressed in similar formalisms - this is one of the good features of the time-correlation function approach. The main differences between the two types of correlation function are in the indices of the Wigner functions and, as will become clear below, in the non-zero values of $\underset{\sim}{v}_\ell$ and $\underset{\sim}{t}_\ell$ for a given normal mode.

We can now derive explicit expressions for the spectral correlation functions. As suggested previously, we simplify the problem by considering various parts of C_M separately, and begin with the "pure rotational" correlation (which is usually associated with far infrared or microwave absorption). This term is

$$C_{far}(t) = \mu_o^2 < \sum_{i,j} \underset{\sim}{n}_i(0)\cdot\underset{\sim}{n}_j(t) > \tag{3.23}$$

The self-terms (i=j) make a large contribution to the summation, but it can happen that mutual angular correlations may be significant, especially for strongly non-spherical molecules at high density. Consider only the self term, and drop the index i, for simplicity. Then

$$C_{far} = N\mu_o^2 \sum_{m=-1}^{+1} < [n_m(0)]_{space} [n_m^*(t)]_{space} > \tag{3.24}$$

where the subscript m now indicates a spherical vector component. Eqs. (3.16) and (3.17) are substituted into eq. (3.24), and an average over orientation at time zero is performed to give

$$C_{far}(t) = N\mu_o^2 \sum_{m,m'} [n_m]_{body} [n_{m'}^*]_{body} < D_{mm'}^1(\delta\Omega) > \tag{3.25}$$

In many cases of interest, the molecular dipole moment is parallel to the symmetry axis, which means that the only component of $\underset{\sim}{n}]_{body}$ that is non-zero is $n_0 (\equiv 1)$; eq. (3.25) then simplifies to

$$C_{far}(t) = N\mu_o^2 < \cos\delta\beta > \tag{3.26}$$

where $\delta\beta$ is the angle between $\underset{\sim}{\mu}(t)$ and $\underset{\sim}{\mu}(0)$.

In addition to the "self" part of $C_{far}(t)$ given in eq. (3.25), it is possible to at least write out the "mutual" contribution to this quantity. The same argument as above yields

$$C_{far}(t) = C_{self} + (N\mu_o)^2 \sum_{mm'm''r} [n_{m'}]_{body} [n_{m''}^*]_{body}$$

$$< D_{mm'}^{*1}(\Omega_1(0)) D_{mr}^1(\Omega_2(0)) D_{rm''}^1(\delta\Omega_1) > \tag{3.27}$$

where 1 and 2 now denote two different molecules in the fluid and the average is taken over all orientations and separation distances of 1 and 2. It is not possible to reduce eq. (3.27) further without drastic approximation; in particular, the reorientational motion of molecule 1 should depend upon its initial proximity to molecule 2, so the averaging cannot be split. Of course, the formal equation similifies somewhat if the dipoles are parallel to a symmetry axis, but the essential difficulty with the coupling remains.

Before discussing vibration-rotation band shapes, it is appropriate to consider "pure rotational" Raman spectra in dense fluids (often called inelastic light scattering). The argument closely parallels that for i.r. absorption, but with the added complication of polarized and depolarized scattering. In the depolarized case, one has a "self" term which is:

$$C_{\perp,\ell s}(t) = N < \alpha_{xz}(\Omega(0))\alpha_{xz}(\Omega(t)) >$$

$$= \frac{N}{4} < [\alpha_{+1}(\Omega(0))-\alpha_{-1}(\Omega(0))]_{space} \cdot [\alpha_{+1}(\Omega(t))-\alpha_{-1}(\Omega(t))]_{space} > \qquad (3.28)$$

After inserting eqs. (3.19) and (3.17) into (3.28) and averaging over $\Omega(0)$, one obtains

$$C_{\perp,\ell s}(t) = \frac{N}{10} \sum_{m,m'} [\alpha_n]_{body} [\alpha_m]_{body} < D_{mm'}{}^2(\delta\Omega) > \qquad (3.29)$$

The "self" part of the depolarized scattering is obtained in a similar way; one has

$$C_{\parallel,\ell s}(t) = N < [\bar{\alpha} + \sqrt{\tfrac{2}{3}}\, \alpha_o(\Omega(0))]_{space} \cdot [\alpha + \sqrt{\tfrac{2}{3}}\alpha_o(\Omega(t))]_{space} > \qquad (3.30)$$

$$= N\, \bar{\alpha}^2 + \frac{2}{15} \sum_{m,m'} [\alpha_m]_{body} [\alpha_{m'}]_{body} \cdot < D_{mm'}{}^2(\delta\Omega) > \qquad (3.31)$$

For a molecule with a three-fold or higher symmetry axis, only $[\alpha_o]_{body}$ is non-zero and eqs. (3.29) and (3.31) simplify considerably. Note that the term in $\bar{\alpha}^2$ in eq. (3.31) has no time dependence; its Fourier transform is a delta function at zero frequency which corresponds to Rayleigh scattering. Of course, the Rayleigh peak does have a finite width; however, it is due to translational motions that have been omitted from the present treatment. Finally, if the molecules in the fluid are sufficiently non-spherical, correlations between orientations of neighboring molecules will exist that give a "mutual" contribution to the inelastic light scattering. The argument follows closely that for the infrared absorption and thus will not be discussed further here.

We now need to take up the vibrational contributions to the absorption and the Raman correlation functions. Ordinarily, the motion associated with a normal mode is not appreciably coupled either with the orientation of the molecule, or with other normal modes in the same molecule or in other molecules. (Treatments of this coupling do exist, but are too advanced for present purposes.) When these assumptions are made, the spectral time-correlation functions simplify greatly. For example, for the infrared case, the pure rotational part is augmented by a series of terms, one for each normal mode. We can consider these separately since, as mentioned above, they usually correspond to vibration-rotation bands which in favorable cases are isolated spectral features that can be Fourier transformed or otherwise analyzed independently from the other bands. The time-correlation function for the infrared absorption associated with the ℓ'th normal mode is thus written as

$$C_{\ell,ir}(t) = N < \underset{\sim}{v}_\ell(0) \cdot \underset{\sim}{v}_\ell(t) > < Q_\ell^2(0) > V_\ell(t) \quad (3.32)$$

A factor of $\exp(-2\pi\, i\, \nu_\ell t)$ has been omitted, which means that the Fourier transform of $C_{\ell,ir}(t)$ will give the spectral intensity at a frequency calculated relative to ν_ℓ. (It can happen that the value of ν_ℓ itself will be significantly altered to the presence of neighboring molecules; in this case, a more careful treatment of $Q_\ell(t)$ is needed.)

In eq. (3.32), the factor in $Q_\ell^2(0)$ is merely the mean square amplitude of motion in the ℓ'th mode and is readily evaluated using the vibrational wave function; after Fourier transformation, $V_\ell(t)$ gives the "vibrational broadening" and the term in v_ℓ gives the rotational width of the vibration rotation band. We consider the rotational term in more detail. The goal here is to write it in terms of the reorientational Wigner functions. This is readily done by following the same argument used previously; one uses the body→space transformation and then averages over all orientations at time zero. The result is essentially the same as in eq. (3.25):

$$< \underset{\sim}{v}_\ell(0) \cdot \underset{\sim}{v}_\ell(t) > = \sum_{m,m'} [v_m]_\ell [v_{m'}]_\ell < D_{mm'}^1(\delta\Omega) > \quad (3.33)$$

where $[v_m]_\ell$ denotes the m'th spherical component of the dipole moment derivative for the ℓ'th normal mode, measured in the body-fixed frame. It is important to realize that one can use standard group theoretical techniques to determine which $[v_m]_\ell$ are non-zero for a given normal mode in a molecule of given point group. In standard character tables, the symmetries of the Cartesian variables x,y,z are shown; of course, v_0 has the symmetry of z; v_1, $-v_{-1}$ have the symmetry of x; and iv_1, iv_{-1}) have the symmetry of y. Thus, if one is interested in a normal mode of given symmetry, one looks in the character table to find whether x,y or z correspond to that symmetry; if not, the mode is inactive ($\underset{\sim}{v}_\ell \equiv 0$); if so, the non-zero spherical components of $\underset{\sim}{v}_\ell$ are ascertained by determining which cartesian coordinates have the symmetry of the ℓ'th mode. In principle, the vibration-rotation bands will have a different width for each symmetry because the time dependence of the $< D^1_{km}(\delta\Omega) >$ changes as the subscripts are changed. However, if one is to extract rotational dynamics from an ir band, it is necessary to assume either that $V_\ell(t)$ is independent of normal mode or that it can be obtained from the Raman band shape (see below).

Once again, the derivation for vibration-rotation Raman band shapes follows a similar path to the infrared case. Ordinarily, only the "self" term is important and thus one needs the transformation equations for $[t_k]_\ell$, the k'th spherical component of the polarizability derivative for the normal mode ℓ. In fact, one can show that (14)

$$C_{\perp,Ram}(t) = \frac{N}{10} \sum_{m,m'} [t_m]_\ell [t_{m'}]_\ell < D^2_{mm'}(\delta\Omega) > V_\ell{}'(t) \qquad (3.34)$$

$$C_{\parallel,Ram}(t) = \left\{ N\ \bar{t}^2 + \frac{2}{15} \sum_{m,m'} [t_m]_\ell [t_{m'}]_\ell < D^2_{mm'}(\delta\Omega) > \ V_\ell{}'(t) \right\} \qquad (3.35)$$

where $V_\ell{}'(t) = < Q^2_\ell(0) > V_\ell(t)$.

The final step in this theory is to use group theory to determine the non-zero $[t_m]_\ell$ for a given point group.

Note that there is a constant $\bar{t}^2$ in the expression for polarized band intensity; this yields a delta function when the Fourier transform is taken. The resulting spectral feature is the narrow peak often observed in polarized Raman spectra and has a finite width due to vibrational relaxation function $\dot{V}_\ell(t)$. In experimental studies, this peak shape is often analyzed to obtain $V_\ell(t)$, which is then used to extract the rotational contributions from the shapes of the other bands, While the transfer of $V_\ell(t)$ from one band to another is by no means rigorously correct, it is perhaps the best available approach to the difficult problem of separating vibrational relaxation from the rotational contribution to the band shape.

As an example, let us briefly consider the case of benzene, a D_{6h} molecule. By reference to the character table, one sees that the eleven spectrally active different vibrational modes of the molecule (eight doubly degenerate, three non-degenerate) have rotational side-bands whose width and shape are governed by the following time-dependent functions:

$$
\begin{array}{lll}
2\ A_{1g} & \text{(Raman)} & \begin{cases} D_{00}^{\;2} & \text{(depolarized)} \\ \bar{t}^2 + t_0^{\;2} D_{00}^{\;2} & \text{(polarized)} \end{cases} \\
A_{2a} & \text{(infrared)} & D_{00}^{\;1} \\
3\ E_{1u} & \text{(infrared)} & (D_{11}^{\;1} + D_{-1-1}^{\;1}) \\
E_{1g} & \text{(Raman)} & (D_{11}^{\;2} + D_{-1-1}^{\;2}) \\
4\ E_{2g} & \text{(Raman)} & (D_{22}^{\;2} + D_{-2-2}^{\;2})
\end{array}
$$

Some vibrational width will be convoluted into the rotational parts of all these bands; estimation of this contribution to the overall band shape is a serious complication in this problem. By comparing the polarized and depolarized A_{1g} bands, one can obtain reasonably unambiguous information about the vibrational relaxation for that mode only. It is doubtful that vibrational relaxation will be the same for all modes, but this appears to be the only approximation that can be made at present.

Symmetry arguments show (15) that the only non-zero $D_{mm'}{}^{j}(\delta\Omega)$ for a molecule with an n-fold symmetry axis are those with $m' = m \pm kn$. For $j < 6$, this means that only $D_{mm}{}^{j}(\delta\Omega)$ describe the rotation of a benzene molecule, as indicated above. The presence of the functions $D_{mm}{}^{j}$ and $D_{-m-m}{}^{j}$ in pairs merely gives a real function of $\delta\Omega$.

Note that the lower index of the D-function is a measure of the importance of rotation around the symmetry axis in affecting these band shapes, since the functions contain factors of $\cos[m(\delta\alpha + \delta\gamma)]$. To obtain some idea of the behavior of these functions, one might start with the rotational diffusion model which predicts that (16)

$$< D_{mm}{}^{j}(\delta\Omega) > = \exp[-t\{j(j+1)\mathcal{R}_x + m^2(\mathcal{R}_z - \mathcal{R}_x)\}] \qquad (3.36)$$

If $\mathcal{R}_z$, the diffusion constant for motion <u>around</u> the symmetry axis, is larger than $\mathcal{R}_x$, the constant for rotation of the axis, one expects a faster decay (and a wider band) as m increases. We emphasize that benzene need not obey the rotational diffusion model and indeed, it is likely that the axial rotation is sufficiently free to cause serious deviations from any diffusional model. One additional factor needs to be dealt with in a careful analysis of the band shapes of the perpendicular bands of the benzene molecule (i.e., those bands whose shape is correlated with the D-functions having $m \neq 0$). This is the first order Coriolis coupling effect.

Briefly described (17), a Coriolis force is present when translation occurs in a coordinate system which is rotating. If the mass and velocity of the translating particle are $m, \underset{\sim}{v}$, the force is equal to $2m(\underset{\sim}{v} \times \underset{\sim}{\omega})$ where $\underset{\sim}{\omega}$ is the rotational velocity of the frame. In the context of vibration-rotation spectra, translation arises from the vibrational motion of the nuclei in a rotating molecule. The resulting force causes a slight oscillatory distortion in the molecular frame if the vibrational motion is non-degenerate. However, this force can have a significant effect on the direction of the dipole moment derivative or the polarizability derivative if one has a degenerate vibrational mode. Degenerate modes signify two (or more) transition moments pointing in different directions in the molecule. The Coriolis force can then cause a rapid change in the moment from one degenerate direction to another, and is best explained by means of an example.

Consider a degenerate (E) vibrational mode in benzene; the dipole or polarizability derivative then lies in the plane of the hexagon and the normal modes correspond to in-plane translational motion of the nuclei. Rotation around the molecular symmetry axis will produce a Coriolis force which is in the plane of the molecule (but perpendicular to the direction of translational motion). As a consequence, the dipole or the polarizability derivative (which would otherwise be fixed in the molecular frame) will rotate rapidly in the plane. Specifically, it will have a rotational velocity equal to $\zeta\omega_z$, where ζ is the Coriolis coupling constant and can range between 1 and -1. Values close to the limits are observed (in gas phase spectra) if the nuclei involved are heavy. This motion can have a very large effect on the spectrum. This is seen most clearly by considering the second moment M_2 defined by

$$M_2 = \frac{\int_{\text{band}} \omega^2 I(\omega)\,d\omega}{\int_{\text{band}} I(\omega)\,d\omega} \tag{3.37}$$

where ω here is $2\pi\nu$, with ν equal to the spectral frequency. It is possible to derive rigorous expressions for M_2, at least in the absence of mutual orientational correlations and induced contributions to the spectrum. In the presence of Coriolis coupling, the expression for a symmetric top second moment is (2)

$$M_{j,m}^{(2)} = \frac{kT}{I}\left(\ell(\ell+1) + \eta m^2 + (1+\eta)(\zeta+2|m|)\right) \tag{3.38}$$

where $\eta = (I/I_{zz} - 1)$. It is evident that for perpendicular bands ($m \neq 0$), one should take careful account of this Coriolis coupling in quantitative analyses of the second moment in particular and of band shapes in general.

Finally, we present a brief discussion of the comparison of simulation (19) and experiment (18) for liquid benzene at room temperature. Experimental data (18) is actually limited to the correlation times ($\tau_c \simeq \tau_r$) for $D_{00}^2(\delta\Omega)$ of the symmetry axis (Raman and light scattering) and for an axis in the molecular plane (n.m.r. relaxation). Simulations will yield the correlation functions for these vectors as well as the correlation times, and show clearly this system does not exhibit the simple exponential decays that are characteristic of the orientational random walk of a symmetric diffuser. To analyze the results, it is helpful to write down the equation relating the correlation functions for two different molecule-fixed vectors i, j with relative orientations Ω_{ij}. For the case where only the $D_{mm}^j(\delta\Omega)$ exist, one has

$$D_{kk}^{j}(\delta\Omega_i) = \sum_m D_{mk}^{j}(\Omega_{ij}) D_{mk}^{*j}(\Omega_{ij} D_{mm}^{j}(\delta\Omega_j) \tag{3.39}$$

In our case, the two vectors are the molecular z axis and an axis perpendicular to z, so that $\beta_{ij} = 90^o$, α_{ij}, γ_{ij} arbitrary. Also the index j=2, so that

$$|D_{mk}^{2}(\Omega_{ij})|^2 = \begin{vmatrix} \frac{1}{6} & \frac{1}{4} & \frac{3}{8} & \frac{1}{4} & \frac{1}{16} \\ \frac{1}{4} & \frac{1}{4} & 0 & \frac{1}{4} & \frac{1}{4} \\ \frac{3}{8} & 0 & \frac{1}{4} & 0 & \frac{3}{8} \\ \frac{1}{4} & \frac{1}{4} & 0 & \frac{1}{4} & \frac{1}{4} \\ \frac{1}{6} & \frac{1}{4} & \frac{3}{8} & \frac{1}{4} & \frac{1}{16} \end{vmatrix} \tag{3.40}$$

Consequently, the relationship between the in-plane $D_{00}^2(\delta\Omega)$ and the reorientation of the molecule-fixed symmetry axis is

$$D_{00}^{2}(\delta\Omega_{in}) = \frac{3}{8}[D_{22}^{2}(\delta\Omega_s) + D_{-2-2}^{2}(\delta\Omega_s)] + \frac{1}{4}D_{00}^{2}(\delta\Omega_s) \tag{3.41}$$

where in, s denote an in-plane vector and the symmetry axis, respectively. Since the correlation times are given just by the time-integral of this equation, it is evident that

$$[\tau_{c,s}]_{22} = \frac{8}{3}[\tau_{c,in}]_{00} - \frac{2}{3}[\tau_{c,s}]_{00} \tag{3.42}$$

Simulation (19) gives $[\tau_{c,s}]_{0,0} = 2.3$ picosec, $[\tau_{c,in}]_{0,0} = 1.3$ picosec; the experimental data indicates 2.8 and 1.2, respectively. Thus, one estimates $[\tau_{c,s}]_{2,2} = 1.9$ (simulation) or 1.4 (experiment). Clearly, the rotation of this molecule is quite anisotropic. In view of this and the fact that rotational diffusion is not obeyed, additional information, such as estimates of $[\tau_{c,s}]_{1,1}$ and $[\tau_{c,s}]_{2,2}$ based on the appropriate Raman line widths, would be very helpful in progressing toward a complete understanding of the orientational dynamics of this molecule.

REFERENCES

1. Berne, B. J. and Harp, G. D. 1970, Adv. Chem. Phys. 17, p. 63.
2. St. Pierre, A. G. and Steele, W. A. 1981, Molec. Phys. 43, p. 123.
3. Evans, M. W. and Davies, G. J. 1976, Adv. Molec. Relax. Proc. 9, p. 129; Evans, M. W. (Spec. Per. Report, Chemical Society, London, 1977) "Dielectric and Related Molecular Processes, Vol. 3"
4. Steele, W. A. 1981, Molec. Phys. 43, p. 141.
5. Gordon, R. G. 1966, J. Chem. Phys. 44, p. 1830.
6. Chandler, D. 1974, J. Chem. Phys. 60, p. 3508.
7. For a discussion of several of these calculations, see Hansen, J. P. and McDonald, I. R. (Academic, New York, 1976) "Theory of Simple Liquids," Chap. 8.
8. Keyes, T. and Kivelson, D. 1972, J. Chem. Phys. 56, p. 1057; Kivelson, D. and Madden, P. 1975, Molec. Phys. 30, p. 1749.
9. Evans, M., Evans, G. J., Coffey, W. T. and Grigolini, P. (Wiley, New York, 1982) "Molecular Dynamics."
10. Keyes, T. 1972, Molec. Phys. 23, p. 737.
11. Rose, M. E. (Wiley, New York, 1957) "Elementary Theory of Angular Momentum."

12. Gordon, R. G. 1968, Adv. Mag. Res. 3, p. 1.

13. Placzek, G. 1931, Z. Physik. 20, p. 84.

14. Steele, W. A. (Plenum, New York, 1979) Ed. S. Bratos and R. Pick "Vibrational Spectroscopy of Molecular Liquids and Solids."

15. Lynden-Bell, R. M. 1980, Chem. Phys. Lett. 70, p. 477; Steele, W. A. 1980, Molec. Phys. 39, p. 1411.

16. Berne, B. J. and Pecora, R. (Wiley, New York, 1976) "Dynamic Light Scattering," Chap. 7.

17. Herzberg, G. (Van Nostrand, New York, 1945) "Infrared and Raman Spectra," Chap. IV.

18. Steele, W. A. 1976, Adv. Chem. Phys. 34, p. 1.

19. Steinhauser, O. 1982, Chem. Phys. 73, p. 155.

ROTATIONAL-VIBRATIONAL CORRELATIONS IN LIQUIDS AND SOLUTIONS

S.Bratos and G.Tarjus
Laboratoire de Physique Théorique des Liquides, Université
P. et M. Curie, 4,Place Jussieu, 75005 PARIS, France

1. INTRODUCTION

Spectral effects due to correlations between rotational and vibrational motions in liquids have been studied, the last decade, by several authors. This correlation arises from a variety of position and angle dependent intermolecular forces such as dipole - dipole, dipole-induced dipole, dispersion, hydrogen bond or charge transfer forces. Alternatively, it may originate in intramolecular forces such as centrifugal and Coriolis forces. The spectral effects generated by these forces are different for liquids and for diluted solutions as they differ for non-degenerate and for degenerate vibrations. The previously published work permits a partial understanding of this problem. However, it has never been reviewed and the purpose of the present paper is to fill this gap.

2. GENERAL CONSIDERATIONS

It is convenient to start the description of the theory by constructing the basic Hamiltonian of the problem. This can be done by (i) writing the Hamiltonian of each individual vibrating - rotating molecule in a form proposed by Watson (1), (ii) summing over all molecules of the liquid sample and (iii) introducing the potential energy of molecular interaction. The resulting Hamiltonian of the liquid sample can then be written :

$$H=\left[\frac{1}{2}\sum_{i\alpha}\sum(p_{i\alpha}^{2}+\lambda_{i\alpha}n_{i\alpha}^{2})+\ldots\right]+\left[\frac{1}{2}\sum_{i}\sum_{\beta\gamma}\mu_{i,\beta\gamma}(J_{i\beta}-j_{i\beta})(J_{i\gamma}-j_{i\gamma})\right]+$$

A. J. Barnes et al. (eds.), Molecular Liquids - Dynamics and Interactions, 151–161.

$$+\left[\frac{1}{2}\sum_{i\delta}\sum \frac{P_{i\delta}^2}{M_i} + V(n,\theta,R)\right] \qquad (1)$$

The quantities $\mu_{i,\alpha\beta}$ entering in Eqn(1) are related to the components of the effective moment of inertia tensor of the molecule i and depend on its normal coordinates n_i. All other symbols have their usual meaning. It is recalled that $\vec{J}_i$ is the overall angular momentum of i'th molecule and $\vec{j}_i$ its n_i-dependent vibrational angular momentum. $V(n,\theta,R)$ denotes the potential energy of molecular interaction and depends on the center of gravity coordinates $R=\{\vec{R}_1,\vec{R}_2,...,\vec{R}_N\}$ of N molecules forming the liquid sample, on their Eulerian angles $\theta=\{\theta_1,\theta_2,...,\theta_N\}$ as well as on their normal coordinates $n=\{n_{11},n_{12},...,n_{1p};...;n_{N1},n_{N2},...,n_{Np}\}$. There are p normal modes by molecule.

A glance on Eqn(1) shows the nature of various rotation - vibration coupling mechanisms. The terms contained in the second square bracket describe the rotations of a non-rigid rotator coupled to its vibrations ; they describe, between other, the effects of the centrifugal and Coriolis forces. In turn, the potential energy $V(n,\theta,R)$ realises an intermolecular coupling between vibrations, rotations and translations. One concludes that, both, intermolecular and intramolecular forces contribute to the vibration-rotation correlations. These correlations are termed as intramolecular or intermolecular according to whether the rotator and the vibrator under study do, or do not, belong to the same molecule.

3. INTRAMOLECULAR ROTATION-VIBRATION CORRELATIONS IN DILUTED SOLUTIONS. NON-DEGENERATE VIBRATIONS.

3.1. Generalities

This class of processes is the simplest to analyse and it is convenient to begin by studying them. Both, intramolecular and intermolecular forces have been considered. The effect of centrifugal forces was first examined by Gordon (2,3). Systematic investigation of all other aspects of the problem was undertaken later. The basic theory is described in papers by Alekseiev and Sobel'man (4,6), Alekseiev, Grasiuk et al. (5), Bratos and Chestier (7), Brueck (8), Temkin and Burshtein (9-11), Tarjus and Bratos (12) and by Leicknam and Guissani (13). In the discussion below, the papers are grouped according to the theoretical method employed.

3.2. Theory

The first theory which has been elaborated to explain the effect of centrifugal forces on non-degenerate vibrations is the moment analysis by Gordon (2,3) ; it rests on the following assumptions. (i) Vibrational motions are treated by the help of quantum mechanics whereas rotational-translational motions are described classically. (ii) Internal vibrational states are not affected by the intermolecular potential energy. (iii) Collision-induced effects are neglected as are neglected internal field effects. (iv) Resonant energy transfer in the liquid is not considered. Starting from these premises the theory predicts the lowest-order rotation-vibration interaction changes in frequency and intensity. Although formulated to describe pure liquids, the assumption (iv) makes it really applicable only to solutions. Moreover, all rotation-vibration correlation effects due to intermolecular forces are eliminated by the assumption (ii). Then, the main effect which is accounted for is that due to the change of the moment of inertia. It varies when going from the ground to the excited vibrational state.

The second group of theories are due to Alekseiev and Sobel'man (4,6), Alekseiev, Grasiuk et al. (5) and Temkin and Burshtein (9-11). Based on the impact theory of spectral shapes they differ in several important aspects. The theory by Sobel'man et al. implies the following assumptions. (i) The perturbation experienced by the system arises from binary collisions. (ii) These collisions are instantaneous. (iii) The collision process is described by an S matrix. Its diagonal elements depend on rotational, but not vibrational, quantum numbers and its non-diagonal matrix elements vanish. (iv) Only a limited number of rotational sublevels are explicitely considered. In turn, the theory by Burshtein et al. conserves the assumptions (i,ii) but employs a different description of the collision process. The S matrix method is replaced by kinetic theoretical methods. The theory contains a collision strength parameter caracterizing the nature of the collision process. Starting from these premises these theories explain successfully the collapse of the Q-branch of the isotropic Raman spectrum through the motional narrowing (Fig.1). On the contrary, the Q-branch of the anisotropic Raman spectrum broadens monotonously until it merges with the O-and S-branches (11). The major restriction of these theories is that they are essentially gas phase theories ; however, this restriction is less stringent for the theory by Burshtein et al. Moreover, the rotation-vibration coupling effects due to intermolecular forces are neglected here again. This is done in the Sobel'man theory by neglecting the v-dependence of the S matrix.

The third theory of this group is the recent theory by Tarjus and Bratos (12). It describes the rotation-vibration correlation effects arising from intermolecular forces but neglects the role of the centrifugal interaction. This contribution is thus complementary to those reviewed earlier. The basic assumptions of the theory are as follows. (i) Active diatomic molecule is executing

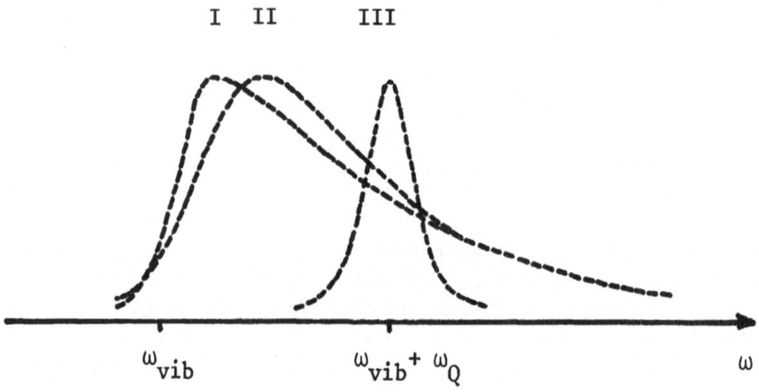

Fig.1. Isotropic Raman band shape transformation with increasing density, for strong collisions : (I) $\Gamma = 0.1$, (II) $\Gamma = 0.3$, (III) $\Gamma = 10.$, where $\Gamma = 1/\omega_Q\tau_J$ (see Ref.(9)).

anharmonic vibrations coupled to the remaining degrees of freedom of the system through the angle and position dependent intermolecular forces. (ii) The active molecule is assimilated to a rigid rotator executing random reorientations. Centrifugal interactions are thus absent. (iii) Vibrations are described quantum-mechanically with the help of the Heisenberg equation of motion. (iv) Rotations and translations are described classically with the help of a simple Langevin equation with a constant transport matrix and a Gaussian random force. The infrared and anisotropic Raman correlation functions are then written in the form of a series. Its leading term is a product of rotational and vibrational correlation functions whereas its higher-order terms give successive corrections of this simple approximation. These corrections describe fine rotation-vibration correlation effects missing in the product approximation. The calculation is simplified by considering the isotropy of the liquid sample. The spectral effects predicted by the theory are small and by no means conspicuous (Fig.2). Its main limitation is that the simple Langevin equation with a Gaussian random force may not be, in general, accurate enough. It does not lend itself easily to incorporate inertial effects but can conveniently be generalized to describe pure liquids.

The last theory is that due to Bratos and Chestier (7). It considers, both, intra-and intermolecular forces. The theory rests on a model similar in many respects to that of the Tarjus-Bratos model. However, the active molecule is not assumed to be rigid. Rotational motions are described in terms of a single stochastic variable and translational motions are not considered explicitely.

A number of physically interesting situations have been analysed. The main limitation of this theory resides in the difficulty to extend it to pure liquids. Moreover, only slowly modulated vibrational relaxation processes are considered.

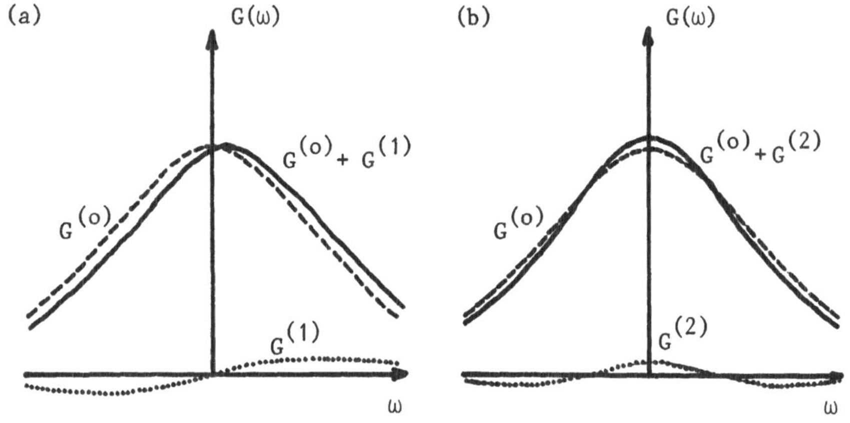

Fig.2. Effect of the first-order correction (a) and the second-order correction (b) on the infrared spectrum of a representative Van der Waals solution ; $G^{(o)}$ represents the simple product approximation (see Ref.(12)).

3.3. Discussion

This Section may be concluded by giving a brief estimate of the actual situation in this area. It is probably fair to say that the spectral manifestation of the intramolecular rotation-vibration coupling are qualitatively understood. Nevertheless, no quantitative study has yet been published. It is probable that the use of computer simulation techniques will accelerate the progress. Unfortunately, the treatment of diluted solutions is technically difficult.

4. INTRAMOLECULAR ROTATION-VIBRATION CORRELATIONS IN DILUTED SOLUTIONS ; DEGENERATE VIBRATIONS

4.1. Generalities

Spectral effect of the rotation-vibration correlation is particularly important for degenerate transitions. The Coriolis interaction, frequently envisaged in this context, was studied in a long series of papers due to Müller and Kneubühl (14,17,19), Müller, Etique et al. (15,16) and Müller (18). Other important papers were published by Eagles and Mc Clung (20), Lévi, Marsault et al. (21), Lynden-Bell (22), Gilbert and Drifford (23), Dreyfus, Berreby et al. (24), Gilbert, Nectoux et al. (25) and by St Pierre and Steele (26). Once again, the papers are grouped according to the theoretical method on which they are based.

4.2. Theory

The first group of papers are relative to the study of band moments in presence of the Coriolis interaction. The main papers are due to Gilbert, Nectoux et al. (25) and to St Pierre and Steele (26). Symmetric tops, spherical tops and linear molecules are carefully examined. These theories are elaborated much in the same spirit as other band shape theories and contain similar restrictions. The processes such as collision-induced absorption or scattering, vibrational relaxation due to intermolecular forces, etc, are not considered. However, the remaining calculation is exact. The four lowest- order moments are determined in this way ; an approximate expression for the memory function is presented too.

Another interesting proposal is due to Lynden-Bell (22). Its objective is to examine the profiles of degenerate infrared or Raman bands for a prolate symmetric top in presence of a Coriolis coupling. This theory is based on the following assumptions. (i) The reorientation of the long axis of the symmetric top is strongly impeded and is only slowly diffusing in angle. On the other hand, the motion about this axis is practically free. (ii) The effect of collisions on the Coriolis rotation is described by two different models. In the first model the collisions are strong. Thus the vibrational angular momentum after a collision is uncorrelated to that before. This is a one-dimensional version of the extended diffusion models. The second model employs the Langevin equation motion. In this model, the vibrational angular momentum changes by small steps and its extremity performs a one-dimensional random walk. All other effects such as vibrational relaxation due to intermolecular forces, etc, are neglected. The band shapes predicted by the theory depend on the correlation time τ_k of the vibrational angular momentum as well as on the value of the Coriolis coupling constant ζ. If τ_k is sufficiently small, the profile is lorentzian; if it is not, the band shape depends on the model. Leaving the restrictions arising from the omission of processes such as vibrational relaxation, etc, aside, the Lynden-Bell theory is mainly limited by the very specific assumption (i). Nevertheless, several results of this paper are reproduced by the later work.

A third group of papers, probably the central one, present an analysis of spectral effects of the Coriolis interaction with the help of the J- and M- extended diffusion models. The main papers are due to Müller and Kneubühl (14,17,19), Müller, Etique et al. (15,16), Müller (18), Gilbert, Nectoux et al. (25), Eagles and Mc Clung (20) and to Gilbert and Drifford (23).

Although basically similar, they calculate the free rotor correlation function, the key element of the theory, in several different ways. (i) Kneubühl, Müller and Etique calculate the Coriolis coupling in degenerate infrared transitions by introducing an apparent angular velocity which differs from the true angular velocity by the factor $1-\zeta$ where ζ is the Coriolis coupling constant. (ii) Drifford, Gilbert and Nectoux determine the Coriolis coupling in degenerate Raman transition systematically, without introducing any ad hoc assumption. They find that some properties, but not all, can be obtained by introducing and apparent angular velocity. (iii) Finally, in their semi-classical theory, Mc Clung, Drifford et al. calculate the free rotor correlation function with the help of quantum mechanics. This can be done by using the observed, or calculated, frequencies and intensities of individual rotation-vibration lines. Once the free rotor correlation function is calculated by one of these methods, the correlation function of the J- model may be determined by introducing an appropriate substitution into its Laplace transform. The study of the M- model employs a similar procedure although it is slightly more elaborate. Symmetric, spherical and linear molecules have been treated in this way (Fig.3). It is difficult to appreciate the limitations of these theories with some precision. However, comparisons of this model with computer simulations indicate that it is not particularly accurate for liquids in absence of Coriolis coupling and there is no reason to believe that the inclusion of this effect would improve the situation. However, these theories

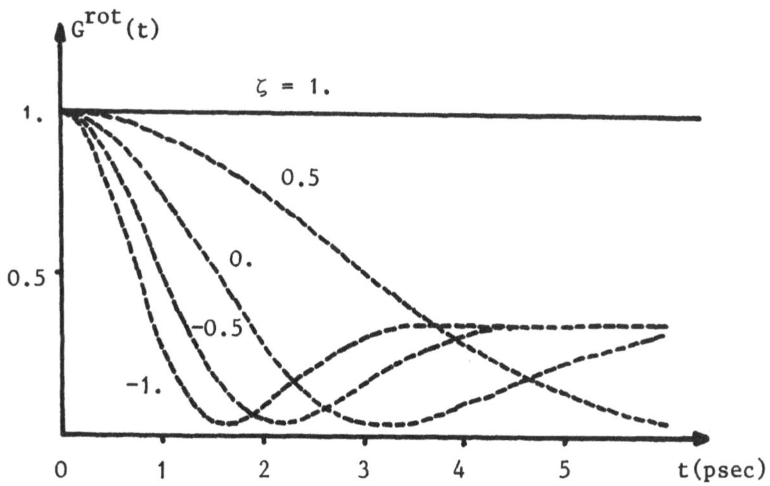

Fig.3. Influence of ζ on the rotational relaxation function $G^{rot}(t)$ of molecules in the gas phase (see Ref.(15)).

are the only theories available at the present time to describe the Coriolis interaction in liquids.

All these papers neglect the rotation-vibration correlation affects arising from molecular interaction. This effect was considered in the paper by Lévi, Marsault et al. (21). In their theory, the collisional events due to the change of angular momentum of the molecule are supposed to bring about changes in the orientation of the transition moment vector. Other premises are similar in all respects to those of the extended diffusion models. The theory describes the experimental data in a satisfactory way. It is difficult to give a reliable appreciation of its value, even more so than for the theories of the preceeding group. One of its limitations certainly is the omission of collision induced processes which modify the length and the direction of the vibrational transition vector. Moreover, the limitations inherent to the use of extended diffusion models have also to be considered. Nevertheless , it should be emphasized that this problem has not yet been examined systematically by any other method.

4.3. Discussion

It is probably fair to say, in appreciating the actual situation in the field, that spectral manifestations of the Coriolis interaction are, in essence, adequately interpreted. Still, the description remains entirely quatitative. From the other side, the effects of intermolecular rotation-vibration correlations on degenerate vibrations are, at the present time, practically unexplored ; this problem certainly merits attention. Here again, the computer simulation techniques will be useful. However, their use can only be envisaged once the quantum-mechanical part of this problem has been solved. This problem refers to the study of degenerate vibrations perturbed by random solvent-solute interactions.

5. INTER-AND INTRAMOLECULAR ROTATION-VIBRATION CORRELATIONS IN PURE LIQUIDS.

5.1. Generalities

The rotation-vibration correlation effects in pure liquids have been much less explored than those occuring in diluted solutions. The early work is due to Van Woerkom, de Bleyser et al. (27) and to Lynden-Bell (28). Recent progress of the theory is described in papers by Wang and Mc Hale (29), Mc Hale (30), Levesque, Weis and Oxtoby (31,32). See also Bratos and Tarjus (33). These papers are discussed in what follows.

5.2. Theory

The first paper in which the separability of the total infrared correlation function into its vibrational and rotational components was questioned is the paper by Van Woerkom, de Bleyser et al. (27). Their theory is based on use of the generalized cumulant expansion theorem for non-commuting quantum mechanical operators and involves the following assumptions. (i) The Hamiltonian of the liquid sample is written in the form H = E + F + G where E is the Hamiltonian for vibrational degrees of freedom, F is the bath Hamiltonian for rotational and vibrational degrees of freedom whereas G is the interaction Hamiltonian. (ii) G is small with respect to E + F. An interaction representation is used in which G is considered as a perturbation. (iii) The averaged ordered exponential is developped into a truncated cumulant expansion series. Only the first- and the second-order cumulants are retained. (iv) The correlation between intra- and intermolecular terms of G is neglected. It is then shown that vibrational relaxation due to intermolecular interactions depends on the reorientational behavior of molecules in the liquid. Writing the correlation function in the form of a product of vibrational and rotational correlation functions introduces errors arising from the fact that the correlations of a given oscillator at several time points are not correctly accounted for. Unfortunately, the assumption (iv) is not realistic and affects the quality of the theory. Nevertheless, the major merit of the authors is that they formulated the problem properly as early as in 1974.

The second group of papers are due to Wang and Mc Hale (29) and to Mc Hale (30) ; see also the paper by Bratos and Tarjus (33). The objective of these theories is to calculate the two lowest spectral moments of the isotropic Raman, infrared and anisotropic Raman spectra. No separability of infrared and anisotropic Raman correlation functions is postulated. The theory is quantum-mechanical in principle athough, in practice, the non-vibrational degrees of freedom are taken to be classical. Its basic assumptions are as follows. (i) The Hamiltonian of the system is written H = E + F + G as in the Van Woerkom paper ; once again G collects all coupling terms. (ii) The moments are calculated by expressing the time derivatives with the help of the Heisenberg equation of motion and by evaluating the commutators. (iii) Collision induced and internal field effects are neglected. The main prediction of this theory is that the first moments of the three spectra do not coincide ; this effect is called the non coincidence effect. It is interesting to mention, however, that the major part of it is obtained even if the separability of the vibrational-rotational correlation function is assumed (33,34). The main limitation of the theory is probably its assumption (iii). Moreover, the analysis of higher-order moments would be necessary.

The last group of papers are relative to the computer-simulation study of the rotation-vibration correlations in pure liquids. The main authors are Levesque, Weis and Oxtoby (31,32) who examined liquid N_2 and HCl in much detail. A molecular dynamics simulation was carried out using 500 molecules for N_2 and 256 for HCl ; periodic boundary conditions were imposed. It was found that, in the isotropic Raman spectrum of liquid N_2, intermolecular rotation-vibration interactions are mainly due to short range repulsion and dispersion forces ; the role of centrifugal forces seems secondary which is an unexpected result. The presence of important interference effects makes any partition illusory. Only intermolecular rotation-vibration correlation effects were examined in the case of infrared and anisotropic Raman spectra of liquid HCl. It results from this calculation that the simple product correlation function is indistinguishable from the total correlation function within the uncertainty of the simulation. This conclusion is similar to that reached by Tarjus and Bratos (12). It should not be forgotten, however, that the latter theory applies to diluted solutions whereas the function determined by Levesque, Weis and Oxtoby is an approximate correlation function of a pure liquid.

5.3. Discussion

The theory of the rotation-vibration correlation in pure liquids is much less advanced then that relative to diluted solutions. All difficulties mentioned earlier appear here again complicated by the presence of important interference effects. Their physical characteristics are entirely unknown at the present time. An additional effort will be needed to settle all these problems definitively.

REFERENCES

1. Watson, J.K.G. 1968, Mol. Phys. 15, p. 479.
2. Gordon, R.G. 1964, J. Chem. Phys. 40, p. 1973.
3. Gordon, R.G. 1964, J. Chem. Phys. 41, p. 1819.
4. Alekseiev, V.A. and Sobel'man, I.I. 1968, Acta Physica Polonica 34, p. 579.
5. Alekseiev, V., Grasiuk, A., Ragulsky, V., Sobel'man, I. and Faizulov, F. 1968, IEEE J. Quant. Electr. 4, p. 654.
6. Alekseiev, V. and Sobel'man, I. 1969, Sov. Phys. JETP 28, p. 991.
7. Bratos, S. and Chestier, J.P. 1974, Phys. Rev. A 9, p. 2136.
8. Brueck, S.R.J. 1977, Chem. Phys. lett. 50, p. 516.
9. Temkin, S.I. and Burshtein, A.I. 1979, Chem. Phys. Lett. 66, p. 52.
10. Temkin, S.I. and Burshtein, A.I. 1979, Chem. Phys. Lett. 66, p.57.
11. Temkin, S.I. and Burshtein, A.I. 1979, Chem. Phys. Lett. 66, p.62.
12. Tarjus, G. and Bratos, S., to be published.

13. Leicknam, J.Cl. and Guissani, Y., to be published.
14. Müller, K. and Kneubühl, F. 1973, Chem. Phys. Lett. 23, p. 492.
15. Müller, K., Etique, P. and Kneubühl, F. 1973, Chem. Phys. Lett. 23, p. 489.
16. Müller, K., Etique, P. and Kneubühl, F. 1974, in Molecular Motions in Liquids, Lascombe, J. Editor, Reidel Publ. Comp., Dordrecht, p. 265.
17. Müller, K. and Kneubühl, F. 1975, Chem. Phys. 8, p. 468.
18. Müller, K. 1976, Chem. Phys. Lett. 40, p. 508.
19. Müller, K. and Kneubühl, F. 1976, Helv. Phys. Acta 49, p. 702.
20. Eagles, T.E. and Mc Clung, R.E.D. 1974, J. Chem. Phys. 61, p.4070.
21. Lévi, G., Marsault, J.P., Marsault-Hérail, F. and Mc Clung, R.E.D. 1970, J. Chem. Phys. 63, p. 3543.
22. Lynden-Bell, R.M. 1976, Mol. Phys. 31, p. 1653.
23. Gilbert, M. and Drifford, M. 1976, J. Chem. Phys. 65, p. 923.
24. Dreyfus, C., Berreby, L., Dayan, E. and Vincent-Geisse, J. 1978, J. Chem. Phys. 68, p. 2630.
25. Gilbert, M., Nectoux, P. and Drifford, M. 1978, J. Chem. Phys. 68, p. 679.
26. St Pierre, A.G. and Steele, W.A. 1981, Mol. Phys. 43, p. 123.
27. Van Woerkom, P.C.M., De Bleyser, J., De Zwart, M. and Leyte, J.C. 1974, Chem. Phys. 4, p. 236.
28. Lynden Bell, R.M. 1977, Mol. Phys. 33, p. 907.
29. Wang, C.H. and Mc Hale, J.L. 1980, J. Chem. Phys. 72, p. 4039.
30. Mc Hale, J.L. 1981, J. Chem. Phys. 75, p. 30.
31. Levesque, D., Weis, J.J. and Oxtoby, D.W. 1980, J. Chem. Phys. 72, p. 2744.
32. Levesque, D., Weis, J.J. and Oxtoby, D.W., to appear in J. Chem. Phys.
33. Bratos, S. and Tarjus, G. 1982, in Raman Spectroscopy, Lascombe, J. and Huong, P.V. Editors, Willey, J. and Sons, New York, p. 317.
34. Bratos, S. and Tarjus, G. 1981, Phys. Rev. A 24, p. 1591.

DYNAMIC LIQUID STRUCTURES THROUGH FAR INFRARED AND MICROWAVE SPECTROSCOPIES

Alain Gerschel

Chimie Physique des Matériaux Amorphes, Université Paris-Sud, 91405 Orsay, France.

I. LIQUID STRUCTURES : STATIC AND DYNAMIC

The picture we present hereafter is an experimental approach. However it is not meant to present the experimental setups nor to discuss the accuracies of spectra. On the opposite it is accepted that ability of measuring and reliability of the results are adequately fulfilled conditions. In the following we will select some actual effects being currently detected and measured in liquids, and we will look at them under the specific angle of liquid structures, so as to include them into the more general scheme of understanding the relationship between liquid static structure (geometrical) and dynamic structure (motional).

More precisely, we will take here as a principle that there exists a correspondence between static and dynamic structures. Dynamical events will be considered as images of static structural aspects and conversely : local order, angular correlations, local or less local density inhomogeneities, may be related to time-dependent observations such as librational coherence, orientational fluctuations, or sometimes rotational coherence, showing either individual or collective characters. For simple molecular liquids such a correspondence appears to be a principle general enough that whenever a static structural detail is found it may be searched for the dynamical constraint that it carries with it, or whenever a specific dynamical effect is detected it can be looked for the geometrical arrangement accompanying it. Precise examples of correspondence will illustrate the following picture of the rotational dynamics, as measured experimentally, through dipolar absorption-dispersion spectroscopy, ranging from far infra-red (FIR) down to microwave and lower frequencies.

A. J. Barnes et al. (eds.), Molecular Liquids - Dynamics and Interactions, 163–199.

As a further distinctive feature our approach does not cover a wast variety of chemical species in the fluid phase. Rather we enter the EMLG's view of considering that much can be learnt through selecting a few model systems, and therefore we restrict the illustration of the general picture hereafter to some compounds made up of a few atoms arranged in simple dipolar molecules. The liquid phase occupies a density domain vast enough that varying parameters of the equation of state within the limiting gaseous and solid phases will allow us to push the investigations deep enough, still keeping clear reference to physically meaningful quantities such as density and molecular kinetic energy.

Last, a definite bias is taken of explaining physics with words rather than through mathematics, these being reserved for specifying definitions in the frame of statistical mechanics, for the sake of linking macroscopic observables with molecular properties.

II. GENERAL PICTURE

To begin with, let us locate the processes investigated in dipolar absorption-dispersion spectroscopy. We can represent these processes with regard to their place in the overall angular trajectories, as steps into a sequence of interrelated dynamical events. Figure 1 gives a schematic representation of such a chain.

The prime mover of the dynamics is the thermal energy that sets on the molecular motion by activating rotational and translational degrees of freedom. The medium within which thermal energy is setting impulses is a locally ordered one with anisotropy settled by molecular symmetry : molecular shape and distribution of electric charges, and altogether specific characters of the intermolecular forces, are responsible of the local structure own symmetry. Whenever different moments of inertia and shape factors favour one or another angular or linear momentum transfers, these also enter the symmetry of "collisions". From the onset of coherent local oscillations, or librations, we can further define these collisions as being correlating events. Their duration is finite and can be estimated as being in the 10^{-14}-10^{-13}s range, this being the first time-scale settled at the intermolecular level. The duration of the oscillatory regime produced by the back-scattering or "cage" effect is variable, depending on molecular geometry, still generally constrained between 10^{-13} to 10^{-12} s, which defines a second time-scale.

Since here the angular oscillatory regime is set on by local torques whose symmetry and amplitudes are intermolecular properties, the correspondence between static and dynamic structure is straightforward. Somewhat more subtl^ is the feed-

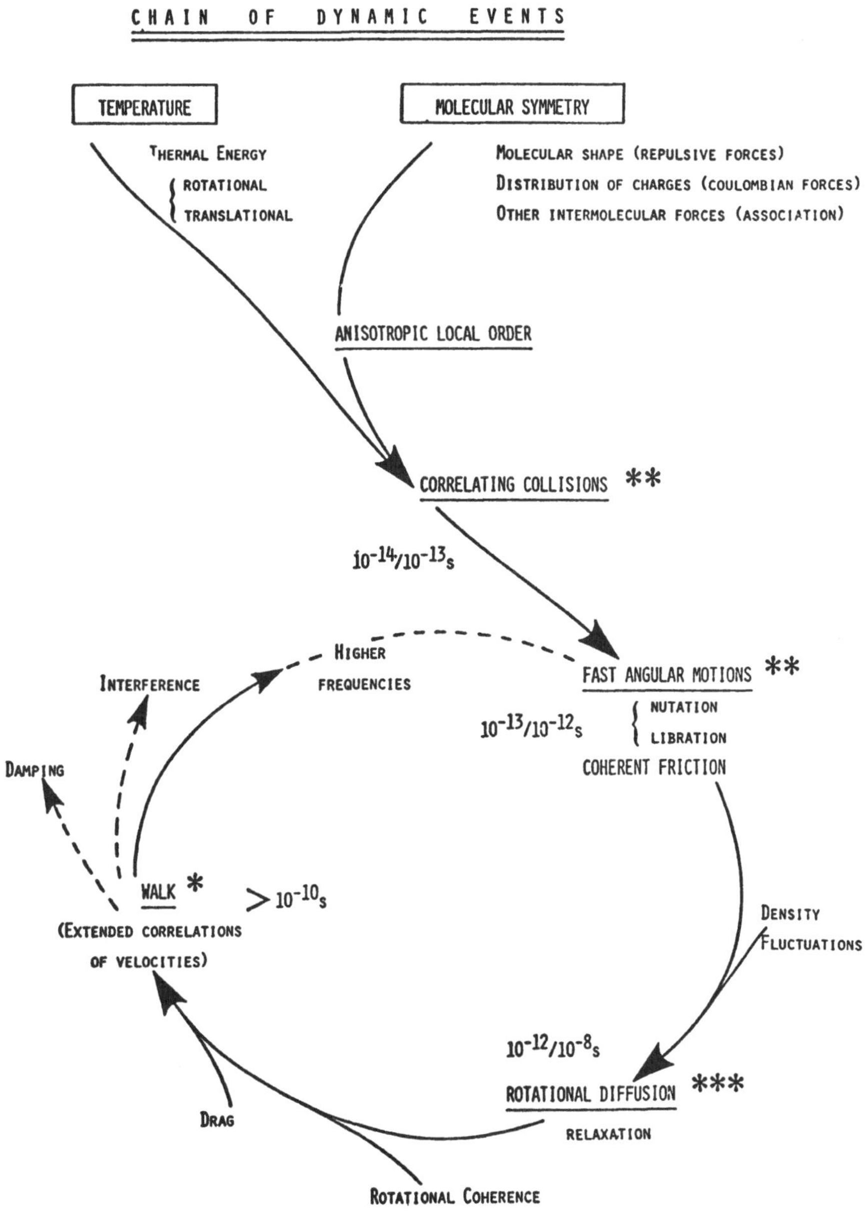
CHAIN OF DYNAMIC EVENTS
TEMPERATURE
MOLECULAR SYMMETRY
THERMAL ENERGY
ROTATIONAL
TRANSLATIONAL
MOLECULAR SHAPE (REPULSIVE FORCES)
DISTRIBUTION OF CHARGES (COULOMBIAN FORCES)
OTHER INTERMOLECULAR FORCES (ASSOCIATION)
ANISOTROPIC LOCAL ORDER
CORRELATING COLLISIONS **
10⁻¹⁴/10⁻¹³s
HIGHER
FREQUENCIES
INTERFERENCE
FAST ANGULAR MOTIONS **
10⁻¹³/10⁻¹²s
NUTATION
LIBRATION
COHERENT FRICTION
DAMPING
WALK *
> 10⁻¹⁰s
(EXTENDED CORRELATIONS
OF VELOCITIES)
DENSITY
FLUCTUATIONS
10⁻¹²/10⁻⁸s
ROTATIONAL DIFFUSION ***
RELAXATION
DRAG
ROTATIONAL COHERENCE

Figure 1.

back influence of the librational motion upon static structure. It calls upon the concept of active friction through which the alignment of molecules is favoured according to their own symmetry characters. The concept of generalized friction will be developped into more detail hereafter in section IV.3.

The next angular displacement of molecules is the rotational diffusion of larger amplitudes, that we usually label "spinning" or "tumbling" motions, following the symmetry of the rotation involved. Spinning is the motion around the principal axis of inertia - which generally happens to be the axis supporting the dipole moment - and tumbling is the motion reorienting this same axis.

Distinct characters of the static structure enter the specification of this more complex motion. Same as for librational oscillations, the symmetry of tumbling reorientations proceeds from both the symmetry of the intermolecular torques and the symmetry of the molecules themselves through the values of their distinct molecular moments of inertia. The most prominent role of inertial factors, and more recently that of torques too, have been made especially clear by W.A. Steele (1). Apart from these molecular and local factors, collective properties of the liquid enter the dynamics through the rate of reorientation. This rate itself is a combination of two factors, generally undistinguishable in the observed mean value : these are the velocity of the angular rotation and the rate of occurence of elementary rotations. Hereafter in section V we will come back to experimental approaches specifying the rate of occurence dependence upon the liquid equation of state (i.e. its temperature and density dependence).

Through transverse viscous coupling the rotational diffusion is able to generate slower and more collective modes on which we are lacking detailed knowledge, and that we just call "walk". The finding out of such modes through molecular dynamic simulations and theoretical essays (2) suppose that they possess a cyclic character, and we may best imagine them as microscopic vortices. As for the experimental manifestation of walk it must be stressed that direct evidence is still lacking, and for that reason we cannot comment very much on this most interesting process. Its actual existence is still controversial (3). Ferhaps a convincing enough argument may be found in the fluctuation-dissipation correspondence principle that links the fluctuating forces at high frequencies with the particle velocities over long periods of time (4). If constraints exist on the high-frequency side, such as oscillational coherence, they shall induce a counterpart on the hydrodynamic motion in which we expect correlated motions to appear. Conversely molecules undergoing extended drags share their momentum with neighbours inducing high-frequency motions. As for static structures, a walk being an extended in time process, its static parallel would be an extended in space correlation. Such are known to exist in somewhat pathological situations such as the approach of critical points. However there is poor evidence from our current experimental knowledge that long range order extends throughout the normal fluid phase.

In the schematic representation of figure 1 the interrelation of dynamic processes at different time scales is clearly signified through a chain. In the picture of this chain we have labelled with three stars those processes that can be considered as well documented - although not necessarily completely understood - while two stars only are assigned to processes where desirable progress is needed, both in experimental and theoretical descriptions. Eventually one star alone is the mark of a poorly documented process from which we have still to learn even about its main characters : symmetry, space and time extension, creation and damping out.

Now we proceed to study into more detail librational interaction-induced motions. Relaxation processes will be more succintly treated in section up, since they are adressed in greater detail elsewhere in this course. In our last section we will touch on some coherent features of the rotational diffusion arising from strong dipolar interactions. No attempt is done to speculate on manifestations of the walk process since it seems unlikely that dipolar spectroscopy will ever be a convenient tool to detect slow currents as long as we dont know how the collective characters of slow motions could be isolated from the total absorption pattern.

III. LIBRATIONAL ABSORPTION

We are using the term librational absorption as the name for a complex spectroscopic feature that displays concomitant dispersion, and that arises from damped librations as the dominating process but still incorporates altogether features from additional processes. Other authors are employing the more general term of "interaction-induced" absorption, however here it is preferred to be more specific by putting emphasis upon the angular oscillatory regime. The term "Poley" absorption is also current practice, still, again, we prefer to keep refering to the distinctive physical mechanism implied.

III.1. Experimental method

To introduce the presentation of some main spectral features and their significance, let us briefly sketch the experimental method. Conceptually it is certainly one of the most simple measurements that can be devised. A liquid sample is being confined between two windows. From one side electro-magnetic radiation is shined, either monochromatic or not, depending on the type of spectrometer used (conventional or Fourier-transform). After having travelled through the sample, the e.m. radiation is focussed onto a detector. Loss of energy in the sample is characterized through the absorption coefficient $\alpha = -(1/d)\ \mathrm{Ln}(P/P_0)$. Here P is the power transmitted through the sample of thickness d, while P_0 is the power transmitted at d equal to zero. Shifts in the wave pattern are also measured against different sample thicknesses : if Δx is the shift corresponding to samples differing in thickness by Δd, the value of the refractive index is $n = 1 + (\Delta x/\Delta d)$. Such conceptually simple experiment can be performed as well in the far infra-red (FIR) and in the microwave "free space" spectroscopies.

It must be clear that in such experiments the e.m. radiation never exercises a driving force setting the molecules in rotation, but just acts as a probe coupling with the natural spontaneous fluctuations of orientation of the molecules. The external field always brings a weak perturbation as compared to the intermolecular torques at work, whatever the intensity shined with conventional sources : the time-dependent response do not vary with fields intensities. Under such conditions, if non linear effects occur they are resulting from intermolecular forces themselves : e.g. the forces at work to create local alignment effects in the field of neighbouring molecules, involving molecular hyperpolarizabilities.

Below we proceed to review the main features of librational absorption bands with some examples taken to illustrate the meaning of band frequencies, band widths, band structures and band intensities.

III.2. Band frequencies

Bands are peaking at frequencies lying generally between 30 and 100 cm^{-1}. As an immediate observation the bands look reminiscent of the well-documented librational absorption in the solid phase. Getting deeper into the study confirms the damped librational character of the molecular oscillations within the liquid's local structure. Thus the spectrum can be understood as a manifestation of the elastic limit in the viscoelastic behaviour of liquids, with the bands central frequencies being ruled by the oscillator strength in the local environment. A low-frequency limit is given by the free rotator frequency $\omega = (2kT/I)^{1/2}$ which is approached at the dense gas limit, while a high-frequency limit is the crystalline mode's frequency observed in the solid phase. Between these two extremes, the central frequency can be identified and studied against the parameters of the equation of

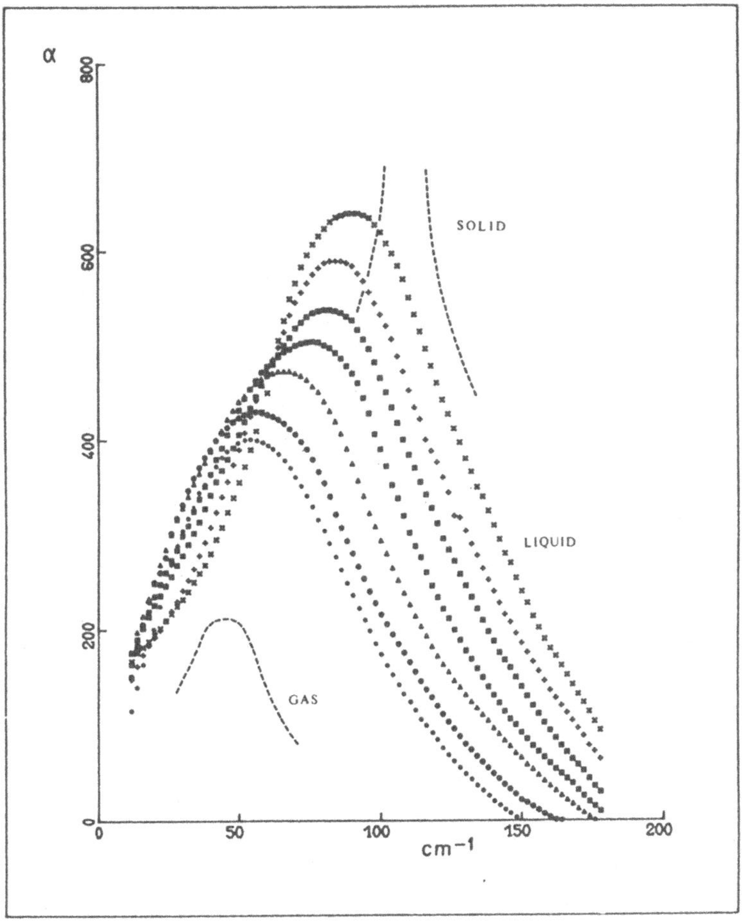

Figure 2. FIR spectra of liquid methyl fluoride. Temperatures and densities follow orthobaric conditions from 133 K (right) up to 293 K (left).

state such as temperature and density. Another possibility of varying the frequency is given by mixing with different solvents, the resulting band shifts depending crucially on the extent of the local perturbations produced through dilution.

Many illustrations of the band frequency dependence upon local environment have appeared in the literature (5). Some examples are reported below.

On figure 2, from reference (6), a shift towards low frequencies is observed as the temperature of the liquid methyl fluoride is increased gradually along the liquid-gas coexistence curve (i.e. under orthobaric conditions). Here the density effect is essentially responsible for the frequency shift following the loosening of local structures.

Dilution of acetonitrile in different solvents have been reported in references (7), demonstrating both temperature effects and dilution effects on the band frequencies. In these examples the replacement of CH_3CN molecules by softer solvent molecules decreased the amplitude of the intermolecular torques, weakening correspondingly the oscillator strength and producing a shift towards low frequencies.

Eventually, through applying gradually increasing pressures up to several kilobars upon the liquid samples, the opposite situation has been observed (8). Tightening of the local structure produced appreciable shifts towards high frequencies in both liquids chlorobenzene and carbon disulfide.

III.3. Band widths

According to the specific nature of distinct liquids very different patterns are observed being generally an interplay between librational absorption and other processes.

An example is given in reference (9), where a broad absorption in H-bonded alcohols is compared with the band of chlorobenzene. In such extreme situations as those displayed in liquids alcohols, there is little chance of ever extracting successfully the librational features from the total broad absorption patterns. But for simplest liquids such as chlorobenzene, and still more for other compounds possessing higher volumic dipole moments and not involved in complexation nor association, bandshapes are fairly attributable to librational motion. In such favourable cases the analysis of bandshapes puts forward two broadening processes arising from both spatial and temporal fluctuations of local order.

Spatial fluctuations are local inhomogeneities of structure distributed in the bulk. As the local structures are more or less loose, the oscillator strength varies correspondingly, just in the same way as it varies when the bulk density is changed in experiments. These inhomogeneities last for longer than a few libratory periods, introducing a distribution in the librations frequencies about the central frequency.

Temporal fluctuations originate from the damping out of the librations after the molecules had oscillated a few times within their environment. This lifetime plays a role in the line-width of the absorption, a narrow line meaning a long duration while a broad line means a short-lived process.

It is important to understand more deeper these two effects and to analyze their respective contributions, since the duration of oscillations is very likely to correspond to the lifetime of the local microstructure itself. This is an example where research adresses at the same time static structures and dynamic structures problems.

III.4. Band structures

In most of the FIR absorption patterns it will be quite difficult to distinguish between different processes occuring in the frequency range. In some particular situations however a structure is clearly visible and in some others the main contribution will dominate so largely over others that it can be adressed most of the absorption features. The most widespread case is of course the intermediate one, but since it cannot be analyzed in any general form we dont attempt to discuss it here. As examples of well-enough resolved structures many examples may be found in solutions. Figure 3 refers to HCl dissolved in fluid argon (10). Other examples of band structures are found in liquid crystals (11). In liquid acetonitrile self-complexation has been reported to exhibit well-resolved FIR band structures (12). Eventually more complex molecules may possess internal modes (torsional vibrations, ring inversions) with spectral densities in the FIR adjoining librational absorption. A typical example is shown for cyanopropyne (13) on figure 4.

Apart from these particular additional processes their are at least two mechanisms that always contribute in the absorption spectra. The "collision-induced" contribution arises from the rotational modulation of transient dipoles during collisions. The charge distribution of one molecule is distorted by the electric field from the charge distribution of its collision partner. The resulting induced dipole moment is varying within the time scale of collisions, that is much faster than the molecular motions themselves, so that it contributes higher frequency absorption to the permanent moment's rotational absorption spectrum.

Another induced contribution arises from environmental

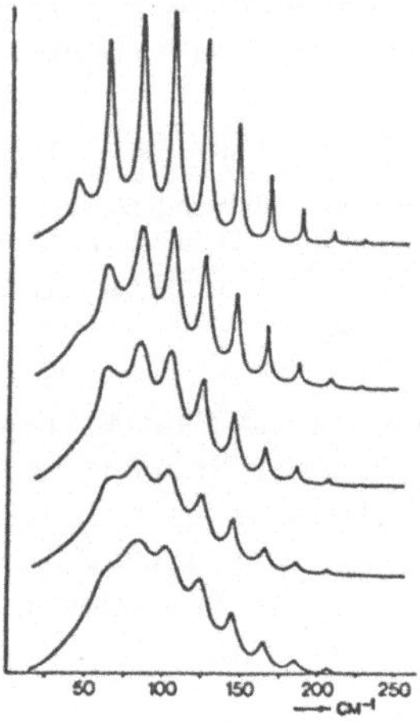

Figure 3. FIR spectra of HCl in Ar at increasing densities from upper to lower curves (from(10)).

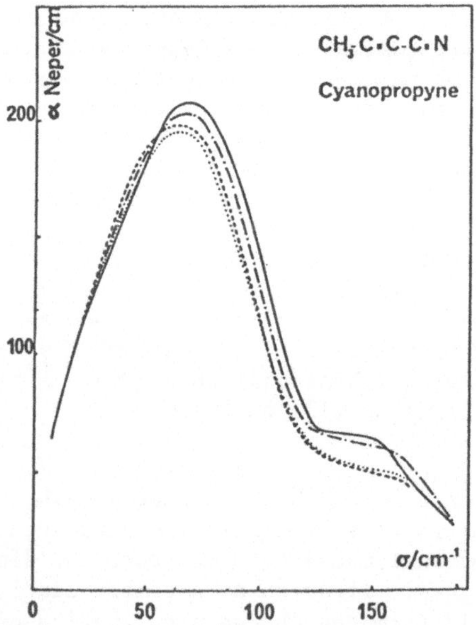

Figure 4. FIR absorption of liquid cyanopropyne (13). Four temperatures shown from 293 K (upper curve) to 343 K (lower curve).

anisotropy which we term "structure-induced" absorption. The induced moments, or induced increments to the permanent moments, behave actually as would permanent moments for the duration of local structures lifetimes. In this way the structure-induced moments follow the librational motions, contributing the FIR absorption spectrum in the same way as the permanent moments do.

It follows that in the case of dipolar compounds, and especially with highly dipolar molecules, we have to deal with experimental profiles including three terms : permanent dipolar absorption, structure-induced dipolar absorption and collision-induced dipolar absorption. As their separation is always something of a puzzle it should tentatively be avoided : this may be indeed possible if one absorbance dominates strongly over the others. As far as the demand of rotational dynamics studies is to deal with an information clear from non-rotational processes, librational bandshapes may carry such information provided collision-induced absorption remains a negligible part in the total absorption.

Examples of spectra dominated by permanent dipolar absorption have been reported above for liquids methyl fluoride (6) and acetonitrile (7) , or can be found in the literature for other small highly polar molecules (6).

III.5. Band intensities

According to the relevant sum rules the total integrated intensity of dipolar absorption bands is proportional to the squared dipole moment (14). The experimental zeroth moment of the dipolar absorption per molecule measured in FIR is :

$$A_{exp} = 9\, n\, (V/N)/(n^2 + 2)^2 \int_0^\infty \alpha(\nu)\, d\nu \qquad \text{III.5.1}$$

including the internal field correction, while the sum rule predicts :

$$A_{th} = 2\pi\mu^2 /3c^2 I \qquad \text{III.5.2}$$

Here n is taken after completion of the relaxation process ($n = \varepsilon_\infty^{1/2}$), V/N is the ratio of volume to number of particles, ν is the frequency, μ is the dipole moment, c is the light velocity and I is the moment of inertia, $I = I_x = I_y$ for symmetric top molecules with the dipole moment lying along Oz.

Comparison of these two expressions applies to experimental absorption provided it has been measured over the whole frequency range, from zero frequency up to beyond the FIR bands. Generally, appreciable differences appear between experimental values from (III.5.1) and theoretical estimates from (III.5.2), posing the problem of the correct value of μ to insert in the sum rule. It seems the most sensible to take μ as being the sum of permanent and structure-induced moments. However since there exist no systematic way of separating induced moments from either structural or collisional origins, that problem may remain unresolved, unless collision-induced absorption be negligible as compared with the two other contributions. Thus we are sent back to such examples as already quoted at the end of former subsection III.4.

Careful examination of measurements carried out with highly polar compounds demonstrated that the effective moments entering the sum rules may differ from the permanent moments measured in vacuum by a few percent up to 20 % (6,15,16). Still it isn't possible to assess exactly how long-lasting are the increments of moments entering the dipolar absorption : we just know that these are not intrinsic molecular properties but structural properties. More progress is very much needed to understand better this interplay of effective dipoles and local structure.

An employment of effective dipoles determined from FIR bands has been proposed to enter the computation of static angular correlation factors, the Kirkwood "g" factors. In the modified formula the integrated band intensity enters instead of the usual dipole moment value determined in vacuum (17). There, a dynamical property is being used to understand more closely the static structure, as controled by the angular position of neighbouring molecules with respect to each other.

III.6. Band moments

Both the shape and the intensity of absorption enter the computation of band moments which are defined as :

$$M(n) = \int_{-\infty}^{\infty} (\omega)^n \, I(\omega) \, d\omega \qquad \text{III.6.1.}$$

Here the index n refers to the order of the moment, $I(\omega)$ is the absorption lineshape, proportional to ε''/ω, $\omega = 2\pi\nu$ is the frequency and ε'' is the susceptibility related to the absorption coefficient by $\alpha(\omega) = \omega\varepsilon''/nc$. As a consequence of $\alpha(\omega)$ being proportional to ω^2 times the intensity entering the definition (III.6.1), the nth moment of $I(\omega)$ is in fact the $(n-2)^{th}$ moment of the FIR band $\alpha(\omega)$.

The importance of computing the moments of experimental bands comes from their connection with molecular properties through statistical averages, as was initially remarked by Gordon (18). The fluctuation-dissipation theorem establishes the correspondence between absorption lineshapes $I(\omega)$ and the angular positions time-correlation function $C_1(t)$:

$$C_1(t) = \langle u(0).u(t)\rangle = \int_{-\infty}^{\infty} e^{i\omega t} I(\omega) d\omega = \Sigma_{n=0}^{\infty} \frac{(it)^n}{n!} \int_{-\infty}^{\infty} \omega^n I(\omega)\, d\omega$$

The Taylor expansion of $C_1(t)$ is

$$\langle u(0).u(t)\rangle = \Sigma_{n=0}^{\infty} (t^n/n!) \left| \frac{d^n}{dt^n} \langle u(0).u(t)\rangle \right|_{t=o} \qquad \text{III.6.2.}$$

Thus the moments of the frequency spectrum may be identified as coefficients in the time power series of $C_1(t)$. Results for the first coefficients have been given for different molecular symmetries (1c). If $\langle N^2\rangle$ is the mean squared torque associated with rotation of the symmetry axis, one has :

for linear molecules : $C_1(t) = 1 - (\frac{2kT}{I})\frac{t^2}{2} + \left\{8(\frac{kT}{I})^2 + \frac{\langle N^2\rangle}{I^2}\right\}\frac{t^4}{24} + \ldots$

for symmetric tops : $C_1(t) = 1 - (\frac{2kT}{I})\frac{t^2}{2} + \left\{8(\frac{kT}{I})^2 + 2(\frac{kT}{I_z})^2 + \frac{\langle N^2\rangle}{I^2}\right\}\frac{t^4}{24} + \ldots$

for spherical tops : $C_1(t) = 1 - (\frac{2kT}{I})\frac{t^2}{2} + \left\{10(\frac{kT}{I})^2 + \frac{N^2}{I^2}\right\} t^4/24 + \ldots$

III.6.3

Computation of correlation functions through theoretical estimates of $\langle N^2\rangle$ is impractical due to the number of terms in the power series required to get accurate functions over the desired range of times. However the inverse procedure of computing experimental mean squared torques values can be achieved through identification of the coefficient of t^4 in expressions (III.6.3) with the fourth derivative of $C_1(t)$ at time zero, computed from Fourier inversion of experimental spectra (19). In this respect FIR lineshapes show a definite advantage as compared with IR (or Raman) spectroscopic lineshapes, since measuring directly the coefficient of absorption $\alpha(\omega)$ weights heavily the high-frequency range, being proportional to ω^2 times the IR lineshapes. The experimental information relevant to $\langle N^2\rangle$ determinations is therefore contained in the 2nd moment of IR lineshapes, which improves appreciably their accuracy.

The importance of measuring torques has been recognized recently (19,20) on the grounds that when real molecules are in close contact, the impulses are finite and exerted over periods of time of a duration comparable with the time interval between them. In such collisions angular momenta are partially conserved through successive rebounds in the local structure. Torques appear to act as correlating forces, exercised for times long enough that the molecules gain dynamical coherence, playing thus the initial role in the dynamic structure. On the opposite, there are good reasons to distrust instant-collisions pictures assuming hard potential models, since the impulse in such collisions would consist in infinite torques exercised over infinitesimal periods of time.

Whenever the time-dependence of the torques indicate correlating collisions, no instant impact model could account for the dynamics, whether for weak or strong randomizing collisions. Generality of these observations has been secured by the large liquid domain investigated in FIR experiments : back-scattering and damped librations remain still detectable at the vicinity of critical temperature (6).

Experimental determinations of mean squared torques values are few. Typical values have been reported in references (19b) and (20).

A remark applies to the experimental values of $\langle N^2 \rangle$ deduced from experimental bandshapes. Since we deal with total (multimolecular) correlation functions, their moments might also contain contributions arising from the intercorrelations of different molecules - or multimolecular effects - since cross terms in the statistical averages do not necessarily vanish. On this account it is necessary to estimate the extent of high-frequency collective effects, if any, whenever torques are to be computed from the spectral moments.

Now, since we are getting increasingly involved in time-correlation functions, we are already in timely position to survey a few aspects of the time-domain representation.

IV. TIME-DOMAIN REPRESENTATION OF ROTATIONAL DYNAMICS

The general formalism of correlation functions has been settled elsewhere in this course. Moreover some very good reviews and textbooks are already available (21). So that we shall keep being specific of just a few statistical functions relevant to discuss experimental power spectra of dipolar rotational origin, especially those functions connected with high-frequency features, i.e. librational absorption.

IV.1. Statistical functions relevant to short-times dynamics

Let us consider the popular "angular positions" autocorrelation function that describes the evolution in time of the angular trajectory of a molecule within the liquid. If the angular positions are defined through unitary vectors $\vec{u}_i$ oriented along the axes carrying the molecular permanent moments, the function is defined through the average :

$$C_1(t) = <\vec{u}_i(0).\Sigma_i\ \vec{u}_i(t)> \qquad \text{IV.1.1.}$$

As far as spontaneous fluctuations of orientation are concerned, that is within the frame of linear response, this function can be expressed in terms of measured susceptibilities respective to any external field coupled with the orientational degrees of freedom. Such is the case for hertzian electromagnetic radiation that couples with the molecular electric moments. Suitable expressions in terms of the susceptibilities have been proposed, duly incorporating a convenient internal field correction. One among them reads :

$$C_1(t)=(2/\pi)(9kTV/4\pi N\mu^2)(\varepsilon_\infty+2)^{-2}\int_0^\infty (\varepsilon''/\omega)\left[\varepsilon_\infty^2/(\varepsilon'^2+\varepsilon''^2)+2\right] \times \cos\omega t.d\omega \qquad \text{IV.1.2}$$

where the Fatuzzo-Mason internal field correction has been retained.

The second derivative of this function is also of great interest with respect to its physical meaning. Its definition is :

$$\ddot{C}_1(t) = C_{rv}(t) = <\dot{\vec{u}}_i(0).\Sigma_i\ \dot{\vec{u}}_i(t)> \qquad \text{IV.1.3}$$

and in terms of susceptibilities :

$$C_{rv}(t)=(2/\pi)(9kTV/4\pi N\mu^2)(\varepsilon_\infty+2)^{-2}\int_0^\infty (\omega\varepsilon'')\left[\varepsilon_\infty^2/(\varepsilon'^2+\varepsilon''^2)+2\right] \times \cos\omega t.d\omega \qquad \text{IV.1.4}$$

It describes the evolution in time of rotational velocities. It also very closely approaches the angular velocities correlation function C_ω and the memory function $K_1(t)$, or generalized friction coefficient of the diffusive rotational molecular motion, a property constituting our main concern in section IV.3 below.

The relationships between $C_{rv}(t)$, $C_\omega(t)$ and $K_1(t)$ deserve some additional comments.

As for their respective definitions, the angular velocity correlation function is

$$C_\omega(t) = <\vec{\omega}(0).\ \Sigma_i\ \vec{\omega}_i(t)> \qquad \text{IV.1.5}$$

where ω is the angular velocity vector.

The function $C_{rv}(t)$ is defined through (IV.1.3) and refers to the vector $\dot{\vec{u}} = \vec{\omega}\times\vec{u}$. Since changes in $\vec{u}$ occur only from changes in $\vec{\omega}$, it may be anticipated that both functions C_ω and C_{rv} shall display common features.

The orientational memory function $K_1(t)$ associated with $C_1(t)$ is defined through

$$\frac{d}{dt} C_1(t) = -\int_0^t K_1(\tau)\, C_1\, (t-\tau)\, d\tau \qquad \text{IV.1.6}$$

The conditions for $K_1(t)$ to coïncidate with $C_\omega(t)$ are generally correctly fulfilled in dense fluids, since they demand that close interactions be frequent (the "fast modulation" condition) and torques be stronger enough than the mean thermal energy : following (22) that condition reads

$$< N^2 > /I^2 >> (2kT/I)^2$$

The fit of both functions together have been checked through careful experimental and numerical experimental works, such as the results shown in figure (5), from reference (23).

The function $C_{rv}(t)$ reduces to any of $C_\omega(t)$ or $K_1(t)$ provided that the rotational velocities decorrelate fast enough as compared to orientations, that is that $C_{rv}(t)$ decays fast enough as compared to $C_1(t)$ (demonstrations appeared in references (14,24). How fast is "fast enough" demand to be examined with real examples, since current rotational dynamics studies adopt either one or the other functions according to their specific approach. For instance real experiments proceed by Fourier inversion of spectroscopic $\alpha(\omega)$ lineshapes, giving directly $C_{rv}(t)$ at the output. On the other hand numerical experiments, model calculations and theoretical approaches adopt preferentially $C_\omega(t)$ since it carries a more staightforward physical meaning.

Comparisons of C_ω or K_1 with C_{rv} have been achieved through separate computations of the functions derived from FIR experimental data covering extended liquid domains (19b). Figure (6) illustrates such comparison. It can be observed that both functions $K_1(t)$ and $C_{rv}(t)$ merge at high densities and move gradually away from each other as the low-density limit is approached, still keeping identical oscillatory periods. The conclusion drawn from such studies is that under fast modulation conditions all three functions are either indistinguishable or very close to each other at short times, and in any case they always carry the same information with concern to the molecular angular velocities.

IV.2. Connection with experiment

Fourier inverting spectra following (IV.1.2) and (IV.1.4) is the route towards drawing time-dependent representations of molecular orientations and rotational velocities. Still

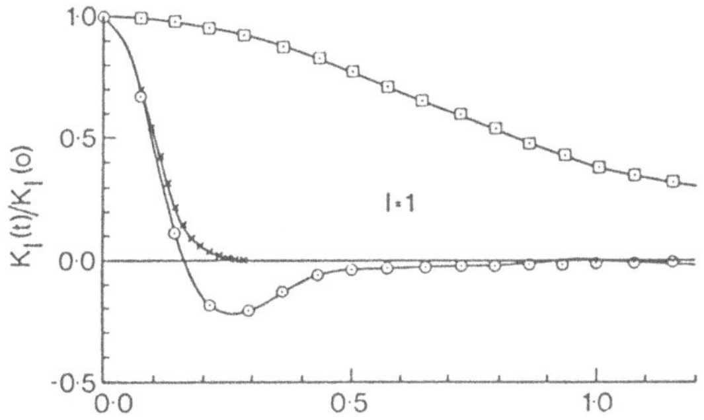

Figure 5. Reduced memory function $K_1(t)/K_1(0)$ (dots) compared to $C_\omega(t)$ (full lines) from simulation of liquid CS_2 (lower curve) and linear free rotor (upper curve).

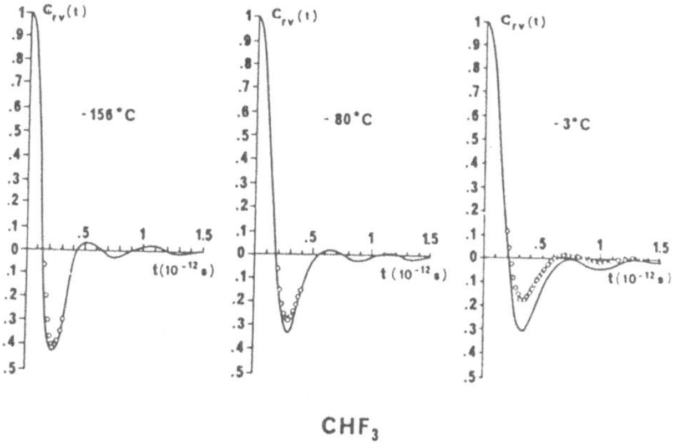

Figure 6. Rotational velocities correlation function $C_{rv}(t)$ (full lines) compared to $K_1(t)$ (open circles), computed from experimental FIR bandshapes in liquid fluoroform.

two questions arise before looking condifently at such functions :

1) Are actually experimental results from macroscopic sample representative of motions at the molecular scale ?
2) What are the demands that experimental spectra must satisfy so as to ascertain that they probe purely rotational dynamics ?

The first question is of course that of the dielectric theory's reliability. It is a delicate question and the appreciation on it can differ according that the approach proceeds from either a theoretical or an experimental viewpoint. As stated in the first sentence of this review, our own approach is experimental, and this is the moment to adhere firmly to our initial choice. Therefore controversial discussions on the preference for one or another local field corrections, or on the validity of imbedded cavities approaches as compared to lattice approaches, are not examined here. We just stress that very interesting advances have been attained in the last years and still others are in progress at present, admittedly as a consequence of the definite impulse given to liquid theories by numerical simulations. References (25) to (27) are examples of such advances.

Our experimental approach rests on down to earth considerations. For instance, both Cole-Glarum and Fatuzzo-Mason internal field derivations converge to the same expression in the FIR frequency range, as a simple consequence of ε'' being small and ε' close to ε_∞ in that range (14). Elsewhere, the actual differences in correlation functions that might indeed arise from applying either of these two theories have been computed, plotted and compared together (19b). In the most stringent tests, which apply to intense absorptions, deviations were found to affect merely the amount of damping of the oscillatory features in $C_{rv}(t)$ and $K_1(t)$. Elsewhere again, an additional effect such as the retardation expected from dielectric friction is within the limits of experimental accuracies (28). There is little hope to separate it clearly from other processes taking place at the high frequency extremity of the relaxation pattern.

Therefore we consider that careful theories following the Onsager cavity approach are trustworthy as much as current experiments can decide. Besides, indications from recent simulation vork (25) lead to prefer the derivation by Mason and Fatuzzo. Such a preference also meets the outcome of a critical examination from theoretical arguments (26).

As regards the second question, precise criteria are listed below to which spectra must conform so as to be Fourier inverted into rotational time-correlation functions. Strictly, the following requirements are much demanding indeed. They end in limitating severely the choice of suitable liquid compounds in order to fit all the conditions. However sensible corrections are available to deal with actual situations, so that a certain choice is already available from Nature. Moreover, when a definite compound appears to satisfy most of the criteria, it can be exploited under extended conditions, both in the pure liquid domain as a function of temperature and density, and in solution with distinct solvents, under various concentrations. So that eventually the wealth of information comes to be significant.

Requirements for useful spectra are as follow :

i) Band structures should not display others than intermolecular modes from rotational degrees of freedom.

In section III.4 above we have mentioned cases of overlapping contributions from non-purely dipolar origin : association or complexation, molecular internal modes, collision-induced contributions. Occasionally specific corrections can be found to reduce the spurious contributions down to negligible amplitudes in the total absorption.

ii) The frequency range should extend without discontinuity from a few Hz up to completion of the librational absorption process, about 5.10^3 GHz.

The profound meaning of this "completeness requirement" is the built-in interdependence of fast and slow modes in the liquid dynamics. As well, when we call for models to guess plausible rotational mechanisms, the main difficulty is to reconcile conceptual simplicity with the demand that the model be reliable at any frequency of the spectral density, or equivalently at any time of the functions in the time-domain. Whenever the absorption patterns are incompletely fitted no comprehensive understanding of the liquid dynamics can be attained.

iii) Especially when dealing with strong absorption-dispersion amplitudes, both absorption and dispersion patterns must be determined over the frequency range.

This point has been made particularly clear by Fulton (29), who demonstrated both qualitative and quantitative changes that take place when the correlation functions are calculated by taking the refractive index to be constant instead of including the actual variation of refractive index with frequency in the calculation. Furthermore, some statistical functions such as the rotational velocities memory functions alluded below (IV.3.3), have analytical expressions that depend critically on terms like $(\varepsilon'-\varepsilon_\infty)$ which enter the internal field corrections. These terms are very sensitive to the variations of real permittivity since, within the high-frequency range, ε' is very close to and sometimes above, sometimes below ε_∞. This is particularly so for highly absorbing molecular species.

iv) A sensible choice of molecular symmetry leads to select axially symmetric molecules so as to avoid anisotropic reorientations that would interplay in the overall dipole reorientation. Further, small molecules not too much elongated will be preferred, since such geometry is expected to minimize rotation-translation coupling, a process very difficult to account for in the analysis of molecular motions.

Eventually a remark applies to the four conditions above. They refer to simple liquid systems where short-time dynamics play a dominant role wide-spread over all the chain of dynamics events. Some other systems are more sensitive to cooperative processes, plurimolecular and with longer time-scales, and for those systems the short-time behaviour looses part of its importance. Examples are found with heavier molecules or in supercooled liquids. In such systems relaxation studies gain more insight from understanding cooperative dynamics than from focussing on short-times processes.

IV.3. Features of generalized friction

Figure 6 is typical of correlation functions $C_{rv}(t)$ derived from spectra of small highly-polar molecules, such as halogenated methane derivatives. Functions are shown at three temperatures : one close to the solidification, one intermediate, and one close to the critical liquid-gas transition. These functions show oscillating tails which come out as the first manifestations of dynamical order : we have termed this property "local coherence", being still confined to a local surrounding compared to the collective librations observed in crystalline phases. Local coherence manifests that close encounters in a liquid medium produce angular momentum retention, or memory, throughout collisions.

Correspondingly the orientational decorrelations show characteristic pits in the time-dependent patterns at times corresponding to the librational frequency (figure 7).

Now we proceed to examine a consequence of the hig-frequency motion upon the overall dynamics. We shall begin with the remark that functions like $C_{rv}(t)$ play the part of a generalized friction coefficient in a Langevin formalism of rotational diffusion, and therefore the spectrum measured in the FIR is nothing but the spectral density of this rotational friction coefficient.

For a comprehensive account of Langevin formalism we recommend one of the recent textbooks (21). In the rotational diffusion case we recall briefly the relationship

$$\vec{N} = - \xi\vec{\omega} + \vec{n}(t) \qquad \text{IV.3.1}$$

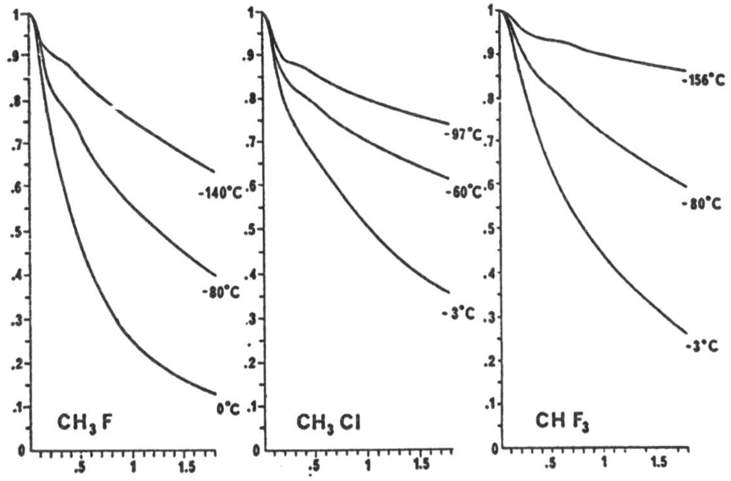

Figure 7. Orientational correlation functions for three liquids at different temperatures of the orthobar (19b).

that links the torque $N = I\dot{\vec{\omega}}$ to the angular velocity ω, $\vec{n}(t)$ being a random torque with a supposedly white noise, the systematic components being included in the first r.h.s. term. This term introduces the friction coefficient ξ as the proportionnality factor between torques and velocities. Physically it possesses a more profound signification as will be seen now.

First, it is necessary to allow for a time-dependence of this coefficient to include the memory effects that are currently observed. The generalized Langevin equation reads then

$$\vec{N} = -\int_{o}^{t} K_1(t-\tau)\ \dot{\vec{\omega}}(\tau)\ d\tau + \vec{n}(t) \qquad \text{IV.3.2}$$

where the kernel of the integral is the memory function $K_1(t)$ very close indeed to $C_{rv}(t)$. The power spectrum of $K_1(t)$ coïncides as well with that of $C_{rv}(t)$, which corresponds to the dipolar absorption spectrum $\alpha(\omega)$.

Thus the spectrum that we designated above as librational absorption finds another physical sense in the time-domain interpretation of molecular motions : it represents the spectrum of molecular friction. To help in visualizing the concept of friction, let us imagine the rotational motion about the dense gas limit. In this particular situation the motion can be depicted as a "collision-limited" process proceeding through successive rotational jumps of significant amplitudes. Collisions themselves do not randomize the angular momenta, on the opposite there is convincing evidence that the angular momentum orientations are well conserved down to dilute gas collisions (30). Getting towards liquid densities strenghtens the local structure's cohesion which in turn produce the splitting of rotational tumbling motion into relaxational plus frictional components. Time-dependent friction appears naturally so as an accompanying process of relaxation. As regards its connection with local structure characters, time-dependent friction acts as a transducer of local structural constraints into dynamical ones, by the bias of forces and torques setting on the molecular velocities along distinct channels.

The relation of friction with diffusion is however a complex one and closely dependent on the relative location of the Fourier components of both frictional and diffusional modes, a situation depending specifically on the liquids under consideration. Fast rotators will show close domains, even merging together at the low-frequency extremity of the liquid phase. More viscous liquids will show well-separated diffusional and librational domains, the coupling of both modes being correspondingly looser. Such a phenomenological diversity explains that the exact mechanism by which friction drives into their way displacements of various amplitudes cannot be ascertained by the sole analysis of high-frequency motions.

As an illustration, let us consider two indications from the analysis of such functions as $C_{rv}(t)$ or $K_1(t)$ showing the time-dependence of friction.

i) Friction is not to be considered as a purely resistive process slowing down displacements. It clearly acts differently according to the frequency of the motion it accompanies.

ii) The oscillating molecular process is activated by thermal energy, but it is also entertained by the feed back collective - low frequency - motions into high-frequency modes.

The first of these remarks is an obvious consequence of the generalized friction patterns to be sometimes positive - corresponding thus to a resistive action - and sometimes negative, with an action that pushes and accompagnies the travelling molecules. Since both of these actions are anisotropic, they put a constraint on the diffusional processes, not only on the limited reorientational or translational steps, but further on the slower currents that may originate from transverse coupling. Here, fluctuation-dissipation laws impose that symmetry constraints ruling high-frequency local fluctuations be further met again in the low-frequency collective processes (4).

Some positive indication of the coupling of slow motions with fast ones has been found by pushing the analysis of memory functions one degree further (19a). The second order memory function $K_{rv}(t)$ is closely related to the torque correlation function (31). It is defined through :

$$\frac{d}{dt} C_{rv}(t) \doteq - \int_0^t K_{rv}(t) . C_{rv}(t-\tau) . d\tau \qquad \text{IV.3.3}$$

This function, although a more complicated object whose derivation from experiment and subsequent interpretation are not straightforward, displays a "long time" component superimposed on the short-time one, revealing the belonging of molecular motions to diverse time-scales.

The physical insight contained in features such as a plateau or a long-time tail in memory functions is uncertain since it only means that slow variables are coupling with the fluctuating "random" torques, the last terms in the r.h.s. of Langevin equations (IV.3.1) or (IV.3.2). Therefore further assignments of a precise nature to the long-lived forces and torques at work are generally difficult to make. Nevertheless the fact itself is an evidence that fast and slow motions are interdependent and it overrules such theories assuming a gaussian white noise for the fluctuating torques. On the other hand attempts to incorporate the coupling through non-linear Langevin equation are rather unorthodox and likely to bear logical mistakes (as discussed in reference 21b). So that it seems most sensible to restrict the memory function analysis rather than to get involved into meaningless methods, despite of the appeal of a red-shifted noise.

V. TUMBLING RELAXATION OF SIMPLE MOLECULES

V.1. Outline

In the chain represented on figure 1 rotational diffusion occupies a central place and is labelled as a well-documented process. Both on theoretical and experimental grounds it has been extensively investigated for more than 50 years. However, despite such a respectable history and of the names of Debye, Onsager and Kirkwood being attached to its description, the problem of dielectric theory is still basically unresolved. Theoreticians have to face the extreme complexity of N-body interactions including both short-range and long-range forces and they cannot expect to receive much help from the considerable wealth of experimental information available. The reason for that lies in the insensitivity of the Markov processes occuring in rotational diffusion as regards the exaçt molecular motions. Experimental data cannot provide but relaxation rates and activation energies, which are very indirect measurements of the detailed microscopic motions.

At present classical textbooks offer wide-spread panoramas of the relaxation characters of solids, liquids and gases, as well as tools for their interpretation at the molecular level (32). However the detailed mechanism through which reorientations occur is still a matter of investigation, requiring the combination of different experimental methods. Furthermore there is no general description of one precise mechanism that would cover as well relaxation in liquids of fast molecular rotators (e.g. simple quasi-spherical molecules) and more complex processes involving cooperative reorientations as encountered in viscous liquids, or supercooled liquids, or liquid crystals, or more generally whenever the constituent molecules display pronounced anisotropies. The role of dynamical intercorrelations prevent to ascertain such a general mechanism : the degree of rotation-translation coupling and that of cooperativity implied in molecular reorientations both contribute the rate of occurrence of rotational steps, the amplitude of the steps and their preferential development. Interesting attempts have been made to design suitable parameters that would take into account the anisotropy and the multimolecular aspects, through specifying molecular boundary conditions of "stick" or "slip" (33,34), and relating "single particle" to "many particles" correlation times(27,35,36).The original Debye approach, which emphasizes the roles of microviscosity and molecular inertia, has also received further support from the recent discovery of the extended validity of the hydrodynamic description down to molecular scale (2,33). Still we are not in a position to succeed in predicting either the details or the rates of the relaxation process in a definite liquid, given the molecular parameters and relevant intermolecular potential details, such as the torques.

To illustrate the screening of molecular reorientation details, as a consequence of relaxation obeying Markov stochastic laws, we give hereafter the example of a well-behaved relaxation obeying Debye equations all over the liquid phase domain, in spite of its proceeding from different mechanisms according to the temperature and density conditions.

V.2. Rotational steps along an orthobaric curve

So as to find out the rotational step amplitudes, it is possible to take advantage of probes which are sensitive to either the dipolar axis reorientation or the polarizability tensor reorientation. The order of the Wigner functions of the Eulerian angles being respectively l=1 and l=2, it is common practice to specify the corresponding correlation times as τ_1 and τ_2. Since rotations of 68° and 41° are required for the decay functions to fall by 1/e, τ_1 and τ_2 refer to the time it takes a molecule to turn through 68° and 41° (37).

For jump processes the ratio τ_1/τ_2 approaches unity while for small angles rotational diffusion the limit amounts to 3. For free diffusion the ratio approaches the value 1.7 corresponding to free rotation.

Actually, values ranging from 1 to 3 has been found in different liquids. Moreover, varying continuously the orthobaric conditions of the liquid state for one selected compound may allow to observe a definite change, so that assignments to different mechanisms can be made. As an example the comparison of τ_1 with τ_2 for liquid OCS under orthobaric conditions is shown on figure 8.

The ratio τ_1/τ_2 evolves from about unity at the vicinity of the triple point up to values of 2.5 at intermediate densities, and then decreases again at the approach of the liquid - gas critical transition. Such an evolution suggests that the mechanism itself undergoes a gradual change from rather large-angle reorientations to more frequent moderate-size rotations and finally to quasi-free rotational diffusion. In that frame the loosening of static structure is the companion process of the gradual change in rotational steps. Large-angle reorientations at the low-temperature end are reminiscent of a solid-like process, not surprisingly since OCS is a linear molecule presenting sterical hindrance to rotation. However since density fluctuations implied in such large steps are relatively rare, that process begins to compete with rotational diffusion by small angles as soon as the mobility of molecules increases in the fluid phase. After rotational diffusion has become the dominant process, the size of angles begin to increase while the density decreases, as also inferred from the FIR bands shift to low frequencies . Eventually a situation comparable to the dense gas collision-interrupted rotation is set at the close approach of the liquid-gas critical transition.

Anyone who is familiar with comparisons of correlation times from different Wigner functions orders understands that such comparisons are subject to certain restrictions, since gene-

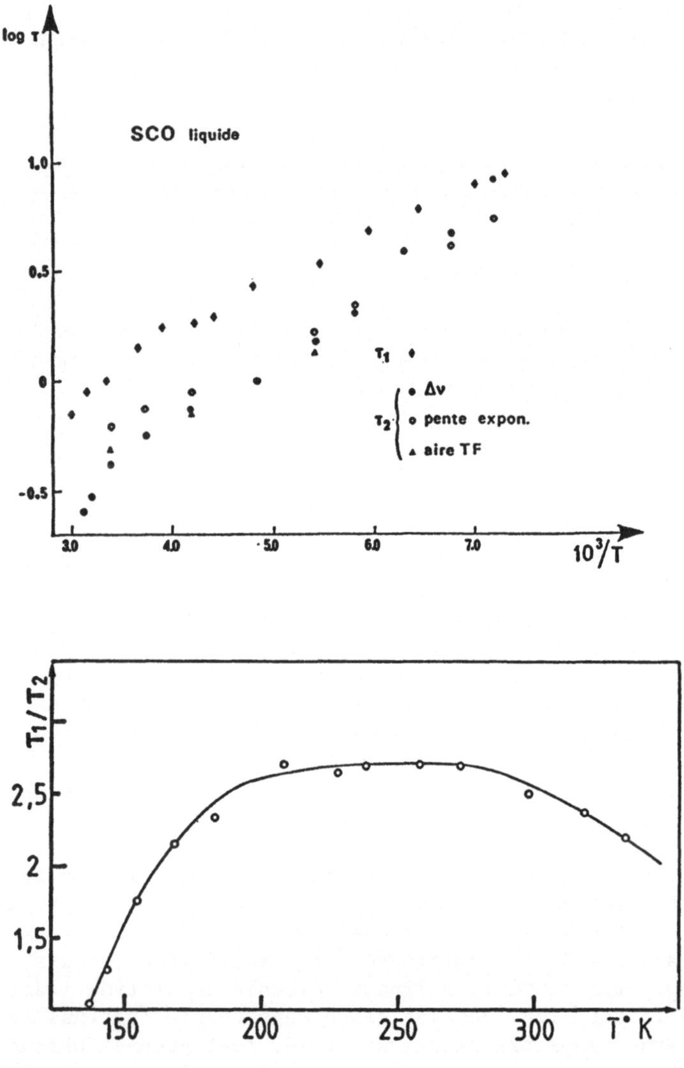

Figure 8. Microscopic correlation times τ_1 and τ_2 and ratios τ_1/τ_2 for liquid OCS. τ_1 from dielectric relaxation (38), τ_2 from Raman (J.P.Perchard and A.Gerschel, unpublished, and (39)).

rally they cannot be obtained through a single experiment, nor event with the same experimental probe. To our current concern, while dipolar absorption gives the multimolecular time τ_1, Raman rotational broadening give the single-particle time τ_2. Correspondingly the absolute values of τ_1/τ_2 considered above should be corrected for short-range cross correlations entering τ_1. It is likely however that in liquid OCS such corrections would barely alter the values of the ratios τ_1/τ_2,and even less their evolution along the orthobar : measurements of the Kirkwood correlation factor under the same conditions have given values almost constant ranging between .8 and .9 (40). Although we do not accept this factor g as an exact measure of the dynamic correlations, recent works have established that the degree of the motion's cooperativity seems to parallel the g factor's departure from unity (23,25).

Most remarkable is perhaps the simple observation that the rotational process preserves its definite relaxational plus librational character over all the temperature and density range. Still more, the plot of Lnτ versus 1/T is a straight line under extended conditions ranging up to a few degrees beneath the critical temperature. It is interesting to note that such behaviour appears to be quite general. Experimental work measuring relaxation rates along the liquid orthobars always conducted to similar observations, whatever the own geometry or polarity of the constituent molecules (6,40). The liquid-gas critical transition has also been investigated in detail without showing rotational absorption anomalies, neither in the microwave nor in the FIR spectral ranges. So that the inflexions in the plots of Lnτ versus 1/T appearing close to the critical transition may be ascribed to a density effect, since when the orthobar critical extremity is approached small changes in temperature induce increasingly large variations in the density.

Before leaving this brief survey let us bring forward a somewhat more general appraisal. Dielectric relaxation studies in simple liquids are difficult and lengthy to achieve, especially when dealing with small and highly polar molecules. But relaxation plays such a central and regulating role in liquid dynamics that its understanding is essential to further advances. In other respects τ_1 multimolecular determinations from dielectric relaxation have the merit of being unequivocal, not convoluting with vibrational relaxation as in spectroscopic bandshapes, nor mixing vith translational information as in neutron inelastic scattering. They also are necessary to understand some high-frequency features which are measured through dipolar absorption-dispersion spectroscopy, such as FIR librational absorption and a more recently discovered high-frequency coherent effect. A brief presentation of this new feature will now conclude this course.

VI. COHERENT ROTATIONAL MODE

Processes occuring at the frontier between relaxation and libration are a challenge to experimentalists. As for dielectric theory, their accurate knowledge would help a lot discriminating between different effects originating from short-range or long-range forces. For instance, the power spectra of dielectric friction and dipolar plasmons are expected to fall in that range, while it is well-known that Debye description of relaxation fails there, not accounting for inertial effects and librational modes (32d, 41,42).

The difficulty resides in the analysis of the absorption pattern when it incorporates mixed contributions from various processes, none of which being ascribed to a precise spectrum in that range. In the examples of liquid chloroform and chlorobenzene that have been for long studied by different authors with the best available accuracies (9,43,44), librational absorption begins before the relaxation spectrum had completely decayed on the high-frequency side. Moreover collision-induced absorption contributes significantly. So that any analysis is bound to be tentative, and further no clearcut additional effect can be detected in the range. Fortunately, systematic investigations on liquids of small highly polar molecules turned out to be rather good so as to analyze the distinct processes. Intense librational absorptions are peaking at high frequencies. On the low-frequency side relaxation processes conform Debye equations without distribution parameters all over the orthobaric domain. In that frame a careful study of the intermediate range has displayed an additional absorption with characters of a resonant process. Since the progress is a recent one it deserves some detailed specification. In counterpart the interpretation is still in its infancy, so that clever explanations can compete to enter the final solution. Hereafter are reported the facts concerning liquid methyl chloride.

Investigations consist of three stages.

First comes the determination of the dielectric relaxation parameters entering the Debye equations :

$$(\varepsilon^{*}-\varepsilon_{\infty})/(\varepsilon_{0}-\varepsilon_{\infty}) = \left[1 + (i\omega\tau)^{1-\beta}\right]^{-1},$$

where

$$\varepsilon^{*} = \varepsilon' - i\varepsilon''$$

Measurements give ε_0, and as many couples of $(\varepsilon',\varepsilon'')$ values as set-up operating frequencies available. Extrapolations allow to infer ε_{∞} values taking a lot of precautions (see references (17), (45) or (46)). Relaxation times τ can be calculated either from the complex plane representations $\varepsilon'' = f(\varepsilon')$ by taking an average value from the determination at each frequency, or through fitting the absorption and dispersion curves to the experimental points. As a result the Debye equations are seen to fit very well the measurements from low to medium frequencies up to about 50 GHz. Beyond that, experimental points move away from the theoretical lines.

Then come the absorption coefficient determinations from FIR spectroscopy. There the absorption pattern is directly obtained as a continuous line against frequency between 15 and 200 cm^{-1} (450 to 6000 GHz). Since refractive index determinations would require a special assembly a fitting procedure is being used so as to extract the refractive index variation along the absorption band (47). The same fit allows to extrapolate the absorption from purely librational origin down below 15 cm^{-1}. It is accepted that in liquid methyl chloride librational absorption from permanent dipolar origin dominates over collision-induced absorption by two orders of magnitude, ensuring that the basic process entering bandshapes is actually a librational one.

Last, a millimeter wave interferometer fills the gap between 50 to 600 GHz, operating with continuous coverage of the frequency range. Both absorption coefficient and refractive index determinations are available with very good accuracies (44,48).

Departure from simple relaxation is illustrated on figures 9 and 10, refering to absorption and dispersion under normalized representations, at three temperatures selected along the orthobar. Two additional processes are seen to occur successively, the first being located beyond $\mathrm{Ln}\nu \simeq 1.8$ (ν= 63 GHz) as well in absorption and dispersion patterns, and the second being the librational absorption lying in the FIR about $\mathrm{Ln}\nu \simeq 3.5$ (ν=3000 GHz, $\bar{\nu}$=100 cm^{-1}). In the following we refer to the first feature as a rotational coherent process since it implies correlated rotations.

Contrasting with the temperature dependent relaxation process, both correlated rotations and librations are located at almost invariable frequencies, which is the signature of a resonant process. In the complex plane representations of $\varepsilon''(\nu)$ versus $\varepsilon'(\nu)$ (Cole-Cole plots) and of $\alpha''(\nu)$ versus $\alpha'(\nu)$ (Scaife polarizability plots (49) the high-frequency lines get still more clearly separated from relaxation (figure 11 and (46)).

Similar evidence has been attained in liquid fluoroform (CHF_3), another compound of small, quasi-spherical, highly polar molecules, while experiments failed to detect any additional features in the spectra of less polar species : CH_2Cl_2, $CHCl_3$, CH_3I and C_6H_5Cl. In this respect correlated rotations appear as a definite outcome of strong dipolar interactions.

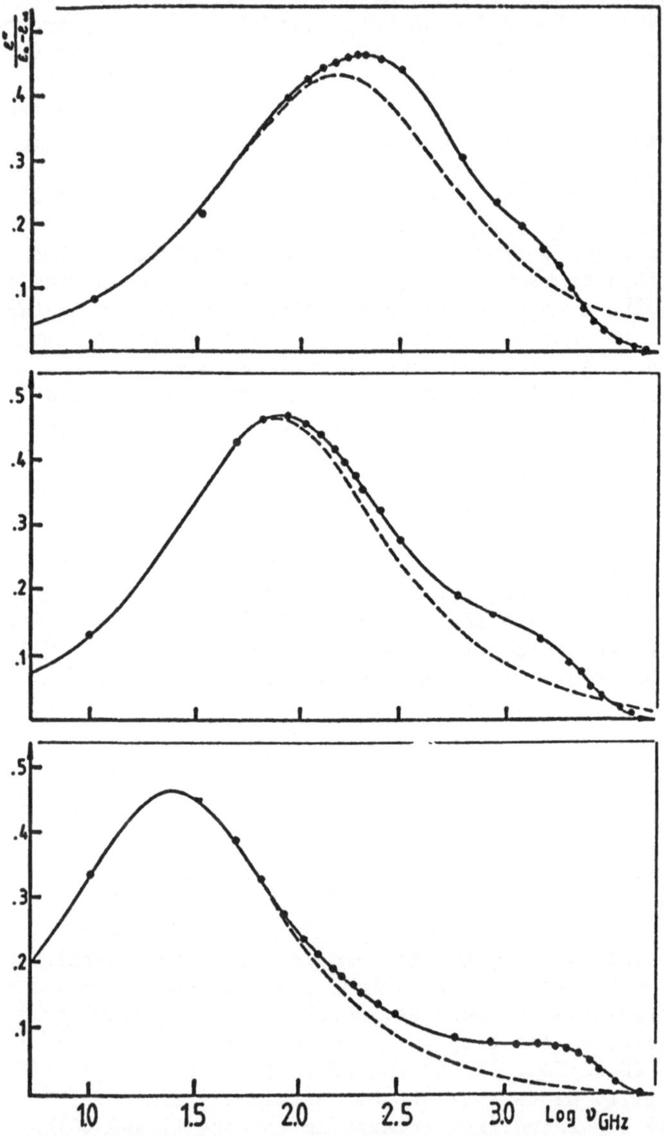

Figure 9. Reduced experimental absorption spectra at three temperatures of liquid CH_3Cl : from top to bottom 333 K, 273 K, 193 K. Full lines, experimental data. Dashed lines, Debye equations.

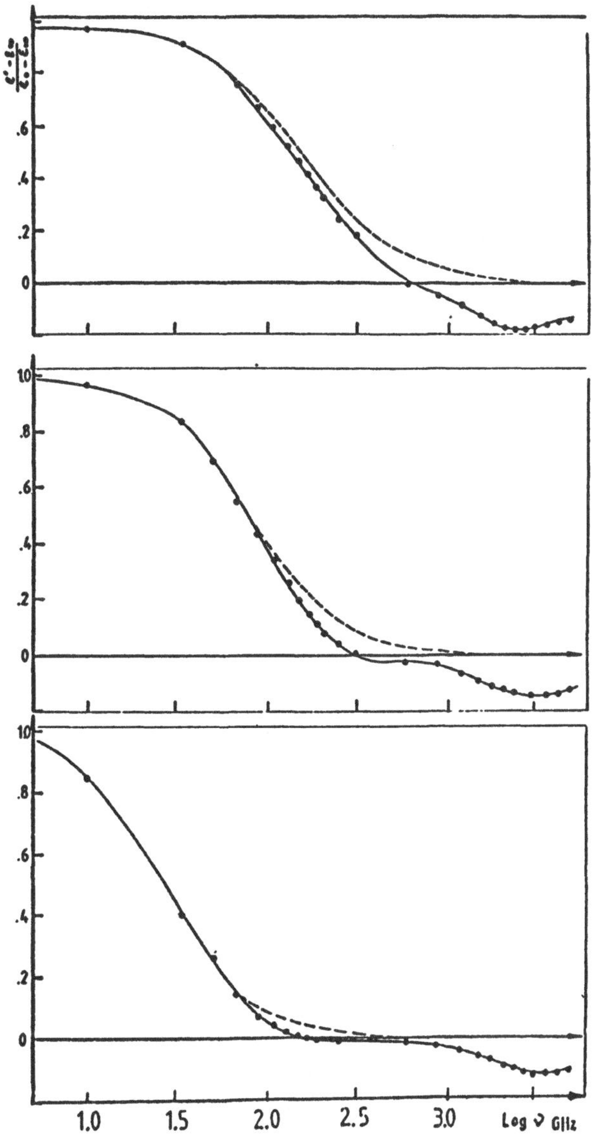

Figure 10. Same as figure 9 for experimental dispersion spectra.

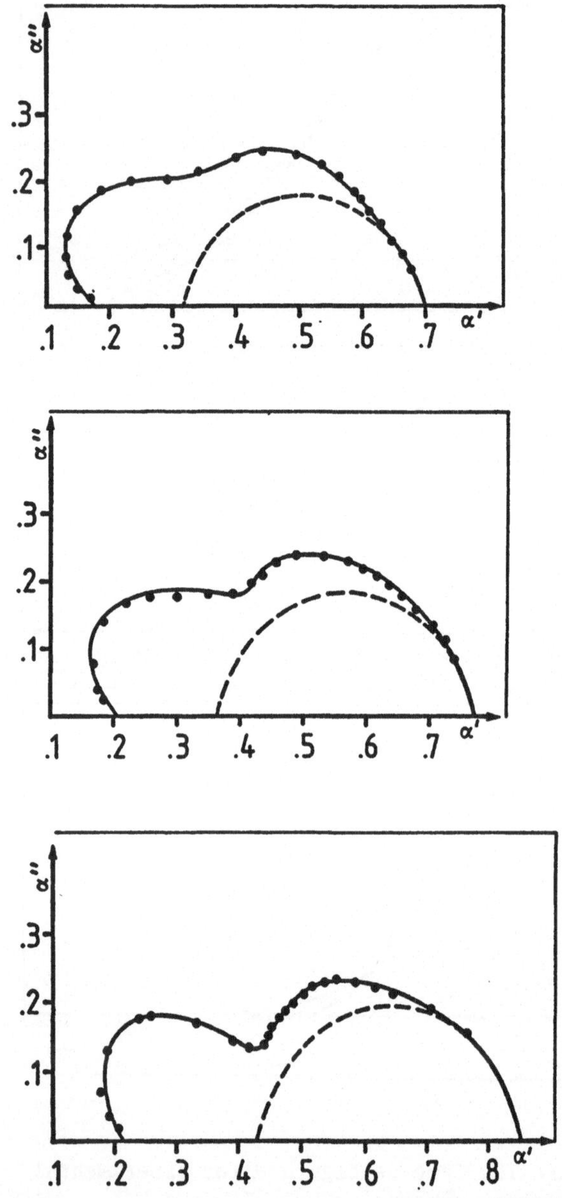

Figure 11. Scaife plots of the complex polarizabilities for three temperatures of liquid methyl chloride. From top to bottom : 333 K, 273 K, 193 K. Full lines, experimental data. Dashed lines, Debye equations.

The next stage is the recognition of the molecular origin of correlated rotations, whether from short-range or long-range interactions. No definite assignment has been established yet. Two different explanations may be put forward on the basis of structural arrangements or dipolar coupling forces.

A structural explanation can be given by analogy with a model that has been proposed for a long time for liquid water, labelled as the "mixture model". In such a model the liquid is considered as comprising both isolated molecules and small clusters. In water, clusters keep molecules tight together through hydrogen bonding. As for small and highly polar molecules it has been observed that the dipolar interaction energy between next neighbouring molecules may amount to values as much as twice that of thermal energy kT, thus representing quite a strong structurating force (17). As a result static correlation factors show noticeable departures from unity, and computed values of the mean squared torques are much higher than would be expected on the basis of purely geometrical shape factors (19). In this frame building up of small clusters that would possess specific dynamics of their own is possible. It would meet two interesting outcomes of numerical simulation work on liquids of dipolar hard spheres (25, 50). Levesque et al. observed high orientational correlations among the molecules located on the first coordination shell about a central molecule. Further, these correlations compensate the field of the central dipole, screening the dipolar interaction in the outer liquid beyond the first shell. From a different approach Brot et al. found ordering processes induced by dipolar interactions, such as oscillating tails of rotational velocities correlation functions, and an accompanying electrostriction effect monitored by the strength of the molecular dipolar field.

Going back to the interpretation of correlated rotations, the issues resumed above give us a picture of short-range order where a central molecule is surrounded by those of its first coordination shell, reminiscent of molecular aggregates that are known to occur about electrons, radicals or ions, and thus refered by analogy as a "dipolar solvation" process. At present, the way the dynamics would be affected by dipolar self-solvation deserves further research concerning mechanisms that entail rotational coherence.

On the other hand, dipolar forces are known to extend much further than hard core repulsive forces that dominate liquid static structures. There is a possibility that a screening process, such as that one alluded just above, ends in reducing the range of effective dipolar forces. However, recent comprehensive theories that include the long-range effect of dipolar forces predict a distinct contribution in the susceptibility absorption patterns (27,28,51). In this respect such liquids as CH_3Cl or

CHF_3 in which the new resonant feature was found are ideal systems to develop collective propogating modes such as dipolarons. However in real experiments it is not clear how far the geometry of classical absorption measurements, such as reported for the rotational coherent features above, would be adequate to detect dipolaron oscillations or even indirect manifestations of these.

Getting deeper into the interpretation is open to imagination and computational skill. Still it demands particularly more experimental data,under enlarged conditions, and if possible from other sources. If the "structural" approach developped in the mixture model is correct, solvophobic effects should appear ; therefore dilution of polar liquids bearing rotational coherence into non-polar solvents should demonstrate persistency of the self-solvated dipoles. But it should cut off modes from long range dipolar interactions. Other techniques such as Rayleigh diffusion can be expected to reveal correlated rotations, provided that the experimental profile be multiplied by the frequency, since in the ε''/ω representation from experiment the corresponding features fall exactly about the rounded inflexion within the central peak and the far wings. New time-resolved laser spectroscopies, in the picosecond and femtosecond time-domain- as explained by Professor Kenney-Wallace in this ASI - should also be quite sensitive to rotational coherence effects.

REFERENCES

1. Steele, W.A., (1963) J. Chem. Phys., 38, p. 2404
   (1976) Adv. Chem. Phys. 34, p.1-104
   "Intermolecular Spectroscopy and Dynamical Properties of Dense Systems" (1980) Proc. Int. School. Phys. E. Fermi LXXV, (North Holland), p. 325-374
2.a Alder, B.J., (1973) Ann. Rev. Phys. Chem. 24,p.325 ; (1981) Ber. Bunsenges Phys. Chem. 85, p.944
  b Pomeau, Y., and Resibois, P., (1975) Physics Reports, 19,p.63
3. Fox, R.F., (1983) Phys. Rev. A, 27, p.3216 ;
   (1983) Physica, 118 A, p. 383
4. London R.E. (1977), J. Chem. Phys., 66, p.471
5.a Chantry , G.W., *"Submillimeter Spectroscopy"* (Academic Press, London 1971)
  b Evans, M., Evans, G.J., Coffey, W.T., Grigolini, P., "Molecular Dynamics" (Wiley, 1982)
6. Gerschel, A., Dimicoli, I., Jaffre, J. and Riou, A., Mol. (1976) Phys., 32, p.679
7.a Arnold, K.E., Yarwood, J. and Price, A.H., (1983) Mol. Phys., 48, p. 451
  b Kroon, S.G. and van der Elsken, J., (1967), Chem. Phys. Lett., 1, p. 285

8. Bradley, C.C., Gebbie, H.A., Gilby, A.C., Kechin, V.V. and King, J.H., (1966) Nature, 211, p.839
9. Chamberlain, J., Afsar, M.N., Davies, G.J., Hasted, J.B. and Zafar, M.S., (1974), I.E.E.E. Trans. Microwave Theor. Tech., 22, p. 1028
10. Frenkel, D., Gravesteyn, D.J. and van der Elsken, J., (1976), Chem. Phys. Lett., 40, p.9
11. Evans, M.W., Davies, M. and Larkin, I.W., (1973), J. Chem. Soc. Faraday Trans.II, 69, p. 1013
12. Evans, G.J., (1983), J. Chem. Soc., Faraday Trans.II, 79, p.547
13. Bossis, G., (1976), Mol. Phys., 31, p. 1897
14. Gerschel, A., Darmon, I. and Brot. C., (1972) Mol. Phys. 23, p. 317
15. Sato, K., Okhubo, Y., Moritsu, T., Ikawa, S. and Kimura, M., (1978), Bull. Chem. Soc. Japan, 51, p. 2493
Ikawa, S., Yamazaki, S. and Kimura, M., (1981), Chem. Phys. 58, p. 275
16. Bossis, G., (1977), Physica, 88A, p. 535 ; (1982), Physica, 110A, p. 408
17. Gerschel, A., (1976), Mol. Phys., 31, p. 209
18. Gordon, R.G., (Academic Press, 1968), Adv. Magnetic Resonance 3, p. 1-42
19.a Gerschel, A., *"Newer Aspects of Molecular Relaxation Processes"* For Symp. N°11, (1977) Chem. Soc. p.115
b Gerschel, A., Brot, C., Dimicoli, I. and Riou, A., (1977) Mol. Phys., 33, p. 527
20. Yarwood, J., (1979), Ann. Rep. C, Royal Soc. Chem. (London), p. 99-130
21.a Mc Quarrie, D.A., *"Statistical Mechanics"*, (Plenum, 1976)
b van Kampen, N.G., *"Stochastic Processes in Physics and Chemistry"*, (North Holland, 1981)
c Hansen, J.P. and Mc Donald I.R., *"Theory of Simple Liquids"*, (Academic Press, 1976)
d Berne, B.J., *"Statistical Mechanics"*, (Plenum, 1977)
e Dupuy, J. and Dianoux, A.J., *"Microscopic Structure and Dynamics of Liquids"*, (1977) NATO ASI, (Plenum, 1978)
22. Desplanques, P. and Constant, E., (1971), C.R.Acad. Sc. Paris, 272B, p. 1354
23. Tildesley, D.J. and Madden, P.A., (1983), Mol. Phys. 48, p.129
24. Anderson, J.E. and Ullman, R., (1971), J. Chem. Phys. 55, p. 4406
25. Brot, C., Bossis, G. and Hesse-Bezot, C., (1980), Mol. Phys. 40, p. 1053 ; Hesse-Bezot, C., Bossis, G. and Brot, C., (1984) J. Chem. Phys., to appear
26. Deutch, J.M., (1973), Ann. Rev. Phys. Chem., 24, p. 301 ; Adalmon, S.A. and Deutch, J.M., (1975), Adv. Chem. Phys. 31, p. 103
27. Kivelson, D. and Madden, P., (1975), Mol. Phys., 30, p. 1749 ; Kivelson, D. and Madden, P., (1984), to appear

28. Mc Mahon, D.R.A., (1980), J. Chem. Phys., 72, p. 1359 ; (1980), J. Chem. Phys., 72, p. 2411
29. Fulton, R.L., (1971), J. Chem. Phys., 55, p. 1386
30. Kato, H., Clark, R. and Mc Coffery, A.J., (1976), Mol. Phys. 31, p. 943 ; Caughey, T.A. and Crosley, D.R., (1977), Chem. Phys., 20, p. 467
31. Kivelson, D. and Keyes, T., (1972), J. Chem. Phys., 57, p.4599
32.a Debye, P., *"Polar molecules"*, (Dover Publ. Inc.,N.Y., 1929)
b Frohlic, H., *"Theory of Dielectrics"*, (2 nd ed. Clarendon Press, Oxford, 1968)
c Smyth, C.P., *"Dielectric Behaviour and Structure"*, (Mc Graw Hill, 1955)
d Hill, N., Vaughan, W., Price, A. and Davies, M., *"Dielectric Properties and Molecular Behaviour"*, (Van Norstrand, 1969)
e Bottcher, C.J.F. and Bordevijk, P., *"Theory of Electric Polarization"*, (2 nd ed., Elsevier, 1978)
f Mc Connell, J., *"Rotational Brownian Motion and Dielectric Theory"*, (Academic Press, 1980)
33. Hu, C.M. and Zwanzig, R., (1974), J. Chem. Phys., 60 , p.4354 Kivelson, D., *"Newer Aspects of Molecular Relaxation Processes"*, Far. Symp. 11, (Chem. Soc., 1977), p.7 Peralta-Fabri, R. and Zwanzig, R., (1979), J. Chem. Phys. 70, p. 504
34. Masters, A.J. and Madden, P.A., (1981), J. Chem. Phys. 74, p. 2450
35. Deutch, J.M., *"Newer Aspects of Molecular Relaxation Processes"*, Far. Symp. 11, (Chem. Soc., 1977), p. 26
36. Bordewijk, P., Naturforsch, Z., (1980), 35a, p. 1207
37. Bartoli, F.J. and Litowitz, T.A., (1972), J. Chem. Phys. 56, p. 413
38. Gerschel, A., to appear
39. Hegemann, B. and Jonas, J., (1983), J. Chem. Phys., 79, p.4683
40. Gerschel, A., (1972), Ber. Bunsenges. Phys. Chem., 76, p.254
41. Davies, M., Pardoe, G.W.F., Chamberlain, J. and Gebbie, H.A., (1968). Trans. Far. Soc., 64, p. 847
42. Hill, N.E., (1971), J. Phys. C., 4, 2322
43. Goulon, J., Rivail, J.L., Fleming, J.W., Chamberlain J. and Chantry, G.W., (1973), Chem. Phys. Lett., 18, p. 211 Rosenthal L.C. and Strauss, H.L., (1983), J. Chem. Phys. 64, p. 282
44. Kisiel, Z., Leibler, K. and Gerschel, A., (1983), J. Phys. E, 16
45. Davies, M., Pardoe, G.W.F., Chamberlain J. and Gebbie, H.A., (1970), Trans. Faraday Soc., 66, p. 273
46. Gerschel, A., Grochulski ., T., Kisiel, Z., Pszczolkowski, L. and Leibler, K., (1984), Mol. Phys, to appear
47. Hill, N.E., (1969), J. Phys. A, 2, p. 398
48. Grochulski , T., Pszczolkowski, L., Checinski, K., Leibler, K. and Gerschel A., (1982), J. Phys. E., 15, p. 304
49. Scaife, B.K.P., (1963), Proc. Phys. Soc., 81, p. 124

50. Levesque , D., Patey, G.N. and Weiss, J.J., (1977), Mol. Phys., 34, p. 1077 ; Patey, G.N., Levesque , D. and Weiss, J.J., (1979), Mol. Phys. 38, p. 219
51. Lobo, R., Robinson, J.E. and Rodriguez J., (1973), J. Chem. Phys., 59, p. 5992

LIGHT SCATTERING SPECTROSCOPY IN LIQUIDS

Th. Dorfmüller

Universität Bielefeld, Fakultät für Chemie,
D-4800 Bielefeld, Federal Republic of Germany.

CONTENTS:

A. J. Barnes et al. (eds.), Molecular Liquids - Dynamics and Interactions, 201–238.

1. INTRODUCTION

Light is scattered by a liquid sample as a consequence of the structural disorder of liquids. Thus the whole of the effect observed in a light scattering experiment is due essentially to the property of liquids of being in a disordered state. This contrasts light scattering spectroscopy to most other spectroscopic techniques which probe essentially molecular properties that are modified to a certain degree by the nature of the liquid environment interacting with the molecules used as spectroscopic probes. Thus the scattered light reflects the structure and the dynamics of the liquid state directly, whereas in other spectroscopic techniques the liquid state properties appear in higher approximation only. In contrast to light scattering the classical scattering techniques like X-ray scattering operating with a radiation, whose wavelength is of the same order of magnitude as the average distance of the scatterers, are commonly used to probe basically the ordered structure of a regular lattice, disorder being manifested only as a secondary feature of the reflexes observed in the experiment.

In light scattering spectroscopy we exploit the two important properties of laser light sources : monochromaticity and coherence. A good laser is monochromatic to approximately 1 MHz and coherent i.e. displaying a definite phase relation in space over the relevant macroscopic sample dimensions. Quantitatively these properties of laser light are reflected in the small deviation of the frequency ω and the wave vector $\underset{\sim}{k}$ from their respective mean values.

In the following introductory chapter the special features of dynamic light scattering will be introduced which make this technique so fertile in studying the dynamics of liquids and more generally of amorphous systems.

2. THEORETICAL BACKGROUND

Figure 1 illustrates the geometry of a typical laser light scattering experiment within a laboratory-fixed frame of reference x,y,z.
The incident light is characterized by the unit polarization vector $\underset{\sim}{n}_i$ parallel to the z-axis and the wavevector $\underset{\sim}{k}_i$ along the direction of propagation. The scattering plane, defined by the two intersecting wavevectors $\underset{\sim}{k}_i$ and $\underset{\sim}{k}_s$, lies in the scattering plane. The scattered light is described by its wavevector $\underset{\sim}{k}_s$ and its polarization vector $\underset{\sim}{n}_s$. The scattering volume is located at the origin O of the frame of reference.

The theoretical basis and the most important applications of dynamic light scattering have been extensively treated in a number of books and a review article (1-9). For this reason only the most important principles will be summarized here and will be used

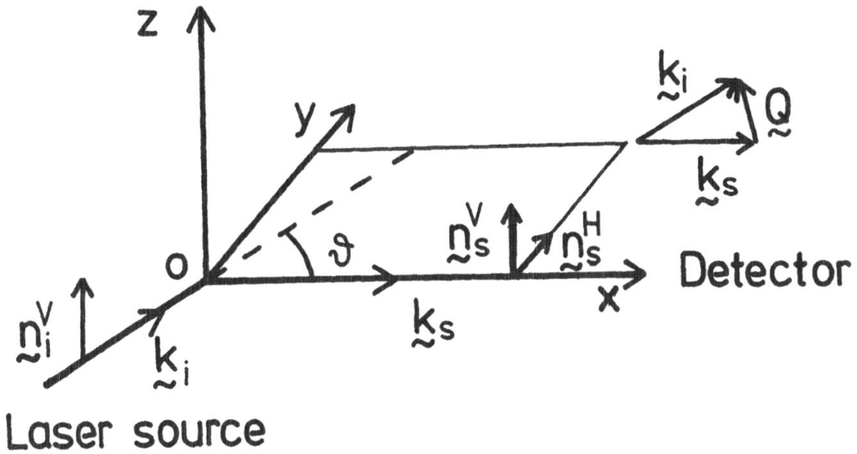

Fig. 1 Geometry of a laser light scattering experiment

subsequently to illustrate the possibilities of dynamic light scattering spectroscopy in dealing with a small number of characteristic applications.

A: The scattered light intensity observed on a detector is the result of the interference of the partial electric fields emitted from the scattering centers within the scattering volume.

The electric field vector of the incident light beam at the time t at a place determined by the vector $\underset{\sim}{r}$ is given by:

$$\underset{\sim}{E}_i(\underset{\sim}{r},t) = \underset{\sim}{n}_i E_o \exp i(\underset{\sim}{k}_i \underset{\sim}{r} - \omega t) \qquad (1)$$

The vector $\underset{\sim}{r}$ locates a scattering center emitting a secondary radiation at a frequency ω_s (quasi elastic scattering approximation). If we place a detector at a scattering angle ϑ and a location $\underset{\sim}{R} = |\underset{\sim}{R}|\underset{\sim}{x}_o$ the scattered field at the detector can be calculated by summing the partial scattered fields $\underset{\sim}{E}_s(\underset{\sim}{r},\underset{\sim}{R},t)$ over the contributions of the whole scattering volume. The phase relation between the partial fields can be taken into account in a very simple way since the excitation in the laser field is coherent. This is performed by multiplying the partial fields by the phase factors $\exp(i\underset{\sim}{Q}\underset{\sim}{r})$:

$$\underset{\sim}{E}_s(\underset{\sim}{R},t) = \int_V dr^3\, \underset{\sim}{E}_s(\underset{\sim}{r},\underset{\sim}{R},t) \exp(i\underset{\sim}{Q}\underset{\sim}{r}) \qquad (2)$$

The integration is carried out over the scattering volume V. Q is the scattering vector defined as follows:

$$\underset{\sim}{Q} = \underset{\sim}{k}_i - \underset{\sim}{k}_s \qquad (3)$$

The magnitude of $\underset{\sim}{Q}$ is an important experimental parameter and it can be expressed as a function of the scattering angle :

$$|\underset{\sim}{Q}| = 4\pi n/\tau \ \sin(\vartheta/2) \qquad (4)$$

n and λ represent the index of refraction of the sample and the incident wavelength respectively.

B: The scattering centers can be defined as optically homogeneous volumes characterized by a local dielectric permittivity $\varepsilon(\underset{\sim}{r})$.

Depending upon the nature of the sample and the experimental conditions it might be useful to identify the scattering centers with the molecules, with molecular "segments" or with properly definded regions of the liquid. For the purpose of simplifying the pertinent equations it is useful to describe the thermal optical inhomogeneity of the liquid by the local deviations of the dielectric permittivity from the macroscopic mean $\bar{\varepsilon}$ for each scattering center:

$$\varepsilon = \varepsilon' - \bar{\varepsilon}$$

Thus a scattering center is characterized by the space and time dependent quantity $\underset{\approx}{\varepsilon}(\underset{\sim}{r},t)$ which is a second rank tensor. The amplitude of the scattered field from a given center observed under a given polarization geometry $\underset{\sim}{n}_i, \underset{\sim}{n}_s$ is determined as far as the sample is concerned by the expression:

$$\underset{\sim}{n}_s \cdot \underset{\approx}{\varepsilon}(r,t) \cdot \underset{\sim}{n}_i \qquad (5)$$

By means of the electromagnetic theory we can now derive an expression for the partial scattered fields $\underset{\sim}{E}_s(\underset{\sim}{r},\underset{\sim}{R},t)$ and hence with equ. 4 for the total scattered electric field at the detector:

$$\underset{\sim}{E}_s(\underset{\sim}{R},t) = P/R \ \ \mathrm{expi}(\underset{\sim}{k}_s \underset{\sim}{R} - \omega t)$$
$$\underset{\sim}{n}_s \left[\int_V dr^3 \exp(i\underset{\sim}{Q}\underset{\sim}{r}) \ \underset{\approx}{\varepsilon}(\underset{\sim}{r},t) \right] \cdot \underset{\sim}{n}_i \qquad (6)$$

The constant P is inversely proportional to λ^2. The integration is over the scattering volume.

The important point which emerges from this formalism is that the scattered field and hence the scattered intensity recorded by the detector depend on two factors:

a) The properties of the light source and the scattering geometry $(\omega_i, \underset{\sim}{k}_i, \underset{\sim}{n}_i, \underset{\sim}{R}, \underset{\sim}{k}_s, \underset{\sim}{n}_s)$.

b) The properties of the liquid sample contained in the dielectric tensor $\underset{\approx}{\varepsilon}(\underset{\sim}{r},t)$.

In an ideal experiment the incident field is monochromatic and perfectly coherent over the scattering volume. The characteristic inhomogeneity of the material expresed by $\underset{\approx}{\varepsilon}(\underset{\sim}{r},t)$ introduces a spatial incoherence in the scattered field reflecting the structure of the liquid, which is furthermore modulated in time by the time evolution of the ε-tensor. Thus both informations, the local distribution and the time evolution of the optical inhomogeneity, arrive at and are recorded by the detector and can in principle be recovered from the signal.

A scattering volume, in which the scattering centers are arranged in a perfect crystalline order, has an ε-tensor which does not depend on $\underset{\sim}{r}$ on a scale comparable to Q^{-1} for scattering angles differing significantly from zero. Under these conditions the integral in equ. 6 is equal to zero except for forward scattering. Any scattering at an angle $\vartheta \neq 0$ is caused by a dielectric tensor which varies locally in such a way that the integral in equation 6 over the sattering volume changes significantly over distances comparable to Q^{-1}, so as to destroy the cancellation of the phase factors $\exp(i\underset{\sim}{Q}\underset{\sim}{r})$ over the whole volume.

We can come to a qualitative understanding of the light scattering formalism and to the underlying physics as expressed in equ. 6 as follows. The value of $\underset{\sim}{E}_s(\underset{\sim}{R},t)$, as far as the sample is concerned, depends upon the value of the integral:

$$\underset{\approx}{\varepsilon}(\underset{\sim}{Q},t) = \int_V dr^3 \exp(i\underset{\sim}{Q}\underset{\sim}{r}) \cdot \underset{\approx}{\varepsilon}(\underset{\sim}{r},t) \qquad (7)$$

which is a Laplace transform of the dielectric fluctuation tensor $\underset{\approx}{\varepsilon}(\underset{\sim}{r},t)$ from the r-space to the Q-space giving the quantity $\underset{\approx}{\tilde{\varepsilon}}(\underset{\sim}{Q},t)$. If $\underset{\approx}{\varepsilon}(\underset{\sim}{r},t)$ is only weakly dependent upon $\underset{\sim}{r}$ relative to Q^{-1} within the scattering volume then the integral reduces to:

$$\underset{\approx}{\varepsilon}(\underset{\sim}{Q},t) = \underset{\approx}{\varepsilon}(\underset{\sim}{r},t) \int_V dr^3 \exp(i\underset{\sim}{Q}\underset{\sim}{r}) = 0 \qquad (8)$$

The situation can be best visualized by noting that in equation 7 the integrand of $\underset{\approx}{\varepsilon}(Q,t)$ can be described by a vector in the complex plane having the module $\underset{\approx}{\varepsilon}(\underset{\sim}{r},t)$ and the phase factor $\exp(i\underset{\sim}{Q}\underset{\sim}{r})$. If the module is approximately constant over the whole phase angle $0-2\pi$, then $\underset{\approx}{\varepsilon}(Q,t) = 0$ over all the scattering angles but for the forward scattering. Generally this is not the case and for a given Q the phase angle changes from 0 to 2π via the space coordinate $\underset{\sim}{r}$. The "liquid structure", i.e. the lack of order manifests itself as a variation of $\varepsilon(Q,t)$ over this range thus leading to a scattering for scattering vectors other than $Q = 0$.

For a given scattering and polarization geometry the scattered light is distributed over the scattering angles according to the Q-dependence of $\underset{\approx}{\varepsilon}(\underset{\sim}{Q},t)$ which reflects the $\underset{\sim}{r}$-dependence of the $\underset{\approx}{\varepsilon}$-tensor. The expression:

$$\underset{\sim}{n}_s \cdot \underset{\approx}{\varepsilon}(\underset{\sim}{Q},t) \cdot \underset{\sim}{n}_i \qquad (9)$$

in equ. 6 reflects the influence of the polarization geometry on the recorded signal. We define the V-polarization normal to the scattering plane, i.e. parallel to the z-axis and the H-polarization within the scattering plane as displayed in figure 1. With these definitions we can further define the scattering intensities under the following polarization geometries: $I_{VV}, I_{VH}, I_{HV}, I_{HH}$. If the dielectric tensor is isotropic, then $I_{VH} = I_{HV} = 0$ and for a scattering angle $\vartheta = 90^{\circ}$ then $I_{HH} = 0$. Generally this is not the case because either the molecules or their arrangement are not optically isotropic or the scattering mechanism is not (collision induced scattering). Under these conditions we observe the depolarized spectrum I_{VH}. In this case we have for optically inactive molecules $I_{VH} = I_{HV}$.

C: The time dependence of a random variable $A(t)$ can be usefully described by the statistical average:

$$\langle A \rangle = \lim_{T \to \infty} \frac{1}{T} \int_0^T A(t')\, dt' \qquad (10a)$$

and the time autocorrelation function (CF):

$$\langle A(0)\, A(\tau) \rangle = \lim_{T \to \infty} \frac{1}{T} \int_0^T A(t')\, A(t'+\tau)\, dt' \qquad (10b)$$

In light scattering we deal with the following quantities:

a) The scattered field: $\underset{\sim}{E}_s(t)$
b) The scattered intensity: $I(t) = |\underset{\sim}{E}_s(t)|^2$

We can define the ensemble averages of these two quantities, but the formalism is generally simpler, if we use the deviations from the mean as the pertinent time dependent functions:

$$\underset{\sim}{E}_s(t) = \bar{\underset{\sim}{E}}_s(t) - \underset{\sim}{E}'_s(t)$$

$$I_s(t) = \bar{I}_s(t) - I'_s(t)$$

where the primed quantities represent the actual values at the time t. With these definitions the averages $\langle \bar{\underset{\sim}{E}}_s(t) \rangle$ and $\langle \bar{I}_s(t) \rangle$

are equal to zero.

The power density $I_A(\omega)$ of the time correlation function of the quantity $A(t)$ is defined as follows by a Fourier transform:

$$I_A(\omega) = \frac{1}{2\pi}\int_{-\infty}^{+\infty} \exp(-i\omega t) \langle A^*(0)\, A(t)\rangle\, dt \qquad (11a)$$

This Fourier transform can be inverted:

$$\langle A^*(0)\, A(t)\rangle = \int_{-\infty}^{+\infty} \exp(i\omega t)\, I_A(\omega)\, d\omega \qquad (11b)$$

Applying these relations to the dynamical variables pertinent for light scattering, we can define the following CF's:

The scattered field CF:

$$C^{(1)}(t) = \langle \underset{\sim}{E}_s^*(\underset{\sim}{Q},0)\, \underset{\sim}{E}_s(\underset{\sim}{Q},t)\rangle \qquad (12a)$$

The scattered intensity CF:

$$C^{(2)}(t) = \langle |\underset{\sim}{E}_s(\underset{\sim}{Q},0)|^2\, |\underset{\sim}{E}_s(\underset{\sim}{Q},t)|^2\rangle \qquad (12b)$$

The normalized scattered field CF:

$$g^{(1)}(t) = C^{(1)}(t) \,/\, C^{(1)}(0) \qquad (12c)$$

The normalized intensity CF:

$$g^{(2)}(t) = C^{(2)}(t) \,/\, C^{(2)}(0) \qquad (12d)$$

The indexes "1" labelling the scattering field CF's is omitted wherever there is no danger of ambiguity.

The two normalized correlation functions are related by the Siegert equation:

$$g^{(2)}(t) = a\,[\,1+b\, g^{(1)}(t)^{\,2}\,] \qquad (12e)$$

which is valid in the case of Gaussian light, i.e. when the scattered field is a stochastic variable with a Gaussian distribution. Since this is not always the case one should be very cautious when comparing interferometric and photon correlation data by means of equation 12e.

The central issue in dynamic light scattering spectroscopy is to extract the information about the molecular dynamics of the liquid sample which is contained in the spectral features of the scattered light. In order to achieve this we have to abandon the phenomenological point of view exposed up to this point and give an interpretation of the properties of the dielectric tensor in terms of the molecular properties of the liquid. In other words we must connect the dielectric tensor with molecular properties the most important in this context being:

a) The mean particle density deviation, i.e. the deviation of the momentary, local particle density $\rho(r,t)'$ from its thermodynamic mean $\bar{\rho}$: $\rho(r,t) = \bar{\rho} - \rho'(r,t)$

b) The molecular polarizability tensor $\underset{\approx}{\alpha}(t)$.

Accordingly we use the following time CF's which represent the fluctuation kinetics of these two molecular quantities:

$$C_\rho(Q,t) = \langle \rho(-Q,0)\cdot \rho(Q,t) \rangle \qquad (13a)$$

$$C_\alpha(Q,t) = \langle \alpha_{is}(-Q,0)\cdot \alpha_{is}(Q,t) \rangle \qquad (13b)$$

α_{is} is defined in equations 14. These CF's reflect the contributions of a) molecular translation and b) molecular rotations as well as possible conformational changes to the time dependence of the dielectric tensor and hence to the light scattering spectrum. The properties of the polarizability tensor reflecting the structure of the molecules determines the nature of the contribution of $C_\alpha(Q,t)$ to the observed spectrum. It is instructive to compare the procedures leading to an expression for the power spectrum in the phenomenological and the molecular approach:

PHENOMENOLOGICAL APPROACH:

Sample property: $\underset{\approx}{\varepsilon}(\underset{\sim}{r},t)$

$$\underset{\approx}{\varepsilon}(\underset{\sim}{Q},t) = \int_V d^3r\ \exp(i\underset{\sim}{Q}\underset{\sim}{r})\cdot \underset{\approx}{\varepsilon}(\underset{\sim}{r},t) \qquad (14a)$$

$$\varepsilon_{i,s}(\underset{\sim}{Q},t) = \underset{\sim}{n}_s\, \underset{\approx}{\varepsilon}(\underset{\sim}{Q},t)\ \underset{\sim}{n}_i \qquad (14b)$$

$$C^{(1)}(t) = \varepsilon_{is}(\underset{\sim}{Q},0)\ \ \varepsilon_{is}(\underset{\sim}{Q},t) \qquad (14c)$$

$$I(\omega) = \frac{1}{2\pi}\int_{-\infty}^{+\infty} \exp(-i\omega t)\ C^{(1)}(t)\ dt \qquad (14d)$$

MOLECULAR POLARIZABILITY APPROACH:

molecular property: $\underset{\approx}{\alpha}(t)$

$$\alpha_{is}(t) = \underset{\sim}{n}_s\cdot \underset{\approx}{\alpha}(t)\ \underset{\sim}{n}_i \qquad (14e)$$

$$\alpha_{is}(\underset{\sim}{Q},t) = \sum_j \alpha^j_{is}(t)\exp(i\underset{\sim}{Q}\underset{\sim}{r}^j(t)) \qquad (14f)$$

$$C^{(1)}(t) = \langle \alpha^j_{is}(\underset{\sim}{Q},t)\cdot \underset{\sim}{\alpha}_{is}(\underset{\sim}{Q},0) \rangle \qquad (14g)$$

$$I(\omega) = \frac{1}{2\pi}\int_{-\infty}^{+\infty} \exp(-i\omega t)\ C^{(1)}(t)\ dt \qquad (14h)$$

The integration in 14a is over the volume elements of the

scattering volume. The summation in 14f is over the discrete molecules j at $\underset{\sim}{r}^j$ in the scattering volume at the time t. Both transforms serve to relate the distribution of polarizable matter in the scattering volume to the experimental parameters contained in the scattering vector Q.

The polarizability tensor $\underset{\approx}{\alpha}$ has a symmetry which is connected to the molecular symmetry. The cases which are theoretically most tractable are molecules with a spherical or with axial symmetry. Whereas in both cases the α -tensor is diagonal, in a tensor with spherical symmetry its three elements are equal to $\alpha = 1/3\, \mathrm{Tr}\, \underset{\approx}{\alpha}$. Molecules with axial symmetry have along the diagonal of the tensor a component parallel to the symmetry axis and a normal component. The optical anisotropy is defined as the difference: $\beta = \alpha_{\parallel} - \alpha_{\perp}$. The integrated intensity of the light scattered at a given angle is related to the value C(O) of the pertinent CF at the time shift t = O. The isotropic component I_{iso} is the area under that part of the light scattering spectrum which is due to the local optical density fluctuations and is proportional to the square of the trace of the α -tensor as shown in equ. 17c.

$$I_{iso} = \langle N \rangle \cdot \alpha^2 \qquad (15a)$$

This quantity is not directly measurable, but it can be calculated from the I_{VV} and the I_{VH} components. Spectra obtained with the VH-polarization geometry are due to pure anisotropic scattering and the integrated intensity of the VH-spectrum is proportional to the square of the optical anisotropy :

$$I_{VH} = \langle N \rangle \cdot \beta^2/15 \qquad (15b)$$

The integrated intensity of the VV-spectrum contains components both of the isotropic and the anisotropic scattering:

$$I_{VV} = \langle N \rangle (\alpha^2 + 4 \cdot \beta^2/45) \qquad (15c)$$

In other words, the time dependence of the trace of the polarizability tensor reflects the translational motion of the molecules and is observed in the isotropic scattering.The time dependence of the polarizability anisotropy is due to reorientational motions and is observed in the VH-scattering. The VV-scattering contains both kinds of motion which, however, can be separated by means of equations 15a-c.

Taking into account the static orientational CF f, we obtain the approximate relation:

$$I_{VH} = \chi\, \beta^2 (1+fN) \qquad (16)$$

which relates the VH-intensity with the anisotropy and the concentration χ of the scatterers and with the orientational correlation f between them. It is important to note that the equation 16 reflects the contribution of the static orientational correlations of anisotropic molecules to the VH scattering. Orientationally parallel correlated molecules ($f > 0$) formally increase the effective concentration of the scatterers. The equation 16 ,in spite of its approximate character, has been very useful in many cases because it might give a hint as to the nature of the scattering moieties which is not always prima facie obvious from the composition of the sample.

3. DYNAMICAL PROCESSES IN LIQUIDS:

3.1. Translational motions

The position of the center of molecular polarizability of each molecule enters the formulae through the phase factors $\exp(i\underset{\sim}{Q}\underset{\sim}{r}(t))$. The phase factors vary as the molecules wander within the scattering volume and/or move in and out this volume. In the case of larger coils, like long flexible molecules, we have to consider the additional possibility that the center of polarizability is displaced by intramolecular conformational fluctuations without a corresponding displacement of the center of mass.

In the case of spherical rigid molecules, neglecting also interaction effects, the α -tensor is spherical and time independent and we can separate in equ. 14e-g the time dependence of the phase factors from the molecular polarizability

$$\alpha_{is} = \alpha\, \underset{\sim}{n}_s \cdot \underset{\sim}{n}_i \quad (= \alpha\,,\ \text{for VV-scattering}) \qquad (17a)$$

$$\alpha_{is}(\underset{\sim}{Q},t) = \alpha \sum_j \exp(i\underset{\sim}{Q}\underset{\sim}{r}^j(t)) \qquad (17b)$$

$$C^{(1)}(t) = \langle N\rangle \alpha^2 \langle \sum_j \exp(i\underset{\sim}{Q}\underset{\sim}{r}^j(t)) \cdot \sum_j \exp(i\underset{\sim}{Q}\underset{\sim}{r}^j(0)) \rangle \qquad (17c)$$

In this case the molecular polarizability translation is the only factor affecting the time dependence of the CF. Because of the factor $\underset{\sim}{n}_s \cdot \underset{\sim}{n}_i$ the depolarized component of the spectrum is absent. What remains is the effect of the fluctuations in number density within the scattering volume and/or of the relative positions of the molecules on I_{VV} which both affect the CF via the phase factors. We can also express this situation by means of the density-density CF of equation 13. This is based upon the relation:

$$\rho(\underset{\sim}{Q},t) = \sum_j \exp(i\underset{\sim}{Q}\underset{\sim}{r}^j(t)) \qquad (18)$$

At this stage we must introduce some physically plausible hypothesis about the nature of the motion of the scattering centers which is the line shaping mechanism of the light scattering spectra. If we use the diffusion model, which is a very obvious choice as a first approximation, especially for large scatterers moving in a fluid medium, we can set up a diffusion equation for the Van Hove CF containing a material parameter, the diffusion coefficient D. At a later stage we have to devise an experimental test whether we can identify D with the macroscopic diffusion coefficient. With these assumptions we obtain the following equations for the translational CF and the I_{VV} spectral shape:

$$C(\underset{\sim}{Q},t) = \langle N\rangle \exp(-Q^2 \cdot D \cdot t) = \langle N\rangle \exp(-t/\tau) \qquad (19a)$$

$$I(\underset{\sim}{Q},\omega) = \langle N\rangle/\pi \cdot Q^2 D/(\omega^2 + (Q^2 \cdot D)^2) \qquad (19b)$$

These equations describe fairly well the translational motion of molecules and of large scatterers like diluted polymers or micelles in a liquid. The extension of these relations to the situation in molecular liquids is an approximation which has to be justified. The main cause of deviations is due to the fact that in light scattering we basically observe collective modes. The single particle motion cannot be extracted from the measured correlation functions without either having recourse to other methods or introducing further simplifying assumptions.

3.2. Rotational motions

If the molecular polarizability tensor is not spherical, then the factorization in equation 17c is not possible without an additional assumption. Owing to the molecular rotation $\underset{\sim}{\alpha}(t)$ is now a function of time as well as the phase factor. If we assume, however, that the orientation of the molecules is independent of their position, then the factorization can be performed without the appearance of cross terms and the CF separates into a translational and a rotational component as follows:

$$\overset{(1)}{C}(t) = \langle \sum_j \exp(i\underset{\sim}{Q}\underset{\sim}{r}^j(t) \cdot \sum_j \exp(i\underset{\sim}{Q}\underset{\sim}{r}^j(0)\rangle$$

$$\langle \alpha^j(t) \cdot \alpha^j(0)\rangle \qquad (20a)$$

$$= \overset{(1)}{C}_{TR}(t) \quad \overset{(1)}{C}_{ROT}(t) \qquad (20b)$$

The independent treatment of the translational and the orientational modes can be justified in the case where the time scales of these two modes are sufficiently different. This is for example done, if one can reasonably assume that the translation is slow enough, so that the reorientation takes place in an environment which does not change over the time required for a significant reorientation. This approximation is generally used in liquids consisting of small quasi spherical molecules reorienting more easily than molecules with a very elongated shape. Conversely, in systems like liquid crystals the reorientation of the main molecular axis is so slow, that the translational motion along this axis can be assumed to take place without interference from the reorientation. Assuming furthermore that a) the α -tensor is diagonal, as this is the case for a linear molecule or a symmetric top, and that b) the reorientation of the molecular symmetry axis can be described by the Debye rotational diffusion model, i.e. that the orientational probability distribution obeys a diffusion equation, then we obtain the following expression for the polarizability correlation function:

$$C(t) = \langle N \rangle \left[\exp(-Q^2 Dt)\, \alpha^2 + 4/45\, \beta^2 \exp(-6\Theta t) \right] \quad (21)$$

Θ is the rotational diffusion coefficient. The polarizability CF can be expressed for the case of the anisotropic rotor in terms of the second order Legendre polynoms as follows:

$$C(t) = \langle P_2(\cos\vartheta(t)) \rangle \quad (22)$$

where $\vartheta(t)$ is the angle of rotation of the molecular symmetry axis in the time interval between 0 and t. The equation 21 gives an expression for this CF for the Debye orientational diffusion model. Although this is generally a good description of the reorientation in a viscous liquid, other models, the so called extended diffusion models, have been devised in the attempt to describe the reorientation in simple liquids.

It is customary to describe the reorientational kinetics by a time reflecting the rate at which the CF decays to its long time limit. In the case of a purely diffusional motion the best choice is the relaxation time τ_R which appears in the exponent of the equation 19 and 21 for the translational and the rotational CF's. For CF's which are not exponential we define a correlation time τ_c as the area under the normalized CF:

$$\tau_c = \int_0^\infty g(t)\, dt \quad (23)$$

These two characteristic times are identical for an exponential CF. In the cases, usually met with complex liquids, where the CF does not have an exponential shape we have the possibility to use an empirical fit function which generally involves an additional parameter characterizing the deviation from the exponential decay. Since in most cases the origin of this effect, which manifests itself as a spreading of the CF over a wider time scale than the exponential, is not well known we do not have at our disposition a theoretical function. In some cases the width is attributed to polydispersity effects, in other to an inhomogeneity of the environment of the scatterers. Probably we also have to deal in some cases with more than one unresolved relaxation processes which furthermore may be physically coupled so that a resolution is not feasible. Even in the polydispersity model of the distribution of relaxation times, which is the one which can be best visualized, the problem of extracting unambiguously a distribution parameter, with a definite physical meaning from data obtained over a limited time range, with a given number of experimental points, and with a given signal-to-noise ratio, is by no means solved. The general problem can be best formulated with the equation:

$$g^1(t) = \int_0^\infty G(\Gamma)\exp(-\Gamma t)\,d\Gamma + \delta(t) \qquad (24)$$

In this equation the CF is given as a distribution $G(\Gamma)$ of relaxation times $\Gamma = \tau^{-1}$. Additionally the inversion of this integral equation involves the noise $\delta(t)$. This mathematical problem is ill-posed in the sense that there exists a set of possible solutions satisfying equation 24 within experimental error. The deviations of these solutions from the "true" $G(\Gamma)$ are unbounded. This latter property is the essential restriction as to the information which can be obtained from experimental correlation functions, given a certain time range and a certain number of channels (see articles 28-33 in (7)).

3.3. Collective motions

At this point it should be noted that in introducing the molecular polarizability tensor to replace the phenomenological dielectric tensor the assumption was made that the molecular α-tensors could be treated independently from each other or, in other words, that the time dependence of the phase factors is independent of the time dependence of the α-tensors. This is actually a serious restriction which amounts to neglecting the contribution of collective reorientational motions to the light scattering spectrum. An exact treatment of the collective motions is not possible, but an approximate treatment (10) has yielded a useful relation between the single particle correlation times τ_s and a collective correlation time τ_{cl} in terms of the static orientational CF f and the dynamic orientational CF g:

$$\tau_{cl} = \tau_s\,(1+Nf)/(1+Ng) \qquad (25)$$

For isotropic liquids consisting of small molecules we generally assume g to be equal to zero, and f is generally found to be a number not very different from 1. A reliable separation of the single particle and the collective reorientation times is only possible, if we have recourse to other methods not complicated by collective motions or by modifying the system so as to exclude collective motions for example by diluting it with an isotropic solvent. On the other hand, the sensitivity of dynamic light scattering spectroscopy to collective motions, which it does not share with any other spectroscopic method, makes it so valuable for the study of the liquid state whose essential dynamic feature is cooperativity. From the point of view of theory, however, the solution of the n-body problem, which is required to reproduce the experimental results, can only be approached indirectly by means of physically plausible and mathematically tractable models.

3.4. Structural relaxation

The relaxation processes observed in complex liquids can be often attributed to intermolecular conformational changes the system undergoes under the influence of external perturbations. This is basically a structural effect, and the main problem is to find out what is the geometry of the conformations involved in the process. From the dynamic point of view, which is the light scattering view, we have to ask ourselves what is the molecular mechanism of such transitions, what are the intermediate stages and what can we learn about these from the measurement of relaxation times.

Liquid systems which undergo structural relaxation processes display mechanical and other dispersion effects. In contrast to vibrational relaxation, which can be observed in the same frequency range, structural relaxation is strongly temperature dependent, the relaxation times changing by several decades in a narrow temperature range. This temperature dependence approximately parallels the changes of shear viscosity with temperature which is especially drastic in the neighbourhood of the glass transition. The mechanical properties of glass forming liquids can be studied by mechanical strain-stress relaxation measurements, by ultrasonics, or by high frequency Brillouin scattering hypersonic measurements which are described in more detail in section 5. The combination of such methods will result in a wide frequency range which is very useful for the interpretation of the data. Brillouin light scattering measurements have thus been used to extend the frequency range on the high frequency side.

In interpreting the hypersonic data of this kind one should keep in mind the connection of sound velocity to the elastic moduli of the medium. At the high frequency involved in Brillouin scattering, the frequency dependence of these moduli is important. The Brillouin shift and linewidth depend on the real and the imaginary parts of the complex longitudinal modulus $M(\omega)$ repectively (11). The longitudinal modulus is connected to the compressional modulus $K(\omega)$ and to the transversal modulus $G(\omega)$ by the equation :

$$M(\omega) = K(\omega) + 4\,G(\omega)/3 \qquad (26a)$$

whose frequency dependence is given by the equations:

$$K(\omega) = K_o + S\omega^2\tau_b/(1+ i\omega\tau_b) \qquad (26b)$$

$$G(\omega) = T\omega\tau_{sh}/(1+ i\omega\tau_{sh}) \qquad (26c)$$

The quantity K_o is the static modulus of compression, S and T are the corresponding relaxation amplitudes. These equations describe the frequency dependence of the moduli in a liquid in which a single relaxation process is effective, characterized by the shear and the bulk relaxation times τ_{sh} and τ_b and the corresponding amplitudes T and S. If there are more relaxation processes, we sum over their respective contributions. The connection between the elastic moduli and the Brillouin shift and linewidth is given by the two equations for the real and the imaginary part of the longitudinal modulus as a function of the frequency:

$$M'(\omega) = \rho u^2(\omega) \qquad (26d)$$

$$M''(\omega) = 2\rho\alpha u^3(\omega) \qquad (26e)$$

where $u(\omega)$ is the sound velocity at the frequency ω, ρ the density and α the attenuation coefficient.

Another aspect of the situation is illuminated by describing the dissipation by the viscosity of the liquid. In a purely viscous liquid, i.e. in which absorption is due only to the shear viscosity, we have:

$$\alpha_{class}/\omega^2 = 2\eta_s/3\rho u^3(0) \qquad (27)$$

indicating that the expression α/ω^2 for classical absorption is frequency independent. Absorption in excess of α_{class} is attributed to the volume viscosity which is frequency dependent and related to the imaginary part of the compressional modulus as follows:

$$\eta_v = \lim_{\omega\to 0} K''(\omega)/\omega \qquad (28a)$$

This equation is similar to the relation connecting the shear viscosity to the imaginary part of the transversal modulus:

$$\eta_s = \lim_{\omega\to 0} G''(\omega)/\omega \qquad (28b)$$

The combination of equations 27 and 28 gives the expression of the attenuation of a relaxing liquid in terms of the shear and the volume viscosity:

$$\lim \alpha/\omega^2 = (\eta_v + 4\eta_s/3)/2\cdot\rho\cdot u^3(0) \qquad (29)$$

the whole of the relaxation effect being hidden in the frequency dependent volume viscosity. Unfortunately, the experimental access to this quantity is not very clear up to now.

A phenomenological model of structural relaxation is obtained by using viscoelastic theory. Actually, the structural relaxation processes always entail a viscoelastic behaviour of the system under some appropriate stress. With the concept viscoelastic liquids one generally understands materials which, under the appropriate conditions, are able both to store energy in elastic deformation and to dissipate energy as heat.

4. EXPERIMENTAL TECHNIQUES

The physically immediately relevant quantity in molecular dynamics is the CF of a molecular dynamical variable which in light scattering is the molecular polarizability. We can either gain access to this quantity by making a measurement in real time by photon correlation analysis, or we can make the corresponding measurement in the Fourier transform of time, i.e.in the frequency space. This latter procedure is the basis of dynamical light scattering spectroscopy where we have to use an instrument with the appropriate frequency resolution in order to be able to measure the lineshapes with the accuracy required to extract the time information contained in the corresponding correlation function.

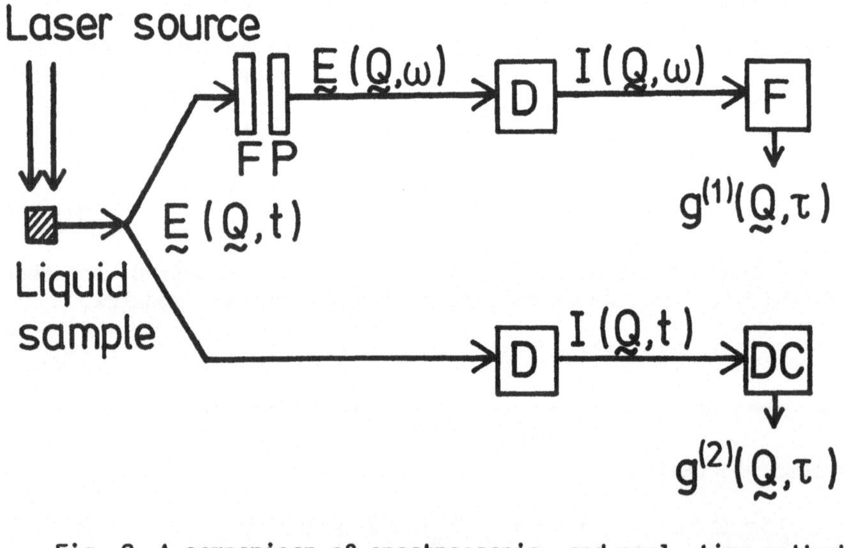

Fig. 2 A comparison of spectroscopic and real time methods

This point needs to be stressed because of the limitations in time resolution imposed by the free spectral range on the one hand, and the spectral resolutions of the available instrumentation on the other. The two principles, by which time correlation functions are currently obtained from the analysis of quasi elastic light scattering, are illustrated in figure 2: The sample is illuminated by the laser source and the scattered light can be analyzed in two different ways:

a) The spectroscopic way involves a scanned frequency filter, for example a Fabry Perot interferometer FP. The output of the instrument is a monochromatic field $E(Q,\omega)$ with the frequency ω which impinges at the detector cathode D, resulting in an electric current whose intensity is proportional to the square of the field. The output of the recording system is Fourier transformed to give the electric field CF $g^{(1)}(Q,t)$.

b) On the real time way the signal $E(Q,t)$ falls immediately on the detector. The output of the recording system follows the fluctuating signal in real time thus, giving the intensity $I(Q,t)$. The fluctuating signal is analyzed in a digital correlator which forms the intensity CF $g^{(2)}(Q,t)$ by multiplying the signal at successive increment delay times Δt ,and summing the products stored in a number of channels until the statistics are sufficiently good to form a reliable statistical average.

The interferometric techniques are described in some detail in (1) and (2). The books (1,3,5,7) are mainly devoted to photon correlation.

Although both methods give us in principle the same information, they are complementary mainly in two respects:

a) An important difference between the two experimental approaches comes from the fact that, whereas interferometric spectroscopy gives us the field CF $g^{(1)}(t)$, the photon correlation analysis results in the intensity CF $g^{(2)}(t)$. These two quantities are related by the equation 12e which, however, is only valid under the assumption that the scattered light is Gaussian which by no means is generally the case.

b) Due to the restrictions in optical instrumentation the free spectral range of a Fabry Perot interferometer can be varied in the range of 1 to 200 GHz corresponding roughly to a time range of 200 to 1 ps. On the other hand, the restrictions of the commercially available hardware computers, like a digital correlator, impose a lower limit to the shortest measurable relaxation time of little less than one microsecond. At the other end of the time scale the mechanical and thermal stability of a light scattering system permit the use of sample times at most of the order of one second. This time range is very large compared to many other methods, since it enables us to measure currently times between a few tenths of a microsecond to one second. The large time range of photon correlation spectroscopy makes possible the study of distributions of relaxation times, which are very often observed in complex liquids, and in some cases the measurement of two distinct relaxation processes with different relaxation times in the same experiment. This makes photon correlation so valuable for complex systems where single exponential correlograms are not the rule. Interferometry, on the other side, is at its best in a very interesting time range but the general successful use of Lorentzian fits indicates that complex situations like distributions of relaxation times are not easily observed due to the narrow frequency range of the instrument at a given free spectral range. This drawback, which is due to the property of the interferometer to dump the information about faster processes into the baseline, is in some respects an advantage, since it simplifies sometimes the situation to the essential process. If used under optimal conditions, light scattering can exploit the whole time range from the subpicosecond scale to minutes, keeping in mind the additional possibility of monochromator spectral analysis. However, there are but a few if any examples in the literature which have demonstrated the feasibility of this cooperation of experimental techniques in light scattering.

5. BRILLOUIN LIGHT SCATTERING

The appearance of the Brillouin doublet, consisting of two shifted lines symmetrical to the central Rayleigh line in the VV-light scattering spectrum, can be understood in the phenomenological picture, when we take into account that the local index of refraction is a function of the local density. The density fluctuations in a liquid can be separated in a system of non propagating fluctuations, and a system of longitudinal propagating fluctuations/modes having the form of a grating of periodic optical density maxima and minima. We can obtain under certain angles Bragg reflections of the scattered light from this optical density grating propagating at hypersonic speed. Thus the scattered light observed under a particular angle filters a particular wavelength/frequency out of the ensemble of the propagating fluctuations. The Brillouin doublet, which is observed at the detector, is best characterized by the frequency shift ω_B with respect to the central line and by its linewidth $\Delta\omega_B$. Disregarding for the moment the question of the exact shape of the Brillouin lines, the shift can be related to the velocity of the sound wave propagating at the particular frequency picked out by the experiment, and the linewidth to the attenuation of this wave. Since the frequency accessible to the Brillouin light scattering experiment is usually in the GHz range, and since by varying the scattering angle we can vary this frequency roughly by one decade, this enables us to observe dispersion phenomena occuring in this interval of the hypersonic frequency range.This has proven to be a useful technique to follow relaxation processes whose relaxation times lie between 100 and 1000 ps. The complete line shape theory of Brillouin light scattering has been developed among others by Mountain (12,13). An important point in connection with this is the theoretical prediction and the observation of an additional unshifted relaxation line whose linewidth gives us directly the inverse relaxation time of the same process as the one responsible for the hypersonic dispersion. Thus a common theoretical treatment of both the unshifted relaxation line and the Brillouin doublet and, whenever possible, a data analysis including both is very helpful in increasing the reliability of the results.

Two quite different phenomena have been mainly examined with the aid of Brillouin light scattering. These are vibrational relaxation phenomena in liquids and structural relaxation processes. As the time window of the Brillouin experiment is not much larger than a decade, the observation of vibrational relaxation has been restricted to such substances whose vibrational degrees of freedom exchange energy with the translational heat reservoir at a rate of several hundred GHz. On the other hand, structural relaxation is strongly temperature dependent, and thus it has been often possible to shift the relaxation time into the Brillouin time window by changing the sample temperature. However, this advantage has the drawback that we have to use a temperature-frequency shift relation which is not rigorously known and which introduces a model dependent bias into the experimental data. However, in spite of this the method, if used with sufficient caution, has proved very useful.

5.1 Vibrational relaxation of simple liquids

The vibrational relaxation of the three liquid tetrachlorides CCl_4, $SiCl_4$, and $SnCl_4$ has been studied extensively (14,15) and some of the results will be reported here because they illustrate the possibities of Brillouin spectroscopy applied to this problem.

Figure 3 displays the hypersonic speed dispersion curves of CCl_4 at four different temperatures. The complete dispersion curve contains additionally an experimental point corresponding to the ultrasonic velocity, i.e. lying at a comparatively low frequency which is located on the ordinate axis. The fit has been made with the classical dispersion relation with the relaxation time, and the amplitude of the dispersion curve as free fit parameters according to the equation:

$$u(\omega)^2/u(0)^2 - 1 = (u(\infty)^2/u(0)^2 - 1)\omega^2\tau^2/(1 + \omega^2\tau^2) \qquad (30)$$

The amplitude of the dispersion curve is connected to the relaxing energy reservoir which is a valuable indicator as to what is actually the relaxing quantity. In the case of CCl_4 it is found that the relaxing energy is equal to the sum of the energies of the vibrational levels of this molecule. This is a confirmation of the assumption that the relaxation process manifested in the hypersonic dispersion is the exchange of energy between the whole of the internal degrees of freedom of this molecule and the external energy reservoir.

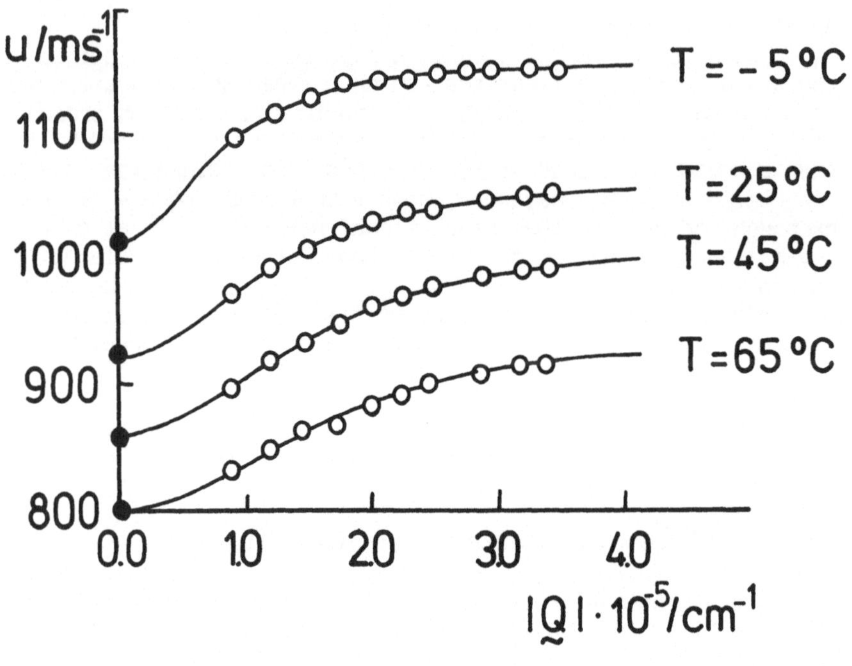

Fig.3: Hypersonic dispersion curves of CCl_4

In the case of the other tetrachlorides $SnCl_4$ and $SiCl_4$, however, the relaxation strengths indicate clearly that the relaxing energy is that of a subset of the internal degrees of freedom. This finding is not easy to rationalize at first sight. It should be noted, however, that, as shown in figure 4, the vibrational levels, which do not participate to the observed process, are the lower ones which furthermore lie below 200 wavenumbers, the thermal energy level at room temperature. This might indicate that the levels below 200 cm^{-1} relax at a faster rate which is not observable within the time-window of the Brillouin spectrum. The problem which remains is, however, that of the mechanism that leads to the coupling of all the levels in CCl_4 and to the decoupling in the two other tetrachlorides. The problem of the existence or non-existence of two distinct relaxation times in vibrational relaxation is not a new one, and much work has been devoted to this (see for instance (16)). It is very probable that we should include symmetry considerations in this discussion.

The relaxation times observed in vibrational relaxation have been related to the rate of "collisions" Z between molecules in

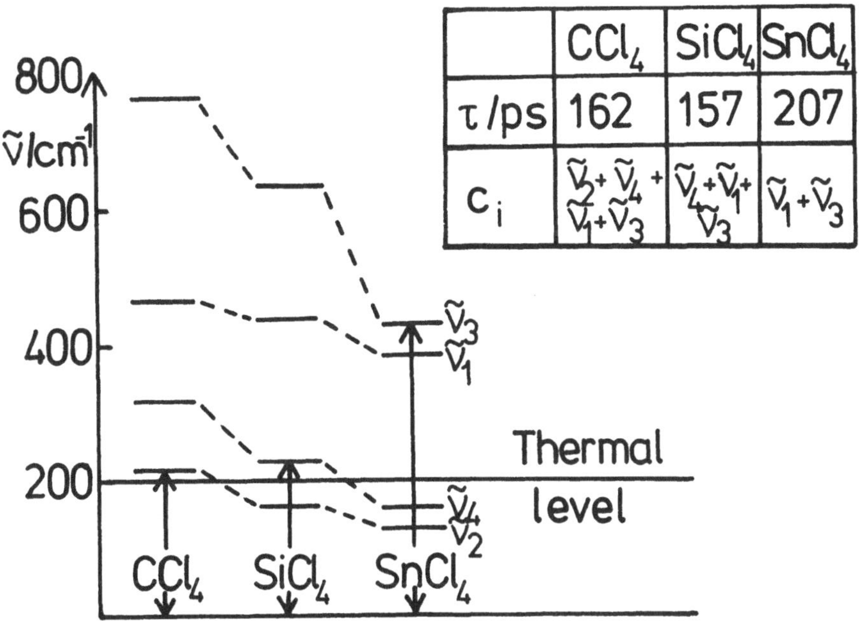

Fig.4: Relaxing energy levels in the three tetrachlorides c_i is the internal energy corresponding to the indicated vibrational modes

the liquid and the probability P of the energy transition on a collision of a certain type (16,17).

$$1/\tau_R = P\,Z\cdot\varphi$$

where φ is a factor correcting for the population of the modes. The rate of collisions, for example in CCl_4, has been calculated by a molecular dynamics simulation (18), and thus we are in the position to calculate the probability factor and to compare it to the theoretical predictions. It has been found that the commonly used calculations based upon the breathing sphere model of the internal degrees of freedom are too rough and cannot reproduce the observed temperature dependence of τ_R. If we use instead a mode matching model, we obtain a much more realistic picture of the relaxation which becomes a function of the configuration of approach of the molecules and of the symmetry of the vibrations involved as compared to the symmetry of the collision (19).

Measurements, which have been performed on liquid mixtures (20,21), enable us to separate the relaxation times for the relaxation of a molecule of the one species undergoing a collision with another molecule of the same species or with a molecule of the other. The conclusion, which can be drawn from these measurements, is that the relaxation times depend very sensitively on the properties of the collision partner and not only on its mass, as the simple collision theory suggests (20-22).

5.2 Hypersonic dispersion in hydrogen bonded liquids

An important issue of the study of high frequency viscoelastic phenomena by Brillouin spectroscopy is to come to a molecular interpretation of the viscoelastic parameters obtained in the experiment. An example of such a study is given by the study of two isomeric pentanediols, 1,5- and 2,4-pentanediol, in the liquid and the supercooled state (23). As already mentioned, instead of varying the scattering angle and thereby observing the dispersion curve at different frequencies, it is more convenient to change the temperature of the sample and, by an appropriate frequency-temperature shift, to obtain a dispersion curve over a much wider range. The experimental data are displayed in fig.5:

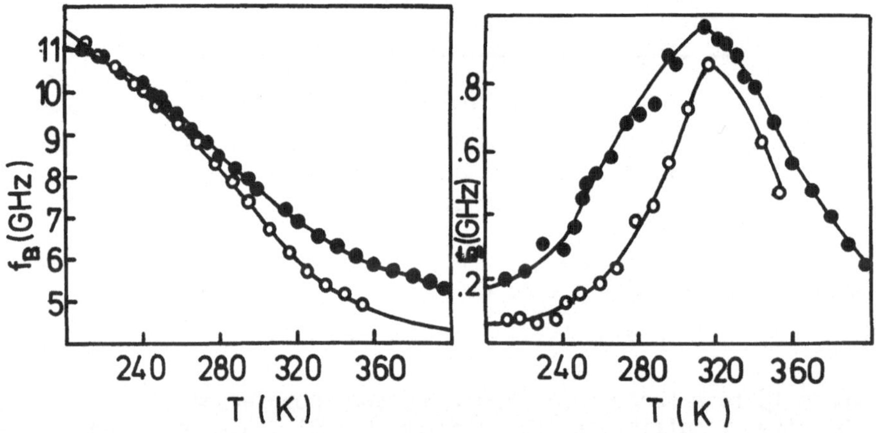

Fig.5: Dispersion of the Brillouin shift f_B and linewidth Γ_B for the two pentanediols as a function of temperature

Two observations can be made offhand by inspection of the data. First, we can see a clear dispersion in the frequency shift and a maximum in the linewidths of the spectra. Second, there are significant differences in the behaviour of the two isomeric substances.

In order to give a molecular interpretation of these results we have to use an appropriate model for the structural relaxation process. Quite generally, we can assume that the propagating longitudinal density fluctuations giving rise to the Brillouin doublet are due to rearrangements of molecules or molecular segments in the direction parallel to the wavevector Q. Lin and Wang (24, 25) treat this model by means of a Mori type theory assuming a memory function which decays exponentially with a relaxation time τ_s. The relations they obtain contain among others the longitudinal elastic modulus which they relate to the longitudinal component of the total stress induced by the sound wave, as it distorts the molecules away from their equilibrium positions. τ_s is the time required for this induced strain to relax away, as the molecules rearrange themselves to a new equilibrium position. The temperature dependence of τ_s for both diols is Arrhenius over the temperature range studied, and we obtain two different activation energies as displayed in figure 6.

In section 6.2, the depolarized light scattering results for the two alcohols will be exposed and the mechanism of the observed structural relaxation will be discussed on the basis of both classes of data.

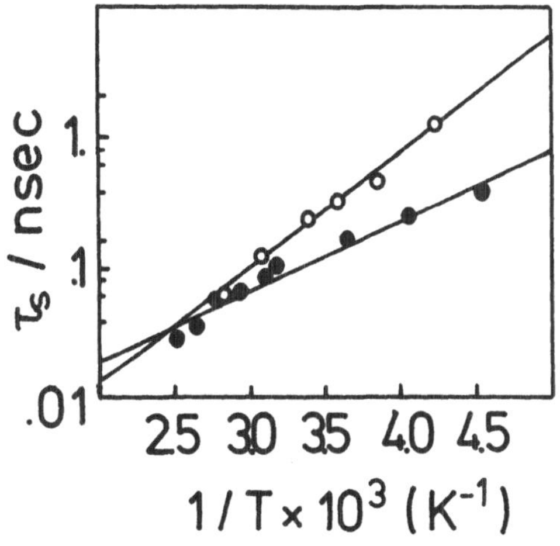

Fig.6: Arrhenius least square fit of the structural relaxation times of 1,5- and 2,4-pentanediols

6. DEPOLARIZED LIGHT SCATTERING SPECTROSCOPY

The study of the unshifted central line in light scattering is a useful tool for the study of reorientational dynamics, since the integrated intensity of this component contains only the dynamics of the anisotropic part of the molecular polarizability tensor. Consequently any translational effect is, at least in a first approximation, absent in these spectra. The frequency range, over which depolarized light scattering can be observed, is very large, but we do not dispose of any single device by which the whole range can be observed in one experiment. For this reason we have to use different techniques which are mainly:

a) Grating monochromator spectroscopy

b) Interferometric spectroscopy

c) Photon correlation analysis

The wavenumber range of the grating monochromator, which can be currently utilized in the central line light scattering spectroscopy, extends from approximately 1 to 200 cm^{-1} corresponding to a frequency range of 30 to 6000 GHz. However, the Rayleigh wings above 50 cm^{-1} are largely due to interaction induced effects which will not be treated here. By means of the interferometric method we can study a frequency range from 1 to 500 GHz. Finally, by means of photon correlation analysis techniques we can study the frequency range between 1 Hz and 10 MHz.

In order to study the liqud dynamics in such a way as to avoid to separate artificially processes, which are coupled and which lie in the time ranges of different experimental techniques, it would be useful to be able to adjust the spectra obtained in these frequency ranges to each other. Unfortunately, this is not feasible as such a procedure requires a very good accuracy, as far as absolute intensities are concerned. As a consequence of this restriction, the time window of each experimental method determines largely the process/processes we are able to analyse. For this reason only a few studies have been devoted to the attempt to join the different frequency ranges into a common picture of liquid dynamics. A solution of this problem would give us the possibility of separating uniquely processes like rotations, induced effects, or collective motions and possibly coupling effects between these. Despite the complexity of the situation, some examples will be presented in the following sections, where simple effects could be separated and interpreted in a relatively unambiguous way.

6.1 The reorientational dynamics of a large molecule: Diphenyl ether

The shape of this molecule is that of an asymmetric rotor, but in view of its geometry it can be approximately treated as an elongated symmetric top. In doing this we reduce the number of components of the central line from five to two. If additionally the I_{VH} intensity is determined by the rotation about a single axis, the reorientation about the other axis being either too slow or too fast for the instrumental range, we can describe the spectrum by a single Lorentzian (26). Figure 7 shows the spectra obtained with a Fabry Perot interferometer with a free spectral range of 25 GHz for three temperatures.

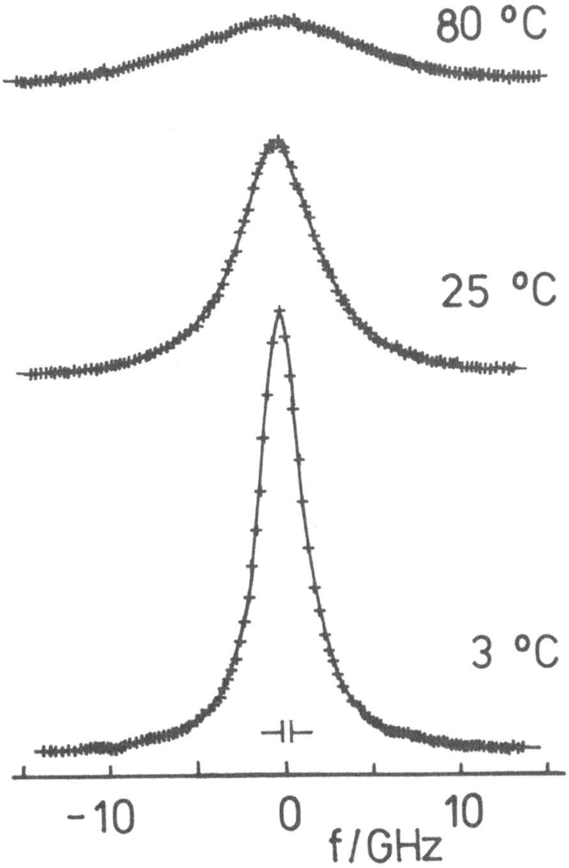

Fig.7 Depolarized light scattering spectra of liquid diphenyl ether. The instrumental linewidth is shown at the bottom. A Lorentzian lineshape function was fitted to the experimental spectra

The figure shows that the Lorentzian fits represent the spectra very well at all temperatures, thus justifying the assumption that the rotation about a single axis basically shapes the spectral line. This justifies also the assumption that the motion is essentially a diffusive reorientation of this axis. Under these conditions we can calculate the reorientation time τ_{OR} from the linewidth Γ_B by means of the equation:

$$\Gamma_B = 1/\tau_{OR} \tag{32}$$

The question arises as to the nature of this reorientation, i.e. if we observe a collective reorientation or a single particle reorientation. We can answer this question by measuring the reorientation time in diphenyl ether which has been diluted with CCl_4.

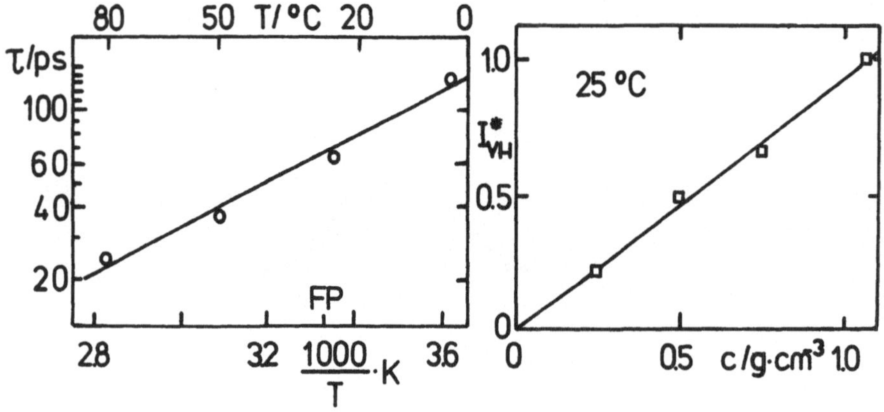

Fig.8: a) Arrhenius plot of the reorientational time
b) Effect of dilution on the reduced intensity

Figures 8 a) and b) display an Arrhenius plot and a plot of the concentration dependence of the reduced intensity of the solutions and the pure liquid. We see that we obtain a common activation energy in all cases and that the reduced intensity increases proportionally to the concentration of the molecules, indicating that the single molecules are the scatterers according to equation 16.

The assumption of a diffusive motion entails also the validity of hydrodynamics for the reorientation. If this is the case, then we are justified to some extent to use the Debye-Stokes-Einstein equation:

$$\tau_{OR} = V\eta_s/kT \tag{33}$$

which has been derived for the diffusive reorientation of a sphere in a viscous continuum. The extension of this model to the molecular scale was sucessful in many cases. The reorientation of a molecule with a non spherical shape, sweeping a hydrodynamic volume V in the process of reorientation under two limiting boundary conditions (stick and slip), has been fitted successfully to many experimental data. The value of this model is, that it gives a connection between the molecular size/shape and the reorientational relaxation time τ_{OR} provided we can assume that the effective viscosity at this level is identical with the macroscopic shear viscosity. Since we have independent data on molecular sizes and shapes, the information we can obtain on the nature of the effective viscosity on the molecular level seems to be the most fruitful application of this equation in connection with the molecular relaxation data as obtained by light scattering.

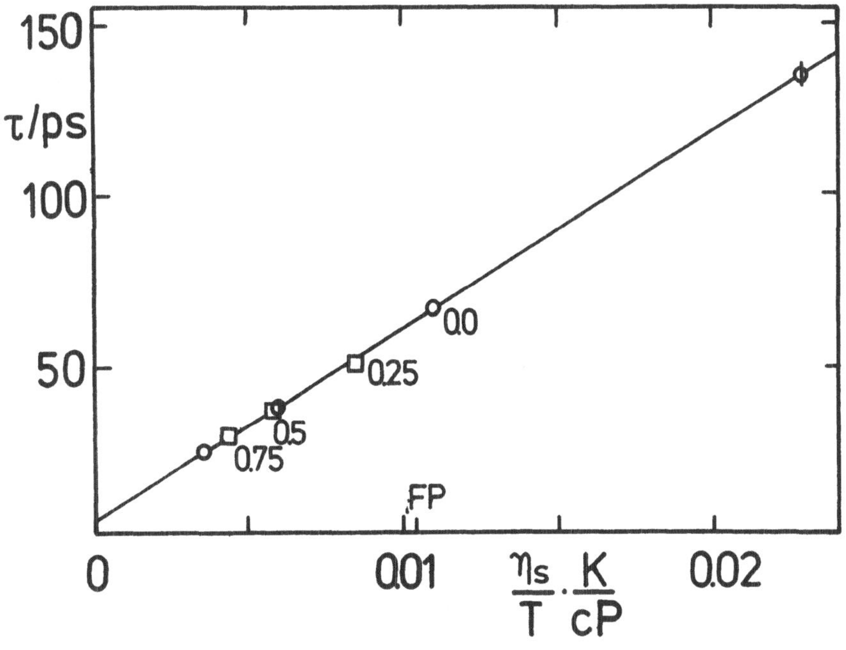

Fig.9: Plot according to equation 33. The viscosity is varied by changing the temperature or the composition of the solution.
O : Variation of T in neat diphenyl ether
□ : Variation of the concentration at 298 K
The numbers indicate the concentrations in g/cm^3

Figure 9 shows a plot of the experimental reorientation time vs. the ratio η_s /T of the solution. The plot contains the temperature dependence as well as the concentration dependence of the solution, since the viscosity is a function of both parameters. The agreement with the equation 33 is excellent. The slope of this straight line, which gives the hydrodynamic volume of the rotational motion in connection with the stick boundary conditions, shows that the reorientation observed is the one of the axis passing through the centers of the phenyl rings. The hydrodynamic volume for reorientation of the molecule is found equal to 79 A^3 which is in good agreement with the size calculated from the Van der Waals radii. This is an indication that in the case of liquid diphenyl ether the molecular motion is subject to a viscous drag which can be indeed described by the macroscopic shear viscosity. This is not the case in all liquids as will be seen in section 5.2 for the structural relaxation of hydrogen bonded liquids.

6.2 Structural relaxation of hydrogen bonded liquids

The differences observed in the structural relaxation of the two diols, 1,5-pentanediol and 2,4-pentanediol, by means of Brillouin light scattering, exposed in section 5.2, can be also fruitfully studied by depolarized light scattering in the interferometric as well as in the photon correlation frequency ranges (27,28). Since these substances are known to display an appreciable amount of hydrogen bonding, as suggested for example by the strong increase of the macroscopic shear viscosity at low temperatures, the question as to the nature of the reorienting moieties observed in depolarized light scattering deserves some inquiry.

The integrated intensity of the depolarized spectra obtained in the interferometric analysis are, according to equation 16, a measure, both of the scattering power and the concentration of the scattering centers. It is important to note that, since we integrate over the spectrum observed in the interferometric experiment, the integrated intensity can give us a clue as to the nature of the moieties which are undergoing reorientation at the particular rate given by the Lorentzian half width of the spectrum, and not of any appreciably larger or smaller clusters which might occur in the liquid. Thus the total intensity of the scattered light is generally larger, since it contains contributions both, from faster motions which contribute to the scattering observed in the baseline of the spectra obtained with the interferometer, and sometimes from slow motions which manifest themselves as narrow spikes at the top of these spectra.

Figure 10 displays the integrated reduced intensities defined as $I^*_{VH} = I_{VH}/\rho$ of the two pentanediols as a function of the temperature. Assuming that the molecular optical anisotropy of the two isomers is essentially similar, the lower values of I^*_{VH} for 2,4-pentanediol indicate that the configuration of this molecule is such as to compensate the anisotropy. This is

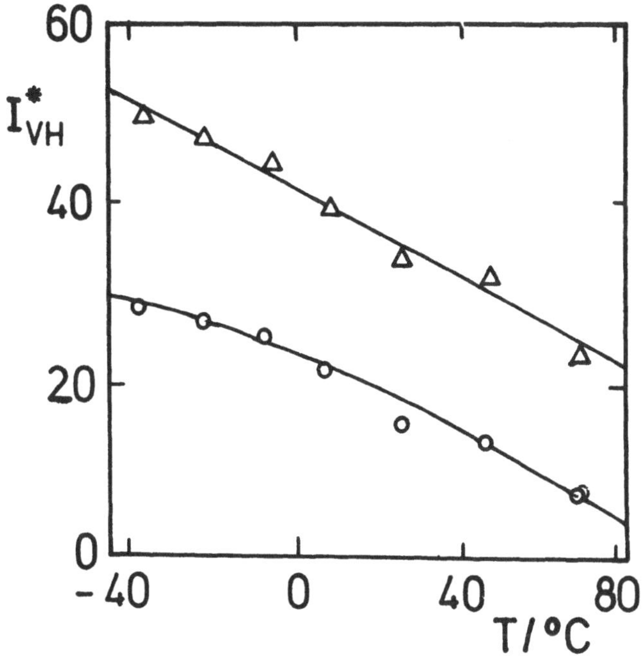

Fig 10: Reduced intensities of the depolarized light scattering
Δ : 1,5-pentanediol
o : 2,4-pentanediol

indeed plausible, since the position of the two OH-groups facilitates the formation of a six member ring by intramolecular hydrogen bonding. Thus, the scattering in this liquid is either due to molecules in a less anisotropic conformation or to those molecules which are in an anisotropic conformation, but whose concentration is now lower than the total concentration of the alcohol molecules due to the intramolecular hydrogen bonding. Simplifying the situation, we can describe it by two structures of different anisotropy between which we observe a temperature dependent equilibrium, so that the total intensity can be described by the relation:

$$I^* = X_1\beta_1^2 + X_2\beta_2^2 \qquad (34)$$

The indices 1 and 2 indicate the two structures with their respective mole fractions and anisotropies.

Since the temperature dependence will be mainly due to the shift of this equilibrium, figure 10 indicates an increase in intramolecular bonding at low temperatures in 2,4-pentanediol.

This is in contrast to the situation in 1,5-pentanediol where intramolecular hydrogen bonding is less favored by the molecular structure. These results can provide an estimate of the free energy of the formation of the relevant structures. In a more realistic approximation, however, the anisotropic scattering in the alcohols can be described by a distribution of the local aniso- tropy rather than to well defined structures with an equally well defined anisotropy. Dilution experiments with $CHCl_3$ have confirmed this picture and have shown that the reduced intensity of 2,4-pentane-diol decreases with the solvent content. We can understand this, as the solvent will probably decrease the ratio of intermolecular to intramolecular hydrogen bonding in the same manner, as the increase of the temperature does.

The study of the dynamic properties can be helpful in understanding the differences between the two alcohols. The relaxation times have been obtained by interferometric spectroscopy and by photon correlation spectroscopy in order to exploit the full range of the two methods from 10^{-1} to 10^{-10} s. Before attempting a common approach to the two spectra it should be noted that, whereas the interferometric spectra of the two alcohols can always be fitted with a single Lorentzian, the correlograms obtained by photon correlation analysis are wider than single exponentials. This is a common observation in such systems (29,30) and the problem has been handled by using a variety of empirical fit functions which introduce an additional parameter characterizing the width of the distribution. In the case of the alcohols, we have used the Williams-Watts distribution function (31) which yielded excellent fits. This equation has the form:

$$g^1(t) = \exp(-t/\tau_0)^{\beta} \qquad (35)$$

where β is the width parameter and τ_0 the primary relaxation time. A mean correlation time is defined as:

$$\bar{\tau} = \int_0^{\infty} dt \, \exp(-t/\tau_0)^{\beta} = \tau_0 \cdot \Gamma(\beta^{-1})/ \qquad (36)$$

$\Gamma(\beta^{-1})$ is the gamma function. Figure 11 displays the fit of a correlogram obtained with a digital clipped correlator with 94 channels by superimposing four successive correlograms using incremental delay times $\Delta\tau$ increased by a factor of four each time.

The temperature dependence of a structural relaxation process is in some cases of the Arrhenius type like in the case in 1,5-pentanediol. Very often we observe, however, a much stronger dependence especially at low temperatures. This has been successfully described by the Vogel-Fulcher-Tamann (VFT) equation involving an additional parameter, the T_0 temperature, which is related to the glass temperature T_g:

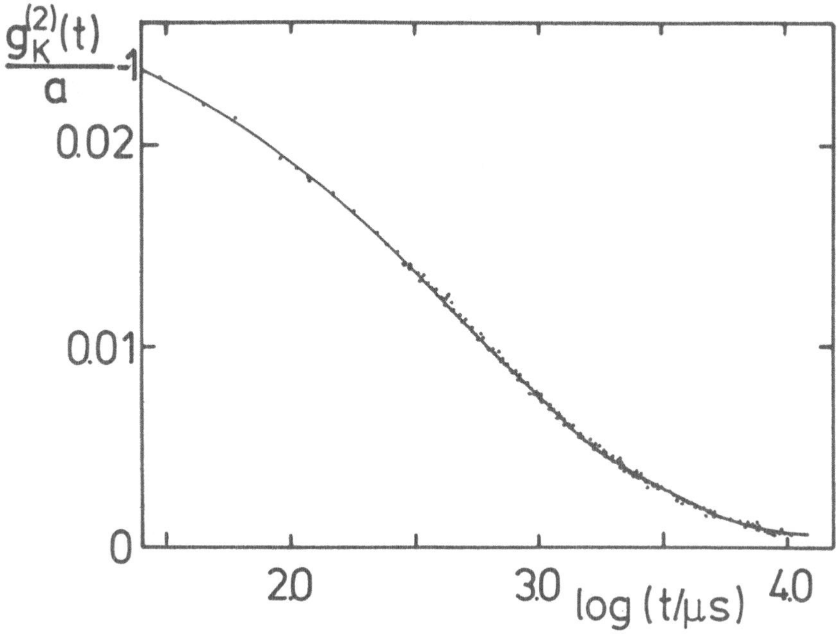

Fig.11: Correlogram of 2.4-pentanediol at -54 4^{0}C
The ordinate is $g^{1}(t)$ according to equ. 12e

$$\bar{\tau} = A \exp(B/(T-T_o) \qquad (37)$$

The model underlying this equation is based upon the assumption that the free volume available to the motion under study determines the relaxation time. The decrease of the temperature reduces this volume near to and below a critical volume. At temperatures lower than the glass point or the T_o temperature the displacement of the molecules or of molecular segments is inhibited by the potential walls of the surrounding molecules or atoms. The effect of temperature is, contrary to the activation complex model, not to provide to the moving molecule enough energy to pass over a constant barrier but rather to induce rearrangements which lower this barrier. This mechanism can explain qualitatively the extremely steep change of the relaxation time on approaching the T_o point.

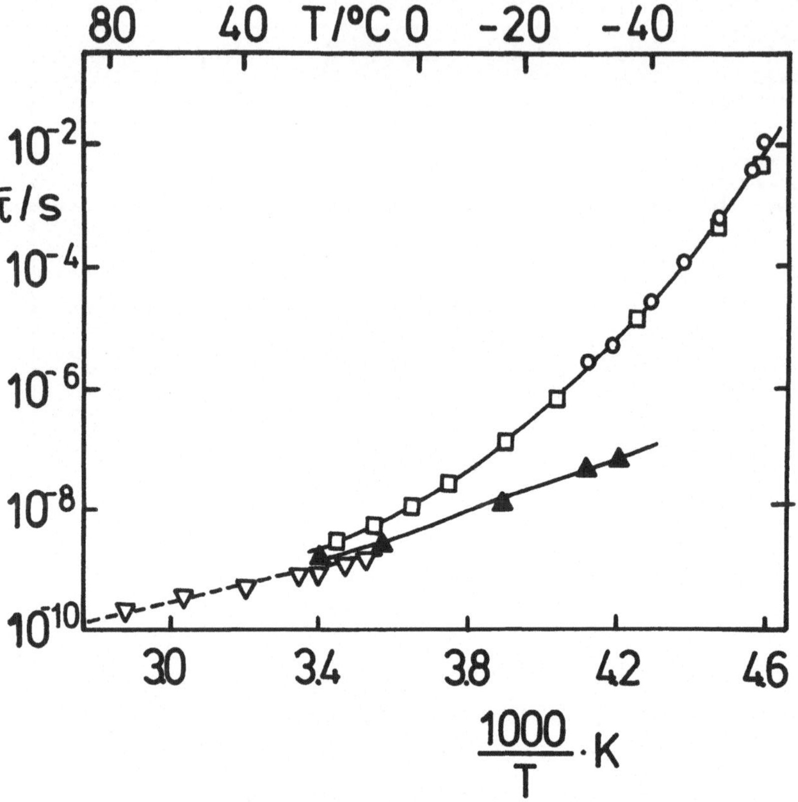

Fig.12: Temperature dependence of the relaxation times of the pentanediols:
2,4-pentanediol; o: Photon correlation,
□: Dielectric relaxation
▽: Interferometric spectroscopy
1,5-pentanediol ▲: Interferometric spectroscopy

The same equation has been successfully used to describe the temperature dependence of the viscosity of glass forming liquids and polymer melts near and above the glass point (32-34). The constant B has been related to some kind of activation energy, although neither the model itself nor the form of the equation 35 lends itself easily to the use of this concept. Formally, the VFT equation takes the form of the Arrhenius equation when $T \rightarrow T_o$, thus justifying an identification of B with the high temperature limit of the slope of the log $\bar{\tau}$ vs. 1/T plot. Figure 12 displays the temperature dependence of the light scattering relaxation times of the two pentanediols.

We can see that the shape of the curve for 2,4-pentanediol does not conform to the Arrhenius equation, and the fit with the VFT equation shows a very good agreement. The dielectric relaxation times, which have been included in the fit, lie on the same curve. The extrapolation from the long time to the short time interferometric data is not good but, as the gap between the photon correlation and the interferometric relaxation times is as large as three decades, which are only bridged by the dielectric times, the significance of this mismatch is not clear. Two other alcohols, glycerol and 1,2,6-hexanetriol, have given a perfect match of the corresponding data (30,34), whereas in the case of some polymers as polysiloxanes (35), the photon correlation data lie on a completely different curve compared to the interferometric data, indicating that different relaxation mechanisms are observed in these time ranges. The high temperature interferometric relaxation times of both alcohols are identical.

It should be noted that in the alcohols, as well as in other glass forming liquids, like o-terphenyl (32) and a number of polymers (36), the correlograms of the I_{VH} component and of the I_{VV} component are very similar, yielding identical correlation times. This is generally interpreted as an indication of a strong coupling of the reorientational and the translational motion of the scattering centers. As a consequence of this observation, the CS measurements have been made on the I_{VV} component which, as far as the CF is concerned, is identical with the I_{VH} component in the case of the pentanediols. However, one cannot exclude the possibility that at higher temperatures the two modes responsible for the two components gradually decouple. This would explain the mismatch of the two light scattering curves in figure 12.

The correlation time data can be further rationalized by the use of the Debye-Stokes-Einstein equation 35, as displayed in figure 13.

The viscosity has been varied both by changing the temperature and the composition in the latter case by diluting the alcohol with $CHCl_3$. The correlation times lie, in agreement with the equation 34, on a straight line. This is in favour of a model by which the reorienting moieties are similar in volume over the whole temperature and concentration range. An additional argument in favour of this model comes from the intensity data of the mixtures. Although the total intensity increases slightly on dilution due to concentration fluctuations, the relaxation amplitude of the correlogram is reduced by approximately 25% in the concentration range studied, and the relaxation time decreases by two orders of magnitude. This is consistent with a model in which the same or very similar groups are reorienting in an environment which dissipates the energy by a viscous mechanism similar to the relief of a macroscopic shear stress. However, in taking a closer look at the effect of the dilution by means of the values of $\bar{\tau} T/\eta_s$ in fig. 13 we observe that this quantity which should be constant, if the only effect of dilution were a change in viscosity, does indeed increase with dilution.

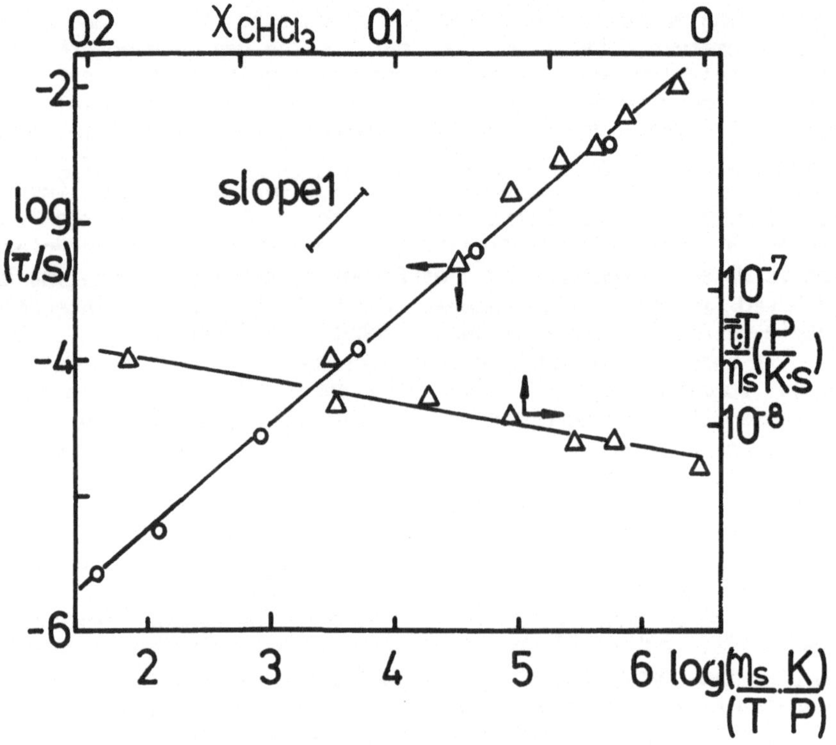

Fig.13: The influence of temperature and dilution on the nature of the reorienting structures
o: undiluted 2,4-pentanediol.
Δ: diluted with $CHCl_3$ at -56.3 K

This can be explained by giving up the picture of the same groups reorienting independently from the composition, but it can also mean that the effective viscosity is concentration dependent not to the same extend as the macroscopic viscosity. The first alternative is not supported by the intensities. The second seems more plausible because it fits better into the picture, according to which the chloroform breaks the intermolecular hydrogen bonds leaving the intramolecular hydrogen bonds largely unaffected. This leads to a decrease of the macroscopic viscosity, however, the reorienting moieties, which are set free by this process, are much more affected than the process by which a macroscopic stress is relieved and which is dominated by the structure of the overall hydrogen bonded network. Thus the dilution with chloroform introduces some kind of freely dangling "side chains" in the otherwise connected network of the bulk substance, which contribute to the light scattering correlation time, but not, or to a lesser degree, to the macroscopic shear viscosity. This picture is still largely hypothetic, but it is consistent with all the experimental findings.

This short discussion of the light scattering results of the pentanediols demonstrates the complexity of the situation in structural relaxation processes as opposed to the reorientation of well defined molecules in a liquid. The problems awaiting a solution in the glass forming liquids are numerous, and a coordinated discussion of results obtained by various methods is necessary.

6.3. High pressure light scattering studies of structural relaxation

In a study of the photon correlation light scattering spectroscopy of o-terphenyl at pressures up to 1250 bar (37) the mean relaxation times were obtained as a function of temperature and pressure as well. This kind of study enables us to measure the activation energy and the activation volume, thus obtaining one more clue as to the molecular mechanism of the structural relaxation process. In the particular case of o-terphenyl it was found that the height of the activation barrier increases with increasing pressure. The activation volume is found approximately equal to the molecular volume of the liquid suggesting that the structural relaxation observed involves an activation which is connected with the motion of a single molecule on the average. At lower temperatures, this volume increases indicating a larger degree of cooperativity. The results of this study could be well described by means of an extended VFT-equation involving a linear pressure dependence of the activation energy and of the glass point temperature.

$$\bar{\tau} = \bar{\tau}_0 \exp((B+\alpha P)/(T-T_0+ bP)) \qquad (37)$$

7. OUTLOOK

The recent introduction of high pressure techniques to the light scattering spectroscopy (37,38) provides a good example of the still unexploited possibilities of dynamic light scattering in the study of liquid state dynamics. There are more promising extensions of light scattering spectroscopy, like the use of resonance enhanced light scattering spectroscopy by which we can study the dynamics of labeled molecules at high dilution (39-41). These developments can greatly increase the usefulness of light scattering spectroscopy and, if applied systematically, this technique can develop into one of the most promising tools for the study of the liquid state.

In spite of the wide dynamic frequency range accessible to light scattering spectroscopy it would be very useful to extend the frequency range of the different techniques in order to close the existing gaps and ,if possible to come to some overlap. This would be very helpful in increasing the reliability of the results. A possibility to extend the range of photon correlation techniques to shorter times is, for example, provided by the measurement of the time of arrival of photoelectron pulses. A rewiev of this and of some other technical developments by B. Chu (42) shows that the possibilities of dynamical light scattering spectroscopy are far from being exhausted.

REFERENCES

1) Berne, B.J., and Pecora, R., Dynamic Light Scattering, J. Wiley, 1976, New York.
2) Chu, B., Laser Light Scattering, Academic Press, 1974, New York, San Francisco, London.
3) Cummins, H.Z., and Pike, E.R. (eds.), Photon Correlation and Light Beating Spectroscopy, Plenum Press, 1974, New York, London.
4) Degiorgio, V., Corti, M., and Giglio, M. (eds.), Light Scattering in Liquids and Macromolecular Solutions, Plenum Press, 1980, New York, London.
5) Cummins, H.Z., and Pike, E.R. (eds.), Photon Correlation Spectroscopy and Velocimetry, Plenum Press, 1977, New York, London.
6) Chen, S.-H., Chu, B., and Nossal, R. (eds.), Scattering Techniques Applied to Supramolecular and Nonequilibrium Systems, Plenum Press, 1980, New York, London.
7) Schulz-DuBois, E.O. (ed.), Photon Correlation Techniques in Fluid Mechanics, Springer, 1983, Berlin, Heidelberg, New York.
8) Earnshaw, J.C., and Steer, M.W. (eds.), The Application of Laser Light Scattering to the Study of Biological Motion, Plenum Press, 1983, New York, London.
9) Fleury, P.A., and Boon, J.P., Laser Light Scattering in Fluid Systems, in: Prigogine, I., and Rice, S.A. (eds.), adv. Chem. Phys. 24, 1-93 (1974).
10) Keyes, R., and Kivelson, D., J. Chem. Phys. 56, 1057 (1972).
11) Harrison, G., The Dynamic Properties of Supercooled Liquids, Academic Press, 1976, London, New York, San Francisco.
12) Mountain, R.D., 1966, J. Res. natn. Bur. Stand. A 70, 207.
13) Mountain, R.D., 1968, J. Res. natn. Bur. Stand. A 72, 95.
14) Samios, D., Relaxationsprozesse bei der Energieübertragung in flüssigem CCl_4, $SiCl_4$, $SnCl_4$ und in deren Mischungen, Dissertation, 1979, Bielefeld.
15) Samios, D., and Dorfmüller, Th., Mol. Phys. 41, 637 (1980).

16) Lambert, J.D., Vibrational and rotational relaxation in gases, in: Rowlinson, J.S. (ed.), The International Series of Monographs on Chemistry, Claredon Press, 1977, Oxford.
17) Herzfeld, K.F., and Litovitz, T.A., Absorption and Dispersion of Ultrasonic Waves, Academic Press, 1959, New York, London.
18) Samios, J., and Dorfmüller, Th., J. Chem. Phys. 76, 5463 (1982).
19) Samios, D., Samios, S., and Dorfmüller, Th., Mol. Phys. 49, 543 (1983).
2o) Dorfmüller, Th., and Samios,D., Mol. Phys. 43, 23 (1981).
21) Samios, D., Dorfmüller, Th., and Asenbaum, A., Chem. Phys. 65, 305 (1982).
22) Dorfmüller, Th., and Samios, D., Vibrational Relaxation in Molecular Liquids, in: Capellos, C., and Walker, R.F. (eds.), Fast Reactions in Energetic Systems, pp.445-459, D. Reidel Publishing Company, 1981, Dordrecht, Boston, London.
23) Lempert, W., Wang, C.H., Fytas, G., and Dorfmüller, Th., J. Chem. Phys. 76, 4872 (1982).
24) Lin, Y.-H., and Wang, C.H., J. Chem. Phys. 69, 1101 (1978).
25) Lin, Y.-H., and Wang, C.H., J. Chem. Phys. 70, 681 (1979).
26) Fytas, G., Lilge, D., and Dorfmüller, Th., J.Chem.Soc.Faraday Trans., 1983.
27) Fytas, G., and Dorfmüller, Th., Ber. Bunsenges. Phys. Chem. 85, 1064 (1981).
28) Fytas, G., and Dorfmüller, Th., Mol. Phys. 47, 741 (1982).
29) Dux, H., Photonenkorrelationsspektroskopische Untersuchung lokaler Relaxationsprozesse in unterkühlten Alkoholen, Dissertation, 1977, Bielefeld.
30) Dux, H., and Dorfmüller, Th., Chem. Phys., 40, 219 (1979).
31) Williams, G., and Watts, D.C., J. Chem. Soc. Faraday Trans. 66, 8C (1970).
32) Fytas, G., Wang, C.H., Lilge, D., and Dorfmüller, Th., J. Chem. Phys. 75, 4247 (1981).
33) Wang, C.H., Fytas, G., Lilge, D., and Dorfmüller, Th., Macromolecules, 14, 1363 (1981).
34) Dorfmüller, Th., Dux, H., Fytas, G., and Mersch, W., J.Chem. Phys. 71, 366 (1979).
35) Fytas, G., Dorfmüller, Th., Lin, Y.H., and Chu, B., Macromolecules 14, 1088 (1981).
36) Fytas, G., Meier, G., Patkowski, A., and Dorfmüller, Th., Colloid & Polymer Sci. 260, 949 (1982).
37) Pressure and Temperature Dependent Homodyne Photon Correlation Studies of Liquid o-Terphenyl in the Supercooled State, to appear in 1983.
38) Fytas, G., Meier, G., Dorfmüller, Th., Patkowsky, A., Macromolecules, 15, 214 (1982).

39) Anglister, J., and Steinberg, I.Z., Chem. Phys. Lett. 65, 50, (1979).
40) Stanton, S.G., and Pecora, R., J. Chem. Phys. 75, 5615(1981).
41) Stanton, S.G., Pecora, R., and Hudson, B.S., J. Chem. Phys. 78, 3365 (1983).
42) Chu, B., Pure & Appl. Chem. 49, 941 (1977).

LOW FREQUENCY DIELECTRIC SPECTROSCOPY AND DYNAMIC KERR-EFFECT OF MOLECULAR LIQUIDS.

Graham Williams

Edward Davies Chemical Laboratories,
University College of Wales,
Aberystwyth SY23 1NE,
Dyfed, U.K.

The reorientational motions of molecules in the supercooled liquid state and in liquid-crystal-forming materials just above the clearing point may occur on a time-scale which is far longer than that encountered for normal low viscosity liquids. Such 'slow motions' are conveniently studied using dielectric and dynamic Kerr-effect techniques in the ranges 10^{-4} to 10^{8} Hz and 10^{-8} to 10^{3}s respectively. The present account describes the behaviour of representative systems, indicates the molecular factors which are involved and, where possible, gives a discussion of mechanisms for reorientation in specific liquid systems.

I. INTRODUCTION

The low frequency dielectric properties of liquids may be studied using transient-current methods (10^{-4} to 10 Hz), low frequency bridge methods (10^{-2} to 10^{5} Hz) and high frequency bridge or resonant-circuit methods (10^{6} to 10^{8} Hz). Much equipment is available commercially and is capable of giving the complex permittivity $\varepsilon(\omega) = \varepsilon'(\omega) - i\varepsilon''(\omega)$ of a sample to high precision and accuracy over wide ranges of frequency ($\omega = 2\pi f/\text{Hz}$) and temperature. These methods have been described in detail (1,2) and involve, in the main, measurement of the real and imaginary parts of the complex impedance of a sample at fixed frequencies. We note that measurements in the range 10^{2} to 10^{5} Hz may be made quickly and with high precision using either a General Radio 1620-A or 1621 capacitance measuring assembly or a Hewlett Packard 4274-A-multifrequency LCR meter.

In this low frequency range relaxation behaviour may be

A. J. Barnes et al. (eds.), Molecular Liquids - Dynamics and Interactions, 239–273.

observed for a variety of systems, notably polymers in solution or in bulk, rotator-phase molecular crystals and liquid crystals. For amorphous solid polymers, studies up to about 1965 had clearly established that α, β and (αβ) dielectric relaxations generally occurred (1, 3-6). For example, for amorphous polyethylene terephthalate a well-defined α relaxation is observed above the apparent glass transition temperature (T_g) and a small, broad β relaxation is observed below T_g. This polymer has dipoles only in its main chain thus chain-backbone motions give rise to both processes. For flexible side-chain polymers, e.g. poly(vinylacetate), α and β processes are again observed (1). The two processes have the following characteristics:-

α process. (i) carries with it most of the available relaxation magnitude $\Delta\varepsilon$.

(ii) the frequency of maximum loss, f_m, is strongly dependent upon sample temperature and has a temperature-dependent apparent activation energy $Q_{app} > 150$ kJ mol^{-1}.

(iii) the loss-factor (ε'') - vs - log f plots are far broader than that for the single relaxation time expression

$$\varepsilon(\omega) = \varepsilon'(\omega) - i\varepsilon''(\omega) = \varepsilon_\infty + \frac{\Delta\varepsilon}{1+\omega^2\tau^2} - \frac{i\Delta\varepsilon\omega\tau}{1+\omega^2\tau^2} \qquad (1)$$

where ε_∞ is the limiting high frequency permittivity for relaxation and τ is a dielectric relaxation time. For the α relaxation in such polymers as polyethylene terephthalate and poly(vinylacetate) the total half width, Δlogf, of the ε''_α - vs - log f plot is in the range 1.8 - 2.2 and the curves are asymmetrical in the Cole-Davidson sense (1,2).

β process. (i) carries with it only a small part of $\Delta\varepsilon$

(ii) the plot of log f_m - vs - $(T/K)^{-1}$ is linear (i.e. Arrhennius form) with Q_{app} in the range 50 - 100 kJ mol^{-1}.

(iii) the ε''_β - vs - log f plots are extremely broad with Δlog f in the range 4-6.

The frequency-temperature locations for α, β and (αβ) processes in all amorphous solid polymers are shown schematically in figure 1. For $T < T_g$ only the β process is observed. For $T > T_g$ α and β processes coexist for a limited range then at higher temperatures coalesce to form the (αβ) process where the latter process is considered to be the α process, principally, continued to higher temperatures only now it relaxes all of $\Delta\varepsilon$ (see refs. 4-6 for discussion of this pattern of behaviour). The curvature of the plot log f_m - vs - $(T/K)^{-1}$ for the α and (αβ) processes is usually represented in terms of the Vogel equation

$$f_m = f_m^o \exp\left[-\frac{B}{T - T_o}\right] \qquad (2)$$

where f_m^o, B and T_0 are constants for a given polymer at a fixed pressure (normally atmospheric).

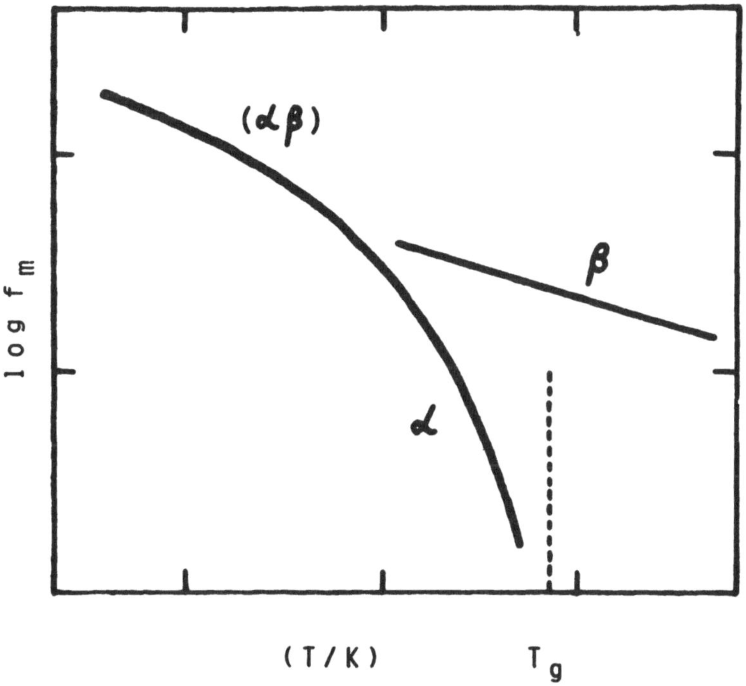

Figure 1. log f_m against $(T/K)^{-1}$ (schematic) for the α, β and (αβ) relaxations in amorphous solid polymers.

A simple phenomenological theory has been given (4-7) which appears to rationalize all the experimental observations for the α, β and (αβ) dielectric relaxations in amorphous solid polymers. It is assumed that a reference dipolar group in the polymer system may partially reorientate via motions in a temporary local environment (β process) but at sufficiently long times may completely reorientate by cooperative rearrangements of that local environment (α process). When the time scales of the two processes are sufficiently different, i.e. for $T < T_g$ or for T just above T_g, the dipole moment vector correlation function may be written as (4-6)

$$\frac{\langle\vec{\mu}(0).\vec{\mu}(t)\rangle}{\langle\mu^2\rangle} = [\sum_r^o p_r q_{\alpha r} + \sum_r^o p_r q_{\alpha r}\phi_{\beta r}(t)]\phi_\alpha(t) \tag{3}$$

where $q_{\alpha r} + q_{\beta r} = 1$, $q_{\alpha r} = [\langle\vec{\mu}\rangle_r^2]/\langle\mu^2\rangle$ and $\langle\vec{\mu}\rangle$ is the net dipole moment residing in the temporary environment 'r' after the 'fast' β_r process has occurred, and $^{o}p_r$ is the probability of obtaining environment 'r'. $\phi_{\beta r}(t)$ is the normalized decay function for the partial reorientation of the dipole vector in environment 'r' and $\phi_\alpha(t)$ is the normalized decay function for the cooperative motions of the dipole with its environment. Since the α process is assumed to be a gross microbrownian process, $\phi_\alpha(t)$ is assumed to be independent of 'r'. The α and individual β_r processes are assumed to be statistically independent for this model. One-sided Fourier transformation of equation (3) gives the complex permittivity according to (8)

$$\left[\frac{\varepsilon(\omega) - \varepsilon_\infty}{\Delta\varepsilon}\right] p(\omega) = 1 - i\omega \mathcal{F}\left[\frac{\langle\vec{\mu}(0).\vec{\mu}(t)\rangle}{\langle\mu^2\rangle}\right] \tag{4}$$

$p(\omega)$ is an internal field factor (8) and $\mathcal{F}$ indicates the Fourier transform. According to equations (3) and (4) we have:-

(i) the α and β processes have magnitudes $\Delta\varepsilon_\alpha$ and $\Delta\varepsilon_\beta$ which are proportional to $\langle\mu^2\rangle\sum_r {}^{o}p_r q_{\alpha r}$ abd $\langle\mu^2\rangle\sum_r {}^{o}p_r q_{\beta r}$, respectively.

(ii) the β process is a weighted sum of individual β_r processes, each having a relaxation function $\phi_{\beta r}(t)$. Thus for limited partial reorientation the overall β process will be a small, broad process.

(iii) For $T < T_g$ only the β process will be obtained. For a limited range above T_g both α and β processes will be obtained then, at higher temperatures, will coalesce into a single (αβ) process which is the continuation of the α process to higher temperatures (i.e. is characterized by $\phi_\alpha(t)$ but now carries with it all of the available relaxation strength $\Delta\varepsilon \propto \langle\mu^2\rangle$).

These predictions of the simple phenomenological model are in accord with experimental dielectric data for amorphous solid polymers (4-7). The model does not specify detailed mechanisms for α and β processes, so, historically, the next stage was to develop such models. Many attempts were made and Table 1 summarizes a number of 'one-body' models and their generalizations to include chain dynamics. Those for chain dynamics incorporate the basic models for one body motion : e.g. the theory of Yamafuji and Ishida (22) is for coupled units each undergoing small-step rotational diffusion, while those of Jernigan (29) and Beevers and Williams (30) are for coupled units each undergoing motion in local (conformational) barrier systems. All the models in Table 1 exclude the short time effects associated with inertial factors and damped librations in a local potential.

Table 1

Models for Molecular Reorientation

A. One-body motions

(i) *Rotational diffusion.*
Debye (9), DiMarzio and Bishop (10), Ivanov (11), Anderson (12).

(ii) *Barrier theories.*
Hoffman (13,14), Adam (15), Brereton and Davies (16).

(iii) *Defect-diffusion/fluctuations.*
Glarum (17), Anderson and Ullman (18), Phillips and coworkers (19), Beevers and coworkers (20).

B. Chain dynamics

Kirkwood and Fuoss (21), Yamafuji and Ishida (22), Stockmayer (23,24), Dubois-Violette and coworkers (25), Geny and Monnerie (26), Hayakawa and Wada (27), Saito and coworkers (28), Jernigan (29), Beevers and Williams (30), Shore and Zwanzig (31), Evans and Knauss (32).

Thus all models only give the long-time rotational dynamics, as is necessary for relaxations observed in the range 10^{-4} to 10^{7} Hz.

For polymer chains a complicating feature is the presence of cross-correlation terms due to the angular correlations of dipole vectors along a chain (1,6,8). The dielectric increment $\Delta\varepsilon$ is a time-averaged (equilibrium) property of a system and may be expressed in terms of the fluctuations of the macroscopic dipole moment $\vec{M}(t)$ of a sphere of volume V according to the Kirkwood-Fröhlich relation (1).

$$\Delta\varepsilon = \frac{4\pi}{3kT}\left[\frac{3\varepsilon_o(2\varepsilon_o+\varepsilon_\infty)}{(2\varepsilon_o+1)^2}\right]\cdot\frac{\langle\vec{M}(0).\vec{M}(0)\rangle}{V} \tag{5}$$

ε_o is the limiting low frequency permittivity, $\Delta\varepsilon = \varepsilon_o - \varepsilon_\infty$, $\langle\vec{M}(0).\vec{M}(0)\rangle$ is the mean square dipole moment of the sphere. For polymer chains having tacticity as a variable, the elucidation of $\langle\vec{M}(0).\vec{M}(0)\rangle$ is complicated but has been done for a variety of chain structures by methods developed by Volkenstein (33) and Flory (34). For the simple case of chains containing only one type of dipole unit (e.g. as in polyethylene terephthalate and polyoxymethylene) we may write (1,5,6).

$$\frac{\langle\vec{M}(0).\vec{M}(0)\rangle}{V} = c_r\left[\langle\mu_r^2\rangle + \sum_{\substack{k' \\ k'\neq k}} \langle\vec{\mu}_k(0).\vec{\mu}_{k'}(0)\rangle\right] \tag{6}$$

$$\equiv c_r\langle\mu_r^2\rangle\, g_1$$

where μ_r is the dipole moment of the repeat unit, c_r is the number of repeat units per unit volume and $\langle\vec{\mu}_k(0).\vec{\mu}_{k'}(0)\rangle$ involve angular correlations between a reference dipole vector k and further dipole vectors k′ along the same polymer chain. For flexible chains the cross-correlation terms decrease rapidly as $|k - k'|$ is increased. Model calculations for polyethers (35,36) show that the cross-correlation terms may be positive or negative, giving a Kirkwood g_1 factor greater or less than unity (see refs. 5 and 36).

For the dynamic situation cross-correlation terms are again important and in place of equation (5) we now have (8,36)

$$\left[\frac{\varepsilon(\omega) - \varepsilon_\infty}{\Delta\varepsilon}\right]p(\omega) = 1 - i\omega\mathcal{F}[\Lambda(t)] \tag{7}$$

$$\Lambda(t) = \frac{\langle\vec{M}(0).\vec{M}(t)\rangle}{\langle\vec{M}(0).\vec{M}(0)\rangle} \tag{8}$$

For the simple case of polymer chains containing only one type of dipole we have

$$\frac{\langle\vec{M}(0).\vec{M}(t)\rangle}{V} = c_r\left[\langle\vec{\mu}_k(0).\vec{\mu}_k(t)\rangle + \sum_{\substack{k' \\ k'\neq k}} \langle\vec{\mu}_k(0).\vec{\mu}_{k'}(t)\rangle\right] \tag{9}$$

so that

$$\Lambda(t) = \frac{\langle\vec{\mu}_k(0).\vec{\mu}_k(t)\rangle + \sum_{\substack{k' \\ k'\neq k}} \langle\vec{\mu}_k(0).\vec{\mu}_{k'}(t)\rangle}{\langle\mu_r^2\rangle + \sum_{k'}\langle\vec{\mu}_k(0).\vec{\mu}_{k'}(0)\rangle}$$

$$\equiv g_1(t)/g_1(0) \tag{10}$$

Equations (7) and (10) generalize the Kirkwood-Fröhlich equilibrium theory to the dynamic situation. The correlation function approach to dielectric relaxation was first made by Glarum (37) and was extended by Cole (38) and by Steele (39). The latter author gave explicit forms for the auto- and cross-correlation functions (in time) in terms of time-dependent conditional probability distribution functions. Later Williams and coworkers (36) applied the correlation function method to amorphous dipolar polymers.

The fact that the dielectric relaxations in amorphous solid polymers was due to the motion of angularly correlated dipoles within a bulk amorphous medium comprising similar dipoles implied that a complete theoretical description of dielectric relaxation for any given polymer would be an extremely formidable task. This was the daunting prospect in the late 1960's and it appeared necessary to construct sophisticated models for chain dynamics, incorporating angular (dipole) correlations, in order to make progress.

We have described this earlier work for amorphous solid polymers in some detail since a large literature had been compiled for polymers by the mid 1960's (not only for dielectric, but also for dynamic-mechanical and NMR relaxation - see refs. 1, 40 and 41) and was in need of explanation in molecular terms. That explanation emerged as a result of studies of small molecule glass-forming systems in the supercooled *liquid* and *glassy* states. The important advance was made by Johari, Goldstein and Smyth (42-47) who showed that dielectric behaviour entirely similar to that described for amorphous solid polymers (1,3-7) could be observed for rigid dipolar solute molecules in non polar solvents or as pure liquids. Their systems were non-polymeric and were not complicated by the possibilities of internal rotation, hydrogen-bonding or of cross-correlation functions. Thus it was possible that the multiple relaxations observed in amorphous solid polymers could be explained on the same 'one-body' basis as that required for the simpler rigid systems. Before we describe the work of Johari, Goldstein and Smyth and subsequent work with supercooled liquids it is important to note earlier dielectric studies with supercooled and other viscous molecular liquids.

II. SUPERCOOLED MOLECULAR LIQUIDS

The dielectric behaviour of viscous molecular liquids has been studied since the early 1930's. Baker and Smyth (48) had observed relaxation in supercooled isoamyl- and isobutyl bromide in the kHz region and had noted the similarity to the relaxations of liquid glucose (49) and glycerol (50). Similar studies were made by Turkevich and Smyth (51) for supercooled isobutyl chloride and 1,2 dichloroisobutane. Denney (52) studied supercooled isoamyl bromide, isobutyl chloride and isobutyl bromide and Winslow and coworkers (53) studied tolyl xylyl sulphone, pentachlorobiphenyl, hexachlorobiphenyl and solutions of tolyl xylyl sulphone in supercooled o-terphenyl. In all studies a low frequency dielectric process, due to dipole reorientation, was observed. Denney (52) had covered a range of relaxation time extending from 10^{-3} to 10^{-11}s by varying temperature. He also showed (54) that supercooled mixtures of two alkyl halides gave a *single* loss process, intermediate in location between those of each pure halide, thus demonstrating that each component reorientates in solution as a part of one general cooperative (α) process. Denney (55) further demonstrated that there was a close parallel between the temperature variation of τ(dielectric) and flow viscosity for isoamyl bromide, isobutyl chloride and isobutyl bromide: for isoamyl bromide the viscosity varied from 10^2 to 10^9 poise. Further comprehensive studies were made by Mopsik and Cole (56) for n-octyl iodide and by Berberian and Cole (57) for isoamyl bromide. These studies of non-hydrogen-bonded viscous liquids had shown that certain small molecules could be made to reorientate as slowly as desired by arranging for them to be in the supercooled liquid state at low temperatures. The similarity in behaviour for the various liquids was noted: (i) curved plots of log f_m - vs - $(T/K)^{-1}$ (see equation (2) above), (ii) asymmetric ε''-vs-log f plots in the sense of Cole and Davidson (58-60). They commented (57) that the various theories for relaxation in supercooled liquids had the common feature that they invoked cooperative interaction effects.

Thus there was an extensive literature for the dielectric behaviour of supercooled molecular liquids before Johari, Goldstein and Smyth (42-47) published their important work. A major contribution made by these latter authors was that they clearly demonstrated that small rigid dipolar molecules gave α and β processes entirely similar to those seen for flexible polymers in the amorphous solid state (1, 3-7) and that their α process was also entirely similar to that seen for flexible short chain molecules (48, 51-57) in the supercooled liquid state. In all cases the α process is observed for $T > T_g$: i.e. in the supercooled *liquid* state and not in the glassy state. Johari, Goldstein and Smyth studied chlorobenzene, 1-chloronaphthalene, o-dichlorobenzene, bromobenzene, tetrahydrofuran and toluene as solutes in the (essentially) non-polar solvent cis-decalin in the supercooled liquid and glassy states and observed α and β processes in the ranges 10^2 to 10^5 Hz, 80 - 140 K. Such small molecules reorientate in normal low viscosity solvents to give dielectric relaxation frequencies in the range 10^{10} to 10^{12} Hz, so the rate of the rotational process was decreased by a factor of 10^5 to 10^{10} by going into the supercooled liquid states. Such 'slow' motions have a different mechanism from that operating in the normal low viscosity liquid, leading to a greatly increased value for Q_{app} ($\sim$200 kJ mol^{-1}) and a broadened loss curve. They also studied mixtures such as 43.5 mol % chlorobenzene in pyridine, 42.5 mol % bromobenzene in pyridine, 36.8 mol % 1-chloronaphthalene in pyridine, 37.5 mol % 1-chloronaphthalene in tetrahydrofuran, 43.4 mol % bromobenzene in tetrahydrofuran and 43.8 mol % toluene in pyridine. They observed α and β processes as before. There was no evidence of structure in the α peaks, indicating that both dipolar components in the mixture move cooperatively in one general α process - a result previously obtained for mixtures of alkyl halides (54). Also the α process was asymmetric in the Cole-Davidson sense, as had been found for the alkyl halidees (52, 54-57). As one example, figure 2 shows the plot of

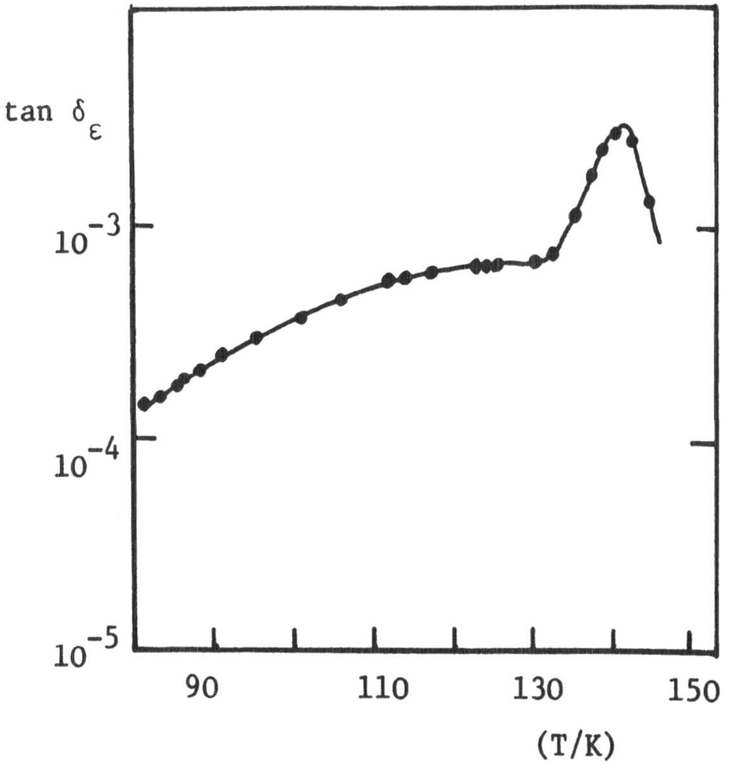

Figure 2. tanδ_ε against T for an 11.5 mol % chlorobenzene/cis-decalin solution at 1 kHz. (after ref. 44 with permission)

$\tan\delta = \varepsilon''/\varepsilon'$ against (T/K) for a chlorobenzene/decalin solution (44). The small broad β process is seen at low temperatures and the larger well-defined α process is seen just above T_g. Heating above 146 K causes crystallization of the supercooled liquid. Johari (46) showed that α and β processes coexist just above T_g, and figure 3 shows his data for a chlorobenzene/decalin

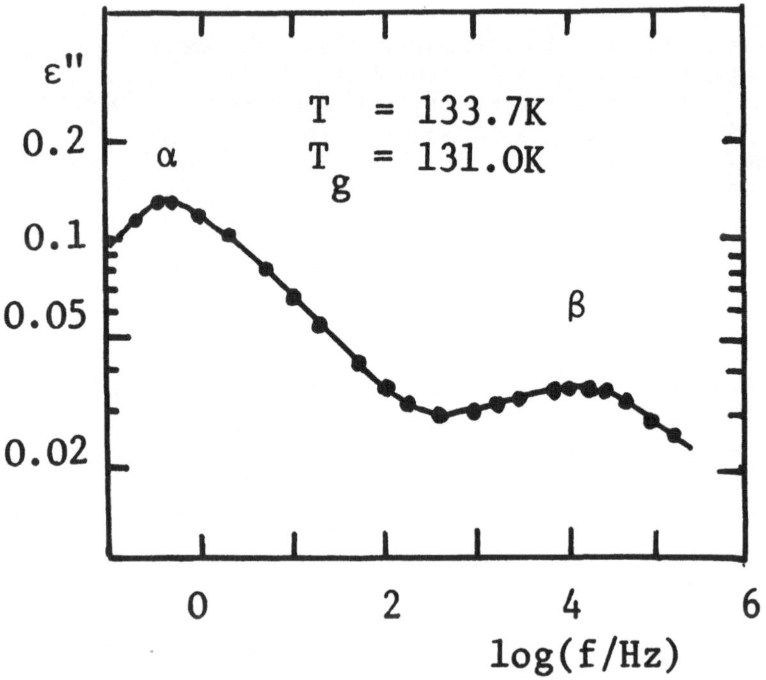

Figure 3. ε'' against log(f/Hz) for a 17.2 mol % chlorobenzene/ cis-decalin solution at 133.7 K. (after ref. 46 with permission)

mixture. The frequency-temperature locations of the α, β and (αβ) processes for the chlorobenzene/decalin solution are shown in figure 4 and the similarity to figure 1 is apparent.

Given such data it was necessary to obtain a general theory for these multiple processes which would accommodate *all* such glass-forming systems. That theory is the general phenomenological one given by equation (3) and which has been discussed above in relation to amorphous solid polymers (see refs 4-6).

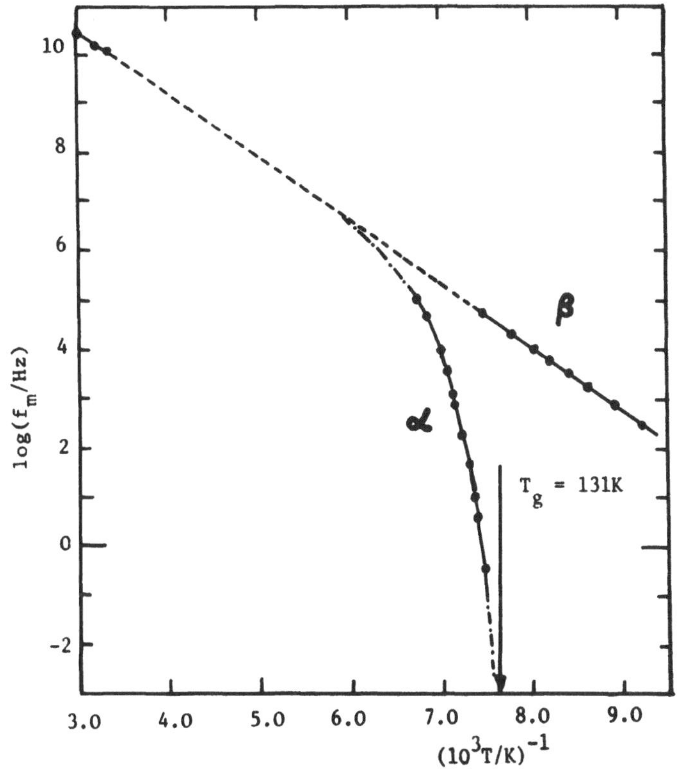

Figure 4. $\log(f_m/Hz)$ against $10^3(T/K)^{-1}$ for a 17.2 mol % chlorobenzene/cis-decalin solution showing α, β and (αβ) relaxation regions. (after ref. 46 with permission)

Thus some progress had been made by the early 1970's towards understanding the multiple relaxations of non-polymeric and polymeric glass-forming systems which were not hydrogen-bonded. As a result of the works of Johari, Goldstein and Smyth, the writer and his coworkers made a quantitative study of several dipolar solutes in the supercooling solvent o-terphenyl. Johari and Goldstein had studied this liquid, which has a small dipole moment, has a T_g near 243 K and freezes at 328.7 K. They had found small α and β processes whose forms were entirely similar to those shown in figure 2. The viscosity-temperature behaviour of o-terphenyl had been studied by Greet and Turnbull (61, 62) and the thermodynamics of liquid and glass by Chang and Bestul (63). McCall and coworkers (64) had studied the α relaxation in o-terphenyl using proton NMR methods. The supercooled liquid range is *ca* 243-272 K so dielectric measurements could be readily made without elaborate cooling arrangements. A variety of dipolar solutes in supercooled o-terphenyl were studied (65-69). Since the solvent loss was small compared with that for the solute, the solute acts as a molecular probe for the cooperative reorientational dynamics (α process) of the supercooled liquid.

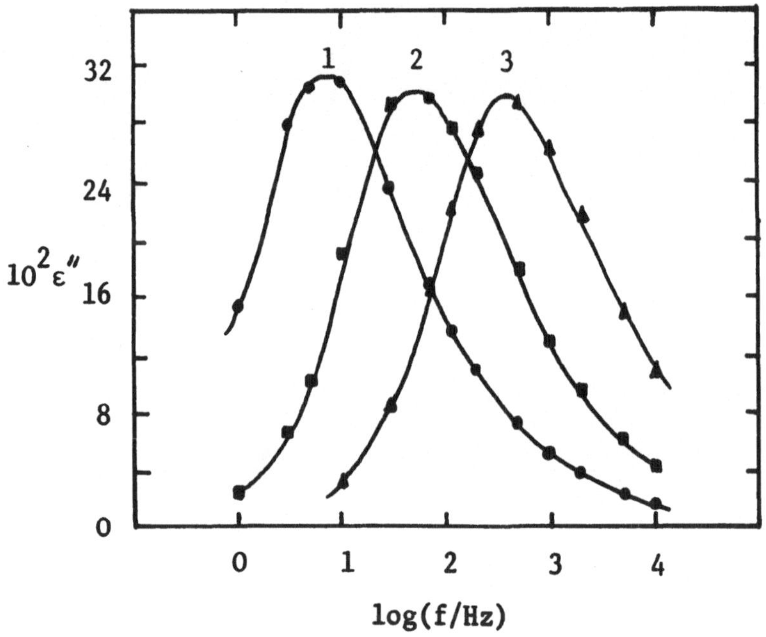

Figure 5. ε'' against log(f/Hz) for a 3.72 % (w/w) solution of phthalic anhydride in o-terphenyl in the supercooled liquid state. Curves 1,2 and 3 refer to 254.8, 258.8 and 263.2 K respectively. (after ref. 66 with permission).

Figure 5 shows the ε''_α - vs - log f plot for phthalic anhydride in o-terphenyl (66), just above the T_g of the *solution.* This process has a large apparent activation energy ($Q_{app} \sim 270$ kJ mol^{-1}) and the loss curve is asymmetric in the Cole-Davidson sense. Williams and coworkers (66-68) observed a similar well-defined α-relaxation for the following solutes: phthalic anhydride, anthrone, nitrobenzene, 1-chloronaphthalene, benzophenone, cyclohexanone, fenchone, camphor and fluorenone. It was shown that the α-process did not relax all of $\langle\mu^2\rangle$. Since ε_0 and $\varepsilon_{\infty\alpha}$ could be measured, and ε_∞ (for all relaxations) could be estimated from

ε_∞ (solvent), the quantity $\Delta\varepsilon_\alpha/\Delta\varepsilon = (\varepsilon_o - \varepsilon_{\infty\alpha})/(\varepsilon_o - \varepsilon_\infty)$ could be determined for each solution. Table 2 lists $\Delta\varepsilon_\alpha/\Delta\varepsilon$ obtained for each solute. The range of values quoted relate to different solutions and include experimental uncertainties. For certain

Table 2
Supercooled o-terphenyl solutions

Solute	$\Delta\varepsilon_\alpha/\Delta\varepsilon$	($\Delta\log(f/Hz)$)	Q_{app}/kJ mol^{-1}
Phthalic anhydride	0.83 - 0.89	2.00 - 2.05	270 - 290
Anthrone	0.91 - 1.00	2.00 - 2.05	260 - 290
Nitrobenzene	0.80 - 1.00	2.2 - 2.3	275 - 285
Camphor	0.38 - 0.46	2.00 - 2.05	245 - 270
1-chloronaphthalene	0.37 - 0.41	2.2 - 2.3	230 - 240
Fluorenone	0.85 - 0.96	2.00 - 2.10	265 - 270
Di-n-butyl phthalate	0.90 - 0.98	2.00 - 2.10	270 - 280
Benzophenone	0.40 - 0.50	2.00 - 2.20	230 - 250
Cyclohexanone	0.46 - 0.59	2.1 - 2.5	230 - 270
Fenchone	0.70 - 0.80	2.0 - 2.3	240 - 250
(o-terphenyl)	-	2.0 - 2.1	263

solutes most of $\langle\mu^2\rangle$ is relaxed by the α-process : e.g. for anthrone, nitrobenzene. For several solutes a substantial part of $\langle\mu^2\rangle$ is relaxed by a further higher frequency process e.g. for camphor, 1-chloronaphthalene. The origin of this variation with chemical structure is largely steric. The camphor molecule is small and nearly spherical thus may, in partial reorientation, reorientate over an appreciable part of 4π solid angle without disturbing the neighbouring o-terphenyl molecules. Fluorenone clearly cannot reorientate appreciably without the cooperation of the surrounding molecules. For molecules such as nitrobenzene, fluorenone, anthrone and phthalic anhydride one may envisage the cooperative motions of solute and solvent as being like the motions of enmeshing cogs. Such motions extend to many molecules beyond a reference molecule, giving a cooperative process which is very different from that for small molecules in low viscosity liquids where the sphere of correlation may only extend to a few molecular lengths around a reference molecule. Surprisingly, the shape of the ε''_α - vs - log f plots were very similar for all solutes. Figure 6 shows a representative plot of normalized loss against $\log(f/f_m)$ for anthrone in o-terphenyl (66). The curve is asymmetric and may be fitted using the macroscopic relaxation function $\gamma(t)$ of the form (70,71)

$$\gamma(t) \simeq \exp[-t/\tau_o]^{\bar{\beta}} \qquad (11)$$

in conjunction with the relation (1,8)

$$\frac{\varepsilon(\omega) - \varepsilon_\infty}{\Delta\varepsilon_\alpha} = 1-i\omega\mathfrak{F}[\gamma(t)] \tag{12}$$

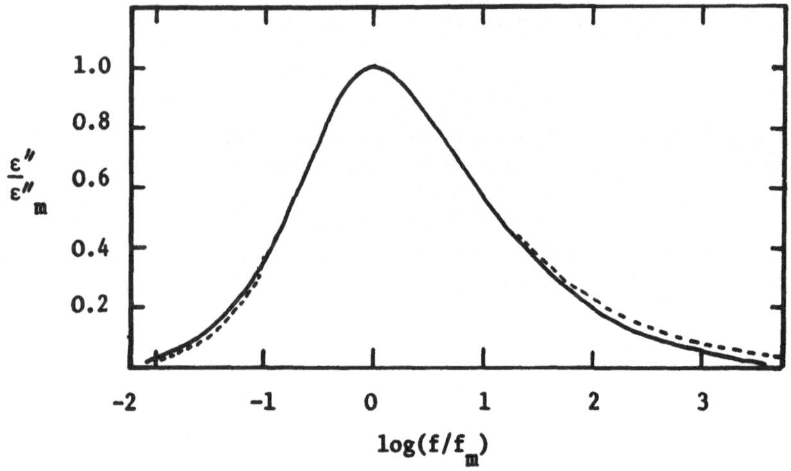

Figure 6. Master loss curve for a 4.2% (w/w) solution of anthrone in o-terphenyl. Dashed line indicates the experimental data, continuous line is that calculated for $\bar{\beta}$ = 0.55. (after ref. 66 with permission).

In figure 6 the dotted line is the experimental curve and the continuous line is for $\bar{\beta} = 0.55$. The agreement is good except at the highest frequencies where the observed values exceed those calculated using equation (11). Equations (11) and (12) give calculated loss curves which are asymmetrical in the Cole-Davidson sense, but are less extended on the high frequency side for a given total half-width (70,71). The half-width of the loss curves, $\Delta\log f$, is shown in Table 2 and its near constancy for the different solutes show that $\bar{\beta}$ may vary only slightly with chemical structure. Systematic changes in $\Delta\log f$ were observed as solute concentration is increased (65-68). In some cases, e.g. for fluorenone, $\Delta\log f$ is approximately constant at 2.05 ± 0.03. Importantly, plots of $\log f_m$ - vs - solute concentration (c) at a fixed temperature showed that $\log f_m$ increased with c for all solutes, showing that the solutes plasticized the solute/solvent system (lowered its T_g value), a phenomenon which is well-known for amorphous polymers (1). However $(\log f_m)_{c\to o}$ was essentially the same for all solutions and was equal to that obtained for the small α-process in o-terphenyl itself (66): i.e. the solute acts as a probe for the cooperative motions of itself with the surrounding molecules. The large values of Q_{app} given in Table 2 are all similar and again reflect the cooperative nature of the process. Note that Q_{app} is in no sense a 'barrier' to reorientation of a representative solute molecule and note that bond-dissociation energies lie in the range 300 - 500 kJ mol^{-1}.

Fluorenone was also studied in mixed solvents (67). Increasing the decalin concentration from 0 - 7.5% in o-terphenyl moved $\log f_m$ by an amount 2.7 to higher values. Increasing the triphenylbenzene concentration from 0 to 15% in o-terphenyl decreased $\log f_m$ by an amount 1.3. In the first case decalin plasticizes the solution (T_g(decalin) = 133K) while in the second case triphenylbenzene antiplasticizes the solution (T_g(TPB) = 333K). The shape of the process did not change appreciably, nor did Q_{app}. Thus the rate of reorientation of the fluorenone molecules could be changed over a wide range by change in solvent composition, but the mechanism for the process is unchanged.

Mixtures of di-n-butyl phthalate (DBP) with o-terphenyl were studied over the entire range of composition (67). This was possible since DBP is a glass-forming liquid. The dielectric α-process in the range 5 - 100 % DBP was mainly due to DBP since it has an appreciable dipole moment. It was shown that the apparent T_g of the mixtures lay on a straight line joining the T_g values for DBP and o-terphenyl (see figure 5, ref. 67). These data show that the α-process in pure DBP is antiplasticized by increasing o-terphenyl content in the mixture. Surprisingly $\Delta\log f$ increased from its value of 1.74 for pure DBP to a maximum value of $\sim$2.55 at 35% DBP then decreased to 2.05 for pure o-terphenyl (see figure 5, ref. 67). Such behaviour may be taken as evidence for concentration variations within a liquid mixture; i.e. at intermediate compositions the DBP molecules favour each other and the o-terphenyl molecules likewise. The experimental loss curves are broader than one expects using a linear relation for $\Delta\log f$ with composition because they are the resultant of superposed component loss curves, reflecting the range of local environments (compositions) experienced by the molecues. The reorientation of a reference molecule 'i' requires the cooperation of a limited number of surrounding molecules and we may draw a sphere around it which contains all those molecules which make an appreciable contribution to $\langle\vec{\mu}_i(0).\vec{\mu}_i(t)\rangle$. Clearly this correlation function is determined by the chemical composition of the sphere. The time-scale in which molecules diffuse through their molecular length will be comparable with, or greater than, the time-scale for molecular reorientation so a reference molecule will reorientate in an environment of essentially fixed chemical composition. DBP molecules in a DBP rich environment relax faster than those in an o-terphenyl-rich environment (T_g(DBP) < T_g(O-T)) thus a range of local compositions leads to a relative broadening of the loss curves in comparison with that for each pure liquid. Shears and Williams (67) quantified this picture. They assumed that each contributing dipole would give a loss curve which was broader than that for a single relaxation time process and that such dipoles would experience a range of local compositions (on average). While the basic loss profile was asymmetric they adopted a simple approach in which it was assumed that the elementary process was represented by the empirical Fuoss-Kirkwood function (see ref. 1), with distribution parameter β, and that there was a rectangular distribution of such processes. For such a case the overall loss-factor was given by (67)

$$\frac{\varepsilon''(\bar{\omega})}{\varepsilon''_m} = \frac{\tan^{-1}\bar{\omega}^{\beta}\alpha - \tan^{-1}\bar{\omega}^{-\beta}\alpha}{\tan^{-1}\alpha - \tan^{-1}\alpha} \qquad (13)$$

where $\bar{\omega} = \omega/\omega_m$, $\alpha = (\tau_b/\tau_a)^{\beta/2}$ and ε''_m occurs at the condition $(\omega_m^2\,\tau_a\,\tau_b) = 1$. Here τ_b and τ_a are the maximum and minimum values of τ in the rectangular distribution for log τ. It followed that the half-width Δlogf of the overall loss-curve was given approximately by (67)

$$\Delta\log f = \frac{1}{\beta}.[0.602 + 2\log F(\alpha)] \qquad (14)$$

where

$$F(\alpha) = \frac{(\alpha - \alpha^{-1})}{\tan^{-1}\alpha - \tan^{-1}\alpha^{-1}} \qquad (15)$$

β is that for a given *average* composition of solution and is obtained using a linear interpolation between the values for the pure liquids. Experimental determination of Δlogf and a knowledge of β gives F(α) and hence α (using equation (15)) for each mixture. Since $\log(\tau_b/\tau_a) = (2/\beta).\log\alpha$ the range of logτ could be estimated. Also using the experimental plot for $\log f_m$ - vs - c_{DBP}, together with $\log(\tau_b/\tau_a)$, the apparent range of c_{DBP} was estimated for each mixture. It was found (67) that $\log(\tau_b/\tau_a)$ rose to a flat maximum value ∿1.8 and the effective *range* of c_{DBP}, Δc, was found to be as much as ±5 % over much of the composition range (67). This analysis, although approximate, showed that extensive concentration fluctuations may occur in a mixture of a dipolar liquid (DBP) and a lowly dipolar liquid (O-T) at low temperatures. Note that dielectric spectroscopy has since been used to study concentration fluctuations in apparently compatible blends of amorphous polymers (72).

In our discussion to this point we have considered that any higher frequency processes for $T > T_g$ would be the β process (equation (3)). Johari, Goldstein and Smyth had shown that the β process (in the glass or just above T_g) was small and broad, just as in amorphous solid polymers. We have no experimental information on the form of the high frequency process for the solute/o-terphenyl solutions of Table 2. We would suggest that for cases where $\Delta\varepsilon_\alpha/\Delta\varepsilon > 0.8$ the β process would be very similar to that observed by Johari just above T_g (figure 4 above). However for the other cases in Table 2 it is possible that the high frequency process may be of a different character, since the local motions relax such a substantial part of $\langle\mu^2\rangle$. Warchol and Vaughan (73) proposed a model of small-step diffusion in a cone to account for limited motions in such supercooled systems. Their model has been extended by Wang and Pecora (74). The essential result of this model (73, 74) is that for motions of a dipole vector limited within a cone of cone-angle θ_o, where $\theta_o < 40°$, we have

$$\frac{\langle\vec{\mu}(0).\vec{\mu}(t)\rangle}{\langle\mu^2\rangle} = A + B\exp[-\nu_1^1(\nu_1^1 + 1)D_R t] \tag{16}$$

where $A = (1 + \cos\theta_o)^2/4$ and $A + B = 1$. D_R is the rotational diffusion coefficient and $\nu_1^1 \equiv \nu_1^1(\theta_o^2)$ is a quantity which is, approximately, inversely proportional to θ_o^2. For small-step diffusion into all 4π solid angle $\nu_1^1 \to 1$, $A \to 0$, $B \to 1$, but for small-step diffusion within the cone a small, 'fast' partial relaxation process is obtained. That part of $\langle\mu^2\rangle$ which is not relaxed by motion in the cone, i.e. A, would be relaxed by motion of the cone-axis (α-process). The formal similarity of equations (3) and (16) is evident but we note that equation (16) specifies a mechanism for the faster process.

Of all the solutes studied in o-terphenyl, only two gave results which were very different to those discussed to this point. Davies, Hains and Williams (69) studied tri-n-butyl ammonium picrate (TnBP) and tri-n-butyl ammonium iodide (TnBI) in o-terphenyl. These solutes exist as ion-pairs, or their aggregates, in non-polar solvents (75). TnBP exists as contact ion-pairs in o-terphenyl and gave (69) a well-defined α process which carried with it essentially all of $\langle\mu^2\rangle$. The process was broad at low temperatures but narrowed on increasing the temperature, approaching the single-relaxation time function at the highest temperatures studied. In addition, the relaxation time for the reorientation of the ion-pairs was about 3 times larger than that for the solvent at the highest temperatures, but this difference decreased with decreasing temperature. These results suggested that at the lowest temperatures, i.e. just above T_g(solvent) the ion-pairs relaxed by a similar mechanism to that for the solutes in Table 2, but as temperature was increased the mechanism gradually changed to that for small-step diffusion. We shall discuss TnBP/O-T further below. For TnBI solutions the α process was rather broad, indicating association of ion-pairs, as was expected from earlier data for this solute in benzene and xylene solutions (75).

The defect-diffusion models of Glarum (17) and Phillips, Barlow and Lamb (19) may be used for the α relaxation for the solutes shown in Table 2. These models give the following normalized relaxation functions

$$\phi_o(t') = \exp(t')\mathrm{erfc}(t'^{\frac{1}{2}}) \quad : \text{Glarum} \tag{17}$$

$$\phi_a(t') = \phi_o(t')[(1-2t')\phi_o(t') + (2t'/\pi)^{\frac{1}{2}}] \tag{18}$$

$$: \text{Phillips, Barlow, Lamb}$$

$t' = t/\tau_d$, where τ_d is a defect-diffusion time. The PBL model extended the Glarum model to include next-nearest-neighbour defects. It was shown (68) that the PBL relation gave a good representation of the α-process for benzophenone, cyclohexanone, fenchone and anthrone in o-terphenyl. Since the empirical function, equation (11) also fitted such data it was apparent that equation (18) could be force-fitted by a suitable choice for $\bar{\beta}$. This was found to be so for $\bar{\beta} = 0.53 \pm 0.01$. While the defect diffusion model was entirely reasonable for rotational motions in viscous liquids, it was still possible to fit our data using *any* of the one-body models in Table 1 : i.e. it was still not possible to distinguish between models on the basis of the dielectric results alone, as will be made clearer in the next section.

III. ORIENTATIONAL DISTRIBUTION FUNCTIONS AND THEIR RELATION TO EXPERIMENT

The basic problem associated with elucidating a relaxation mechanism from experimental data is most clearly seen in terms of the time-dependent conditional orientational distribution function $\rho(\Omega,t/\Omega_o,0)$. The quantity $\rho(\Omega,t/\Omega_o,0)d\Omega d\Omega_o$ is the probability of obtaining the orientation of the body (molecule, chain segment) in the element of solid angle $d\Omega$ around Ω at time t *given* its orientation was in $d\Omega_o$ around Ω_o at t = 0. We may expand the distribution function in terms of the elements $D^J_{K,M}(\Omega)$ of Wigner rotation matrices and the orientational time-correlation functions $C^J_{K,M}(t)$ (76,77) so that

$$\rho(\Omega,t/\Omega_o,0) = \sum_{J,K,M} \left[\frac{2J+1}{8\pi^2}\right] D^{J*}_{K,M}(\Omega) D^{J}_{K,M}(\Omega_o) C^{J}_{K,M}(t) \qquad (19)$$

Using the orthogonality properties of the rotation matrices (78) it may be shown (76,77) that

$$C^{J}_{K,M}(t) = \langle D^{J}_{K,M}(\Omega(t))\ D^{J*}_{K,M}(\Omega_o(0))\rangle$$

$$= \int_{\Omega}\int_{\Omega_o} \rho(\Omega,t/\Omega_o,0)\ D^{J}_{K,M}(\Omega)\ D^{J*}_{K,M}(\Omega_o)\, d\Omega d\Omega_o \qquad (20)$$

A given experiment gives information on a single correlation function or a weighted sum of correlation functions (76,77) and is therefore unable to describe fully the orientational distribution function. Dielectric relaxation, for example, gives $C^{1}_{o,o}(t)$ or weighted sums of $C^{1}_{K,M}(t)$, so is insufficient to describe the distribution function. This is the basic problem for all relaxation studies: it is not possible to obtain the mechanism of reorientation using the results of one experimental method. It is necessary to obtain data using as many complementary experiments as possible and then to seek a consistent interpretation based on all these data - a principle stated clearly by Hildebrand (79). Further insight into the mechanism of the α-process could only be obtained by measuring a complementary correlation function to that for dielectric relaxation. Consequently Dr. M. S. Beevers and the writer built a dynamic Kerr-effect apparatus in the early 1970's at Aberystwyth and carried out complementary dielectric and Kerr-effect studies of several supercooled liquids (80,81).

The static Kerr-constant is $K = \Delta n/[\lambda E^2]$ where Δn is the optical birefringence, measured at a wavelength λ, which is produced by a directing electric field E. K is made up of contributions from induced dipole moments (K_{ind}) and permanent dipole moments (K_{μ}) (82,83). For axially-symmetric molecules

$$K_{ind} = \sigma N \Delta\alpha_o \Delta\alpha_\infty / 5kT \qquad (21)$$

$$K_{\mu} = \sigma N \Delta\alpha_\infty \mu^2 / 15k^2T^2 \qquad (22)$$

σ is a proportionality factor, N is the number of molecules per unit volume, Δα is the anisotropy of molecular polarizability and subscripts 'o' and '∞' refer to 'low' and optical frequencies respectively. For liquids having an appreciable dipole moment $K_{\mu} \gg K_{ind}$. Equations (21) and (22) assume the absence of angular correlations between molecules. When such angular correlations are present the above equations are generalized according to the Kielich relations (84)

$$K_{ind} = \frac{\sigma\Delta\alpha_o\Delta\alpha_\infty}{5kT} \sum_i^N \sum_j^N \frac{<3\cos^2\Theta_{ij} - 1>}{2} \quad (23)$$

$$K_\mu = \frac{\sigma\Delta\alpha_\infty\mu^2}{15k^2T^2} \sum_i^N \sum_j^N \sum_k^N \frac{<3\cos\Theta_{ij}\cos\Theta_{ik} - \cos\Theta_{jk}>}{2} \quad (24)$$

Θ_{ij} is the angle between the principal axes of molecules i and j. K_{ind} is seen to give information on P_2-type angular correlations between molecular axes (P_2 indicates the Legendre polynomial of index 2). K_μ gives information on two and three-particle correlation terms.

For the dynamic Kerr-effect one may consider the responses to 'step-on' and 'step-off' applied electric fields. The rise and decay functions for step-on and step-off are equivalent for the induced part of the Kerr-effect but this is not the case for the dipolar part (85-88). Consider first the case of axially-symmetric molecules. The conditional probability distribution function for a representative molecule is written as

$$\rho(\Omega,t|\Omega_o,0) = \frac{1}{4\pi} + \frac{1}{4\pi}\sum_{m=1} (2m+1)P_m(\cos\Theta)\kappa_m(t) \quad (25)$$

where $P_m(\cos\Theta)$ is the Legendre polynomial of index m, Θ is the polar angle between Ω and Ω_o and the $\kappa_m(t)$ are field-free orientational time correlation functions.

$$\kappa_1(t) = <P_1(\cos\Theta(t))> = <\cos\Theta(t)> \quad (26)$$
$$\equiv C^1_{o,o}(t)$$

$$\kappa_2(t) = <P_2(\cos\Theta(t))> = \frac{<3\cos^2\Theta(t)-1>}{2} \quad (27)$$
$$\equiv C^2_{o,o}(t)$$

For systems of such molecules in the absence of angular correlations between molecules it may be shown (85-88) that the induced part of the Kerr-effect has a step-on and step-off response function given by $\kappa_2(t)$. The dipolar part of the step-off function is also given by $\kappa_2(t)$ but the step-on function is much more involved (85-88). Cole (86) has recently shown that the step-on function for the dipolar part is a weighted sum of $\kappa_1(t)$ and $\kappa_2(t)$ functions so in principle $\kappa_1(t)$ and $\kappa_2(t)$ may be obtained from a comparison of the

Kerr-effect rise (step-on) and decay (step-off) transients for a rectangular pulse in E(t). Rosato and Williams (87-88) have deduced expressions for the rise and decay functions for the Kerr-effect and for dielectric relaxation for the more general case of a body undergoing reorientational motions (as for equation (19)). They show for a molecule having C_{2v} symmetry that the Kerr-effect decay for the induced part is a weighted sum of correlation functions $\langle D^2_{o,m}(\Omega(t))D^2_{o,m}(\Omega(0))\rangle$, and for the dipolar part is a single term involving $\langle D^2_{o,o}(\Omega(t))D^2_{o,o}(\Omega(0))\rangle = \kappa_2(t)$. Fluorenone has C_{2v} symmetry so the Kerr-effect decay transients give information on $\kappa_2(t)$ which complements that for $\kappa_1(t)$ obtained from dielectric relaxation studies.

Note that the steady-state a.c. Kerr-effect is not simply related in the general case to the transient Kerr-effect in the time-domain (86-88), but exact relations can be given if the model for reorientation is specified. For example Benoit gave the necessary relations for the small-step diffusion model for axially-symmetric molecules (85). In this case $\kappa_m(t) = \exp[-m(m+1)D_R t]$ where D_R is the rotational diffusion coefficient (see also refs. 86,89). Thus for the case where the permanent dipole moment contribution to K greatly exceeds the induced dipole moment contribution, the effective relaxation time for step-on response, $\tau_{K,r}$, is far larger than that for step-off decay, $\tau_{K,d}$ and, further, $\tau_{K,d} = (1/3)\tau_\varepsilon$, where τ_ε is the dielectric relaxation time.

IV. COMPARATIVE DIELECTRIC AND KERR-EFFECT STUDIES OF SUPERCOOLED LIQUIDS.

The first such comparative studies were made by Beevers and coworkers (90) for the glass-forming liquid tri-tolylphosphate (TTP) and its mixtures with o-terphenyl. They showed that the Kerr-effect rise and decay functions were equivalent in the range 214-246K and were broad, being characterized by $\bar{\beta}$ in the range 0.75-0.85. The dielectric measurements were made concurrently in the same cell and gave τ_ε and $\bar{\beta}_\varepsilon$ values which were close to $\tau_{K,d}(\simeq\tau_{K,r})$ and $\bar{\beta}_K$, respectively, at each temperature. Converting the Kerr-effect relaxation times into an equivalent $f_m = (2\pi\tau)^{-1}$, taking into account the appropriate correction factor for the empirical relaxation function for t→f conversion, gave the results shown in figure 7. It is evident that $\tau_{K,r} \simeq \tau_{K,d} \simeq \tau_\varepsilon$ at each temperature. Thus the small-step diffusion model does not apply to the reorientational motions of the TTP molecules. These data may be rationalized by saying that when a molecule reorientates it does so completely (i.e. jumps through angles of arbitrary size). This 'fluctuation-relaxation' model (20) gives

$$\rho(\Omega,t|\Omega_o,0) = \delta(\Omega - \Omega_o)\zeta(t) + \frac{(1 - \zeta(t))}{4\pi} \qquad (28)$$

$$\kappa_1(t) = \kappa_2(t) = \kappa_m(t) = \zeta(t) \tag{29}$$

$\zeta(t)$ is a relaxation function due to fluctuations whose functional form can be given, approximately, by the PBL model, equation (18). Studies were also made for fluorenone/o-terphenyl solutions (81) where it was again shown that $\tau_{K,r} \simeq \tau_{K,d} \simeq \tau_\varepsilon$ with $\bar{\beta}_K$ values in the range 0.55 - 0.65. Thus the fluorenone molecules moved not by small-step rotational diffusion but possibly by a 'fluctuation-relaxation' mechanism.

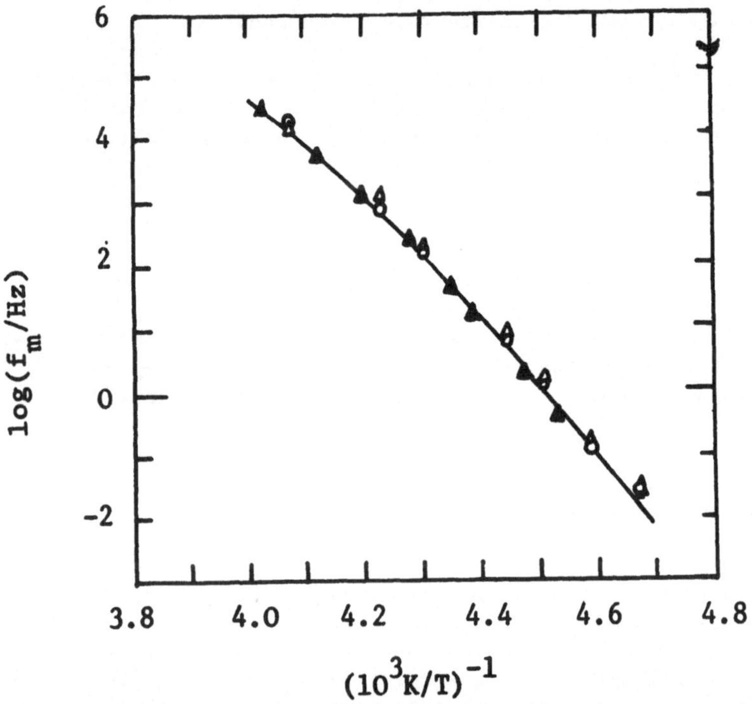

Figure 7. log(f_m/Hz) against $(10^3T/K)^{-1}$ for pure liquid tritolyl phosphate. ▲ *Dielectric data,* ○ *Kerr-effect rise transient data* Δ *Kerr-effect decay transient data (after ref. 90 with permission).*

They also studied TnBP in o-terphenyl (81) and found the following. At low temperatures the Kerr-effect rise and decay transients were nearly equivalent and were fitted with $\bar{\beta} \simeq 0.7$. At the highest temperatures studied the rise and decay transients were different, the rise transient being slower than the decay transient. The plot of $\log f_m$ - vs - $(T/K)^{-1}$ for a 3.29% (w/w) solution of TnBP is shown in figure 8. Note that $\tau_{K,d} \simeq (1/3)\tau_\varepsilon$ at the highest temperatures.

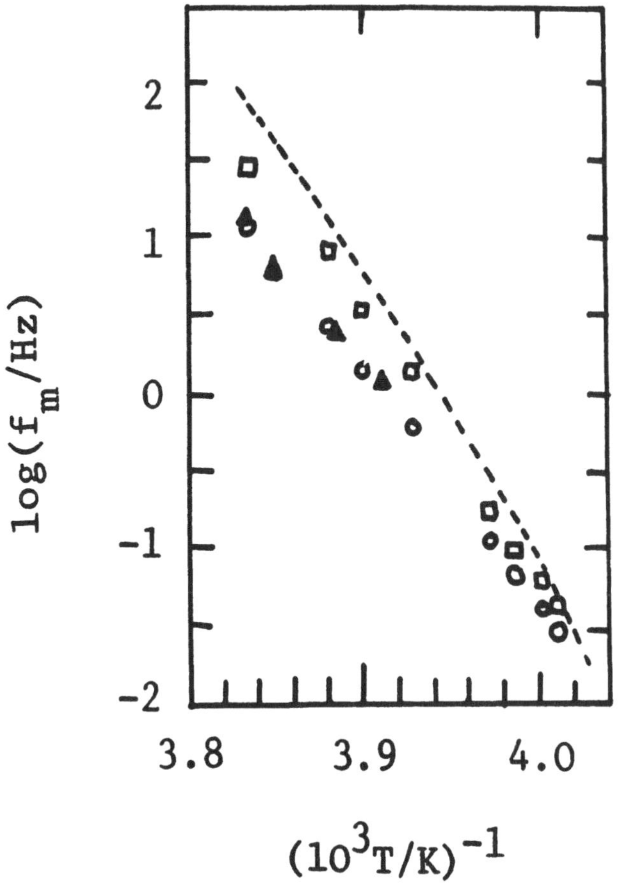

Figure 8. log(f_m/Hz) against $(10^3T/K)^{-1}$ for a 3.3% (w/w) solution of TnBP in o-terphenyl.▲ dielectric data, ○ Kerr-effect rise transient data,□Kerr-effect decay transient data. Dashed line represents dielectric and Kerr-effect data for pure o-terphenyl (after ref. 81 with permission).

It was also shown (81) that $\bar{\beta}_K \approx 0.90$ (near single relaxation time) at the highest temperatures so these results suggest that TnBP ion-pairs undergo small-step diffusion at higher temperatures. The fact that $\tau_{K,r} \to \tau_{K,d} \to \tau_\varepsilon$ at lower temperatures, and tend to the solvent value, and that small $\bar{\beta}$ values are obtained suggest that in this range, the ion-pairs move by the 'fluctuation-relaxation' mechanism as for fluorenone and TTP. This is an example where the mechanism for reorientation changes as temperature is changed. The comparative studies impose constraints on the interpretation, allowing certain mechanisms to be eliminated and others favoured.

TnBP was also studied in a polymeric solvent, poly-(propylene glycol) (PPG) (91). The molecular weight of the PPG liquids were 400, 1000 and 2000. Contrary to expectation, the solute and solvent motions were well-separated with τ_ε(solute) being about 30 times larger than τ_ε(solvent) As a result, two distinct dielectric loss peaks were obtained. The Kerr-constant of PPG was very small in comparison with that for TnBP so the Kerr-effect rise and decay transients were essentially due to solute. Figure 9 shows the derived plot of $\log f_m$ - vs - $(T/K)^{-1}$ for solute and solvent processes.

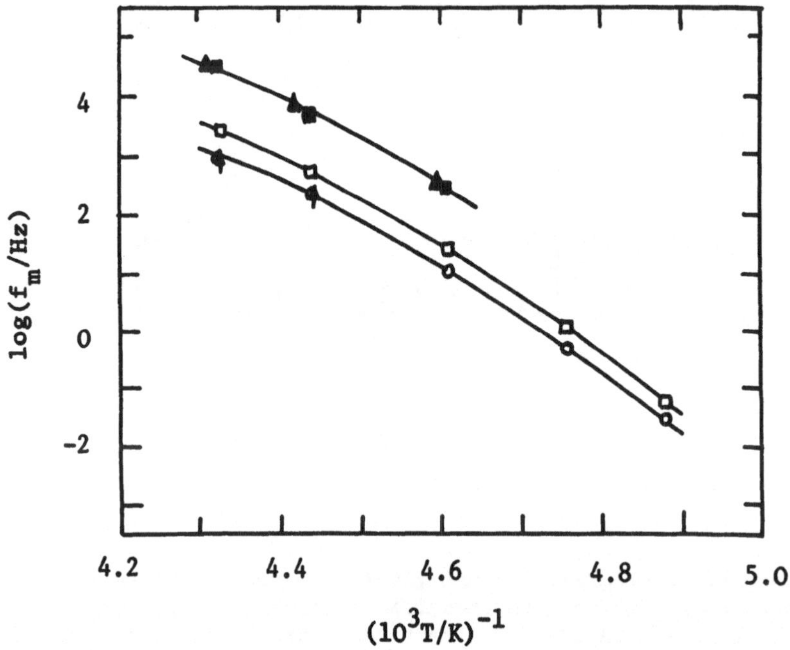

Figure 9. $\log(f_m/Hz)$ against $(10^3T/K)^{-1}$ for a 1.4% (w/w) solution of TnBP in polypropylene glycol 1025. Points marked ■ ♠ and ▲ refer, respectively, to dielectric data for the solvent peak in solution, solute peak in solution and pure solvent. Points marked ○ and □ refer respectively to the Kerr-effect rise and decay data for the TnBP solute (i.e. solute relaxation) (after ref. 91 with permission).

The relaxation for solvent is due to *segmental* motion of the flexible polyether chain. The reorientation of the solute ion-pairs gives dielectric and Kerr-effect relaxations. Inspection of figure 9 shows that $\tau_{K,r} > \tau_{K,d} \approx (1/3)\tau_\varepsilon$, suggesting that TnBP moves by small-step diffusion. Thus in this case the comparative studies give information on the mechanism for motion of solute in a solvent which also gives a low frequency dielectric process.

In our account to this point we have only described cooperative relaxation in non-hydrogen-bonded viscous liquids. The dielectric relaxation behaviour of hydrogen-bonded viscous liquids such as monohydroxyl and polyhydroxyl alcohols is well-documented (for reviews see refs. 2, 92, 93 and refs. therein). Crossley and Williams (92, 93) studied 2-methyl-2,4 pentanediol, 2,4 pentanediol, dipropylene glycol, 2-ethyl 1,3 hexandiol and 6-methyl-3 heptanol using both dielectric and Kerr-effect techniques. The complications introduced by hydrogen bonding (viz. auto + cross-correlation terms) prevent a detailed analysis of these results but it is worth noting that $\tau_{K,r} \simeq \tau_{K,d} \simeq \tau_\varepsilon$ for 2-methyl-2,4 pentanediol, dipropylene glycol and 6-methyl-3-heptanol. The behaviour of 2,4 pentanediol was anomalous, having $\tau_\varepsilon \approx 3\tau_{K,r} \gg \tau_{K,d}$ at all temperatures. For this liquid intra-molecular hydrogen bonding competes with inter-molecular hydrogen bonding so further analysis of the data is not possible.

V. FURTHER CONSIDERATIONS OF THE MECHANISMS FOR COOPERATIVE RELAXATION IN SUPERCOOLED LIQUIDS.

The data for the Kerr-effect and dielectric relaxations for the solutes of Table 1 and for TnBP suggest that the cooperative α-process occurring just above T_g involves many molecules, and their motions are just like those of a system of enmeshing cogs. The 'fluctuation-relaxation' model, where $\zeta(t)$ is described by a defect-diffusion model, rationalizes our observations. However, other models may also give the result $\kappa_1(t) = \kappa_2(t)$: e.g. the diffusional motions of a bond on a tetrahedral lattice (94).

A different approach may be of value, and involves memory functions (76). The correlation function $\kappa_m(t)$ is required to obey the Volterra equation (77).

$$\frac{d\kappa_m(t)}{dt} = -\int_0^t \Gamma_{om}(x)\ \kappa_m(t-x)\,dx \qquad (30)$$

$\Gamma_{om}(x)$ is the first memory function for $\kappa_m(t)$ (89). For 'fast' motions in normal low viscosity liquids this memory function falls to zero in the time-scale 10^{-13} to 10^{-12} and involves inertial and 'recoil' terms. After $\Gamma_{om}(x)$ has decayed to zero, it follows that the remainder of $\kappa_m(t)$ decays exponentially. The α-process we have described occurs in a time-range which is 10^9 to 10^{13} times longer than that for normal low viscosity liquids and $\kappa_1(t)$ and $\kappa_2(t)$ are far removed from the simple exponential form. It is possible that in the supercooled liquids a long-time memory function is operative whose origin is very different from that normally described (76). The form of this memory function is obtained from equation (30). Taking Fourier transforms and using the convolution theorem it readily follows from equation (30) that (76,89)

$$\Gamma_{om}(\omega) = \frac{\kappa_m(0)}{\kappa_m(\omega)} - i\omega \tag{31}$$

where $\Gamma_{om}(\omega) = \mathcal{F}[\Gamma_{om}(t)]$, $\kappa_m(\omega) = \mathcal{F}[\kappa_m(t)]$ and $\kappa_m(0) \equiv \kappa_m(t{=}0)$. Setting the internal field factor $p(\omega)$ equal to unity, equations (4) and (12) give $\kappa_1(t) \simeq \gamma(t)$. Now $\gamma(t) \simeq \exp(-t/\tau_0)^{\bar{\beta}}$ for several of the supercooled solutions so insertion of $\kappa_1(t) = \exp[-t/\tau_0]^{\bar{\beta}}$ into equation (31) gives $K_{om}(\omega)$. For several of the supercooled liquids $\bar{\beta} \simeq 0.5$, and the Fourier transform may be carried out analytically (70,71) giving

$$\kappa_1(\omega) = \frac{1}{i\omega}\left[1 - \frac{1}{2}\left(\frac{\pi}{\tau_0}\right)^{\frac{1}{2}} \cdot \frac{1}{(i\omega)^{\frac{1}{2}}} . \exp\left(\frac{\ell^2}{i\omega}\right) . \mathrm{erfc}\left(\frac{\ell}{(i\omega)^{\frac{1}{2}}}\right)\right] \tag{32}$$

where $2\ell = (\tau_0)^{-\frac{1}{2}}$. Insertion of equation (32) into equation (31) gives $\Gamma_{om}(\omega)$. Inverse Fourier transformation of $\Gamma_{om}(\omega)$ gives $\Gamma_{om}(t)$ and such a form for $\Gamma_{om}(t)$ for $\bar{\beta} = 0.5$ represents the first memory function for $\kappa_1(t)$ and $\kappa_2(t)$ for the cooperative (α) process in several systems (e.g. fluorenone/o-terphenyl). In addition, the analytic form for $\Gamma_{om}(t)$ for $\bar{\beta} = 0.5$ allows one to write a *master equation* for $\kappa_m(t)$, based on equation (30), for the dynamics of certain supercooled liquids, in which the memory function is explicit. Such an approach avoids the concept of a distribution of relaxation times. It would be particularly useful in practice for non-isothermal experiments, as in volume relaxation obtained on cooling towards the glass (40) and thermally stimulated discharge current experiments.

This approach turns the focus of attention from the orientational distribution function $\rho(\Omega,t|\Omega_0,0)$ to the memory function $\Gamma_{om}(t)$. But how may one envisage the physical origin of such a memory function? It arises as a natural consequence of fluctuations which occur in the local thermodynamic variables G, H, S, P, V as envisaged by Landau and Lifshitz (95) and is not due to a distribution of elementary processes. There is a need to have a formal theory for the dynamics of such local fluctuations in liquids from which it would be anticipated that memory functions and relaxation functions similar to those described would emerge. In the absence of such a theory we note that while equation (30) is perfectly general for a classical correlation function (76), the modified Langevin equation for Brownian motion (see ref. 76 p. 610) leads to an equation of the form of equation (30) in which $\Gamma_{om}(t)$ is the correlation function of the 'random force':

$$\Gamma_{om}(t) = \langle F_m(0).F_m(t) \rangle \tag{33}$$

This result is sometimes called 'the second fluctuation-dissipation theorem' (76). Thus within the terms of the modified Langevin equation, $\Gamma_{om}(t)$ gives information on the force-force correlation function which is operative in reorienting the molecule. $\Gamma_{om}(\omega)$ is a measure of the spectral density for these random forces. Attention is thus given to the nature of $\langle F_{om}(0).F_{om}(t)\rangle$. For small-step diffusion the random forces are assumed to be uncorrelated in time and to be of infinitesimal duration, leading to $\Gamma_{om}(t) = \delta(t-x)$ and hence, from equation (30), a correlation function which decays exponentially with t. The fact that $\kappa_1(t) \simeq \exp[-t/\tau_o]^{\frac{1}{2}}$ suggests that the forces acting on the reference molecule are correlated in time : i.e. the molecule is subjected to a continuously variable force whose time variation is similar to that for the reorientation of the molecule, thus giving $\Gamma_{01}(t)$ which decays on a time-scale similar to that for the decay of $\kappa_1(t)$. In this way the correlation function may become a 'slow' function of time; i.e. of a slower functional form than the simple exponential form $\exp(-t/\tau)$. Models for the fluctuations in $\Gamma_{01}(t)$ may be proposed which would lead to $\Gamma_{01}(t)$ and $\kappa_1(t)$ of the forms which are observed for certain supercooled liquids, and these are under consideration.

The above discussion of memory functions was prompted by the remarks made at the ASI by Professor Steele regarding the long-time exponential behaviour of $\kappa_m(t)$ expected from equation (30) when $\Gamma_{om}(x)$ decays rapidly (as in normal low viscosity liquids).

VI. RELATED KERR-EFFECT STUDIES

We have seen that dielectric and dynamic Kerr-effect techniques provide complementary information on the rotational dynamics in liquids in which extensive angular correlations between molecules is absent. For liquids which become liquid-crystalline on cooling (e.g. the nematogens such as benzylidene aniline derivatives and alkyl cyanobiphenyls, or the cholesterics such as cholesteryl esters) extensive angular correlations occur between molecules just above the clearing point T_c. The Kerr-effect provides a sensitive method for detecting the formation of angular correlations and for studying the dynamics of the ensemble

(96-102). For example, Beevers (96) showed that the static Kerr-constant K became large (and negative) in the pretransition region for the nematogen N-(p-methoxybenzylidene)-p-n-butyl aniline (MBBA) and obeyed a relation $(T - T^*)^{-1}$ where $T^*(<T_c)$ is an apparent second-order transition temperature. For the nematogen n-heptyl cyanobiphenyl (HCB) Davies and coworkers (97) showed that K became large (but was always positive) in the pre-transition region and also followed the $(T - T^*)^{-1}$ relation. Equations (23) and (24) indicated that $(K_{\mu} + K_{ind})$ for an angularly correlated system may be positive or negative and will increase in magnitude as the angular correlations develop in the isotropic liquid as T_c is approached. Beevers (96) also studied MBBA in carbon tetrachloride solution and showed that the apparent molecular Kerr-constant only became positive for solute concentrations less than 0.18M: i.e. the strong angular correlations between MBBA molecules persisted down to that concentration. In contrast to the pre-transitional behaviour of the Kerr-constant for pure MBBA (96) and HCB (97), the static dielectric permittivity, ε_0, dielectric relaxation time, τ_{ε}, the optical refractive index and liquid density do not show any unusual variation as T_c is approached from above (97, 102): i.e. they do not exhibit premonitory behaviour. Complementary dielectric and Kerr-effect relaxation studies were made by Schadt (102) for several alkyl cyanopyrimidines and their mixtures. While ε_0 and τ_{ε} gave apparently-normal variations with decreasing temperature for the isotropic liquid, K and τ_K followed the $(T-T^*)^{-1}$ relation, with $T^* < 1.0$ K, indicating the formation of local order in the liquid. While the presence of such order is detected by such Kerr-effect experiments, the individual terms in equations (23) and (24) are not easily extracted from the data. The theory of the dynamic Kerr-effect for an angularly correlated system appears to be extremely complicated, partly because of the difference between step-on and step-off responses discussed above. In the presence of angular correlations, Cole (86) showed that the induced dipole moment term gives rise and decay functions, for step-on and step-off fields, which are equivalent and give information on $\sum_i \sum_j \langle P_{2i}(0).P_{2j}(t)\rangle$. Note that this induced term has a strength factor given by equation (23); i.e. the g_2 equilibrium correlation function for the molecular axes. Such a factor also governs the magnitude of the depolarized Rayleigh scattered light in the pre-transition region, as was discussed by Gierke and Flygare (103) for MBBA. The permanent dipolar contribution to the dynamic Kerr-effect for an angularly correlated system has been discussed by Cole (86) for step-on and step-off fields. The results are complicated. The step-off transient involves $\sum_i \sum_j \langle P_{2i}(0).P_{2j}(t)\rangle$ and the step-on transient is a weighted mixture of P_1 and P_2 auto- and cross-correlation functions (see ref. 86, equation (19)). Note that Pecora (104) has discussed the form of correlation functions arising for the induced dipole term in terms of the conditional probability distribution $\rho(\Omega_j,t|\Omega_i,0)$ between pairs of molecules.

All the experimental data for MBBA, alkyl cyanobiphenyls and alkyl cyanopyrimidines (97-102) show that $\tau_{K,r}$, $\tau_{K,d}$ and τ_K (steady-state a.c.) all increase as T_c is approached, following the $(T-T^*)^{-1}$ relation, but τ_ε shows no unusual variation and $\tau_\varepsilon < \tau_K$. It is well-known that in the *liquid-crystalline* nematic state a well-defined dielectric process is observed (97), which is thought to be due to the 'flip-flop' reorientations of the dipole vector with respect to the (fixed) nematic-director axis. For the isotropic liquid just above T_c the Kerr-effect data suggest that local directors are present so it would be reasonable to assume that the dipole vector may undergo 'flip-flop' motions (P_1 relaxation), leading to dielectric relaxation, but that at longer times the randomization of the director, or its disappearance due to fluctuations, would lead to Kerr-effect relaxation. Hence $\tau_K >> \tau_\varepsilon$ in the pretransition region.

A quite different interpretation may be made in terms of relations derived by Kivelson and coworkers (105, 106). They relate a single particle correlation time τ_{is} and a collective correlation time τ_{ic} by the equation

$$\tau_{ic} = \frac{g_i}{j_i}\,\tau_{is} \tag{34}$$

i(=1 or 2) indicates the order of the polynomial, g_i is the static and j_i is the dynamic orientational correlation factor. For dielectric and induced-dipole moment Kerr-effects we have, respectively

$$\tau_{1c} = \frac{g_1}{j_1}\cdot\tau_{1s} \qquad \text{dielectric} \tag{35}$$

$$\tau_{2c} = \frac{g_2}{j_2}\cdot\tau_{2s} \qquad \text{Kerr-effect} \tag{36}$$

g_1 and g_2 are the equilibrium Kirkwood and Kielich factors discussed above. It is not clear how j_1 and j_2 are to be estimated. Clearly a simulation for a model system of angularly correlated molecules would yield τ_{1c}, τ_{2c}, g_1, g_2, τ_{1s} and τ_{2s}. In the absence of such information one may make progress only by making assumptions, the most reasonable one being that j_1 and j_2 vary only slowly with temperature in the pre-transition range. It would then follow that τ_{1c} scales as g_1 and τ_{2c} scales as g_2. Experimental dielectric data for HCB (97) shows that g_1 changes only slowly but g_2 changes remarkably in the pre-transition range (see also ref. 103 for the variation of g_2 for MBBA). Thus the marked increase in τ_K (and τ(light scattering (103)) and the relatively small increase in τ_ε as T_c is approached can be rationalized, at least qualitatively, by the Kivelson approach. Equations (35) and (36) were derived using the Mori projection operator method. Cheung (107) has discussed this derivation and has derived the following relation

$$\tau_{2c} = \tau_{2s} \frac{(1 + fNh)}{(1 + fN)} \tag{37}$$

where $g2 = (1 + fN)$, $h = (\tau_d/\tau_{2s})$ and τ_d is a distinct correlation time which is defined by the relation:

$$\tau_d = \frac{\int_0^\infty <P_2(\cos\theta^{(1)}(0))P_2(\cos\theta^{(2)}(t))>dt}{<P_2(\cos\theta^{(1)}(0))P_2(\cos\theta^{(2)}(0))>} \tag{38}$$

(1) and (2) refer to different molecules in the liquid. Thus τ_{2c} and τ_{2s} are related in terms of two well-defined microscopic parameters fN and h. Equation (37) is quite general and does not depend on any model or theory of reorientation (107). For further discussion of equations (34) - (38) the reader is referred to ref. 107, p. 162-165. The scaling of two dynamic quantities by a factor involving equilibrium correlations in a liquid, as is the case for equation (34), also occurs for the dynamic scattering of light for a partially-ordered liquid in which the particles undergo translational diffusion (108, 109) (see also ref. 107, p. 164) and I thank Professor Versmold for drawing this to my attention at the ASI.

VII CONCLUDING REMARKS

The similarity of the multiple relaxations observed for small-molecule and polymeric glass-forming systems is striking and implies common underlying relaxation mechanisms for the individual processes. Insufficient information is contained in the results of a single experimental method to allow the mechanism of relaxation to be established, so comparative studies, using dielectric and Kerr-effect relaxation, have been made which put constraints on the use of particular models. Such studies have shown that small-step rotational diffusion does not apply in all cases except for the (large) tri-n-butyl ammonium picrate ion-pairs in o-terphenyl at higher temperatures. While a 'fluctuation-relaxation' mechanism, taken with a model for defect-diffusion, is able to rationalize the data for the cooperative α-process, other mechanisms may well apply. A different approach involving memory functions and the correlation function for the effective random forces is outlined. It is possible that a modelling of the random forces would provide a rationalization for the shape of the α-process of the supercooled liquids.

The presence of extensive local ordering in the isotropic liquid state for liquid-crystal-forming media is readily detected using the equilibrium and dynamic Kerr-effects. Since the observed quantities are complicated weighted sums of angular correlation terms it seems not to be possible to extract the individual terms from Kerr-effect data alone. However if the properties of a model system of angularly-correlated molecules were simulated by computer it should be possible to calculate the dielectric and Kerr-effect quantities in the pre-transition region. The Kerr-effect would provide a good test of such model calculations since, as we have described, K may increase to large positive values (as for HCB) or negative values (as for MBBA).

REFERENCES

1. McCrum, N.G., Read, B.E. and Williams, G., 1967 Anelastic and Dielectric Effects in Polymeric Solids, J. Wiley, London and New York.
2. Hill, N.E., Vaughan, W.E., Price, A.H. and Davies, M., 1969, Dielectric Properties and Molecular Behaviour, Van Nostrand, New York.
3. Wada, Y., 1977, in Dielectric and Related Molecular Processes, ed. M. Davies (Spec. Period Report), The Chem. Soc., London, 3, p. 143.
4. Williams, G. and Watts, D.C., 1971 in NMR Basic Principles and Progress, Vol. IV, NMR of Polymers, Springer, Berlin, Heidelberg and New York, p. 271.
5. Williams, G., 1979, Advances in Polymer Science, 33, p.60.
6. Williams, G., 1982, in Static and Dynamic Properties of Solid Polymers, NATO Advanced Study Institute, ed. R. Pethrick and R.W. Richards, D. Reidel Publ. Holland, p.213.
7. Williams, G., 1972, in J.Chem.Soc., Faraday Symposium No. 6, p.44 of discussion.
8. Williams, G., 1972, Chem.Rev., 72, p. 55.
9. Debye, P., 1929, Polar Molecules, Chemical Catalog Co., New York.
10. DiMarzio, E.A. and Bishop, H., 1974, J.Chem.Phys., 60, p.3802.
11. Ivanov, E.N., 1964, Sov.Phys., J.E.T.P., 10, P.1041.
12. Anderson, J.E., 1972, cited in J.Chem.Soc., Faraday Symp., 6, p.90 of discussion.
13. Hoffman, J.D., and Pfeiffer, H.G., 1954, J.Chem.Phys., 22, p.132.

14. Hoffman, J.D., 1955, J.Chem.Phys., 23, p.1331.
15. Adam, G., 1965, J.Chem.Phys., 43, p.662.
16. Brereton, M.G. and Davies, G.R., 1977, Polymer, 18, p.1764.
17. Glarum, S.H., 1960, J.Chem.Phys., 33, p.639.
18. Anderson, J.E., and Ullman, R., 1967, J.Chem.Phys., 47, p.2178.
19. Phillips, M.C., Barlow, A.J. and Lamb, J., 1972, Proc.Roy. Soc., Ser. A, 329, p.193.
20. Beevers, M.J., Crossley, J., Garrington, D.C. and Williams, G., 1976, J.Chem.Soc., Faraday Trans II, 72 p.1482.
21. Kirkwood, J.G. and Fuoss, R.M., 1941, J.Chem.Phys., 9, p.329.
22. Yamafuji, K. and Ishida, Y., 1962, Koll.Zeit., 183, p.15.
23. Stockmayer, W.H., 1967, Pure Appl. Chem., 15, p.539.
24. Stockmayer, W.H., 1976, in Fluides Moleculaires, ed. Balian, R. and Weill, G., Gordon and Breach, New York, p.101.
25. Dubois-Violette, E., Geny, F., Monnerie, L. and Parodi, O., 1969, J.Chim.Phys., 66, p.1865.
26. Geny, F. and Monnerie, L., 1977, J. Polymer Sci., Polymer Phys. Edn., 15, p.1.
27. Hayakawa, R. and Wada, Y., 1974, J. Polymer Sci., Polymer Phys. Edn., 12, p.2119.
28. Saito, N., Okano, K., Iwayanagi, S., Hideshima, T., 1963, in Solid State Physics, ed. Seitz, F. and Turnbull, D., Academic Press, New York, Vol. XIV, p.343.
29. Jernigan, R.L., 1972, in Dielectric Properties of Polymers, ed. Karasz, F.E., Plenum Publ. p.99.
30. Beevers, M. and Williams, G., 1975, Adv. Mol. Relax. Proc., 7, p.237.
31. Shore, J.E., and Zwanzig, R., 1975, J.Chem.Phys., 63, p.5445.
32. Evans, G.T. and Knauss, D.C., 1980, J.Chem.Phys., 72, p.1504.
33. Volkenstein, M.V., 1963, Configurational Statistics of Polymeric Chains, Wiley-Interscience, New York.
34. Flory, P.J., 1969, Statistical Mechanics of Chain Molecules, Wiely-Interscience, New York.
35. Read, B.E., 1965, Trans.Faraday Soc., 61, p.2140.
36. Cook, M., Watts, D.C., and Williams, G., 1970, Trans. Faraday Soc., 66, p.2503.
37. Glarum, S.H., 1960, J.Chem.Phys., 33, p.1371.
38. Cole, R.H., 1965, J.Chem.Phys., 42, p.637.
39. Steele, W.H., 1965, J.Chem.Phys., 43, p.2598.
40. Ferry, J.D., 1970, Viscoelastic Properties of Polymers, Wiley, New York.
41. Ward, I.M., 1971, Mechanical Properties of Solid Polymers, Wiley, London.
42. Johari, G.P., and Smyth, C.P., 1969, J.Amer.Chem.Soc., 91, p.5168.
43. Johari, G.P. and Goldstein, M., 1970, J.Phys.Chem., 74, p.2034.
44. Johari, G.P. and Goldstein, M., 1970, J.Chem.Phys., 53, p.2372.

45. Johari, G.P. and Smyth, C.P., 1972, J. Chem. Phys., 56, p.4411.
46. Johari, G.P., 1973, J.Chem.Phys., 58, p.1766.
47. Johari, G.P., 1972, in J.Chem.Soc., Faraday Symp., 6, p.42.
48. Baker, W.O. and Smyth, C.P., 1939, J.Amer.Chem.Soc., 61, p.2063.
49. Thomas, S.B., 1931, J.Phys.Chem., 35, p.2103.
50. Kobeko, P.P. and coworkers, 1939, J.Tech.Phys., USSR, 8, p.715.
51. Turkevich A. and Smyth, C.P., 1942, J.Amer.Chem.Soc., 64, p. 737.
52. Denney, D.J., J.Chem.Phys., 27, p.259.
53. Winslow, J.W., Good, R.J. and Berghausen, P.E., 1957, J.Chem.Phys., 27, p.309.
54. Denney, D.J., 1959, J.Chem.Phys., 30, p.1019.
55. Denney, D.J., 1959, J.Chem.Phys., 30, p.59.
56. Mopsik, F.I. and Cole, R.H., 1966, J.Chem.Phys., 44, p.1015.
57. Berberian, J.G. and Cole, R.H., 1968, J.Amer.Chem.Soc., 90, p.3100.
58. Davidson, D.W. and Cole, R.H., 1950, J.Chem.Phys., 18, p.1417.
59. Davidson, D.W. and Cole, R.H., 1951, J.Chem.Phys., 19, 1484.
60. Davidson, D.W., 1961, Canad.J.Chem., 39, p.571.
61. Greet, R.J. and Turnbull, D.J., 1967, J.Chem.Phys., 47, p.2185.
62. Greet, R.J. and Turnbull, D.J., 1967, J.Chem.Phys., 46, p.1243.
63. Chang, S.S., and Bestul, A.B., 1972, J.Chem.Phys., 56, p.503.
64. McCall, D.W., Douglass, D.C. and Falcone, D.R., 1969, J.Chem.Phys., 50, p.3839.
65. Williams, G. and Hains, P.J., 1971, Chem.Phys.Lett., 10, p.585.
66. Williams, G. and Hains, P.J., 1972, J.Chem.Soc., Faraday Symp., 6, p.14.
67. Shears, M.F. and Williams, G., 1973, J.Chem.Soc., Faraday Trans. II, 69, p.608.
68. Shears, M.F. and Williams, G., 1973, J.Chem.Soc., Faraday Trans. II, 69, p.1050.
69. Davies, M., Hains, P.J. and Williams, G., 1973, J.Chem.Soc., Faraday Trans. II, 69, p.1785.
70. Williams, G. and Watts, D.C., 1970, Trans. Faraday Soc., 66, p.80.
71. Williams, G., Watts, D.C., Dev, S.B. and North, A.M., 1971, Trans. Faraday Soc., 67, p.1323.
72. Wetton, R.E., MacKnight, W.J., Fried, J.R. and Karasz, F.E., 1978, Macromolecules, 11, p.158.
73. Warchol, M.P. and Vaughan, W.E., 1978, Adv.Mol.Relax,Proc. 13, p.317.
74. Wang, C.C. and Pecora, R., 1980, J.Chem.Phys., 72, p.5333.
75. Davies, M.M. and Williams, G., 1960, Trans.Faraday Soc., 56, p.1619.

76. Berne, B.J., 1971, in Eyring, H., Jost, W., and Henderson, D.; (Eds.) Physical Chemistry, An Advanced Treatise, Vol. VIIIB, The Liquid State, Academic Press, New York, p.539.
77. Berne, B.J. and Pecora, R., 1976, Dynamic Light Scattering, Wiley-Interscience, New York.
78. Rose, M.E., 1957, Elementary Theory of Angular Momentum, Wiley, New York.
79. Hildebrand, J.H., 1964, Amer.Philosophical Soc., 108, p.411.
80. Beevers, M.S., Crossley, J., Garrington, D.C. and Williams, G., 1977, J.Chem.Soc., Faraday Trans. II, 73, p.458.
81. Beevers, M.S., Crossley, J., Garrington, D.C. and Williams, G., 1976, J.Chem.Soc., Faraday Symp., 11, p.38.
82. Stuart, H.A., Funk, E. and Muller-Warmuth, W., 1967, Moleculstruktur, Springer Verlag, Berlin and New York.
83. Fredericq, E. and Houssier, C., 1974, Electric Dichroism and Electric Birefringence, Oxford U.P., London.
84. Kielich, S., 1972, in Dielectric and Related Molecular Processes, ed. M. Davies, The Chem. Soc., London, Spec. Period. Reports, 2, p.192.
85. Benoit, H., 1951, Ann. Phys. (Paris) 6, p.561.
86. Cole, R.H., 1982, J.Phys.Chem., 86, p.4700.
87. Rosato, V. and Williams, G., 1981, J.Chem.Soc., Faraday Trans. II, 77, p.1767.
88. Rosato, V. and Williams, G., 1982, in Molecular Interactions, Vol. 3., ed. Ratajczak, H., and Orville-Thomas, W.J., Wiley, New York, Ch. 8, p.373.
89. Williams, G., 1978, Chem.Soc. Reviews, 7, p.89.
90. Beevers, M.S., Crossley, J., Garrington, D.C. and Williams, G., 1977, J.Chem.Soc., Faraday Trans. II, 73, p.458.
91. Crossley, J., Elliott, D.A. and Williams, G., 1979, J.Chem. Soc., Faraday Trans. II, 75, p.88.
92. Crossley, J. and Williams, G., 1977, J.Chem.Soc., Faraday Trans. II, 73, p.1651.
93. Crossley, J. and Williams, G., 1977, J.Chem.Soc., Faraday Trans. II, 73, p.1906.
94. Valeur, B. and Monnerie, 1976, J. Polymer Sci., Polymer Phys. Edn., 14, p.11 and p.29.
95. Landau, L.D. and Lifshitz, E.M., 1958, Statistical Physics, Addison Wesley, Mass.
96. Beevers, M.S., 1975, Mol. Cryst.Liq.Cryst., 31, p.333.
97. Davies, M., Moutran, R., Price, A.H., Beevers, M.S. and Williams, G., 1976, J.Chem.Soc., Faraday Trans. II, 72, p.1447.
98. Coles, H.J. and Jennings, B.R., 1976, Mol.Phys., 31, p.571.
99. Coles, H.J. and Jennings, B.R., 1976, Mol.Phys., 31, p.1225.
100. Coles, H.J. and Jennings, B.R., 1978, Mol.Phys., 36, p.1661.
101. Kolinsky, P.V. and Jennings, B.R., 1980, Mol.Phys., 40, p.979.
102. Schadt, M., 1977, J.Chem.Phys., 67, p.210.
103. Gierke, T.D. and Flygare, W.H., 1974, J.Chem.Phys., 61, p.2331.

104. Pecora, R., 1968, J.Chem.Phys., 50, p.2650.
105. Keyes, T., and Kivelson, D., 1972, J.Chem.Phys., 56, p.1057.
106. Kivelson, D. and Madden, P., 1975, Mol.Phys., 30, p.1749.
107. Cheung, P.S.Y., 1977, J.Chem.Soc., Faraday Symp., 11, p.162.
108. Brown, J.C., Pusey, P.N., Goodwin, J.W. and Ottewill, R.H., 1975, J.Phys., A, Math., Gen., 8, p.664.
109. Pusey, P.N., 1975, J.Phys., A, Math., Gen., 8, p.1433.

DEPOLARIZED LIGHT SCATTERING EXPERIMENTS

H. Versmold
Physikalische Chemie
Universität Dortmund
D-4600 Dortmund, W-Germany

1. INTRODUCTION

Depolarized light scattering belongs to the most powerful experimental techniques to study the orientational dynamics and structure in molecular liquids. Since the experimental and theoretical aspects of the light scattering method are well documented in the literature |1, 2, 3|, in this chapter particular emphasis will be given to the different schemes presently used to evaluate depolarized light scattering spectra. By referring to typical experimental results it will then be shown which kind of information concerning the reorientational dynamics and orientational structure of liquids can be obtained with this method.

2. EXPERIMENTAL SPECTRA

In a typical light scattering experiment a laser is used as source for monochromatic linear polarized light. The incident light field can be characterized by its electric field strength $\vec{E}_i$, polarization Vector $\vec{n}_i$, frequency ω_i, and wave vector $\vec{k}_i$. In passing through the scattering cell, which contains the substance under study, the field induces dipoles $\vec{\mu}_{ind}(t)$ in the fluid particles via their molecular polarizability $\overleftrightarrow{\alpha}(t)$. The sum of the induced moments, located

A. J. Barnes et al. (eds.), Molecular Liquids - Dynamics and Interactions, 275–308.

in the scattering volume, scatters light. This scattered light field can similarly be described by $\vec{E}_f$, $\vec{n}_f$, ω_f and $\vec{k}_f$. For a 90° scattering experiment four arrangements, which differ by the polarization of the incident and scattered light, are of interest. Commonly the letter V is used for vertically and H for horizontally polarized light, such that the four combinations VV, VH, HV, and HH refer to the four experimental arrangements. Conventionally the first letter gives the polarization of the incident and the second that of the scattered light. A VV experiment is also called a polarized scattering experiment, since the polarization direction does not change during scattering. Contrary, VH, HH, and HV experiments are termed depolarized scattering experiments.

In order to further characterize the polarized and depolarized light scattering it is convenient to divide the molecular polarizability $\vec{\vec{\alpha}}(A,t)$ of a fluid particle A into a spherical part $\alpha^{(o)}(A)\ \vec{\vec{1}}$ and a non-spherical traceless part $\vec{\vec{\alpha}}^{(2)}(A,t)$ |4, 5|

$$\vec{\vec{\alpha}}(A,t) = \alpha^{(o)}\ \vec{\vec{1}} + \vec{\vec{\alpha}}^{(2)}(A,t). \tag{1}$$

Details of this decomposition will be given below. For the present introductory purpose it is sufficient to know, that the two polarizability components give rise to two independent light scattering components which differ by geometrical factors. The scattering from the spherical polarizabilty $\alpha^{(o)}(A)$ is accompanied by a factor $(\vec{n}_i \cdot \vec{n}_f)^2$, i.e. this so called isotropic scattering component is observable only in the VV experiment. On the other hand, the anisotropic scattering component resulting from the $\vec{\vec{\alpha}}^{(2)}(A,t)$ polarizability is accompanied by the geometrical factor $(3+(\vec{n}_i \cdot \vec{n}_f)^2)$. Thus, this anisotropic scattering component contributes both to the polarized (VV) scattering and to the depolarized (VH, HV, and HH) scattering with an intensity ratio of 4/3.

Finally, if the molecules under consideration vibrate in a particular normal mode v the spherical and the non-spherical part of the polarizability may vary due to such a vibration. To first order this variation can be written as

$$\alpha_v^{(n)}(A,t) = \alpha^{(n)}(A,t) + \alpha_v^{(n)\prime}(A,t)\cdot q_v(t),\ n = 0,2, \quad (2)$$

where $q_v(t)$ is the v-th normal coordinate, $\alpha_v^{(n)\prime}(A,t) = \partial\alpha^{(n)}(A,t)/\partial q_v|_o$ is the polarizability derivative, and $\alpha^{(n)}(A,t)$, as before, the permanent polarizability.

The polarizability components $\alpha^{(o)}(A,t)$ and $\alpha^{(2)}(A,t)$ give rise to the isotropic and anisotropic Rayleigh spectra centered at the laser frequency ω_L. If for a given normal vibration v the $\alpha_v^{(n)\prime}(A,t)$ do not vanish they will lead to Raman scattering. Figure 1 shows typical spectra as observed in a double monochromator experiment. The upper part of the figure represents

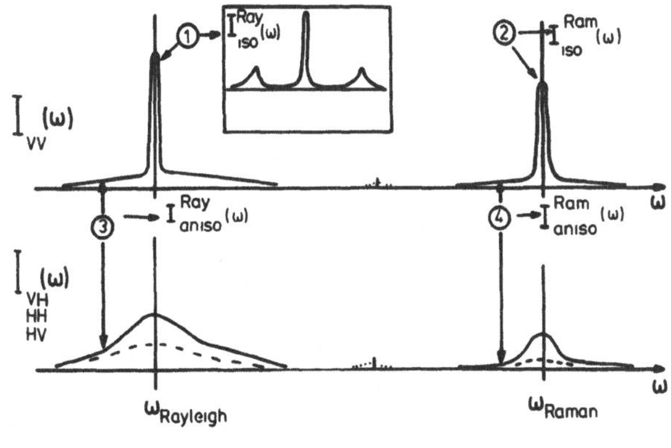

Fig. 1: Polarized and depolarized Rayleigh and Raman spectra

the polarized or VV spectrum $I_{VV}(\omega)$ and the lower part the corresponding depolarized or VH, HH, and HV spectrum. In the vicinity of the laser frequency $\omega_L = \omega_{Rayleigh}$ the polarized and depolarized Rayleigh spectra are observed. The most pronounced feature of the polarized Rayleigh spectrum is a very sharp peak ① at the center of the band resulting from the isotropic permanent polarizability $\alpha^{(o)}(A)$. In a double monochromator experiment this peak usually is not resolved. If, however, a Fabry-Perot interferometer is used for the spectral decomposition this peak is found to consist of a central Rayleigh and two symmetrically shifted Brillouin lines. This isotropic Rayleigh-Brillouin spectrum $I^{Ray}_{iso}(\omega)$ is the Fourier transform of density-density correlations in the fluid |1, 2|.

Beside the central peak just mentioned the polarized Rayleigh spectrum consists of a broad diffuse background ③ resulting from the non-spherical part $\alpha^{(2)}(A,t)$ of the polarizability. This is the anisotropic Rayleigh spectrum $I^{Ray}_{aniso}(\omega)$, which reflects orientational fluctuations but contains also interaction induced contributions to be discussed below. The anisotropic spectrum $I^{Ray}_{aniso}(\omega)$ ③ also occurs with a (by the factor 3/4) reduced intensity in the depolarized spectra. Despite the somewhat smaller intensity, $I^{Ray}_{aniso}(\omega)$ is easier observed in a depolarized experiment since here no interference with the isotropic spectrum takes place. The terms depolarized and anisotropic Rayleigh scattering are often used synonymously.

Isotropic and anisotropic Raman scattering contributions due to non-vanishing $\alpha_v^{(o)\prime}(A,t)$ and $\alpha_v^{(2)\prime}(A,t)$ occur frequency shifted by ω_v from the laser frequency. One such Raman band is included in Fig. 1 The polarized spectrum again consists of two components. The narrow central peak ② is the isotropic Raman band $I^{Ram}_{iso}(\omega)$ resulting from $\alpha_v^{(o)\prime}(A)$. It is the Fourier transform of the vibrational correlation function and can be used to study vibrational dephasing. The broader background ④ is the anisotropic Raman spectrum $I^{Ram}_{aniso}(\omega)$. It results from the polarizability component $\alpha_v^{(2)\prime}(A,t)$ and is the Fourier transform of a compound rotation-vibration correlation function |6, 7|.

The isotropic and anisotropic scattering components of a given band can be determined from the experimental polarized and depolarized spectra simply by using

$$\begin{aligned} I_{iso}(\omega) &= I_{VV}(\omega) - 4/3\, I_{VH}(\omega) \\ I_{aniso}(\omega) &= I_{VH}(\omega). \end{aligned} \tag{2}$$

3. DEPOLARIZED RAYLEIGH SCATTERING CONTRIBUTIONS

Once an anisotropic or depolarized Rayleigh spectrum is obtained experimentally the question remains which information can be gained from it. The evaluation of experimental spectra is by no means a trivial matter. This is due to the fact that not only the molecular polarizability $\alpha^{(2)}(A,t)$ but also intermolecular contributions affect the polarization of the scattering volume and thus contribute to a depolarized Rayleigh spectrum. Before discussing different evaluation schemes in the next section, here we consider in more detail the different scattering mechanisms.

Depolarized Rayleigh spectra are influenced by fluctuations of $\alpha^{(2)}(A,t)$ due to molecular reorientation. Beside this orientational scattering so called interaction induced mechanisms also cause fluctuations of the induced electric moment of the scattering volume. The best investigated mechanism of this type is the dipole induced dipole (DID) mechanism in which a dipole induced by the laser field in a neighbour particle B contributes a fluctuating field at particle A which adds to the laser field at particle A |8-11|. The influence of induced quadrupoles etc. has also been discussed |12, 13|, however, much less is known about such higher order induced mechanisms. For the total moment induced in the scattering volume at time t we can write |14, 15|

$$\vec{\mu}_{ind}(t) = \sum_A \overleftrightarrow{\alpha}(A,t)\cdot\vec{E}_A + \frac{1}{2}\sum_A \overleftrightarrow{\overleftrightarrow{\beta}}(A,t): \vec{E}_A \vec{E}_A + \frac{1}{3}\sum_A \overleftrightarrow{\overleftrightarrow{A}}(A,t) : \overleftrightarrow{E}'_A \tag{3}$$

where $\overleftrightarrow{\alpha}(A,t)$, $\overleftrightarrow{\overleftrightarrow{\beta}}(A,t)$, and $\overleftrightarrow{\overleftrightarrow{A}}(A,t)$ are the (dipole) polarizability, the hyperpolarizability, and the quadrupole polarizability of a particle A at time t. The summations extend over all particles in the scattering volume. $\vec{E}_A$ and $\overleftrightarrow{E}'_A$ are the field vector and the field gradient tensor at particle A:

$$\vec{E}_A = \vec{E}_i(A) + \sum_{B\neq A} \overleftrightarrow{T}^{(2)}(\vec{r}_{AB})\cdot\overleftrightarrow{\alpha}(B,t)\cdot\vec{E}_i(B) - \frac{1}{3}\sum_{B\neq A} \overleftrightarrow{\overleftrightarrow{T}}^{(3)}(\vec{r}_{AB}) : \overleftrightarrow{\Theta}(B,t) - \frac{1}{3}\sum_{B\neq A} \overleftrightarrow{\overleftrightarrow{T}}^{(3)}(\vec{r}_{AB}) : \overleftrightarrow{\overleftrightarrow{A}}(B,t)\cdot\vec{E}_i(B) \tag{4}$$

and

$$\overleftrightarrow{E}'_A = \sum_{B\neq A} \overleftrightarrow{\overleftrightarrow{T}}^{(3)}(\vec{r}_{AB})\cdot\overleftrightarrow{\alpha}(B,t)\cdot\vec{E}_i(B). \tag{5}$$

Here, $\overleftrightarrow{\Theta}(B,t)$ is the quadrupole moment tensor of molecule B at time t and $\overleftrightarrow{T}^{(2)}(\vec{r}_{AB})$ and $\overleftrightarrow{\overleftrightarrow{T}}^{(3)}(\vec{r}_{AB})$ are the dipole-dipole and the dipole-quadrupole tensor |16-18|.

If the expressions (4) and (5) are inserted into (3) a rather intractable result for the induced moment is obtained in general. Molecular symmetry, however, may considerably reduce the number of contributing terms. In order to make this simplification apparent we consider table 1 which gives the number of independent constants of the lowest molecular moments and polarizabilities for some common point groups |16|. It is immediately obvious that no simplification occurs for mole-

Table 1. Number of Constants

Group	$\vec{\mu}$	$\overleftrightarrow{\Theta}$	$\overleftrightarrow{\alpha}$	$\overleftrightarrow{\beta}$	$\overleftrightarrow{A}$
C_{2v}	1	2	3	3	4
C_{3v}	1	1	2	3	3
D_{6h}	0	1	2	0	0
$D_{\infty h}$	0	1	2	0	0
T_d	0	0	1	1	1
sphere	0	0	1	0	0

cules of C_{2v} and C_{3v} symmetry since none of the molecular properties vanishes. For these molecules orientational, DID, and higher order interaction induced scattering contributions plus cross terms between them make up the depolarized spectrum. For molecules with a center of inversion, like D_{6h} and $D_{\infty h}$ molecules, the $\overleftrightarrow{\beta}$ and $\overleftrightarrow{A}$ tensor vanish. Now equations (3) with (4) and (5) simplifies to,

$$\vec{\mu}_{ind}(t) = \overleftrightarrow{A}(t)\cdot\vec{E}_i, \tag{6}$$

where the total polarizability A(t) is given by

$$\overleftrightarrow{A}(t) = \overleftrightarrow{\alpha}(t) + \Delta\overleftrightarrow{A}(t) \tag{7}$$

with

$$\overleftrightarrow{\alpha}(t) = \sum_A \overleftrightarrow{\alpha}(A,t) \tag{8}$$

and

$$\Delta\overleftrightarrow{A}(t) = \sum_A \sum_{B\neq A} \overleftrightarrow{\alpha}(A,t)\cdot\overleftrightarrow{T}^{(2)}(\vec{r}_{AB}(t))\cdot\overleftrightarrow{\alpha}(B,t). \tag{9}$$

Thus, for molecules with a center of inversion the total polarizability $\overleftrightarrow{A}(t)$ is the sum of the collective polarizability $\overleftrightarrow{\alpha}(t)$, equation (8), and a purely DID interaction induced polarizability $\Delta\overleftrightarrow{A}(t)$, equation (9). Keeping in mind that this simplification occurs only for molecules with sufficiently high symmetry we restrict the discussion in the following to DID intermolecular contributions.

Although experimentally anisotropic scattering is most conveniently studied in a depolarized experiment, the theoretical discussion is much more transparent for the anisotropic VV scattering contribution. If we assume that the laboratory Z-axis coincides with the vertical direction defined previously the rank 2 spherical component of the total polarizability $A_{ZZ}^{(2)}(t)$ responsible for anisotropic VV scattering, can be written as |19|

$$A_{ZZ}^{(2)}(t) = \frac{2}{3}\,\gamma\, Q_{ZZ}(t) + \Delta A_{ZZ}(t), \tag{10}$$

where $\gamma = (\alpha_{\parallel} - \alpha_{\perp})$ and $\alpha_{\parallel}$ and $\alpha_{\perp}$ are the polarizabilities parallel and perpendicular to the molecular symmtry axis. In equation (10)

$$Q_{ZZ}(t) = \sum_{A} P_2(\Theta_A(t)) \tag{11}$$

is called a collective orientational variable. Here, P_2 is the second Legendre polynomial and $\Theta_A(t)$ the angle of the symmetry axis of molecule A relative to the laboratory Z-axis. Finally, the collective interaction induced variable $\Delta A_{ZZ}^{(2)}(t)$ is given by |19|

$$\Delta A_{ZZ}^{(2)}(t) = \sum_{A}\sum_{B\neq A} (\overleftrightarrow{\alpha}(A,t)\cdot\overleftrightarrow{T}^{(2)}(\vec{r}_{AB})\cdot\overleftrightarrow{\alpha}(B,t))_{ZZ}^{(2)} \tag{12}$$

$$= \sum_{A}\sum_{B\neq A} \Big\{\alpha_A\alpha_B\, F_{OO}^{(2)}(\vec{r}_{AB})$$

$$+ \alpha_A \gamma_B F^{(2)}_{02} (\Omega_B, \vec{r}_{AB}) + \gamma_A \alpha_B F^{(2)}_{20} (\Omega_A, \vec{r}_{aB})$$

$$+ \gamma_A \gamma_B F^{(2)}_{22} (\Omega_A, \Omega_B, \vec{r}_{AB}) \Big\} \tag{12a}$$

Here, $\alpha = (\alpha_{\parallel} + 2\alpha_{\perp})/3$ and the $F^{(2)}_{m,n}$ $(\Omega_A, \Omega_B, \vec{r}_{AB})$ are functions of the orientations Ω_A and Ω_B of the molecules A and B and of their center-center vector $\vec{r}_{AB}$. Explicit expressions for the $F^{(2)}_{m,n}$ $(\Omega_A, \Omega_B, \vec{r}_{AB})$ have been given by Frenkel and McTague |19|. Without going into further details we note that both the collective orientational variable and the interaction induced variable depend on the molecular orientation and will therefore be correlated.

In order to eliminate this correlation at least at t = 0 the induced polarizability $\Delta A^{(2)}_{ZZ}(t)$ can be decomposed into one part which is correlated with the collective orientational variable $Q_{ZZ}(t)$ and a purely interaction induced or collision induced uncorrelated part $\Delta A^{(2)}_{CI}(t)$ as follows |19, 20|

$$\Delta A^{(2)}_{ZZ}(t) = \frac{2}{3} \Delta\gamma \, Q_{ZZ}(t) + \Delta A^{(2)}_{CI}(t), \tag{13}$$

with

$$\Delta\gamma = \frac{3}{2} \frac{\langle Q_{ZZ}(t) \; \Delta A^{(2)}_{ZZ}(t) \rangle}{\langle Q_{ZZ}(t) \; Q_{ZZ}(t) \rangle} \tag{14}$$

and

$$\Delta A^{(2)}_{CI}(t) = \left\{ \Delta A^{(2)}_{ZZ}(t) - \frac{\langle Q_{ZZ}(t) \; \Delta A^{(2)}_{ZZ}(t) \rangle}{\langle Q_{ZZ}(t) \; Q_{ZZ}(t) \rangle} \cdot Q_{ZZ}(t) \right\}. \tag{15}$$

$\Delta\gamma$ is an intermolecular contribution to the molecular polarizability anisotropy such that the effective molecular anisotropy becomes

$$\gamma_{eff} = \gamma + \Delta\gamma . \tag{16}$$

By inserting equations (13) and (16) into equation (10) we are now able to write down the correlation function for depolarized Rayleigh scattering as

$$G^{(2)}(t) = \langle A^{(2)}_{ZZ}(0)\ A^{(2)}_{ZZ}(t)\rangle \tag{17}$$

$$= \frac{4}{9}\gamma^2_{eff} \langle Q_{ZZ}(0)\ Q_{ZZ}(t)\rangle + \langle \Delta A^{(2)}_{CI}(0)\ \Delta A^{(2)}_{CI}(t)\rangle + \frac{2}{3}\gamma_{eff}\left[\langle \Delta A^{(2)}_{CI}(0)\ Q_{ZZ}(t)\rangle + \langle Q_{ZZ}(0)\ \Delta A^{(2)}_{CI}(t)\rangle\right] \tag{18}$$

The first term on the right hand side of equation (18) contains the collective orientational correlation function $\langle Q_{ZZ}(0)\ Q_{ZZ}(t)\rangle$ weighted by the (generally unknown) effective polarizability anisotropy γ_{eff}. The second term is a pure collision (interaction) induced correlation function and the third term is a cross orientation - interaction induced correlation function, which, due to the definition of the collision induced polarizability $\Delta A^{(2)}_{CI}(t)$, given by equation (15), vanishes at $t = 0$. Accordingly, the depolarized Rayleigh spectrum, which is the Fourier transform of the correlation function $G^{(2)}(t)$, will be composed of three spectral contributions: (a) a collective orientational, (b) a collision induced, and (c) a cross orientation collision induced spectrum.

The integrated anisotropic spectrum, i.e. the total depolarized scattering intensity, has frequently been used by experimentalists to study orientational pair correlations in molecular liquids. This total intensity $\hat{I}$ is proportional to the correlation function

$G^{(2)}(t)$ at t=0, for which one gets

$$G^{(2)}(t=0) = \rho \left\{ \frac{4}{45} \gamma_{eff}^2 < \sum_B P_2(\Theta_{AB})> + N^{-1} <\Delta A_{CI}^{(2)2}> \right\}. \quad (19)$$

Here ρ is the particle number density and N the number of particles in the scattering volume. The quantity $<\sum_B P_2(\Theta_{AB})>$ is an orientational pair correlation factor often referred to as |1, 21, 22|

$$g_2 \equiv 1 + f_2 \equiv < \sum_B P_2(\Theta_{AB})>. \quad (20)$$

Θ_{AB} is the angle between the symmetry axis of a reference particle A and that of a neighbour molecule B and the summation extends over all particles B in the scattering volume including the reference particle A.

From equation (19) it is obvious that the total depolarized intensity can provide information on the pair correlation factor g_2 only if independent information on γ_{eff} and $<\Delta A_{CI}^{(2)2}>$ is available.

4. EVALUATION OF EXPERIMENTAL SPECTRA

Often depolarized Rayleigh experiments are carried out in order to study the reorientational dynamics in liquids. From the previous section it is obvious that a separation of the orientational spectra is required. We note, that the problem of spectra separation is not solved generally in a satisfactory manner at present. Here, we shall discuss two evaluation schemes, which are suited to treat the strongly hindered diffusional reorientation of larger molecules in liquids on one side and the freer reorientation of small molecules on the other.

The first evaluation method, mainly applied to analyse scattering data from larger organic molecules, starts with the assumption that reorientational dynamics can be described by a collective diffusion law |22, 1|. For molecules with a threefold and higher symmetry axis the orientational correlation function $g^{rot}(t)$ results to |22, 1|

$$g^{rot}(t) = g_2 \cdot \exp\left\{- t/\tau_{LS}\right\} \quad (21)$$

with

$$\tau_{LS} = \frac{g_2}{j_2}\tau_S. \quad (22)$$

Here, τ_{LS}, the light scattering correlation time is related to the single particle correlation time $\tau_S = 1/(6D_\perp)$, where $D_\perp$ is the rotational tumbling diffusion constant of the molecule, g_2 the orientational pair correlation factor introduced above and j_2 an angular velocity correlation factor which equals unity if the angular velocities of different fluid particles A and B are uncorrelated |22, 1|.

A second assumption of this method concerns the shape of the collision induced spectrum, which, as suggested by the spectral wings, is taken to be essentially exponential:

$$I^{CI}(\omega) = A\,\omega^n \exp\left\{- \omega/\Delta\right\}. \quad (23)$$

Here, A, n, and Δ are adjustable empirical constants.

Finally, it is assumed that the collision induced variable $\Delta A_{CI}(t)$ fluctuates much more rapidly than the orientational variable $Q_{ZZ}(t)$ |23, 20|. This separation of time scales causes the cross correlation functions in equation (18) to vanish and one is thus left with only two spectral contributions. Examples of this evaluation method will be given in the next section.

In the second method the separation of the spectra is achieved via the second moments. As pointed out in the previous section the experimental spectrum is the sum of an orientational $I^{Or}(\omega)$, a collision induced $I^{CI}(\omega)$, and a cross $I^{Cross}(\omega)$ contribution,

$$I^{exp}(\omega) = I^{Or}(\omega) + I^{CI}(\omega) + I^{Cross}(\omega). \quad (24)$$

Since almost no information is available concerning the two non-orientational spectra, it is assumed that they again can be approximated by a spectrum of the form

$$I^{NO}(\omega) = I^{CI}(\omega) + I^{Cross}(\omega) = A\cdot\omega^{n}\exp\left\{-\ \omega/\Delta\right\}, \qquad (25)$$

where, as in equation (23), A, n, and Δ are adjustable parameters. Often n is set equal to zero. Δ is usually determined from the far wings of the spectrum, where the spectra become exponential, and A is determined by comparing the second moments of the spectra as follows: Since the orientational second moment is |24, 25|

$$M_2^{Or} = 6kT/(I\cdot g_2), \qquad (26)$$

the second moment of the experimental spectrum can be written as

$$M_2^{exp} = M_2^{Or} + M_2^{NO}(A,\ n,\ \Delta). \qquad (27)$$

If M_2^{exp} is derived from the experiment and M_2^{Or} from equation (26), $M_2^{NO}(A,\ n,\ \Delta)$ can be used to calculate A. Once the model spectrum for the non-orientational contributions is determined, this can be used to calculate the orientational spectrum from the experimental spectrum via equation (24).

It should be noted that in particular the second evaluation method is to be considered empirical. Comparison of experimental with recent molecular dynamics results indicates however, that despite the crude estimate of the non-orientational spectrum interesting information on the reorientational dynamics in liquids can be obtained by this evaluation scheme.

Typical experimental investigations illustrating the two evaluation methods are presented in the next section.

5. EXPERIMENTAL EXAMPLES

5.1 Determination of g_2 |26, 27|

(a) Total Intensity

As pointed out in section 3 the total depolarized Rayleigh intensity provides information on the orientational pair correlation factor g_2 defined by equation (20). Equation (19) indicates, however, that the total intensity is made up not only of an orientational but also of a collision induced contribution, which must be removed. As an example as to how g_2 can be determined we consider an experimental investigation of Battaglia, Cox, and Madden |26| on CS_2, C_6H_6, and C_6F_6. With the assumptions of collective rotational diffusion and rapid fluctuation of the collision induced contribution they fitted the depolarized spectra of these substances to a central diffusional Lorentzian and a collision induced exponential. Typical results are shown in fig. 2 for CS_2 at 232 K. The experimental spectrum is drawn as solid line (——). The figure further shows the fitted Lorentzian (...),the data minus the

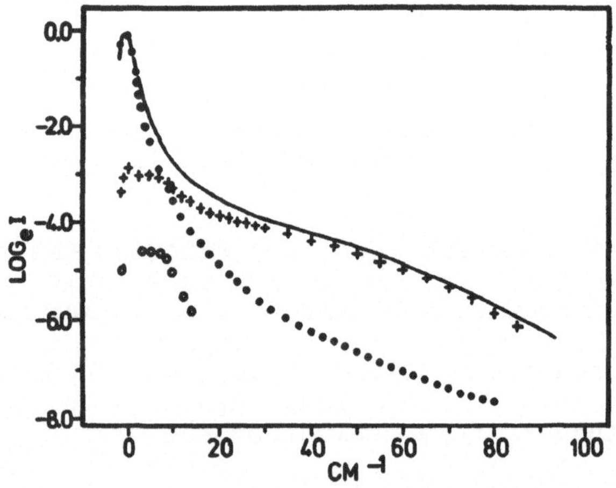

Fig. 2: Depolarized Rayleigh spectrum of CS_2 at 232 K (——). Taken from Ref. 27

Lorentzian(+++), and the residuals of a fit of a Lorentzian and an exponential background (ooo). The orientational integrated intensity $\hat{I}^{Or}$ determined from the Lorentzian provides information on g_2 since according to equation (19)

$$I^{Or} \sim \gamma_{eff}^2 \cdot g_2. \tag{28}$$

Further information is necessary to separate γ_{eff}^2 from g_2. In order to achieve this Madden and coworkers referred to the Cotton-Mouton constant ${}_mC$ which can be written as |26|

$${}_mC \sim \gamma_{eff}\, \Delta\chi \cdot g_2/kT. \tag{29}$$

Here, $\Delta\chi$ is the magnetizability anisotropy of the molecules under study, which is known from the gas phase. Experimental results on g_2 and γ_{eff}/γ at room temperature as determined from depolarized Rayleigh and Cotton-Mouton experiments are reproduced in table 2.

Table 2: g_2 from total intensity and Cotton-Mouton experiments |26|

	CS_2	C_6H_6	C_6F_6
g_2	1.13	1.16	2.75
γ_{eff}/γ	0.59	0.54	0.52

Although the large value of g_2 in C_6F_6 points to orientational pair correlations in this liquid, the values close to unity in CS_2 and C_6H_6 should not be taken as indication that pair correlations are absent here. Cancellation effects or an orientational structure with preferred relative orientation close to the magic angle (54.5°) could also be responsible for the low g_2 values |28, 29|. Certainly surprising is the large change of γ to γ_{eff} for particles in condensed fluids due to interaction induced mechanisms.

(b) g_2 from τ_{LS}/τ_S ratio

According to Keyes and Kivelson |22| the collective orientational correlation function in the diffusion limit is given by equations (21) and (22). It has been argued |22| and to a certain extent verified experimentally |3|, that the angular velocity correlation factor j_2 is close to unity for ordinary liquids. If this is the case, experimental determination of τ_{LS} by Rayleigh scattering and determination of the single particle correlation time τ_S by Raman scattering or nuclear magnetic relaxation allows the approximate determination of g_2 by calculating the ratio

$$\frac{\tau_{LS}}{\tau_S} = \frac{g_2}{j_2} \simeq g_2. \tag{30}$$

In table 3 values of τ_{LS}, τ_S, τ_{LS}/τ_S, and j_2 for CS_2, C_6H_6, and C_6F_6 at room temperature as reported by Madden and coworkers |26| are reproduced. There is, in fact, a strong correlation between the ratio τ_{LS}/τ_S and the g_2 values given in table 2. Within experimental error j_2 equals unity, with perhaps a slight tendency to follow g_2 away from unity as g_2 increases.

Table 3: τ_{LS}, τ_S, τ_{LS}/τ_S, and j_2 for CS_2, C_6H_6, and C_6F_6

	CS_2	C_6H_6	C_6F_6
τ_{LS}/ps	1.7	3.0	14.0
τ_S/ps	1.45	2.7	5.5
τ_{LS}/τ_S	1.17	1.11	2.54
j_2	0.97	1.01	1.08

The temperature dependence of g_2 and j_2 of CS_2 has also been studied by Cox, Battaglia, and Madden |27| with the result that g_2 increases slightly with decreasing temperature.

5.2 Determination of orientational correlation functions

If one is interested to obtain from a spectrum more detailed information on the reorientational dynamics than merely τ_{LS}, the second evaluation procedure mentioned in section 4 is usually applied. Slightly different estimates of the non-orientational spectrum have been used by different authors |30-36|, none of them, however, is really well founded. On the other hand, although molecular dynamics investigations have reached a stage of sophistication that collective and induced properties can be simulated, at present no convergent picture concerning the importance, the form, and the magnitude of the spectral contributions $I^{Or}(\omega)$, $I^{CI}(\omega)$, and $I^{Cross}(\omega)$ is in sight. This is mainly due to the different ways in which the polarizable matter is treated in the simulations |19, 8, 37, 38|. Depending on whether a single molecular polarizability or site polarizabilities are attributed to the molecules under study rather different spectral contributions and total spectra were obtained. Thus despite its empirical character the above mentioned second evaluation scheme remains the presently only viable alternative which receives its legitimation from the reasonableness of the results. In the following we consider several typical experimental investigations.

(a) Molecular reorientation of CO_2 |35, 36|

Molecular rotation of CO_2 has recently been studied at constant temperature 40° C in the density range $\rho = 0.1\ \rho_c$ to $2.35\ \rho_c$ |35|, where ρ_c is the critical density. Due to the large density variation by almost a factor of 25 a wide dynamical range from almost free rotation to liquid-like hindered reorientation was covered with these experiments. Figure 3

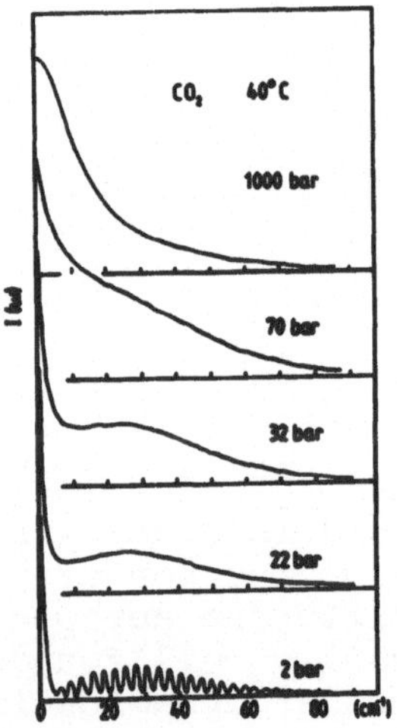

Fig. 3: Depolarized Rayleigh spectra of CO_2 at 40° C and pressures as indicated |35|

shows depolarized Rayleigh spectra obtained at various pressures as indicated and in figure 4 corresponding orientational correlation functions are reproduced.

In figure 4 the time is given in reduced units, $t^* = (kT/I)^{1/2} \cdot t$; 1 reduced unit = 0.408 ps at 40° C.

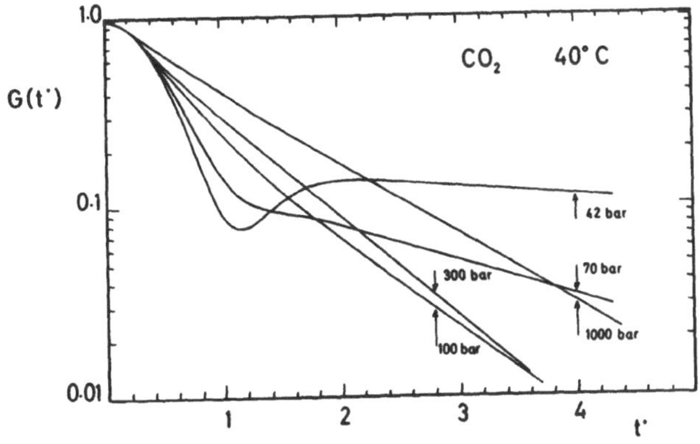

Fig. 4: Orientational correlation functions of CO_2 at 40° C and the pressures indicated |35|

In a second series of experiments for two fixed densities $\rho = 0.47\ \rho_c$ and $\rho = 1.9\ \rho_c$ the temperature dependence of the reorientation of CO_2 at constant density was investigated |36|. In figure 5 and 6 the orientational correlation functions at temperatures as indicated are shown for the low and the high density respectively.

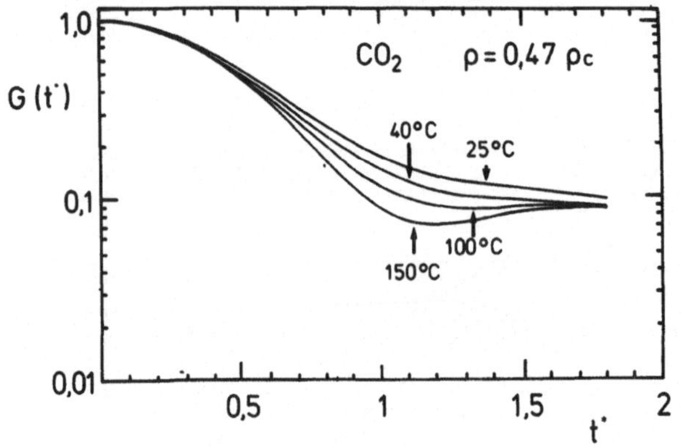

Fig. 5: Orientational correlation functions of CO_2 at constant density $\rho = 0.47\ \rho_c$ and temperatures as indicated

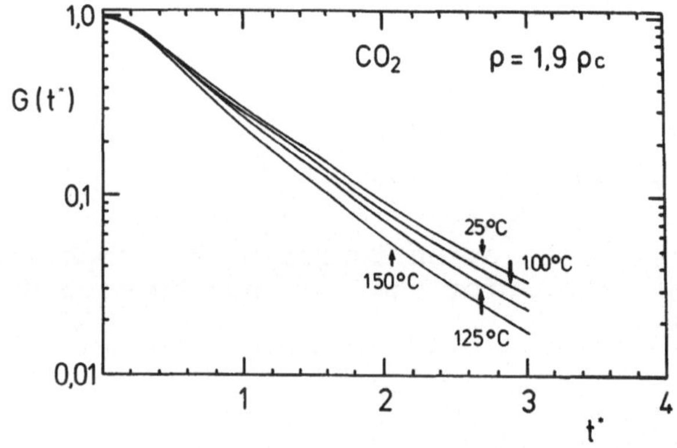

Fig. 6 : Orientational correlation functions of CO_2 at constant density $\rho = 1.9\ \rho_c$ and temperatures as indicated

Several evaluations of the correlation functions were carried out in order to account for the wide dynamic range covered by the experiments. First a comparison with the well known J-diffusion model will be

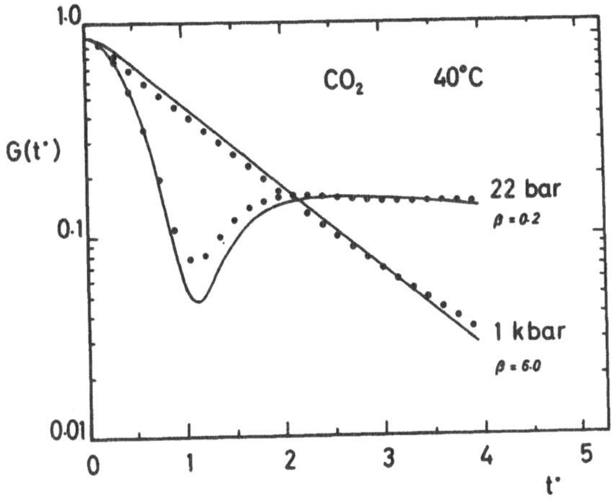

Fig. 7 : Comparison of experimental (....) and J-diffusion (——) correlation functions for CO_2 at 40° C

given. Figure 7 gives typical results obtained by fitting the J-diffusion model (——) to the experimental correlation functions (....) and figure 8 and 9 show

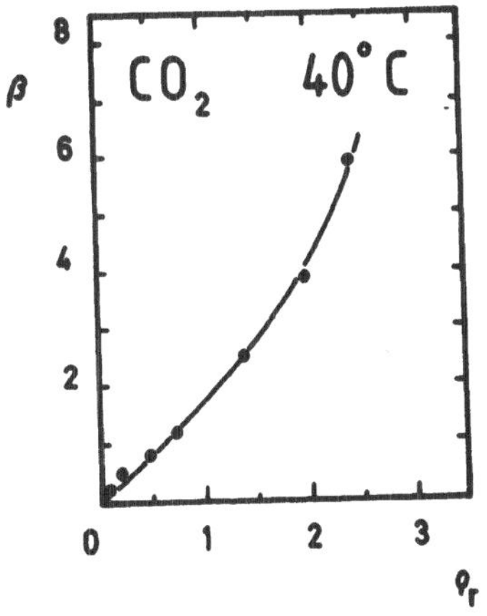

Fig. 8

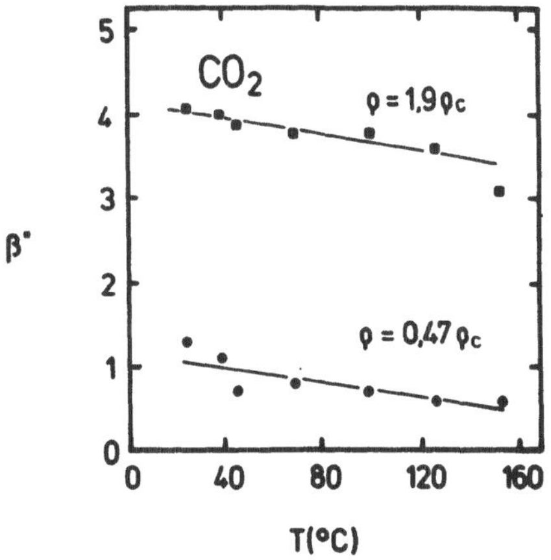

Fig. 9

Density and temperature dependence of the reduced J-diffusion collision frequency β^*

the density and temperature dependence of the reduced J-diffusion collision frequency β^* as obtained from such fits. In figure 8 the reduced density is defined as $\rho_r = \rho/\rho_c$. The more than linear increase of β^* with ρ_r at high density in figure 8 seems to be related to an increase of the local structure with density |35|. Remarkable is the finding that β^* decreases with increasing temperature at constant density.

One further approach to the understanding of orientational correlation functions is the memory function formalism |1, 39|. A memory function $K(\tau)$ is defined by the equation

$$\frac{dG(t)}{dt} = - \int_0^t d\tau\, K(\tau)\, G(t-\tau). \qquad (31)$$

If G(t) is determined by experiment as described above, equation (31) can be solved numerically to obtain an experimental memory function. A typical experimental memory function for CO_2 at 25° and $\rho = 1.9\ \rho_c$ is shown in figure 10 (——). Several approximations to the memory

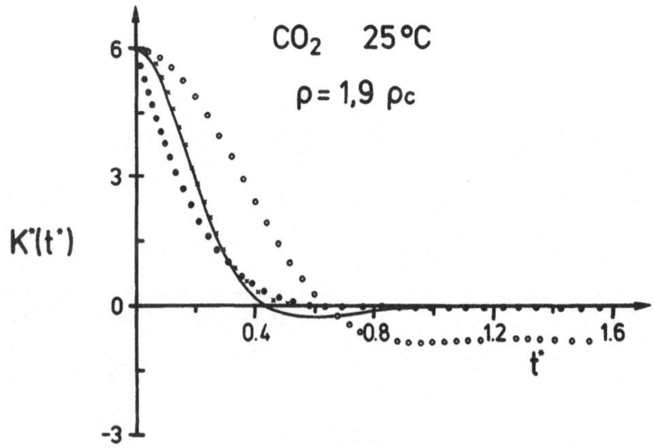

Fig. 10 : Memory functions for CO_2 at 25° C and $\rho = 1.9\ \rho_c$

function have been discussed: (a) The J-diffusion memory function K_J is of the simple form |40, 41|

$$K_J(t^*) = K_f(t^*) \cdot \exp\left\{- \beta^* \cdot t\right\}, \tag{32}$$

where $K_f(t^*)$ is the free rotor memory function. Both, the best fitted J-diffusion memory function (···), and the free rotor memory function (ooo) are included in figure 10. (b) Nee and Zwanzig related the memory function to the angular velocity correlation function $c_\omega(t) = \langle\omega(o)\ \omega(t)\rangle$ |42|,

$$K(t^*) = 6(kT/I) \cdot C_\omega(t^*). \tag{33}$$

(c) Since the right and left hand side short time expansions of equation (33) are inconsistent, Steele |25, 43| introduced the approximation

$$K(t^*) = K_f(t^*) \cdot C_\omega(t^*), \tag{34}$$

which he further approximated by a Gaussian

$$K(t^*) = K(o) \cdot \exp\left\{- \left[\frac{\langle N^2\rangle}{4(kT)^2} + 5\right] \cdot t^{*2}\right\}. \tag{35}$$

Here, $\langle N^2\rangle$ is the mean squared torque and for convenience we shall call $\langle N^2\rangle/4(kT)^2$ the mean squared torque factor. The best fitted Gaussian approximation to the experimental memory function is also included in figure 10 (x x x).

One interesting feature of the experimental memory function is its getting negative much earlier than the free rotor memory function. This behaviour is related to an oscillatory behaviour of the angular velocity correlation function $C_\omega(t)$ and indicates a highly damped librational motion of the CO_2 molecules. The librational pattern of the memory function becomes more pronounced as the density increases |35|.

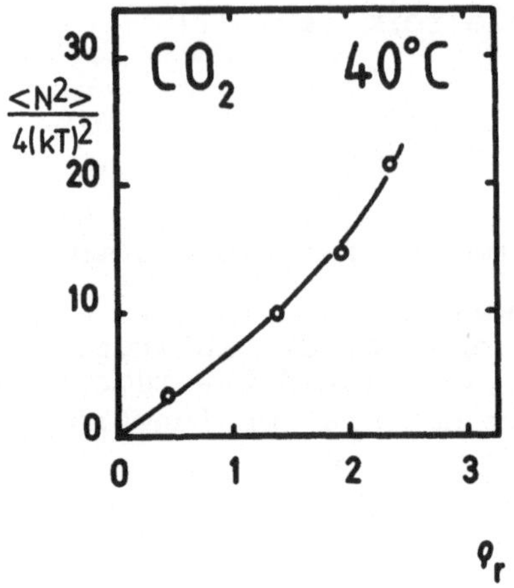

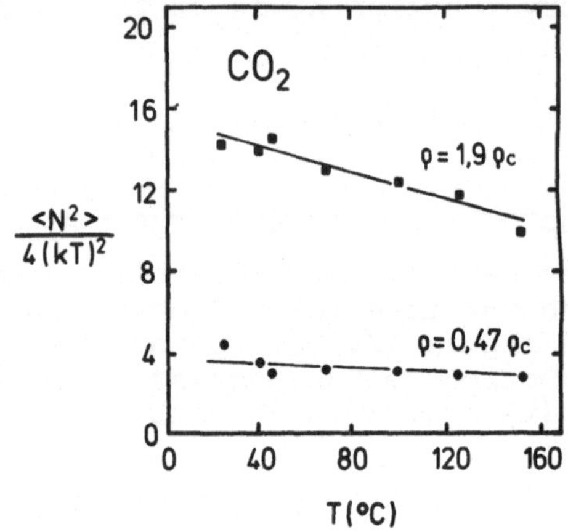

Fig. 11: Fig. 12:

Density and temperature dependence of the mean squared torque factor $\langle N^2\rangle/4(kT)^2$

In figures 11 and 12 the density and temperature of the mean squared torque factor is reproduced which resembles the behaviour of β^* shown in figures 8 and 9. The decrease of the mean squared torque factor with temperature at constant density results mainly from the $(kT)^2$ factor in the denominator. As indicated in table 4

Table 4 : CO_2 at $\rho = 1.9\ \rho_c$

T/K	$\langle N^2\rangle/10^{-40}J^2$	$\langle N^2\rangle/10^{-40}J^2$
298	9.7	10.5 (302 K)
313	10.5	10.4
322	11.7	11.7 (323 K)
348	12.2	
373	13.2	
398	14.4	
423	13.4	

second column the mean squared torque increases slightly with temperature. These results are in very good agreement with recent molecular dynamics results of Singer et al. |44, 45| which are reproduced in the third column of table 4. Although the almost perfect agreement between experiment and MD calculation should not be overemphasized it indicates the essential correctness of the correlation and memory functions obtained by the evaluation procedure.

The main conclusions from these investigations are:

(a) the short time dynamics of CO_2 has a librational character at sufficiently high density.

(b) At constant temperature: With increasing ρ $\langle N^2 \rangle$ increases and the reorientation gets slower.

At constant ρ: Despite increasing $\langle N^2 \rangle$ the reorientation gets faster with increasing temperature.

(b) Molecular reorientation of CH_3J |31|

Dill, Litovitz and Bucaro |31| investigated the pressure dependence of rotational correlation functions of acetone, benzene, and methyl iodide by depolarized Rayleigh scattering. In fig. 13 their orientational

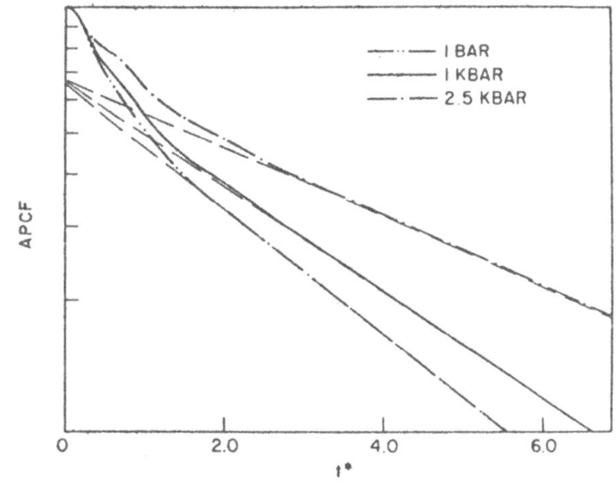

Fig. 13 : Orientational correlation functions of CH_3J at pressures as indicated. Taken from Ref. |31|

correlation functions of CH_3J at 1 bar, 1 kbar, and 2.5 kbar are shown. The correlation functions which all start at very short times with the free rotor type behaviour, show a with pressure increasing oscillatory pattern at intermediate times and an exponential decay at long times. This separation of orientational correlation functions into three parts is typical for small molecules. The intermediate oscillatory range indicates a librational type of motion. As figure 13 indicates the librations become more pronounced with increasing density. A similar behaviour was observed for benzene |30|. The long time part of the correlation function reflects collective reorientational diffusion which gets slower as the density increases.

In order to study the librational character of the correlation functions in more detail Dill, Litovitz, and Bucaro |31| calculated the so called generalized second moments $M_2(t)$,

$$M_2(t) = \int d\omega \exp\left\{i\omega t\right\} \omega^2 I(\omega). \tag{36}$$

For t = o this function is identical to the ordinary second moment. At larger times $M_2(t)$, a cross orientation-angular velocity correlation function, resembles the angular velocity correlation function $C_\omega(t)$ for systems close to the diffusion limit |46-48|. Contrary to the memory functions discussed above, $M_2(t)$ has the disadvantage that the area under this correlation function vanishes, which might lead to an overestimation of the librational character. In figure 14 $M_2(t)$ correlation functions as obtained by Litovitz and coworkers are shown. Again it is evident that the oscillatory character of the correlation functions gets more pronounced as the density is increased.

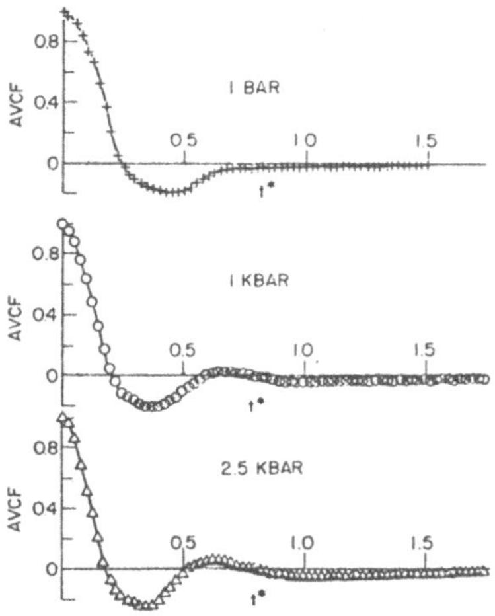

Fig. 14: Cross orientation-angular velocity correlation function for CH_3J at various pressures |31|

(c) Depolarized Rayleigh scattering from benzene

The reorientational motion of benzene has received much attention and has been studied by various experimental techniques. In section 5.1 we discussed results for τ_{LS} and τ_S and the information on orientational pair correlations derivable from such correlation times. Here, we want to consider the depolarized Rayleigh spectrum in a more extended frequency range. Figure 15 shows such a spectrum (——) taken at 21° C |30|. One characteristic feature of the spectra of benzene and other planar ring molecules like pyridine is - beside the common central Lorentzian and the exponential wing spectrum - a pronounced shoulder at about 80 cm^{-1}. This shoulder was first observed and related to librational reorientation by Dardy, Volterra, and Litovitz |30|. These authors found that the shoulder in the spectra and the librational character of the correlation functions increases as the temperature of benzene is lowered or the density (via pressure) increased.

More recently, the dependence of the spectral shoulder and the correlation functions on the mixture

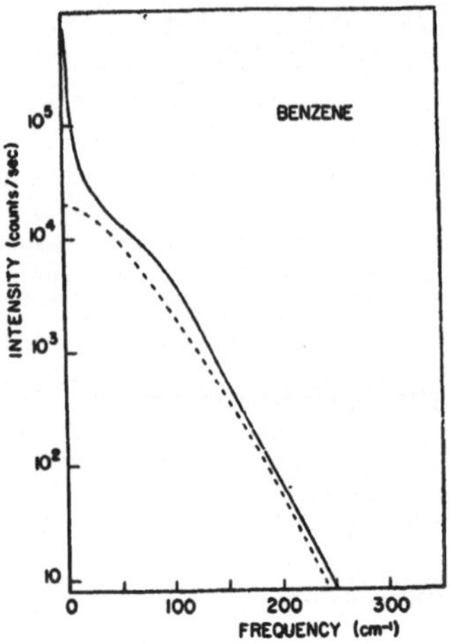

Fig. 15 : Depolarized Rayleigh spectrum of benzene. Taken from Ref. |30|

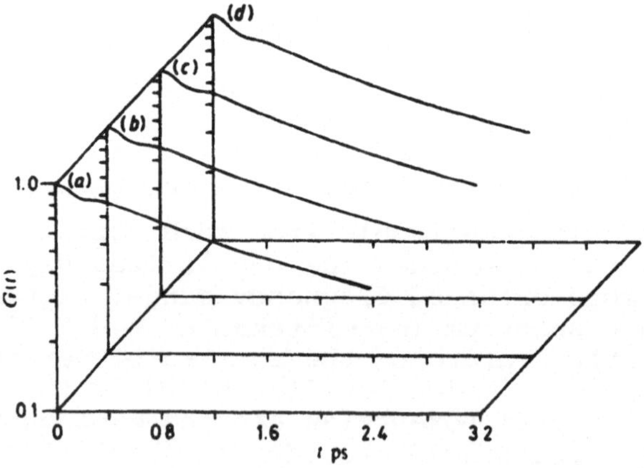

Fig. 16 : Correlation functions of benzene in solutions with CCl_4. (a) 100 mole% benzene, (b) 50 mole%, (c) 30 mole%, (d) 15 mole% benzene composition of mixtures of benzene with the isometric sólvents CCl_4, $Si(CH_3)_4$, and $C(CH_3)_4$ was investigated |49|. Surprisingly, dilution of benzene in CCl_4 has very little effect on the spectrum. Correlation functions for (a) 100% benzene, (b) 50 mole%, (c) 30 mole%, and (d) 15 mole% benzene are reproduced in figure 16. Experimental memory functions K(t) of the same systems are shown in figure 17. Again it is surprising, that

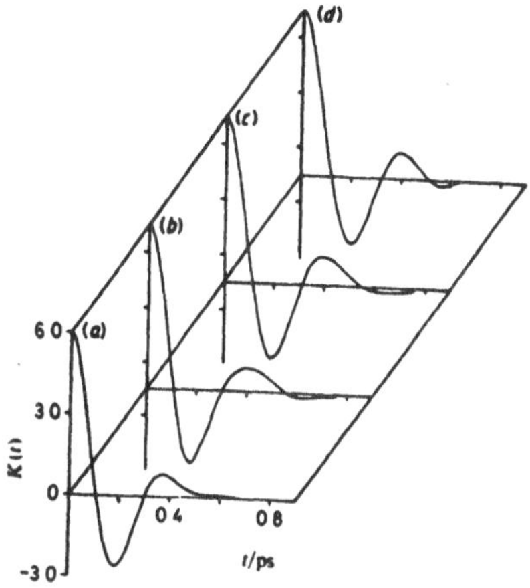

Fig. 17 : Experimental memory functions K(t) for benzene/CCl_4 solutions. Concentrations as in figure 16

the memory functions show almost no concentration dependence. The behaviour changes dastically if benzene is mixed with $Si(CH_3)_4$ or $C(CH_3)_4$. Correlation functions for benzene in solution with $C(CH_3)_4$ at 6° C are shown in figure 18. The solution concentrations were (a) 100 mole% benzene, (b) 75 mole%, (c) 50 mole%, (d) 36 mole%, (e) 16 mole% benzene.

The corresponding memory functions are shown in figure 19. Both the correlation and the memory functions smothen considerably as the solvent concentration increases. A similar behaviour has also been observed for

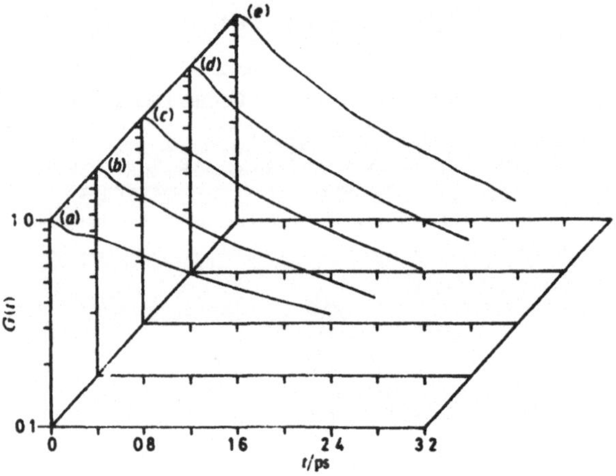

Fig. 18 : Correlation functions G(t) of benzene in solutions with $C(CH_3)_4$ at 6° C.
(a) 100 mole% benzene, (b) 75 mole%, (c) 50 mole%, (d) 36 mole%, (e) 16 mole% benzene

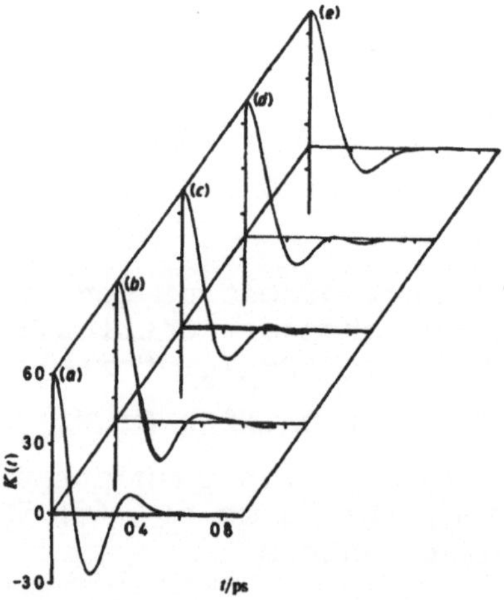

Fig. 19 : Memory functions K(t) for benzene/$C(CH_3)_4$ solutions at 6° C. Concentrations as in figure 18

benzene/$Si(CH_3)_4$ mixtures |49|. Since the three solvents CCl_4, $Si(CH_3)_4$ and $C(CH_3)_4$ are rather similar as far as the molecular size and polarizability are concerned, there is no immediately obvious reason why the strong change of the memory functions, by replacing CCl_4 by, say, $C(CH_3)_4$, should be due to an interaction-induced mechanism. There is thus strong indication that this change is due to a difference of the short-time reorientational dynamics of the benzene molecules in the different solutions. A simple qualitative explanation of this difference can be given in terms of the mass of the solvent particles: If the molecules in the vicinity of a given benzene particle are light and spherical the benzene seems to be able to push them aside with little tendency to reverse its own angular velocity. On the other hand, if a benzene molecule is encaged by heavy or very anisotropic neighbours then it performs a librational motion during the time it takes the local environment to relax.

6. FUTURE DEVELOPMENTS

Depolarized Rayleigh scattering experiments will certainly profit from improvements of laser systems, of spectrometer components and of spectra sampling and data evaluation systems. The most severe difficulty facing the depolarized Rayleigh and Raman technique to study reorientational dynamics in liquids, however, is not technical but is the problem of the separation of the orientational from the non-orientational interaction induced spectral contributions. So far no reliable separation procedure has been developed in general. Recent molecular dynamics calculations even have cast some doubt on the validity of separation schemes so far considered to be valid under rather general conditions.

To clarify this situation it seems to be highly desirable to perform combined experimental and molecular dynamics investigations at identical p-V-T state points. For correct light scattering results from molecular dynamics investigations not only the dynamics must be correct but also the distribution of polarizable matter in more extended molecules must be correctly accounted for. This seems to be another unsolved problem today |37|. Finally, the molecular dynamics calculations should provide orientational, collision-induced, and cross correlation functions and spectra in order to allow a critical revision of the existing evaluation procedures and eventually to give reliable guidance concerning the evaluation of experimental spectra.

REFERENCES

1. B.J. Berne and R. Pecora, Dynamic Light Scattering, Wiley, 1976

2. B. Chu, Laser Light Scattering, Academic Press, 1974

3. D.R. Bauer, J.I. Brauman, and R. Pecora, Ann. Rev. phys. Chem., 27, 443 (1976)

4. M.E. Rose, Theory of Angular Momentum, Wiley, 1957

5. A.R. Edmonds, Angular Momentum in Quantum Mechanics, Princeton University Press, 1957

6. F.J. Bartoli and T.A. Litovitz, J. Chem. Phys. 56, 413 (1972)

7. S. Bratos and E. Maréchal, Phys. Rev., A4, 1078 (1971)

8. P. Madden, this book

9. P. Madden, Mol. Phys., 36, 365 (1978)

10. W.M. Gelbart, Adv. chem. Phys., 26, 1 (1974)

11. A.J.C. Ladd, T.A. Litovitz, and C.J. Montrose, J. Chem. Phys., 71, 4242 (1979)

12. A.D. Buckingham and G.C. Tabisz, Optics Lett., 1, 220 (1977); Mol. Phys., 36, 583 (1978)

13. H.A. Posch, Mol. Phys., 37, 1059 (1979), 1137 (1980), 46, 1213 (1982)

14. T.I. Cox and P.A. Madden, Mol. Phys., 39, 1487 (1980), 43, 287 (1981), 43, 307 (1981)

15. T.I. Cox, Thesis, Cambridge, 1978

16. A.D. Buckingham, Adv. chem. Phys., 12, 107 (1967)

17. A.D. Buckingham in Microscopic Structure and Dynamics of Liquids, edited by J. Dupuy and A.J. Dianoux, Plenum Press, 1977

18. A.J. Stone, this book

19. D. Frenkel and J.P. McTague, J. Chem. Phys., 72, 2801 (1980)

20. T. Keyes and B.M. Ladanyi, Mol. Phys., 33, 1063 (1977), 33, 1099 (1977), 34, 765 (1977)

21. A. Ben Reuven and N.D. Gershon, J. Chem. Phys., 51, 893 (1969)

22. T. Keyes and D. Kivelson, J. Chem. Phys., 56, 1057 (1972)

23. T. Keyes, D. Kivelson, and J. McTague, J. Chem. Phys., 55, 4096 (1971)

24. R.G. Gordon, J. Chem. Phys., 40, 1973 (1964)

25. A.G. St. Pierre and W.A. Steele, Mol. Phys., 43, 123 (1981)

26. M.R. Battaglia, T.I. Cox, and P.A. Madden, Mol. Phys., 37, 1413 (1979)

27. T.I. Cox, M.R. Battaglia, and P.A. Madden, Mol. Phys., 38, 1539 (1979)

28. O. Steinhauser and M. Neumann, Mol. Phys., 37, 1921 (1979)

29. R.W. Impey, P.A. Madden, and D.J. Tildesley, Mol. Phys., 44, 1319 (1981)

30. H.D. Dardy, V. Volterra, and T.A. Litovitz, J. Chem. Phys., 59, 4491 (1973)

31. J.F. Dill, T.A. Litovitz, and J.A. Bucaro, J. Chem. Phys., 62, 3839 (1975)

32. P. van Konynenburg and W.A. Steele, J. Chem. Phys., 56, 4776 (1972), 62, 2301 (1975)

33. P.E. Schoen, P.S. Cheung, D.A. Jackson, and J.G. Powles, Mol. Phys., 29, 1197 (1975)

34. H. Langer and H. Versmold, Ber. Bunsenges. phys. Chem., 83, 510 (1979)

35. H. Versmold, Mol. Phys., 43, 383 (1981)

36. U. Zimmermann and H. Versmold, Mol. Phys., in print

37. B.M. Ladanyi, J. Chem. Phys., 78, 2189 (1983)

38. D. Tildesley, this book

39. B.J. Berne and G.D. Harp, Adv. Chem. Phys., 17, 63 (1970)

40. F. Bliot and E. Constant, Chem. Phys. Lett., 18, 253 (1973)

41. T.E. Eagles and R.E.D. McClung, Chem. Phys. Lett., 22, 414 (1973)

42. T.W. Nee and R. Zwanzig, J. Chem. Phys., 52, 6353 (1970)

43. W.A. Steele, Mol. Phys., 43, 141 (1981)

44. K. Singer, A. Taylor, and J.V.L. Singer, Mol. Phys., 33, 1757 (1977), 37, 1239 (1979)

45. K. Singer and C.S. Murthy, (private communication)

46. A.G. St. Pierre and W.A. Steele, J. Chem. Phys., 62, 2286 (1975)

47. J. Kushik and B. Berne, J. Chem. Phys., 59, 4486 (1973)

48. D. Kivelson, Mol. Phys., 28, 321 (1974)

49. W. Härtl and H. Versmold, J. Chem. Soc., Faraday Trans. 2, 79, 1143 (1983)

NUCLEAR MAGNETIC RELAXATION AND MOLECULAR REORIENTATION

H. Versmold

Physikalische Chemie

Universität Dortmund

F.R. Germany

1. INTRODUCTION

Nuclear magnetic relaxation represents in many respects the most completely developed method to investigate molecular reorientation in liquids. Experimental and theoretical progress has been made for many years and the method is well documented in textbooks |1-3| and review articles |4,5|.

In this chapter a brief introduction to the most important relaxation mechanisms in liquids will be given first. Next orientational correlation functions and effective correlation times will be introduced. Finally, we refer to typical experimental examples demonstrating the unique selectivity of the magnetic relaxation method to study specific reorientational processes.

2. NUCLEAR MAGNETIC RELAXATION MECHANISMS

Let us consider a system of isolated spins I = 1/2 suddenly exposed to a strong external magnetic field B_o. After switching on the field the spins will be found with equal probability in the two Zeeman states with energies E_+ and E_- . No redistribution will take place for an isolated spin system since the Zeeman levels are its stationary states. The behaviour of the spin system just described is in disagreement with

A. J. Barnes et al. (eds.), Molecular Liquids - Dynamics and Interactions, 309–330.

experimental experience which shows that as time passes a redistribution over the levels takes place. The spin redistribution is combined with a build-up of nuclear magnetization $M_z(t)$ which for most liquids follows the simple Bloch equation |1-4|

$$\frac{dM_z(t)}{dt} = -\frac{1}{T_1}(M_z(t) - M_o) \quad , \qquad (1)$$

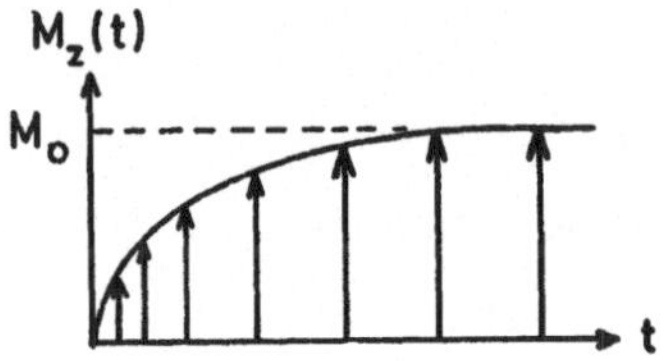

Fig. 1: Definition of longitudinal relaxation time T_1

which can be solved to give (see also figure 1)

$$M_z(t) = M_o(1 - \exp\{-t/T_1\}). \qquad (2)$$

Here, M_o is the equilibrium magnetization, and T_1 the longitudinal or spin-lattice relaxation time. A phase relaxation of the spin system, describable by a transverse relaxation time T_2, also takes place |1-4|. In simple cases T_2 does not provide information beyond the one already given by T_1 and, therefore, we restrict the discussion to T_1.

The relaxation of the spin system shows that our assumption of non-interacting isolated spins contradicts the actual behaviour of the system and that we must include local spin interactions in our description. In fact, local spin interactions, which fluctuate in time due to dynamical processes in the liquid, provide the time dependent pertubation necessary for spin transitions between the stationary Zeeman states.

There are several different spin interactions which can cause spin relaxation. The most important ones to study reorientation processes in liquids are:

(a) Magnetic dipole-dipole interaction $\hat{H}_1^D(t)$

This interaction is most important for nuclei with spin I = 1/2. Since nuclear spins are associated with nuclear magnetic dipole moments $\vec{\mu}$ |1,2|, each spin produces a magnetic dipole field. Let us consider a spin pair I and S belonging to the same molecule. The field of spin S at the site of I provides a local pertubation $\hat{H}_1^D(t)$ for spin I and vice versa. As the molecules rotates, the spin-spin vector $\vec{r}_{IS}(t)$ pointing from spin I to spin S also rotates and thus the dipole-dipole interaction $\hat{H}_1^D(t)$ will be modulated in time.

(b) Electric quadrupole interaction $\hat{H}_1^Q(t)$

If the nucleus under consideration has a spin $I \geq 1$, it will possess both a magnetic dipole $\vec{\mu}$ and an electric quadrupole Q. In such cases usually the interaction of the nuclear quadrupole moment Q with the electric field gradient tensor, caused by the intramolecular charge distribution, is much larger than the previously mentioned dipole-dipole interaction. For nuclei of atoms involved in single or triple bonds the electric field gradient tensor is often found to be axially symmetry, with q being the principal value in the bond direction. As the molecule containing the spin I rotates, the quadrupole interaction $\hat{H}_1^Q(t)$ will be modulated in time since the principal axes fluctuate in space.

(c) Anisotropic chemical shift interaction $\hat{H}_1^{CS}(t)$

The chemical shift of a given spin I is caused by the shielding of the external field B_o via the molecular electronic charge distribution. Only for nuclei at a site with cubic and higher symmetry the shift will be independent of the orientation of the molecule with respect to the external field. For nuclei on sites with lower local symmetry the shift is a function of the orientation of the molecule. Although the chemical shift tensor can be antisymmetric in principle |1,4|, we restrict the discussion to symmetric shift tensors here. The orientation dependent part of the shift provides a perturbation $\hat{H}_1^{CS}(t)$ which again fluctuates as the molecule containing the spin rotates. The $\hat{H}_1^{CS}(t)$ interaction is important for the relaxation of spins I = 1/2 in very strong external fields B_o.

(d) Spin rotation interaction $\hat{H}_1^{SR}(t)$

The last interaction to be considered here differs from the three previous ones since it depends not only on the orientation but also on the angular momentum $\vec{J}$ of the molecule. If a molecule is in a given state of angular momentum the charge distribution moves and produces a magnetic field |4|. In liquids the molecules are not in quantized rotational states and the additional field fluctuates rapidly. Since these fluctuations are primarily due to fluctuations of the angular momentum the spin-rotation interaction provides different information on the reorientational dynamics than the three spin interactions considered previously.

The perturbation Hamiltonians $\hat{H}_1(t)$ can all be written as scalar products of two spherical tensors |4,6,7|. Since the transformation properties of the spin-rotation interaction $\hat{H}_1^{SR}(t)$ are slightly more involved, we merely consider the dipole-dipole, quadrupole, and anisotropic chemical shift (shift tensor assumed to be symmetric) interaction, for which we write |1,2,4,6,7|

$$\hat{H}_1(t) = \sum_{m=-2}^{+2} (-1)^m \; F_m^{(2)}(\Omega(t)) \cdot A_{-m}^{(2)}(\hat{I}). \qquad (3)$$

Here, the $A_{-m}^{(2)}(\hat{I})$ are functions of the spin operator $\hat{I}$ and depending on the specific interaction also functions of the spin operator $\hat{S}$ or the magnetic field B_o. The explicit forms can be found in NMR textbooks |1,2,4|. The coupling functions $F_m^{(2)}(\Omega(t))$ describe the dependence of the local spin interaction on the orientation $\Omega(t)$ of the molecule containing the relaxing spin I. It is interesting to note that both sets $A^{(2)}_{-m}$ and $F^{(2)}_{m}(\Omega(t))$ form irreducible tensors of rank 2, for which the transformation properties under rotations are well known |6,7|. In general, a molecule fixed principal axes system can be found in which the cartesian components of the $F^{(2)}(\Omega(t))$ are diagonal. In this principal system also the spherical components $F^{(2)}_{m}$ of the spin interaction take on a particular simple form which is given in table 1 |1,4|.

Table 1: Coupling functions $F^{(2)}_{m}$ in their principal axes system

$F^{(2)}_{m}$ (P)	Interactions		
	dipole-dipole	quadrupole	aniso. chem. shift
$F^{(2)}_{0}$	1	1	1
$F^{(2)}_{\pm 1}$	0	0	0
$F^{(2)}_{\pm 2}$	0	$\frac{1}{\sqrt{6}}\eta_q$	$\frac{1}{\sqrt{6}}\eta_\sigma$
common factor	$-\frac{1}{2}\gamma_I\gamma_S r_{is}^{-3}$	$\frac{eQq}{4I(2I-1)\hbar}$	$\frac{1}{2}\gamma_I\,\Delta\sigma\,B_0$

Here, γ_I and γ_S are the magnetogyric ratios of the nuclei I and S, $\vec{r}_{IS}$ is the spin - spin vector, $q = V_{zz}$ is the field gradient component in the principal z direction, $\Delta\sigma = \sigma_{zz} - \frac{1}{2}(\sigma_{xx} + \sigma_{yy})$ is the shift anisotropy in the (shift) principal z direction, η_q is the field gradient asymmetry parameter, $\eta_q = (V_{xx} - V_{yy})/q$, and η_σ the chemical shift asymmetry parameter defined similary |1,4|.

Next, the function $F^{(2)}_{m}$ given in their molecule fixed principal axes system must be related to the laboratory system in which the Zeeman interaction $\hat{H}^{Z}_{O}$ dominates over the local perturbations. This relation can be achieved by two successive rotations: A first Ω_p is necessary to relate the principal axes system of the local interaction to the principal axes system of the motional ellipsoid of the molecule. The second rotation $\Omega(t)$ transfers the principal axes system of the motional ellipsoid to the laboratory axes system. We note, that the first rotation Ω_p is time independent, depending only on the local interaction under consideration and the structure of the molecule. On the other hand the second rotation $\Omega(t)$ is time dependent and accounts for the reorientational dynamics. For the two successive rotations we use the spherical tensor transformation properties |6,7|. If we denote with $F^{(2)}_{m''}(P)$, $F^{(2)}_{m'}(A)$, and $F^{(2)}_{m}(\Omega(t))$ the interaction functions in their principal system (P), in the motional ellipsoid system (A), and in the laboratory system ($\Omega(t)$) respectively we obtain

$$F^{(2)}_{m'}(A) = \sum_{m''} F^{(2)}_{m''}(P)\, D^{(2)}_{m'',m'}(\Omega_p) \tag{4}$$

and

$$F^{(2)}_{m}(\Omega(t)) = \sum_{m'} F^{(2)}_{m'}(A)\, D^{(2)}_{m',m}(\Omega(t)), \tag{5}$$

$$= \sum_{m',m''} F^{(2)}_{m''}(P)\, D^{(2)}_{m'',m'}(\Omega_p)\, D^{(2)}_{m',m}(\Omega(t)), \tag{6}$$

where the $D^{(2)}_{m,n}(\Omega)$ are elements of the Wigner rotation matrix of rank two |6,7|.

Next, we define orientational correlation functions |1,2|

$$g(t) \equiv g_m^{(2)}(t) = \langle F_m^{(2)*}(\Omega(o)) \cdot F_m^{(2)}(\Omega(t))\rangle \qquad (7)$$

and the spectral densities

$$I(\omega) \equiv I_m^{(2)}(\omega) = \int \exp(-i\omega t)\, g_m^{(2)}(t)\, dt. \qquad (8)$$

The relaxation rate $1/T_1$ is related to the spectral density $I(\omega)$ at the resonance frequency $\omega_o = \gamma_I B_o$ |1,2|,

$$\frac{1}{T_1} \sim \left\{ I(\omega_o) + \alpha\, I(2\omega_o) \right\}, \qquad (9)$$

where $\alpha = 4$ for dipole-dipole and quadrupole interaction, and $\alpha = 0$ for anisotropic chemical shift relaxation. By inserting equations (6), (7), and (8) into equation (9) a rather intractable result for the relaxation rate $1/T_1$ is obtained. Often, however, the following simplifications hold:

(a) <u>Extreme narrowing ($\omega_o \cdot \tau \ll 1$) :</u>

In ordinary liquids the extreme narrowing condition $\omega_o \cdot \tau \ll 1$ applies, which means that the reorientational fluctuations occur on a much shorter time scale than the magnetic resonance period time. In this case the effective correlation time $\tau_{eff}^{(2)}$ is simply related to $I_m^{(2)}(\omega)$ by

$$\tau_{eff}^{(2)} = \int_o^\infty g_m^{(2)}(t)\, dt = \frac{1}{2} I_m^{(2)}(\omega = o)\,. \qquad (10)$$

Further, equation (9) simplifies to

$$\frac{1}{T_1} \sim \tau_{eff}^{(2)}.$$

(b) $\eta_q,\ \eta_\sigma = 0$

If the site of the nucleus in the molecule is on a threefold or higher symmetry axis the asymmetry parameters η_q and η_σ vanish. With table 1 and equations (10), (7), and (6) we can write

$$\tau_{eff}^{(2)} = \sum_{m',m''} D_{o,m'}^{(2)}(\Omega_p)\ D_{o,m''}^{(2)*}(\Omega_p)\ \cdot\ \tau_{m',m''}^{(2)}, \qquad (12)$$

where the partial correlation times are defined as

$$\tau_{m',m''}^{(2)} = \int_o^\infty < D_{m',m''}^{(2)}(\delta\Omega(t)) > dt. \qquad (13)$$

In order to derive equations (12) and (13) the fluid isotropy has been used. The symbol $\delta\Omega(t)$ denotes the rotation from the initial orientation $\Omega(o)$ of the molecule at time 0 to the final orientation $\Omega(t)$ at time t.

(c) Reorientation axially symmetric

The assumption that the motional ellipsoid is axially symmetric causes m' = m" and simplifies equation (12) to

$$\tau_{eff}^{(2)} = \frac{1}{4}\,(3\cos^2\theta-1)^2\cdot\tau_o^{(2)} + 3\sin^2\theta\cos^2\theta\cdot\tau_1^{(2)}$$

$$+ \frac{3}{4}\,\sin^4\theta\cdot\tau_2^{(2)} \qquad (14)$$

Here θ is the angle between the unique principal axis of the spin interaction and the symmetry axis of the motional ellipsoid.

(d) Reorientation diffusional

The final simplification to be considered here occurs if the reorientation can be taken as diffusional with rotational diffusion constants $D_\perp$ (for tumbling) and $D_\parallel$ (for spinning):

$$\tau_o^{(2)} = 1/(6D_\perp); \quad \tau_1^{(2)} = 1/(5D_\perp + D_\parallel);$$

$$\tau_2^{(2)} = 1/(2D_\perp + 4D_\parallel). \tag{15}$$

We are now able to write down the magnetic relaxation rates $1/T_1$ as functions of the effective correlation times $\tau_{eff}^{(2)}$ for the spin interactions considered above.

1. Dipole-dipole interaction (intramolecular)

$$1/T_1^D = \frac{4}{3}\hbar^2\gamma_I^2\gamma_S^2\, S(S+1)\, r_{IS}^{-6} \cdot \tau_{eff}^D \tag{16}$$

2. Quadrupole coupling (I> 1)

$$1/T_1^Q = \frac{3}{40}\left(\frac{eQq}{\hbar}\right)^2 \frac{2I+3}{I^2(2I-1)} \cdot \tau_{eff}^Q \tag{17}$$

3. Anisotropic chemical shift

$$1/T_1^{CS} = \frac{2}{15}\gamma_I^2 B_o^2 (\sigma_\parallel - \sigma_\perp)^2 \cdot \tau_{eff}^{CS} \tag{18}$$

4. Spin-rotation interaction (linear molecules)

$$1/T_1^{SR} = \frac{4}{3} C^2\, I\, (kT/\hbar^2) \cdot \tau_J \tag{19}$$

In the last relation C is the spin rotation coupling constant |4| and I the moment of inertia of a linear molecule. τ_J is the angular momentum correlation time.

From the discussion given above it is evident that the effective correlation times τ_{eff} of the first three relaxation mechanisms probe the reorientational dynamics of quite different molecule fixed axes. For the dipole-dipole interaction the probe direction is that of the spin-spin vector $\vec{r}_{IS}$. In the case of axially symmetric field gradient and shift tensors the symmetry axis of these tensors defines the probe direction. Combination of relaxation results from different nuclei of the same molecule often allows to determine the different rotational diffusion constants. Examples will be given below.

In the diffusion limit the correlation time of the spin-rotation interaction, the angular momentum correlation time τ_J, is related to the orientational correlation time $\tau_0^{(2)}$ via the Hubbard relation |4,8|

$$\tau_0^{(2)} \cdot \tau_J = I/(6kT). \qquad (20)$$

For nuclei with I> 1 quadrupole relaxation usually dominates the magnetic relaxation behaviour. Contrary, for spin I = 1/2 nuclei the three interactions, dipole-dipole, anisotropic chemical shift, and spin rotation may contribute simultaneously to an experimental relaxation rate

$$\left(\frac{1}{T_1}\right)_{exp} = \frac{1}{T_1^D} + \frac{1}{T_1^{CS}} + \frac{1}{T_1^{SR}} . \qquad (21)$$

By varying the temperature (spin rotation relaxation rates increase whereas the other rates decrease with increasing temperature) and the external field strength B_o (chemical shift relaxation depends on B_o, the other rates do not) it is often possible to separate the three relaxation contributions in equation (21) |4|. The Hubbard relation (20) can in addition be used to interrelate the relaxation rates.

3. EXPERIMENTAL EXAMPLES

3.1 Anisotropic reorientation of acetonitrile

As a first example we consider the anisotropic reorientation of acetonitrile, the first system for which anisotropic reorientation was detected experimentally |9,10|. Beside nmr relaxation a number of other spectroscopic methods have been used since then to investigate the reorientational dynamics of this liquid |11|. Figure 2 shows the deuterated acetonitrile molecule with the probe nuclei ^{14}N (spin I = 1) and ^{2}D (spin I = 1). Both nuclei relax by quadrupole interaction. The principal z directions of the field gradient tensors coincide with the bond directions for both nuclei. These "probe directions" are included in figure 2 as arrows. The motional ellipsoid

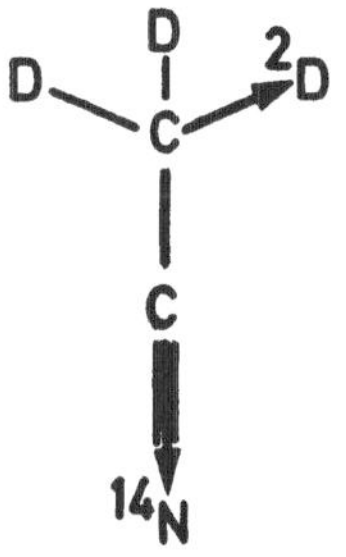

$\theta_D = 109.5^o$

$\theta_N = 0^o$

Fig. 2: Structure of CD_3CN molecule and probe directions for the quadrupole relaxing nuclei ^{14}N and ^{2}D

of the CD_3CN molecule will be axially symmetric. Its symmetry axis coincides with the C_3 axis of the C_{3V} molecule. The two nuclei define different angles θ, for the ^{14}N nucleus $\theta_N = 0^o$ and for the ^{2}D nuclei $\theta_D = 109.5^o$. If the magnetic relaxation rates $1/T_1(^2D,\ \theta_D = 109.5^o)$ and $1/T_1(^{14}N,\ \theta_N = 0^o)$ are determined, equation (17) enables us to calculate two different effective correlation times, $\tau_{eff}(\theta_D)$ and $\tau_{eff}(\theta_N)$. In the limit of rotational diffusion we obtain from equations (14) and (15)

$$\tau_{eff}(\theta) = \frac{1}{4}\,\frac{(3\cos^2\theta-1)^2}{6D_\perp} + 3\,\frac{\sin^2\theta\cos^2\theta}{5D_\perp + D_\parallel} + \frac{3}{4}\,\frac{\sin^4\theta}{2D_\perp + 4D_\parallel} \qquad (22)$$

which, if written down for θ_D and θ_N, provides two equations for the unknown diffusion constants $D_\perp$ and $D_\parallel$ |9,10|.

In figure 3 we show the temperature dependence of $D_\perp$ and $D_\parallel$ of CH_3CN as determined by the nmr method. Some results from other spectroscopic techniques are also included. Let us first consider the two $D_\perp$ curves

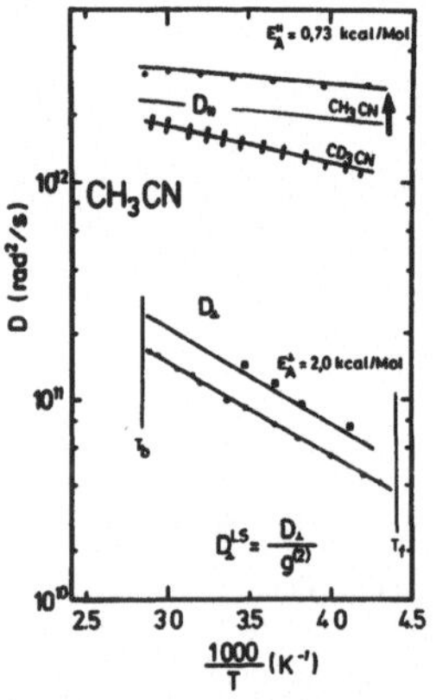

Fig.3: Temperature dependence of the rotational diffusion constants $D_\perp$ and $D_\parallel$ of CH_3CN

given in the lower part of the figure. The upper solidly drawn line corresponds to $D_\perp$ as obtained from ^{14}N re - laxation experiments |9|. These values are subject of some uncertainty due to the poor knowledge of the ^{14}N quadrupole coupling constant in the liquid state |9|. Also included (□ □ □) are $D_\perp$ determined by Raman scattering |12|, which compare well with the nmr results. The lower $D_\perp$ curve (· · · ·) represents depolarized Rayleigh results. As outlined in the chapter on depolarized light scattering this method probes collective reorientation. The light scattering diffusion constant $D_\perp^{LS}$ is related to $D_\perp$ via

$$D_\perp^{LS} = \frac{D_\perp}{g_2} , \tag{23}$$

where we have assumed that the angular velocity correlation parameter j_2, defined in the previous chapter equals one. The deviation of $D_\perp^{LS}$ from $D_\perp$ thus indicates orientational pair correlations in CH_3CN, which is in good agreement with the findings of other experimental methods.

Next we turn our attention to the $D_\parallel$ curves given in the upper part of figure 3; the experimental $D_\parallel$ curve obtained from the ^2D relation |9-11| is marked by crosses (+++++++). A first comment on these results is that the $D_\parallel$ values are extremely large, which means that the axial reorientation is more likely to resemble free rotation than diffusion. Although the diffusion result, equation (22), is no longer strictly applicable, it has frequently been used as a convenient tool to analyse experimental data. Raman bands of non totally symmetric vibrations have also been used to study the fast axial reorientation of methyl halides and acetonitrile |13,14|. $D_\parallel$ results from such investigations of CH_3CN |14| are included in figure 3 (· · · ·). The $D_\parallel$ values from these studies are noticeably larger than the nmr results. The agreement becomes better as one takes into account the different moments of inertia of CH_3CN and CD_3CN, i.e. if one scales the CH_3CN Raman results to CD_3CN. The solidly drawn line in figure 3 corresponds to such a scaling. Still there remains a difference between the results from the two methods which are at present not fully understood |14|.

In figure 4 the pressure dependence of $D_{\perp}$ and $D_{\parallel}$ of CH_3CN as obtained by Bull and Jonas |15| by ^2D and ^{14}N relaxation measurements is reproduced.

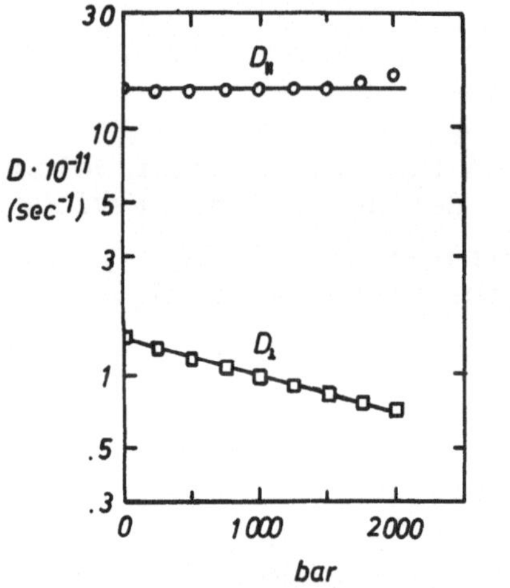

Fig. 4: Pressure dependence of the rotational diffusion constants $D_{\perp}$ and $D_{\parallel}$ of CH_3CN

Again we note that the axial spinning motion of the CH_3CN is faster by roughly a factor of ten than the tumbling. Further, $D_{\parallel}$ is almost independent of pressure, whereas $D_{\perp}$ shows a significant dependence. The activation volumes determined for the spinning and the tumbling motion were $\Delta V^{\ddagger}(D_{\parallel}) \simeq 0$ and $\Delta V^{\ddagger}(D_{\perp}) = 8.5\ cm^3/mol$ |15|. Thus, although certain details are not fully understood, the pressure and temperature experiments clearly show that the acetonitrile molecule performs a rather free axial rotation (spinning) with a low activation energy $E_A^{\parallel} \simeq 0.7$ kcal/mol and a vanishing activation volume. Contrary the tumbling is strongly coupled to the structure and dynamics of the local surroundings. It has a significant activation volume and an activation energy $E_A^{\perp} \simeq 2.0$ kcal/mol which is close to that of the viscosity.

The method just described has found wide application. A number of results for small molecules has been critically reviewed by Griffith |11|. Before leaving this experimental example we want to call attention to a recent molecular dynamics investigation on CH_3CN in which different correlation times and correlation functions were simulated |16|. This investigation basically confirms the results obtained by experimentalists. Beyond that it was found that none of the commonly used models is capable to describe the fast axial rotation correctly. In particular, the small angle diffusion and the J-diffusion model give incorrect orientational correlation functions.

3.2 Temperature dependence of the ^{2}D relaxation in alcohols

As outlined in section 2, normally, magnetic relaxation occurs in the extreme narrowing limit, $\omega_o \cdot \tau \ll 1$, in ordinary liquids. One disadvantage of this situation lies in the fact, that the relaxation rate $1/T_1$ provides information on the effective correlation time τ_{eff} rather than on the spectral density $I(\omega)$ or the full orientational correlation function g(t). The situation improves if the liquid under study can be supercooled which is the case with simple alcohols. With sufficiently long correlation times τ one could study the frequency dependence of $1/T_1$ which would be the most complete information. If a frequency variable nmr spectrometer is not available a very interesting "temperature spectroscopy" can be carried out at constant frequency, which dates back to the early days of magnetic resonance |17|. In figure 5a we consider the frequency dependence of the spectral density $I(\omega)$ of a system performing an isotropic reorientation with a diffusion constant $D_\perp$, for which $I(\omega)$ is a Lorentzian,

$$I(\omega) \sim \frac{\tau_o^{(2)}}{1 + \omega^2 \tau_o^{(2)2}} \quad . \tag{25}$$

Further, we assume that the temperature dependence of $D_\perp$ is given by a simple Arrhenius law

$$D_\perp = D_\perp^\infty \cdot \exp(-E_A^\perp/kT), \tag{26}$$

where $E_A^\perp$ is the activation energy for the isotropic reorientation process. The three $I(\omega)$ curves drawn in figure 5a belong to temperatures $T_a < T_b < T_c$ respectively.

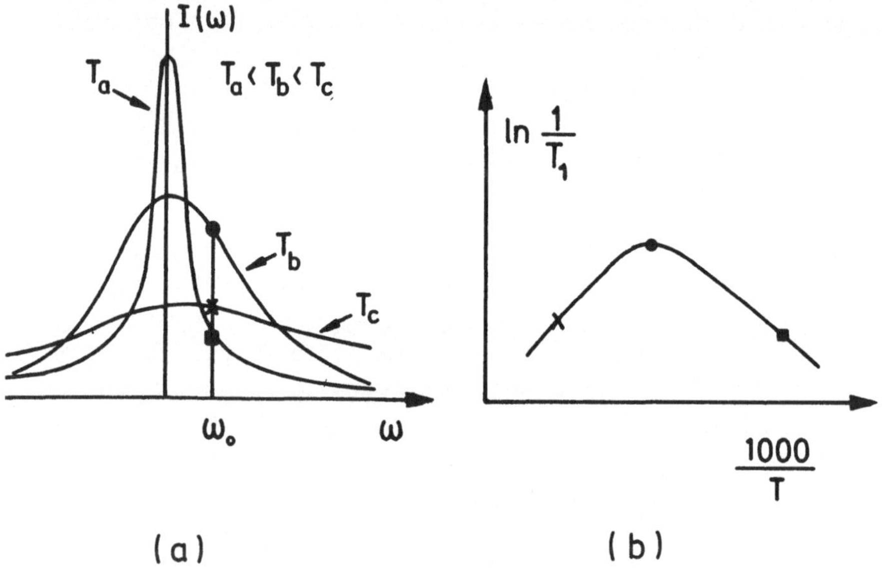

Fig. 5: Temperature dependence of $I(\omega)$ and $1/T_1$ for a simple rotational diffusion process

vely. If the temperature of the system is lowered from T_c to T_a, at a fixed frequency ω_o ($= \gamma B_o$) the spectral density $I(\omega_o)$ first increases and then decreases again. Since $1/T_1$ is essentially proportional to $I(\omega)$ the rate $1/T_1$ follows this behaviour which is shown in figure 5b by a $\ln(1/T_1)$ versus $1000/T$ plot. The maximum in this plot occurs roughly at the temperature at which $\omega_o \cdot \tau_o^{(2)} \simeq 1$.

For an anisotropic rotational diffusion process $I(\omega)$ will be the superposition of three Lorentzians of the type equation (25) with three correlation times $\tau_o^{(2)}$, $\tau_1^{(2)}$, and $\tau_2^{(2)}$. If we also assume an Arrhenius behaviour for $D_{\|}$ with an activation energy $E_A^{\|}$ the three correlation times according to equation (15) will show different temperature dependences. In such a situation we would expect at least three maxima of $1/T_1$ as a function of 1000/T. For a slowly tumbling and fast spinning molecule with decreasing temperature the first maximum occurs at $\omega_o \cdot \tau_o^{(2)} \simeq 1$. It is related to the tumbling and its hight will be proportional to $1/4(3\cos^2\theta-1)^2$. Two further maxima occur at $\omega_o\tau_1^{(2)} \simeq 1$ and $\omega_o\tau_2^{(2)} \simeq 1$ with amplitude factors as given in equation (14).

The relaxation behaviour just described can be used to investigate the interrelation of the overall and intramolecular motion in larger flexible molecules. As an example we consider the relaxation behaviour of ethanol |18| and propanol |19|.

The temperature dependence of the ^2D relaxation rates of selectively deuterated ethanol, (a) CH_3CH_2OD, (b) CH_3CD_2OH, and (c) CD_3CH_2OH are shown in figure 6a. The $1/T_1$ maximum of the OD group relaxation occurs at the highest temperature, the relaxation behaviour of the CD_2 and CD_3 group being very similar. The conclusion is that the hydrogen bonded OD group performs the slowest reorientation. Surprisingly, the methyl group does not show a fast anisotropic rotation behaviour. In order to clarify this finding, mixture studies of ethanol, CH_3CD_2OH and CD_3CH_2OH, with glycerol were performed. Results for 25, 50, and 75 mole% ethanol-glycerol mixtures are shown in figures 6b, 6c, and 6d. In figure 6b the CD_2-rate shows a large shift to higher temperature with a single maximum. Contrary, the CD_3-rate shows a kink in the region where the CD_3-rate has its maximum and increases further to a second relaxation maximum. As the glycerol content is increased

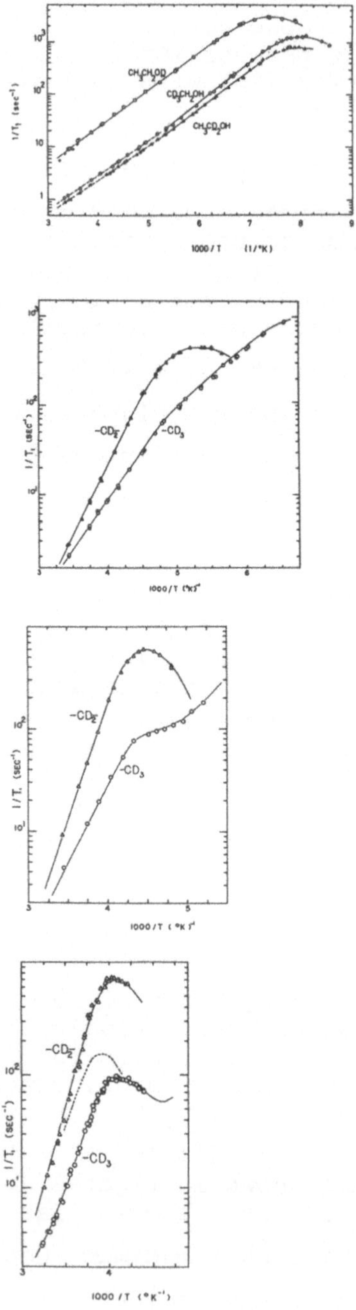

(a) 100% ethanol

(b) 75% ethanol 25% glycerol

(c) 50% ethanol 50% glycerol

(d) 25% ethanol 75% glycerol

Fig. 6 : Temperature dependence of deuteron relaxation rates of the selectively deuterated ethanols CH_3CH_2OD, CH_3CD_2OH, and CD_3CH_2OH (a) in pure ethanol, and (b), (c), and (d) in ethanol/glycerol mixtures (mixture composition as indicated).

(figure 6b and 6c) the kink develops into a separated maximum which occurs at the same temperature as the maximum of the CD_2-rate, however, with a by the factor $1/4(3\cos^2\theta-1)$ reduced amplitude. Obviously, the kink and the maximum of the CD_3-rate belong to the methyl group tumbling which closely follows the CD_2-rotation. That the spinning of the methyl group is faster than its tumbling in the mixtures is apparent from figure 6b and 6c, in which an increase of the CD_3-rate to a second maximum is clearly indicated. Analysis of the data gives an activation energy of E_A = 4.0 kcal/mole for the internal barrier for the CD_3 spinning |18|.

The relaxation behaviour can be understood in the following way: In pure ethanol the internal barrier for the methyl group is higher (4 kcal/mole) than the barrier to rotation of the ethyl group as a whole (3.3 kcal/mole). Therefore, the motion of the ethyl part is faster than the internal rotation and the relaxation behaviour resembles that of a rigid group. In the mixtures the overall motion of the ethyl group is considerably slowed down whereas the internal CD_3 rotation remains untouched. Here, the relaxation behaviour becomes that of a fast spinning methyl group.

As one further example we consider the deuteron relaxation of propanol of selectively deuterated samples (a) $CH_3CH_2CH_2OD$, (b) $CH_3CH_2CD_2OH$, (c) $CH_3CD_2CH_2OH$, and (d) $CD_3CH_2CH_2OH$, as shown in figure 7. The relaxation behaviour of the OD-group and the two CD_2-groups resembles very much the one of these groups in pure ethanol as shown in figure 6a. However, due to the higher viscosity of propanol and the longer alkyl chain the reorientation of the alkyl group is slower than the internal rotation of the methyl group. As a result the CD_3-relaxation rate shows a kink in the temperature range of the CD_2 rate maxima, which, as in figure 6b, indicates a moderately fast methyl rotation in propanol.

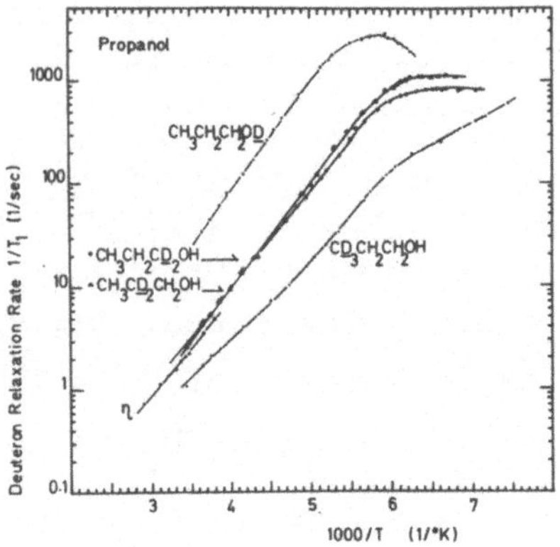

Fig. 7: Temperature dependence of ^{2}D relaxation rates $1/T_1$ in propanol

Attempts have been made to account for the relaxation behaviour of these systems quantitatively |18,20|. So far, however, experiments are far ahead the theoretical understanding of such highly viscous liquids. Light scattering experiments indicate that one is not faced with simple relaxation processes in highly viscous liquids and polymers. Often the relaxation has been found to be formally describable by a William-Watts distribution |21|. The high selectivity to individual motional modes makes the magnetic relaxation method particularly appropriate for more detailed investigations of such systems.

REFERENCES

1. A. Abragam, The principle of nuclear magnetism, Oxford University Press, 1961

2. C.P. Slichter, Principles of magnetic resonance, Harper & Row, 1963

3. T.C. Farrar and E.D. Becker, Pulse and Fourier Transform NMR, Acdemic Press, 1971

4. H.W. Spiess, NMR-Basic Principles and Progress, 15, 55 (1978)

5. J. Jonas and H.S. Gutowsky, Ann. Rev. Phys. Chem., 31, 1 (1980)

6. A.R. Edmunds, Angular momentum in quantum mechanics, Princeton University Press, 1957

7. D.M. Brink and G.R. Satchler, Angular momentum, Oxford University Press, 1962

8. P.S. Hubbard, Phys. Rev., 131, 1155 (1963)

9. T.T. Bopp, J. Chem. Phys., 47, 3621 (1967)

10. D.E. Woessner, B.S. Snowden, Jr., and E.T. Strom, Mol. Phys., 14, 265 (1968)

11. J.E. Griffiths, Vibrational Spectra and Structure, 6, 273 (1977)

12. J.E. Griffiths, J. Chem. Phys., 59, 751 (1973)

13. T. Bien, M. Possiel, G. Döge, J. Yarwood, and K.G. Arnold, Chem. Phys., 56, 203 (1981)

14. J. Gompf, H. Versmold, and H. Langer, Ber. Bunsenges. Phys. Chem., 86, 1114 (1982)

15. T.E. Bull and J. Jonas, J. Chem. Phys., 53, 3315 (1970)

16. H.J. Böhm, R.M. Lynden-Bell, P.A. Madden, and I.R. McDonald, to be published

17. N. Bloembergen, E.M. Purcell, and R.V. Pound, Phys. Rev., 73, 679 (1948)

18. H. Versmold, Ber. Bunsenges. Phys. Chem., 78, 1318 (1974)

19. H. Versmold, to be published

20. H. Versmold, Ber. Bunsenges. Phys. Chem., 84, 168 (1980)

21. T. Dorfmüller, this book

PICOSECOND LASER SPECTROSCOPY AND MOLECULAR DYNAMICS: I PUMP-PROBE SPECTROSCOPY TECHNIQUES AND PHOTODYNAMICS II NONLINEAR LASER SPECTROSCOPY AND MOLECULAR MOTION

G.A. Kenney-Wallace

Lash Miller Laboratories, University of Toronto, Toronto
Canada M5S 1A1

1. INTRODUCTION

The exciting and rapid developments of recent years in the generation and measurement of ultrashort pulses (1-3), with pulse durations now as short as 30 femtoseconds (30×10^{-15}s (4)), are driving the exploration of new physics and chemistry in condensed media in systems as diverse as semiconductors and cytochrome C. One vast and significant area in which the fundamental questions seem well characterised, but their resolution less so, is that of molecular dynamics in liquids. Not only is the microscopic description of the molecular structure and dynamics of liquids one of "the last frontiers", to quote Dr. Yarwood's opening remarks, it is also one of the *new* frontiers for ultrafast spectroscopy, if one views the research potential from an interdisciplinary perspective (5). *Direct* access to the picosecond (ps) and now femtosecond (fs) time domain is a reality that hopefully over the next five years will permit direct, time-ordered optical observations of the dynamical interactions that accompany any molecular motion in a liquid, as the system is perturbed (far) from and returns to equilibrium. Because of the separation of timescales which underlies some events, it will be possible to explore dynamical interactions in the time domain that appear simultaneously in the wings of the frequency domain, spectral scattering data. Furthermore, new types of nonlinear spectroscopy can probe new interactions, complementary to the wealth of data in light scattering for example(6). The EMLG triangular relationship between theory, experiment and simulation is a symbiotic one, confronted by which the many-body problem begins to appear less formidable, although no more tractable in an analytic sense. Nevertheless, the

A. J. Barnes et al. (eds.), Molecular Liquids - Dynamics and Interactions, 331–356.

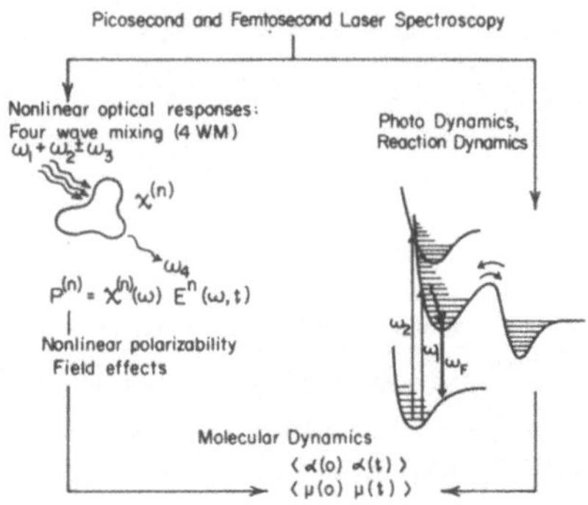

Figure 1: Applications of Ultrafast Linear and Nonlinear Laser Spectroscopy to molecular dynamics discussed in text.

confluence of ultrafast laser spectroscopy and theory and simulation of molecular dynamics in dense media is a very challenging direction for which the *transient*, *optical* and *coherent* interactions have a unique role to play. In these typically dipolar, quadrupolar or polarizability interactions, we are probing both the molecule-laser field interaction and the *a priori* molecule-molecule interaction, as well as their mutual coupling or modulation of the local dynamics (5). Thus it is necessary to study the interactions and structure of the fluid as well the dynamics, since in reality they are inseparable.

The role of the microenvironment around a molecule also plays a major role in the reaction dynamics of the ground state species, or in the photo induced dynamics of the electronically excited state species, which may be undergoing relaxation back to the ground state in competition with reaction via crossing of excited state potential energy surfaces. Molecular dynamics in liquids becomes the necessary underlying description of chemical dynamics in liquids. Figure 1 illustrates the links between laser spectroscopy, molecular dynamics and chemical dynamics that I wish to describe in the following pages. Each field demands a rigorous study in its own right. The purpose of these articles is to focus on the links between these different areas, in order to introduce the reader to the contemporary problems of studying ultrafast phenomena.

One of the central issues in studying reaction dynamics is how you describe the flux over the reaction barrier, whether from simple collision theory, or transition state theory, to modified generalized Langevin theory, Kramer's theory...or even to quantum mechanical tunnelling - This is a selective not exhaustive list of possibilities (7,8). When the timescale of the photo chemical event (be it photodissociation, isomerization, charge transfer) occurs on the time scale of certain pivotal molecular motions, one cannot use separation of timescale arguments anymore, nor mean field theories,nor equilibrium properties to describe what is happening at the *microscopic* level. It is very important therefore to have a quantitative understanding of the rotational, vibrational and translational dynamics in both ground *and* excited states in order to assemble the necessary initial and final state arguments from which the time-evolution of the reactive scattering and chemical dynamics can be constructed. For example, in photoinduced isomerization are the barriers internal (intra molecular) or do they also contain a certain component of viscous drag (frequency-dependent friction term) from the solvent? Do the solvent molecules change the potential and shape of the barrier? What is the competitive timescale of energy relaxation (collision-induced or resonant) of the solvent molecules in competition with the frequency of barrier crossing? Of course, one cannot fit in principle any real molecule with 2 parameters, since we have a multidimensional potential, and yet the intermolecular forces and torques in real fluids seem more amenable to our simplistic notions of short-range repulsive and long range attractive forces than perhaps expected. Of course, if one is testing the least sensitive paramenter, then it will not be sensitive to the finer features of the potential. Thus it is very important to study molecular dynamics and chemical dynamics from several different perspectives and explore for a given molecule different but complementary experimental techniques, each of which can focus on a different labelled property of the molecule (e.g. nuclear spin, dipole, nonlinear (hyper) polarisability) and its relaxation or recovery time in the single particle or many-particle case.

The link between spectroscopy and theory in liquids is through time correlation functions (TCF), on which there are excellent reviews including those in this NATO volume (9,10). The link between the ultrafast laser-induced nonlinear phenomena and interpretation is also through correlation functions, since optically-induced phenomena are usually substantially faster than any electronic or mechanical recording device and therefore the phenomenon must become its own clock. In this chapter I,we focus on the generation and precision measurement of ultrashort pulses and the linear spectroscopy of photodynamics in fluorescing molecules. The chapter following II focuses on the nonlinear spectroscopy of molecules and the very recent and novel applications of four wave mixing (4WM) interactions, stimulated through the nonlinear susceptibility $\chi^{(3)}$, third-order in the applied field, to reveal microscopic details of structure and dynamics in liquids at subpicosecond times through the nonlinear polarization $P^{(3)}$. The origin and time evolution of such nonlinear phenomena (11-13) will be discussed as a background for these experiments. However, it is not possible to comprehensively discuss all aspects of any of the topics selected to appear below, and so these chapters should

$$P^{(3)}_{\omega_4} = \chi^{(3)}_{ijk\ell} \; E(\omega_1)\, E(\omega_2)\, E^*(\omega_3); \; (\omega_4 = \omega_1 + \omega_2 - \omega_3) \qquad [1]$$

serve as a stimulus for further reading. The references have been chosen as general starting points for more detailed investigations of ultrafast spectroscopy.

2. PUMP-PROBE LASER SPECTROSCOPY: FRONTIERS OF TIME

Over the past decade, not only have pulse durations decreased from 10^{-11} to 10^{-13}s but there has been a dramatic increase in the tunability of lasers, such that tunable coherent radiation can now span the VUV to the very long wavelength laser radar. Femtosecond spectroscopy, like most advances, has begun in the *visible* region and considerable research and development is necessary to expand this present spectral range around 600 nm (4). However, it is also the case that for many problems in photo dynamics, for which the state selectivity or the nature of the optically prepared initial state is of paramount importance, the spectral linewidth ($\Delta\nu$) of the pulse must remain narrow. Thus the transform-limited bandwidth relationships ($\Delta\nu\Delta \sim \hbar$) govern the temporal properties of the laser pulse and, for example, a 5 ns pulse of 0.01 cm^{-1} linewidth prepares a different ensemble than a 300 fs pulse of 26 cm^{-1} linewidth at the same wavelength.

In *pump-probe* laser spectroscopy, the initial pulse resonantly *pumps* the atom or molecule into an excited state, following which the excited system is *probed* by a second much weaker laser pulse. Both vibronic and vibrational population relaxation are studied this way. By splitting off a fractional intensity from the *pump* pulse and diverting the newly created *probe* pulse through an optical delay line prior to its arrival at the excited system, we obtain optimal internal synchronization of the two pulses. Since the atoms or molecules are initially prepared by the pump pulse in a non-equilibrium state, whose polarization, absorption or radiative properties are different from the ground state, the system has been "labelled" and the probe pulse monitors the labelled molecules after the initial event at successive delay times τ, obtained by varying the length of the optical delay line. If the initial event is a nonresonant process, such as 4WM, the same pump-probe designs are nevertheless usually utilised, since the precision of the measurement of the relaxation event understudy is entirely controlled by the precision with which we can determine τ. The limits to such measurement which are met at the frontiers of femtosecond spectroscopy are examined later in this section. The "label" varies: it can be absorption or emission intensity, Raman, polarisation dependence, or phase coherence of the individual atom or molecule or the ensemble.

In this brief description of contemporary approaches to ultrafast spectroscopy, we will focus specifically on some of the newer techniques which can produce tunable pump and probe pulses of duration $\geq 10^{-13}$ s. Linear and nonlinear optical pulse compression techniques (14,15) are utilized to compress visible dye laser pulses ~600 nm from 100 fs to 30 fs (4), or from 10 ps to 200 fs (16) in two stages. Both active and passive modelocking techniques have been exploited to generate the initial ps and sub ps dye laser pulses prior to pulse compression, but only the active approach leads to *tunable* pulses. Figure 2 illustrates how tunable ps and fs dye laser pulses can be generated, using an actively mode locked laser as the master oscillator and synchronous pumping techniques; subsequently the pulses can be amplified if *intense and tunable* ultrashort pulses are required. A summary of synchronous pumping techniques is given by Jain in ref (3), where we have also discussed the detailed features of our own oscillator-amplifier system (Kalpouzos et al., in ref. 3) which is based on an acousto-optically mode-locked (AML) argon ion laser. Many of the various applications of synchronously pumped lasers are discussed in (17). Details of the most recent advances in AML CW mode-locked Nd:YAG lasers as master oscillators will be given here since they offer new versatility for the future (18,19). The overall concepts are similar, so Figure 2 is a general schematic for the oscillator-amplifier system. The oscillator could be a CW modelocked argon ion (514 nm) Krypton (614 nm) or Nd:YAG laser (1.064 μ), which

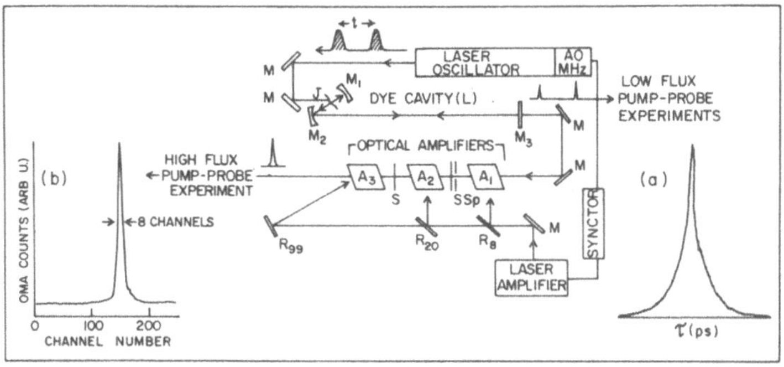

Figure 2:
Scheme for ultrafast spectroscopy. See text for details

generates trains of pulses of typically 100-150 ps in duration spaced at intervals determined by the modelocker frequency. The 100 MHz selected for operating our AML Nd:YAG laser (Quantronix 116, in-house modified) leads to t=10 nanosecond (ns) intervals between the 120 ps pulses. The effective optical cavity length of the Nd:YAG must be matched to within μ distances by the length (L) of the three mirror, folded dye laser cavity ($M_1, M_2 \ldots M_3$) in order to achieve the synchronous pumping condition, which stipulates that the gain modulation in the dye cavity must correspond to the amplitude modulation at the fundamental (and harmonics) of the longitudinal mode separation frequency, c/2L. As a consequence the net gain is not only of very short duration in comparison to the ns spontaneous emission lifetime (τ_f) of the dye molecules but it is independent of τ_f, and as a result a train of ultrashort, single dye laser pulses emerges from M_3 with comparable inter pulse intervals (10 ns) but 10^2 shorter in time, typically 2-10 ps In our dye laser design (19) we have achieved tunable pulses of 300-400 fs duration (or a reduction of 500) as shown in insert (a) of Fig. 2. The physical principles and assumptions on which such ps and fs pulse diagnostics are based will be discussed below, when we introduce the correlation function format for pulse determination.

The emerging dye laser pulses are of nJ (10^{-9}J) energy per pulse and suitable for low flux pump-probe experimentation. If higher energy and power density is required, for multiphoton or some nonlinear experiments for example, then this low energy dye laser pulse can be injected into a series of optical amplifiers (A_1,A_2,A_3) that are independently pumped by a separate amplifier laser,which is locked into a synchronous time relationship with the oscillator , via the RF modulation on the modelocking crystal, for example. For the argon ion dye laser system, we utilised the ns pulses from a Quanta Ray Nd:YAG, frequency-doubled to λ=530 nm with KDP crystals, and designed a low jitter ECL electronic synchronization scheme (SYNCTOR) which controls both the oscillator-amplifier and the streak camera data acquisition techniques for looking at pulse diagnostics and experimental results. Insert (b) in Figure 2 shows the sum of 22 amplified pulses captured on the silicon intensified target (SIT) vidicon tube which is attached to the back of the streak camera (20). We achieve a gain of over 10^5, but the repetition rate necessarily drops to 10 Hz. In ref. (4), Shank and coworkers summarise their pioneering passively mode-locked, colliding pulse model through which pulse durations as short as 70 fs at 600 nm can be achieved and subsequently amplified up to GWatts at 10Hz. However, synchronous amplification of a AML Nd:YAG oscillator *and* AML amplifier, at rates up to 500 Hz, has been described recently by Mourou et al. (18) who report 600 fs pulses of up to mJ energy.

Precision of measurement is particularly important when working at the frontiers of ps and fs spectroscopy because much of our intuitive sense is not naturally attuned to the consequences of wave propagation for a limited number of optical cycles or the effects of dispersion in the wake of intense laser beams. Pulse correlation diagnostics are discussed separately in the next section, but we conclude this tutorial by listing some of the concerns that must be uppermost in designing and intepreting experiments to investigate ultrafast phenomena. Of the many points to mention, the five selected here are a) phase and group velocity, b) linear and nonlinear despersion, c) frequency-time bandwidth relations, d) transit times, e) focusing optics. Although each would prove a stumbling block if ignored, these effects can also offer imaginative routes to new applications of ps and fs spectroscopy when properly utilised (1-4).

While a single plane wave propagates at a phase velocity $v_p=\lambda\nu=\omega k^{-1}$ with wave vector $k=2\pi/\lambda$, $\omega=2\pi\nu$, information travels as a superposition of waves or *wave packet*, with a group velocity $v_g = \lambda\nu'$. Furthermore, the group velocity dispersion (GVD) leads to temporal broadening of the pulses, and self phase modulation (SPM) leads to nonlinear frequency broadening as the pulse intensity increases. The physical consequences of linear dispersion are that low frequencies (e.g. red) travel *faster* than higher frequencies (e.g. green) in the visible and near IR, and such consideration are imperative to the correct measurement and elucidation of time-resolved spectra. (In quartz, dispersion in v_g changes sign at longer IR wavelengths, and this is a principal reason for the emphasis on 1.3μ for optical fibre communications since the transmission dB loss is at a minimum there).

The time delay τ_d between any two optical components of frequency separation $\Delta\omega$ propagating along medium of path length z is given by [2a]:

$$\text{(a)}\ \tau_d = z\,\frac{\delta^2 k}{\partial\omega^2}\,\Delta\omega \qquad \text{(b)}\ \frac{\delta\omega}{\omega} = -n_2\frac{z}{c}\,\frac{\delta\langle E^2\rangle}{\delta t} \qquad [2]$$

The implications of temporal pulse broadening are seen vividly when we point out that a 100 fs pulse (λ=600 nm) passing through 10 cm cell of benzene would emerge ~500 fs in length! Clearly a 10 ps pulse would not be perceptibly affected. However, if linear dispersion occurs via this mechanism, the grating pair technique (14) can be used to compress the pulse again by introducing a dispersive delay. The nonlinear dispersion and frequency broadening is more complex. SPM depends on the nonlinear refractive index n_2, the path and the field intensity. SPM has a quadratic dependence on the pulse duration through the derivative of the time average value $\langle E^2\rangle$ of the instananeous AC field, given in [2]. This is another important route to pulse compression if a dispersive delay can be introduced after SPM, and this area is one of very active research at the present time.

The time-frequency bandwidth relationships underlie the above pulse width considerations but also appear in another light if the nature of the resonant molecule laser interaction is carefully considered. A 1 ps visible pulse carries ~3 cm^{-1} spectral width, and thus it is pertinent to enquire, from *the molecular perspective*, how many optically prepared states are to be found within that bandwidth? This leads to the need to tailor the experiment so that the molecular dynamics probed are indeed intrinsic to the natural molecular system, and not a consequence of specific level coupling in the radiation field whose dynamics are field-driven. Finally, it is useful to remember that in designing an experiment in ps or fs spectroscopy, the image of the laser pulse which is probed in the sample has finite dimensions. Thus by transit times we refer to the fact that since $\nu = c/n$, a focal image of 80μ in n-hexane for example (n_D^{20} = 1.37506) corresponds to 194 fs. Furthermore, there is a spatial separation between the focal points of UV and IR light passing through the same quartz lens, which may impose an effective minimum delay in the time resolution of a pump (UV) probe (IR) experiment.

3. PULSE MEASUREMENTS AND CORRELATION FUNCTIONS

The most widespread technique employed to measure the duration of ultrafast pulses is based on the determination of the second-order autocorrelation function of the intensity. Figure 3 illustrates both the principles and the experimental configuration employed for CW autocorrelation measurements of ps and fs pulses in real time. There are other optically-based methods utilised to measure optical responses over times orders of magnitude faster than electronic detectors, e.g. photodiodes which are currently limited to ~60ps rise time, and even faster than streak camera determinations, which are capable of resolving ≥1 ps. However, the discussions below focuses on the background free ($G_o^{(n)}$) autocorrelation approach as a paradigm for pulse diagnostics. While there are both *fast* and *slow* correlations of optical pulses, where the former results in the retention of the phase information (Ø), in the act of measurement all information regarding phase perturbation is usually lost. For n distinct light pulses having real electric field amplitudes $E_j(t) = \xi(t)\cos[\omega_j t + \emptyset_j(t)]$, the general nth-order, background-free *fast* correlation function $g_o^{(n)}$ is given by [3], such that $g_o^{(n)}(0,0\ldots,0)=1$.

$$g_o^n(\tau_1,\tau_2\ldots,\tau_{n-1})$$

$$= \frac{\int_{-\infty}^{\infty} \{E_1(t)E_2(t+\tau_1)\ldots E_n(t+\tau_{n-1})\}^2\, dt}{\int_{-\infty}^{\infty} \{E_1(t)E_2(t)\ldots E_n(t)\}^2\, dt} \qquad [3]$$

The *slow* nth-order correlation function $G_o^{(n)}$ is given by the optical time average in [3]. The *slow* correlation function thus reflects the time dependent pulse envelopes $\xi_j(\tau_{j-1})$. A detailed discussion of the transition from fast to slow autocorrelation functions, which is accompanied by the loss of phase information

$$G_o^{(n)}(\tau_1,\tau_2.......,\tau_{n-1}) = \langle g_n^{(n)}(\tau_1,\tau_2.....\tau_{n-1})\rangle_{\tau,n} \quad [4]$$

in $\emptyset_i(\tau_{i-1})$, and the differences between the classes of autocorrelation function for given input pulses, has appeared elsewhere (21,17). Pragmatism and space leads to a single case for our study here.

The most frequently measured autocorrelation function is the second-order case $G_o^{(2)}$, in which the pulse is folded into a single replica of itself, and it is this example which is illustrated in Figure 3. The theory and nonlinear optical measurements are shown schematically for a single dye laser pulse (ω), which enters a "real time autocorrelator" or interferometer that operates in both the time and frequency domain. A fraction of the dye pulse intensity is diverted into the autocorrelator via a beam splitter (BS), and this pulse intensity is in turn split into two replicas (I_ω,I_ω') by the 2μ pellicles (P), which direct and reflect the light along the orthogonal arms of an interferometer. At the end of each arm is retroreflector coaxially mounted on the cone of an audio speaker, which can travel in the Z direction for ± 1 cm, thus varying the optical pathlength of the interferometer. The returning pulses are reflected (or transmitted) again by the pellicles into a lens L, from which they are tightly focused into a transparent, nonlinear crystal of KDP. The latter is typically a 100-200μ wafer, cut to accept ω and to optimise the phase matching consideration for ω, 2ω for the frequency-dependent refractive index (n) of KDP. Second harmonic generation (SHG) occurs to produce radiation at 2ω at an angle (dotted) which bisects the diverging fundamental (ω) beams as they exit from the crystal. The SHG alone passes through an aperture and optical filter to a photodector (PMT). SHG occurs because of the second order nonlinear polarization $P^{(2)}$ which is generated in the crystal at the same time as the linear polarization,as the optical waves propagate through the medium. Since a 100 fs pulse is a mere ~30μ in length SGH occurs here *only as a consequence of the precise overlap in space and time* in the KDP of the two travelling ps or fs pulses. $P^{(2)}$ becomes a source term for new radiation waves, which propagate at a phase velocity corresponding to a harmonic frequency, and their resultant constructive interference and thus amplitude reaches a maximum at an optical thickness corresponding to the various phase relationships of these components. Figure 3 shows the three overall and simple mathematical relationships under consideration here; equation [1] indicates the origin of the oscillating cos 2ωt term which appears superimposed on the fundamental ω in the full experssion for nonlinear polarization, second order in the incident electric field E_o, for materials of nonlinear susceptibility $\chi^{(2)}$ which do not posses centrosymmetric symmetry. Otherwise $P^{(2)}$ vanishes. The appropriate nonlinear theory as developed from Maxwell's equations is elaborated upon elsewhere (11-13). The intensity of the second harmonic light $I_{2\omega}$ is proportional to the square of the intensity of the fundamental, maintains the coherence of the incident radiation, and displays polarization properties that depend on the incident polarisation of the fundamental. In the autocorrelator, the SHG light intensity at the photomultiplier is given by the integral in equation [2], Figure 3, where τ is the difference in the arrival time of the two replica pulses at the overlap (L focus) in the KDP crystal. Let the fixed positions of the retroflectors (rest positions of speakers) define an optical pathlength for which the integral in [2] has a maximum value, i.e. $\tau=0$. This corresponds to the central timing diagram in Figure 3 where both ps or fs pulses are fully overlapping in the KDP crystal. When one of the speaker-arms is moving, it is equivalent to varying the optical pathlength of one pulse and, since $\tau=2nz/c$, we essentially are sweeping through values of τ from $-\infty$ to $+\infty$. The timing diagram simulates initially one pulse (shaded) at $\tau=-5$ps (or ~1.5 mm decrease in travel along its optical path) and no overlap, then full overlap at $\tau=0$ as the pulses arrive in the crystal at the same time because they travel equivalent optical paths, and finally $\tau=+$ 7 ps, wherebye one pulse (shaded) travels on optical path ~2.1 mm longer than the other, and once again no overlap occurs. By moving both speaker arms synchronously but 180° out of phase, events from fs up to 300 ps can be recorded.

This overlap integral is the fact the second-order autocorrelation function $G_0^{(2)}(\tau)$ as equation [3] shows. The additional term r(t) is the rapidly varying, phase interference factor which is included for completeness but actually as <r(t)> it is optically averaged to zero. Finally, the output from the PMT is recorded in real time on an XY oscilloscope, and the profile of $G_0^{(2)}$ is captured from which we may deconvolve to obtain the orginal function I(t) and thus determine the pulse width. Three very important points must be made here. First, a second-order autocorrelation

Figure 3: Principles of ps and fs autocorrelation measurements.

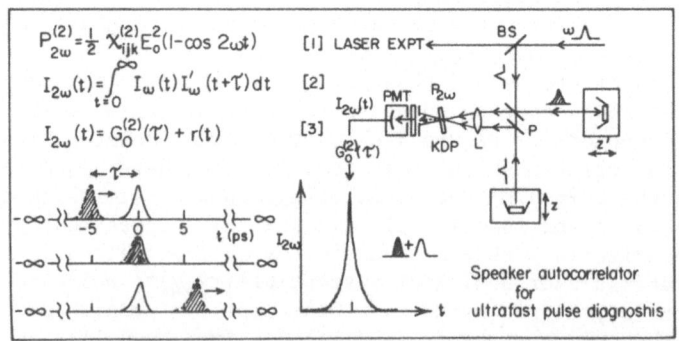

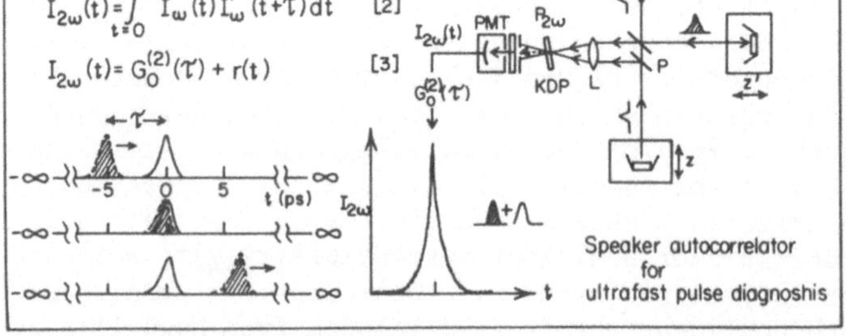

function is fated to be symmetric about τ and thus we cannot obtain information on the skewedness of the pulse in this manner. $G_o^{(3)}$ is the necessary measurement for that information and requires a 3 photon correlation. Second, we must *assume* a pulse shape in order to deduce the pulse width in time based on transform-limited bandwidth considerations, and this choice could lead to a factor of 2 uncertainty in the real value on going from, for example, gaussian where $\Delta\nu\Delta t$=0.4413 to exponential where $\Delta\nu\Delta t$= 0.1103 (21). Third, these are *ensemble* measurements, not single pulse measurements, and the measurement in reality carries information not only on the pulse envelope $G_p(\tau)$ as said before but also on pulse substructure $G_N(\tau)$, which is the Fourier transform of the spectrum of the laser output. The influence of the latter as coherence spikes on pulse determinations and any ps or fs pump-

$$G_o^{(2)}(\tau) = G_p^{(2)}(\tau)[1 + G_N^{(2)}(\tau)] \qquad [5]$$

probe spectroscopic data is beyond the scope of this chapter but should be recognized as a central point of discussion in ultrafast pulse diagnostics.

Finally, it is of vital importance to analyse the shape and duration of single pulse profiles and this requires a streak camera. Comparative pulse diagnostics studies on our laser oscillator amplifier system have been reported in detail elsewhere (22) and illustrate the need to maintain excellent cavity control (to within μ precision) to generate *single* laser pulses, rather than multiple pulse bursts, whose presence is often masked in the broadened wings of the $G_o^{(2)}$ ensemble correlation function.

4. PHOTODYNAMICS VIA LASER-INDUCED FLUORESCENCE

The laser-induced fluorescence (LIF) of molecules can be used as a powerful probe of the photodynamics of the optically prepared molecule participating in intramolecular relaxation processes and in intermolecular processes. Fluorescence *depolarization* (and its ground state analogue, transient dichroism) offers a versatile probe of the orientational relaxation of the excited molecules and has found much application in studies of *slip* and *stick boundary conditions* and in tests of hydrodynamic descriptions (17,23-26). We outline below two examples of contemporary work in photodynamics in which the independent studies both converge towards an acurate picture of the solvation dynamics of larger molecules. Since the fluorescence lifetimes τ_f of most molecules are ~10^{-8}s and the molecular rotation times vary from 10^{-12} to 10^{-7}s, it is important to select a system where $\tau_f > \tau_{rot}$ but τ_f is not too long compared to the correlation time, since complete randomisation of the molecular dipole orientations will occur and the correlation information lost. Hence most studies focus on larger molecules which in

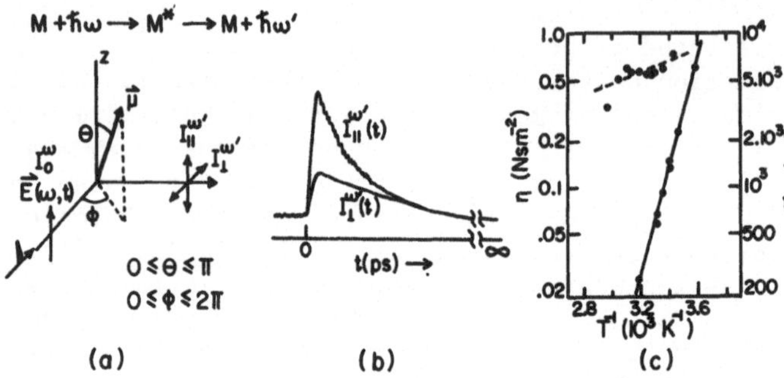

Figure 4: PhotodynamicsI: Fluorescence depolarization, see text

retrospect do not always offer the most stringent or unabiguous tests of hydrodynamic predictions. We will return to this point later after describing the techniques.

(i) Fluorescence depolarisation

Figure 4 illustrates the reference coordinates for a fluorescence depolarization experiment in which a ps laser pulse, of initial polarization direction defined by $\vec{E}_z$, resonantly excites a population of molecules of dipole μ orientated initially in an isotropic distribution (eg. a liquid). A population of excited states is created, with transition dipole moment vector (or major component there of) which now lies along the same z axis in a $\cos^2\theta$ distribution. The laser pulse must be sufficiently intense to burn a hole in the equilibrium orientational probability distribution and create an anisotropy in the system, but not so intense as to induce stimulated emission. The excited state population is thus labelled by a polarization vector and spontaneous emission will occur with the *same* polarization direction as that of the incident exciting pulse I_o, *unless the molecular chromophore changes its orientation before or during emission.* Molecular reorientation thus leads to a loss of population emitting in the z plane or contributing to $I_{\parallel}$, and appearance of a signal at $I_{\perp}$ from molecules emitting light whose polarization now has a major component orthogonal to the incident light. The measurement of $I_{\parallel}(t)$ and $I_{\perp}(t)$ must show a convergence at long times, since the system must return to thermal equilibrium and resume an isotropic orientational distribution. The intepretation of the signals begins with the recognition that we are dealing with an electric-dipole interaction and, assuming one photon absorption, the second rank interaction tensor generates a $\langle P_2[\cos\theta(t)]\rangle$ correlation for the orientational probability density. If the molecule is spherical, it has been shown for diffusive motion that the anisotropy factor r(t) following δ-pulse excitation is given by:

$$r(t) = [I_{\parallel}(t) - I_{\perp}(t)]/[I_{\parallel}(t) + 2I_{\perp}(t)] \qquad [6]$$

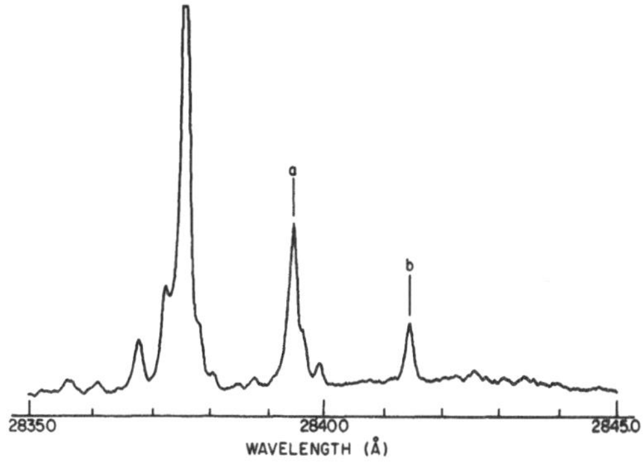

Figure 5: Fluorescence excitation spectrum of jet cooled indole-ethane clusters (29).

$$r(t) = \frac{2}{5} < P_2(\underset{\sim}{u}(o).\underset{\sim}{u}(t)]> = \exp\,[-6D\tau_{rot}] \qquad [7]$$

The anisotropy is sensitive to shape and other factors such as hindered rotation and solvation complexes, particularly in H-bonding fluids. Since the literature in this area is of long standing following pioneering work of Perrin (1926) and has been reviewed, we will confine our remarks to a few recent observations of general importance to the fine-tuning of the microscopic interactions in liquids.

Fluorescence depolarization measures the *excited state* orientational dynamics, which corresponds to the local dynamical solvent-solute interactions arising from the excited state dipole moment. This change in charge density can in some cases lead to significant configurational changes, as reflected by large Stokes shifts in the fluorescence spectrum, so it is not always the case that *ground state* dynamics (observed in transient dichroism or through polarization spectroscopy) should be identical to those in the excited state. Any changes in the anisotropic polarizability between ground and excited states can be also resolved in this way by observing the transient birefringence in the system (26).

The Stokes-Einstein-Debye relationships predict that there is

$$\tau_{rot} = \frac{8\pi\eta R^3}{6kT} = \frac{\eta V}{kT} \qquad [8]$$

a linear correlation between τ_{rot} and η the viscosity for a given T where V is the hydrodynamic volume of the particle of radius R, and hydrodynamic *stick* boundary conditions are assumed. However, even while there is ample evidence of linear correlations in a wide range of dyes and medium size or larger organic molecules, (6,10,23-26) the slope does not always appear consistent with physical reality for V. Microscopic theories of rotational diffusion predict significant diviations from hdyrodynamic behaviour, particularly for small nonpolar molecules (6,10,25). To what extent these are boundary condition problems or symptoms of the intrinsic inability of hydrodynamics to project out molecular parameters is a continuing and lively debate. It is not our intention to comprehensively explore all the recent arguments in this area but instead to note two important maxims that emerge from recent results, and to use as an example the orientation of rhodamine 6G in glycerol (24). The first point is that any correlation between τ_{rot} and η could be an indication that *both* dynamic variables are affected by the solvent molecular structure, which acts as a hidden third partner inthe correlation, thus masking details of the boundary conditions or changing efficiencies of angular momentum transfer. The second point is that in a bath or rotating molecules of disparate size, it becomes a question of *accomodation versus perturbation* (24). In glycerol, the H-bonding provides a broken three dimensional network within which a small molecule such as rhodamine 6G can be accomodated in a "cage" environment and undergo relatively free rotation or libration. A larger molecule, such as myoglobin, perturbs the network so that rotational motion of the quest molecule is indeed governed by the dynamics of the host solvent . The temperature dependent slope of τ_{rot} for rhodamine (○) 6G in Figure 4 gives a rotational barrier of ~9kJ mole^{-1} compared to ~59 kJ mole^{-1} for myoglobin, while the pure glycerol solvent(●) data from many sources on η as f(T) yield 68kJ mole^{-1}, as evidence for these conclusions.

In summary, LIF and fluorescence depolarization are an important tools for determining the dynamical behaviour of molecules but to date have not really been able to address the *fundamental* questions of molecular dynamics as a developing theory. The new ultrashort ps and fs pulses will allow us to observe dynamics and the time-dependent correlation functions at earlier times than 10^{-12}s, when there is expected to be a deviation from hydrodynamic behaviour. To-date, most studies conclude that a simple Debye diffusion model works remarkably well for larger size molecules. However, unless viscosity effects are studied by changing T or P and not always the solvent, which frequently introduces new notions of shape, polarizability, and structure for small molecule systems, it is questionable whether the fundamental questions of molecular motion and angular correlation times can be unambiguously addressed. On the other hand "cage effects," introduced earlier through a multidimensional H-bond solvent, are not restricted to that case and are very important to understand for librational or translational motions of small molecules in pure liquids or solute molecules diluted in large molecule solvents. The persistance of the cage may govern the dynamics of the tagged particle. What are then the pertinant correlation times? Finally, two photon fluorescence from a well-defined optically prepared state in a molecule of specified symmetry may prove to be an interesting test of the P_3 correlations for the orientational probability density.

In concluding this section, we observe that dynamics of molecules on surfaces and in micelles can be studied this way.

Laser-induced fluorescence also offers the possibility of looking at the solvation structure in its isolated form, that is, short-range order without the continuum contribution. This has attracted keen interest in supersonic molecular beam studies of van der Waals complexes (27-28). The successive additions of n solvating molecules B form a solvated species AB_n. Spectral shifts and lineshape changes in the absorption of A as it becomes AB_n, or in the laser-induced fluorescence decay of those consecutively solvated species in given v,v' states, can be monitored. While these

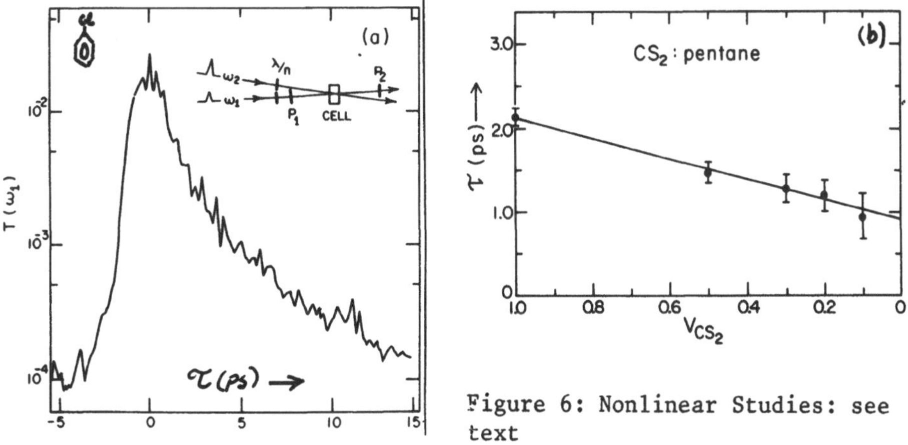

Figure 6: Nonlinear Studies: see text

beam data undoubtedly will assist in the quantitative development of ideas of solvation structures, they represent a special case with respect to studies of dynamics in liquids at warmer temperatures, since there the transitory clusters of molecules can exercise many more degrees of motional freedom, and high frequency fluctuations play a major role in microscopic behaviour. However, those changes in vibrational or vibronic relaxation that are observed are indeed diagnostic of the partially solvated species. Figure 5 shows the fluorescence excitation spectrum of Indole $(C_2H_6)_n$, where peak a refers to n=1, and b to n=2. (The parent peak of bare indole is the one off-scale) These data are drawn from a comprehensive study of indole and many organic and rare gas solvation partners (29). In the excited singlet state probed, only a small change in dipole moment occurs ($|\Delta\mu|=0.14 \pm 0.05D$). While good agreement is seen with the theory of solvent-induced dispersion force interactions, when the spectral shifts in the rare gas complexes are correlated with polarizability, the more subtle details of the interactions with *polyatomic* complexing partners correlate well with gas phase basicity (29). From this work it will be possible to examine the specific signatures of liquid or bulk behaviour (collective effects, band structure), so the onset may be correlated with structural information. While one could argue that liquid behaviour is statistical, it is really *a quest-*

ion of timescales, that is as to whether or not sufficient time elapses during the event under study for all phase space to be sampled.

5. MOLECULAR RELAXATION STUDIED IN THE TIME AND FREQUENCY DOMAIN

Time and frequency domains are rigorously linked through a fourier transform (FT) operation and hence information gained in one domain can be readily interconverted into the other, as shown below for the polarization and light intensity in the medium.

$$P(r,t) = \int P(\omega,k)\, e^{i(k\cdot r-\omega t)} d\omega dk; \quad I(t) = \int I(\omega)\, e^{i\omega t} dt \qquad [9]$$

Theories of light scattering spectra have been developed to explicitly link the lineshapes to the FT of autocorrelation functions, which are chosen as suitable linear combinations of elements of the dielectric tensor (6). Arguments based on separation of timescales permit one to associate the rapidly varying fluctions with interaction-induced phenomena, and the more slowly varying dynamic variables are usually assigned to molecular reorientation and translation. Single particle correlation functions are predicted to have different time behaviour than a two particle correlation functions, and furthermore each experimental determination must be intepreted through the correct-rank (ℓ) of the tensorial interaction. Given all these considerations, why are there now advantages to addressing ultrafast molecular motion directly in the ps and fs domain, when information has been long if indirectly available via FT techniques? While other authors in this volume may present specific arguments for a given technique, I want to merely stimulate some thoughts or points which have important and general significance to all experimental approaches in briefly answering the above question.

In liquids and dense gases where collisions, intramolecular molecular motions and energy relaxation occur on the picosecond timescales, spectroscopic lineshape studies in the frequency domain were for a long time the principle source of dynamical information on the *equilibrium state* of manybody systems. These interpretations were based on the scattering of incident radiation as a consequence of molecular motion such as vibration, rotation and translation. Spectroscopic lineshape analyses were intepreted through arguments based on the fluctuation-dissipation theorem and *linear response* theory (9,10). In generating details of the dynamics of molecules, this approach relies on FT techniques, but the statistical physics depends on the fact that the radiation probe is only *weakly coupled* to the system. If the pertubation does not disturb the system from its equilibrium properties, then linear response theory allows one to evaluate the response in terms of the time correlation functions (TCF) of the *equilibrium* state. Since each spectroscopic technique probes the expectation value of *different* observables and often also different k vectors, it is important to regard all as complementary experimental approaches to the overall problem of molecular motion. Lineshapes analyses, although indirect, are still very accurate sources of dynamical information for those motions contributing towards the central spectral components. *The information from the spectral wings is more complex.* If indeed one observes that a single Lorenztian line shape fits all the data, and thus a single mechanism and exponential relaxation time is deduced, then the spectral wings present little problem. If however there are several components to be fitted to the lineshape, then the non-Lorenztian lineshape has to be fitted by an algorithm that is not model-free. In other words, mechanistic assumptions are built in to the lineshape fitting parameters and thus become a limit to obtaining fundamental answers. If the spectrum is not homogeneously broadened, lineshape analyses are once again difficult to intepret as in any spectroscopic problem.

Theoretically, correlations between rotational and vibrational motion are usually neglected (9) and the overall correlation function thus becomes the product of two individual correlation functions describing these degrees of freedom. Of course such an approach is not model-free since one is assuming the moments of inertia are independent of vibration and that the vibrations are independent of interactions leading to angular-dependent potentials. Increasingly this separation is being questioned. Experimental evidence (30) shows that even in dilute solutions, where single particle rotation is anticipated, depolarized Rayleigh and Raman scattering both second rank rotational tensor, $\ell=2$ do not reveal the same orientation times for CH_3CN in CCl_4. Theoretical studies have shown significant lineshape differences for IR and polarized and depolarized Raman scattering even in the absence of orientational correlations. While other workers have questioned precisely what correlation times are obtained for comparison with theory, and whether certain TCF are indeed even experimentally accessible (31). Clearly more time-domain studies are required to evaluate the rotational-vibration coupling contributions which if not understood or recognized can lead to misintepretation of spectral lineshapes. Further examples of lineshape complications are corrections for hot bands, overtones or isotope bands and perturbations, or refractive index changes through the band, any of which might well lead to a shift of 1 cm^{-1} in IR band going from 1020 to 1021 cm^{-1} peak (30). That very small shift is sufficient to lead to oscillations in the correlation function due to the appearance of an imaginary part. The wings are particularly sensitive to such effects, since the rotational wings and interaction-induced wings often display comparable spectral density in small molecules. While separation of timescale arguments are frequently used to resolve two components unambiguously, it is also clear that concellation effects can be masked by the fact the information is not directly time-resolved.

Such cancellation effects have been seen for higher-body interactions in CS_2 (9b,32).

Thus the picosecond and femtosecond time domain can reveal uniquely this *separation of time-scale argument*, if valid for a given atomic or molecular liquid. In principle the "spectral wings" component will appear first in a time-resolved experiment, before the evolution of the central components, thus removing model-based fitting parameters that are often invoked in frequency domain analyses. Furthermore, the possibility of observing strong field effects and the time-evolution or dynamical variables of a system *perturbed significantly from equilibrium* are major goals not only of our studies but for nonlinear response theory, without which the general dynamical description of molecular liquids will remain incomplete. In any interaction between the *intense* optical field and the molecular liquid, the tenets of linear response theory and links to equilibrium properties clearly must be reexamined and tested, not merely assumed to still apply. Also, with nonlinear spectroscopy we will be selectively examining the higher rank tensor interactions and the TCF relaxations, which are predicted (in a diffusive model) to decay more rapidly by factors of $\ell(\ell+1)$. Such decays in small molecules, if present, will be on the time scale of interaction-induced phenomena and require careful study.

PART II NONLINEAR LASER SPECTROSCOPY AND MOLECULAR DYNAMICS

The interaction of intense, ultrashort laser pulses with an atomic or molecular system can generate a series of transient, nonlinear responses (11-13) that depend on the electronic and molecular structure (σ,π electrons) of the medium. These nonlinear interactions are frequency and phase dependent, and sometimes several effects can occur simultaneously. Dielectric breakdown is catastrophic at very high fields $\geq 10^{11}$ W.cm^{-2} but nonlinear events are studied typically at $\leq 10^{9}$ W.cm^{-2}. As an intense picosecond laser pulse propagates through the medium, one set of nonlinear responses can be envisioned as intensity and frequency changes in the refractive index, n^{λ}, which give rise to self-focusing, self-phase modulation and other effects. The electric-dipole coupling is used for the field-matter interaction and Maxwell's equations to describe propagation of the electromagnetic field. Higher order, field-dependent ($\vec{E}$) terms are added to the nonlinear optics expression for the macroscopic polarization (P) of the material [10], where $\chi^{(n)}$ is the nth order susceptibility, and $n^2=(1+4\pi\chi^{(1)})$. Alternatively, atomic or molecular polarizabilities can be utilised for a molecular description, as in [10a], but these conventionally refer to only electronic not nuclear contributions.

$$P = \chi_{ij}^{(1)}\vec{E} + \chi_{ijk}^{(2)}\vec{E}\vec{E} + \chi_{ijkl}^{(3)}\vec{E}\vec{E}\vec{E}\ldots \qquad \text{macro} \qquad [10]$$

$$\Delta\mu = \alpha\vec{E} + \beta\vec{E}.\vec{E} + \gamma\vec{E}.\vec{E}.\vec{E} \qquad [10a]$$
(linear) (β, γ hyper polarizabilities)

Since λ is large with respect to molecular dimensions, we can assume that the macroscopic polarization (P) depends locally on $E(\mathbf{r},t)$. The real and imaginary parts of $\chi^{(n)}$ are related by Kramers-Konig relationships. Symmetry arguments reduce the 48 contributions to $\chi^{(3)}$ to a few irreducible components of the 4th rank tensor (13). Second harmonic generation arises from the second term, while third harmonic generation (THG), stimulated Raman scattering SRS, CARS, phase conjugation and four wave mixing (4WM) processes are among those which originate in the third term. It is important to realize that, due to local field effects, in an N body system $\chi^{(1)}/N \neq \alpha$.

The dynamical nature of nonlinear responses (33) can be better emphasized by writing down the explicit time-dependent equations describing a given interaction, such as the transient stimulated Raman scattering (34,35) or the optical Kerr effect (36), which reveal the time-dependent memory of the initial coherence, phase memory and orientational alignment imposed on the system during that ps interaction. In concluding this article, we focus on the time-dependence of the nonlinear quantum electronic responses of molecular systems to intense ps and fs polarized laser fields, and the role of the molecular dynamics of the liquid in the persistence of the memory of the nonlinear interaction.

1. $\chi^{(3)}$ spectroscopy

We now described some nonlinear experiments which reveal molecular dynamics and the reasons why we have chosen CS_2 as a reference molecule in the solvents studied. Usually one considers that any two or more particles interact through a field that involves anisotropic overlap, dispersion, permanent and induced multipole interactions, including those via hyper polarizabilities when nonlinear terms are included. It is of paramount importance to *minimize the number of interactions probed* in a given experiment, otherwise there will be too many contributions to the spectrum, and time-dependence of the signal will be difficult to deconvolute with any realism based in the physics of the system. If the molecule posses elements of symmetry, the number of independent coefficients required to describe the mean orientation in space is accordingly reduced. Hence selecting a linear triatomic such as CS_2 or CO_2 of $D_{\infty h}$ symmetry, with only polarizability and quadrupole contributions to the interaction tensor, limits the terms involved. (Of course collision-induced effects are always present in the liquid phase.) Furthermore, in the case of CS_2 it appears as though the *hard core model*, empasing short range effects, effectively reproduces those features of the potential that also seem to be important to the local equilibrium liquid structures (32). Liquid CS_2 has no strong electrostatic multipole moments, so this is perhaps not surprising. Only at lower temperatures is the quadrupole moment expected to become structually influential in a major way. The theory and simulation reproduces the equilibrium g(r) observed in scattering experiments satisfactorily, and in computer simulations CS_2 has been studied (37) using a hard core potential to calculate g_2, which has been deduced from light scattering and Cotton-Mouton experiments (32). While its linear spectroscopy is known, CS_2 is most familiar to ps laser studies as a nonlinear optical system, whose fast relaxation responses have been extensively used in optical Kerr shutters since it has one of the highest nonlinear refractive indices n_2 known, 2×10^{-11} esu (33). Thus the equilibrium properties of CS_2 liquid are well documented.

We designed nonlinear laser spectroscopy experiments to probe the *time-dependent* responses of CS_2 as it was driven far from eqilibrium and as the nature of its nearest neighbour interaction was altered. We chose to study CS_2 and binary systems with alkanes, alcohols and aromatic hydrocarbons, since they appear the most ubiquitous solvents in chemical reactions, and yet can be classified into liquids of varying degrees of interaction for the purposes of our study. Since much of this work has been written up in detail elsewhere (Kalpouzos et al., in ref. (3); 5,38,39,40) we will only give the salient details here. An account of all other work on CS_2 at subpicosecond times from this and other laboratories is given in (5).

We have been studying the optical Kerr effect through both intracavity measurements and in the more conventional external pump-probe configuration. The passage of an intense picosecond polarised laser pulse (ω_2) pulse through a nonlinear medium induces a transient birefringence $\delta n(t)$, which in a liquid will ultimately relax over a timescale τ correlated to the molecular reorientational dynamics in the fluid. A second polarized, but attenuated laser probe pulse (ω_1) will experience a polarization rotation as it passes through the induced birefringence, over pathlength ℓ. The subsequent phase shift $\delta\phi$ appears in the transmission function (T) of ω_1 directed to pass through the sample via two linear polarizers (P_1,P_2), crossed for maximum extinction of ω_1 in the zero pump field case. Figure 6a illustrates the typical response in chlorobenzene at 300K, showing at least two components in the relaxation following 300 fs pump-probe at $\omega_1=\omega_2=615$ nm (22). The explicit relationships are given as follows, where E_2 is the applied laser field and local field effects are neglected. The nonlinear refractive index n_2 can be separated into an electronic and molecular component, n_2^{elec} and n_2^{mol} respectively. The overall change in n, the *total* refractive index, is given by:

$$n = n_o + n_2 \langle E^2 \rangle ;\; n_2 = n_2^{elec} + n_2^{mol} \qquad [11a]$$

$$dn = \tfrac{1}{2} n_2 <E>^2 \sim \frac{1}{n} \chi^{(3)} <E^2>; \quad \delta\phi(t) = \frac{2\pi}{\lambda_1} \ell \, \delta n(t) \qquad [11b]$$

$$\delta n(t) = \underset{\text{fast}}{n_2^{elec} <E^2_{\omega_2}>} + \frac{n_2^{mol}}{\tau_{mol}} \underset{\text{slow}}{\int_{-\infty}^{t} <E_{\omega_2}^2 (t')> e^{-(t-t')/\tau_r} dt'} \qquad [12]$$

$$T_{\omega_1}(t') \sim \int_{-\infty}^{+\infty} I_{\omega_1} (t-t') \sin^2 \left(\frac{\delta\phi(t)}{2}\right) dt \qquad [13]$$

If the phase shift is small, then $\delta\phi(t)$ is directly proportional to $<E^2>$ or the intensity I at ω_2. Thus we can write

$$T_{\omega_1}(t') \propto \int_{-\infty}^{+\infty} I_{\omega_1} (t-t') I_{\omega_2}^2(t)dt \qquad [14]$$

In the event the pulse duration t_p is longer than the molecular relaxation time, the expression in [14] corresponds to a third-order correlation function $G_0^{(3)}(\tau)$. Note also from equation [13] that the transient birefringence has both an instantaneous and a slower response to the picosecond pump pulse, since n_2^{elec} as the electronic component of the nonlinear susceptibility is expected to follow the laser pulse profile but the molecular component n_2^{mol}, a combination of vibrational and orientational parts, carries with it a time-dependence characteristic of each molecular system. It is through the latter that information on orientational correlations can be deduced (39).

We have discussed previously (5) the possibility that a comparison of the *rise time* of the Kerr transients would lead to quantitative answers concerning the influence of strong fields on the orientational motion of a dipolar molecule, in the event *both* the hyperpolarizability (or $\chi^{(3)}$ interactions) and dipole moment contribute to the time-dependent orientational alignment with respect to $\vec{E}$. Asymmetry in the optical Kerr transients would be a consequence of P_1, P_2 *and* P_4 terms contributing to the rise of the induced birefringence under the applied field E but only P_4 and P_2 relaxation terms under field-free diffusion. The simultaneous dipole and polarizability decay functions depend to some extent on the relative short-time momentum correlations and long-time positional correlations in the liquid. We need to know the extent in time of the former, and in studying the Kerr transients information can be obtained on this effect. Recently both theoretical and experimental papers have appeared on these points, for dc Kerr measurements on large polymer systems have indeed shown this to be the case (41). This provides another example of how the expectations of quantum electronics, based largely on *non-interacting* particles, must be modified by the response of the material under circumstances where intermolecular forces and torques can modulate both the magnitude and dynamics of $\chi^{(3)}$ phenomena. For this reason, the corollary holds *that* $\chi^{(3)}$ *phenomena are a sensitive measure of molecular interactions and dynamics in condensed media.* Furthermore, by selecting different pump and probe polarizations, it should be possible to deduce the relationships between different $\chi^{(3)}_{ijkl}$ components of the nonlinear susceptibility tensor (5,38). Theoretically we have developed a formalism to deduce the orientational correlations that emerge through $\chi^{(3)}$ 4WM interactions that is complementary to, but not identical with, depolarized Rayleigh (DRS) light scattering (39). **The Kerr and Rayleigh experiments** operate through different interactions since in the Kerr effect we probe the collective orientation density for only forward scattering (or zero scattering vector) in contrast to DRS, which contains Fourier components for all scattering angles. The TCF appropriate to $\chi^{(3)}$ fourth rank tensor interactions should reveal a P_4 description of the orientational probability density and hence logically g_4/j_4 ratio of static to dynamic correlation functions. DRS of course also reveals interaction-induced effects in the wings and carries rotational-translational coupling as well, which is frozen-out in fs data. If diffusional motion is involved, the P_4 component

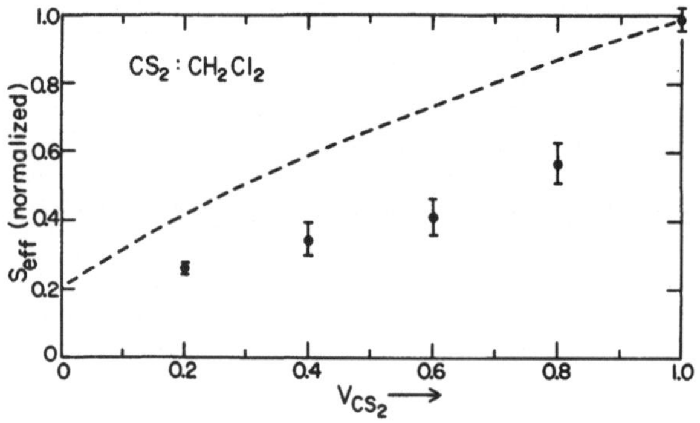

Figure 6(c): 4WM scattered intensity $I_{\omega 4}^{\frac{1}{2}} \equiv S_{eff} \equiv \chi^{(3)} n^{-2}$ for CS_2 in CH_2Cl_2. Dotted line shows calculated value, and data are ♦ from experiments described in ref. (5,39,40).

should delay 3.3 times $\{\ell(\ell+1)\}$ faster than the P_2 component in DRS but the relative weights of the P_4 and P_2 and the role of 3 and 4-particle correlations (to which P_4 is more sensitive) have yet to be established. Clearly these are goals for ps and fs spectroscopy and ones we are addressing in our laboratory via a range of 4WM processes in molecular liquids of carefully chosen molecular symmetries eg. CCl_4. Finally, the *strong-field* effects in Kerr and 4WM experiments beg the question of linear response theory, and further work needs to be completed on field-driven responses. Interestingly enough, consistently *longer* times (1.7-2.3ps) for τ_{rot} in CS_2 have been observed in $\chi^{(3)}$ experiments compared to DRS experiments (1.5-1.7 ps) in contrast to the expected *faster* decays;

these details have been fully discussed in (5). Clearly CS_2 is still a very interesting test case, which is living up to its reputation as the hydrogen atom of nonlinear optics and presents challenges in molecular dynamics for fs nonlinear laser spectroscopy. In order to separate out P_4 effects from interaction-induced responses which occur at times up to generally ~500 fs from computer simulations, it is important to choose molecules (such as the series benzene, chlorobenzene and nitrobenzene) which have interesting interactions but rotate in general a little slower. Otherwise the important goal of observing *a complete set* of P_1, P_2, P_3, P_4 terms for a full orientational probability density (ℓ=1,2,3,4) will not be met in practice, if indeed the $\{\ell(\ell+1)\}$ relationship between P_1 terms and their decay holds as in small step diffusion (9,10).

2. Phase conjugation

Optical Kerr experiments rely on field-induced orientation and are not restricted to optically coherent interactions. In phase conjugation 4WM experiments, we probe the $P^{(3)}_{\omega_4}$ term of eq. [1] which is strictly a coherent interactions (40). These are rotational dephasing experiments in the case of CS_2. We *would not expect to see collision-induced effects in CS_2 (such as observed as spectral wings in the depolarized Rayleigh experiments) in these 4WM coherently driven phase conjugation experiments unless the collisions were very efficient at disrupting phase memory.* Evidence from vibrational dephasing suggests this is not the case, since T_2=20 ±1ps for the ν_1 Raman mode at 656.5 cm^{-1} (32,42). (For details of the phase conjugation or "real-time holography" experiments, see ref. 5 and 40.) Thus we can describe the mechanism probed via phase conjugation in CS_2 as a pure reorientational dephasing time, which at high power densities will evolve and decay in time as a coherent optical Kerr effect. That is, the initially prepared phase coherence in the ensemble will be accompanied by a high degree of orientational alignment. Hence the relaxation times may change as a function of pump intensity if the molecule-molecule relaxation-interactions are shape-dependent, e.g., 2-body and higher angular correlations are present; such experiments are in progress as are computer simulations of this effect (39).

When CS_2 is diluted a nonpolar liquid of vanishingly small $\chi^{(3)}$ such as an alkane (RH), the spectrum of ω_4 changes lineshape to reflect the concomitant changes in the microscopic environment, namely the local intermolecular interactions and thus the reorientational motions. The viscosities of CS_2 (0.36 cp) and n-pentane (0.24 cp) are not very dissimilar at room temperature, and the orientational relaxation time of n-pentane was recently measured to be ~2.2 ps (43), similar to pure CS_2. The CS_2-alkane interactions and structure however are quite different from the pure liquids. The relaxation time (derived from the spectral linewidths of ω_4) is plotted as a function of volume fraction (V) of CS_2 in figure 6b. Even in the most dilute solutions, less than 10% of the signal is solvent background and the dominant source of $P^{(3)}$ are the now isolated CS_2 molecules. Upon substitution of alkane of CS_2 molecules in the solvation shell the local density, volume and g(r) parameters must shift, which is why all data are normalized to a volume fraction rather than number density. For each dilution (0.8<V<0.2) there will be a statistical distribution of n-pentane molecules in the *inner* solvation shell, and this could lead to a distribution of relaxation times within the full ensemble; thus these data represent the most probable times associated with rotation in the first shell solvation structure. Since CS_2 is governed by hardcore, short range interactions, changes in and beyond the second solvation shell can be neglected as a primary influence on the dynamics. At infinite dilution, $\tau \sim 0.9$ ps. We conclude from Figure 6b that the *changing microscopic interactions around CS_2 and the subsequent changes in the orientational relaxation times can be revealed through these 4WM phase conjugation experiments with subpicosecond resolution.* Furthermore the nature of the correlation function changes from the initial many body CS_2-CS_2 correlation function $\sum_{i,j} \langle A_i(0) A_j(t) \rangle$ or *collective* orientational times to the ensemble averaged *single particle* time $\sum \langle A_i(0) A_i(t) \rangle$, corresponding to orientational dephasing of CS_2 in n-pentane.

How can one separate, at least to a first approximation, dynamical effects from structural effects? In an *non-interacting model*, the nonlinear polarizability of CS_2 will not be influenced by its solvation structures and thus no amplitude modulation of $\chi^{(3)}$ is expected. *Strong interactions* will enhance or dimish $\chi^{(3)}$. The assumption of pair-additivity underlies all calculations of intermolecular potentials. Similarly, if the dynamical interactions probed by 4WM are merely angle-dependent pair interactions leading to an effective $\chi^{(3)}$, which is the algebraic sum of the individual $\chi^{(3)}$ (CS_2) and $\chi^{(3)}$ (solvent) components, then a $\chi^{(3)}$.... can be calculated for CS_2 in any liquid. The actual 4WM signals observed when plotted for the various solutions and corrected for refractive index differences in the solutions, Fig. 6c, quite clearly deviate from the non-interacting model. The local dipolar interactions and orientational correlations have substantially modified the nonlinear polarizability of CS_2 and thus the *molecule-field* interactions too. Local field corrections, when incorporated following several different prescriptions, do not alter the degree of deviation of the data from the non-interacting case, because both the calculated and observed $\chi^{(3)}$ are similarly scaled. Once again, ps and fs 4WM experiments currently in progress are determining the time evolution and decay of the conjugate wave and the amplitude of different contributions to $\chi^{(3)}$, namely electronic, vibrational and orientational, to determine which part of the "architecture" of $\chi^{(3)}$ has been modified by the *molecule-molecule* CS_2-solvent interactions. These data will also reveal the dephasing times for specific dipole, induced-dipole and quadrupole interactions for

TCF analysis (39).

Acknowledgments The author gratefully acknowledges the Fellowship support of the Guggenheim Foundation and the continuing research support of the U.S. Office of Naval Research and the Natural Sciences and Engineering Research Council of Canada for the studies involved and in the preparation of this article for the NATO Advanced Study Institute in Florence concluding in July, 1983.

References

1. *Ultrashort Laser Pulses* ed. D.J. Bradley, G.B. Porter and M. Key, The Royal Society, London (1980).
2. *Picosecond Phenomena*, E.P. Ippen, Special Issue of IEEE JQE-19 (1983).
3. *Picosecond Lasers and Applications* ed. L. Goldberg, SPIE vol. 322 (1982). A survey of many recent advances from different laboratories is contained therein.
4. R.L. Fork, C.V. Shank, R. Yen and C. Hirlimann, IEEE J. QE-19 500 (1983) and references therein.
5. G.A. Kenney-Wallace *Nonlinear Optical Spectroscopy and Molecular Dynamics in Liquids* in *Applications of Picosecond Spectroscopy* ed. K. B. Eisenthal, D. Reidel Publishing Company, 1984, p. 139; and in ref. 1, p. 309-319.
6. D. Kivelson and P.A. Madden, Ann. Rev. Phys. Chem. 31, 523 (1980).
7. For example, see reviews by R. Kapral, Adv. Chem. Phys. 48, 71 (1981); S. Adelman, Adv. Chem. Phys. 44, 143 (198)
8. D.G. Truhlar, W.L. Hase and J.T. Hynes, J. Phys. Chem. 87, 2664 (1983).
9a) W.A. Steele, Adv. Chem. Physics 34, 1 (1976) and in this NATO volume; 9b) P. Madeen, this volume.
10. B. Berne in *Advanced Treatise on Physical Chemistry*, vol V111B ed. H. Eyring, D. Henderson and W. Jost, (Academic Press 1971)
11. N. Bloembergen, *Nonlinear Optics*, Benjamin, (N.Y. 1965)
12. P.N. Butcher, *Nonlinear Optical Phenomena*, text publication from Ohio State University (Columbus Ohio)
13. R. Hellwarth, Prog. in Quantum Elec. 5, 1-68 (1977).
14. E.B. Treacy, IEEE JQE-5, 454 (1969).
15. H. Nakatsuka, D. Grischkowsky and A.C. Balant, Phys. Rev. Lett. 47, 910 (1981).
16. B. Nikolaus and D. Grischkowsky, App. Phys. Lett. 43, 2281(1983).
17. G.R. Fleming, Adv. Chem. Phys. 49, 1 (1982) and references therein.
18a) T. Sizer, J. Kafka, I. Duling, C. Gabel and G. Mourou, IEEE JQE-19, 506 (1983) and references therein. b)A.M. Johnson and N.M. Simpson, Optics Letters 8,554 (1983).
19. C. Kalpouzos, J. Lobin, V. Mizrahi and G.A. Kenney-Wallace, to be published.
20. E. Quitevis and G.A. Kenney-Wallace, Rev. of Sci. Inst. (in

press 1983)
21. K. Sala, G.A. Kenney-Wallace and G.E. Hall, IEEE JQE-16, 990 (1980).
22. E. Quitevis, P.M. Kroger, C. Kalpouzos, G.A. Kenney-Wallace Can. J. Chem. 61, 975 (1983).
23a) H.E. Lessing and A. Von Jena, Appl. Phys. 19, 131 (1979); Chem. Phys. 60 245 (1979), b) K.G. Spears and L.E. Cramer, Chem. Phys. 30, 1 (1978).
24. Stephen A. Rice and G.A. Kenney-Wallace, Chem. Phys. 47, 161 (1980).
25a) J.T. Hynes, Ann. Rev. Phys. Chem. 28, 301 (1977). b) J.T. Hynes, R. Kapral and M. Weinberg, J. Chem. Phys. 70, 1456 (1979). c) R. Zwanzig, J. Chem. Phys. 68, 4325 (1978).
26. D. Waldeck, A.J. Cross, D.B. McDonald and G.R. Fleming, J. Chem. Phys. 74, 3381.
27. *van der Waals Molecules*, Faraday Disc. no. 73 (1982).
28a) A. Amirav, U. Even, J. Jortner, J. Chem. Phys. 75, 2489 (1981), b) T.R. Hays, W. Henke, H.L. Selzle and E.W. Schlag Chem. Phys. Lett. 77, 19 (1981).
29. J. Hager and S.C. Wallace, J. Phys. Chem. 87, 2121 (1983) and to be published; M.Smith, J. Hager and S.C. Wallace, J. Phys. Chem. (in press 1983).
30 a) J. Yarwood, this volume for remarks on precision of lineshape data and arguments in present literature. b) J. Gompf, H. Versmold, and H. Langer, Ber. Bunsenges. Phys. Chem. 86, 1114 (1982).
31. R. Lynden-Bell, this volume and references therein.
32. T.I. Cox and P.A. Madden, Mol. Phys. 39, 1487 (1980); ibid, Mol. Phys. 38, 1539 (1979).
33. D.H. Auston reviews these phenomena in *Ultrashort Light Pulses*, ed. S. Shapiro (Springer-Verlag, 1977).
34. A. Laubereau and W. Kaiser, Rev. Modern Phys. 50, 608 (1978).
35. W. Zinth, H-J. Polland, A. Laubereau and K. Kaiser, Appl. Phys. B26 77 (1981).
36. K. Sala and M.C. Richardson, Phys. Rev. A12, 1036 (1975).
37. W.B. Street and D.J. Tildesley, in *Structure and Motion in Molecular Liquids*, Faraday Disc. No. 66, Chem. Soc. (London 1978); R.W. Impey, P.A. Madden and D.J. Tildesley, Mol. Phys. 44, 1319 (1981).
38. J. Etchepare, G.A. Kenney-Wallace, G. Grillon, A. Migus and J-P Chambaret, IEEE JQE-18, 1826 (1982).
39. M. Golombok, G.A. Kenney-Wallace and S.C. Wallace (in press, 1984) and to be published.
40. G.A. Kenney-Wallace and S.C. Wallace, IEEE JQE-19, 719 (1983)
41a) R.H. Cole, this volume b) G. Williams, this volume.
42. J.P. Heritage, App. Phys. Lett. 34, 470 (1979).
43. G.D. Patterson and P.J. Carroll, J. Chem. Phys. 76, 4356 (1982).

EXPERIMENTAL DETERMINATION OF CORRELATION FUNCTIONS FROM INFRARED AND RAMAN SPECTRA

J. Yarwood

Department of Chemistry, University of
Durham, Durham City, DH1 3LE, U.K.

The object of this chapter is to present an overview of the determination of the reorientational and vibrational correlation functions from infrared and Raman spectroscopic band profiles. I shall concentrate on the objectives of such measurements, on the experimental and computational difficulties and on the possible shortcomings of the rather well-established [1-4] methodology.

I. OBJECTIVES

The accurate determination of time correlation (relaxation) functions for a variety of molecular motions and relaxation processes is of crucial importance for detailed study of the liquid state. Such response functions [1,5] are currently used for a wide variety of purposes.

(i) To provide information on the rates of (a) rotational and translational motions and (b) energy and phase relaxation, in molecular liquids.

(ii) To probe the effects of environment (in particular intermolecular torques or intermolecular geometry) on the motions and relaxation modes of fluid phase molecules.

(iii) To allow a check to be made on the intermolecular potentials used for molecular dynamics simulations (see chapters in this book on M.D. methods and results).

(iv) To aid theoreticians developing models for molecular rotation or vibrational relaxation - since the essential features of the correlation function (especially at short times) must eventually be reproduced by such models.

In my opinion, combined use of experimental, theoretical and

A. J. Barnes et al. (eds.), Molecular Liquids - Dynamics and Interactions, 357–382.

computer simulation work must hold the key to making real progress towards understanding molecular liquids.

II. METHODOLOGY

The problem is conveniently set in context by reference to figures 1 and 2 which show the IR and Raman bands of ν_3 of CH_3I

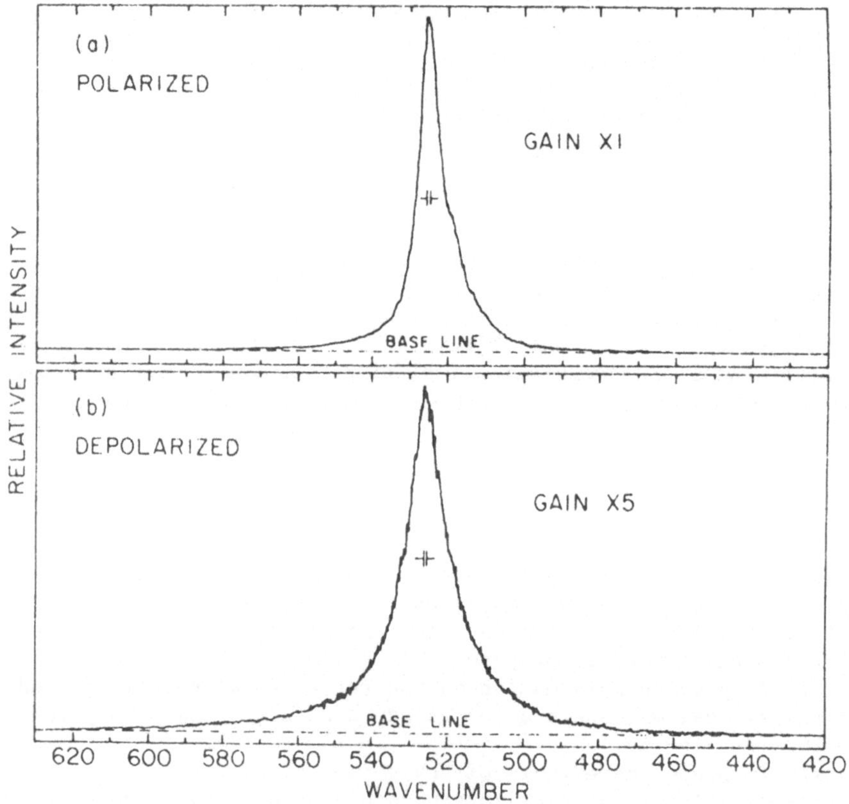

Fig. 1. Raman $I_{VV}(\omega)$ and $I_{VH}(\omega)$ spectra for ν_3 of liquid CH_3I (Reproduced by permission from J. Chem. Phys., 61, 346 (1974)).

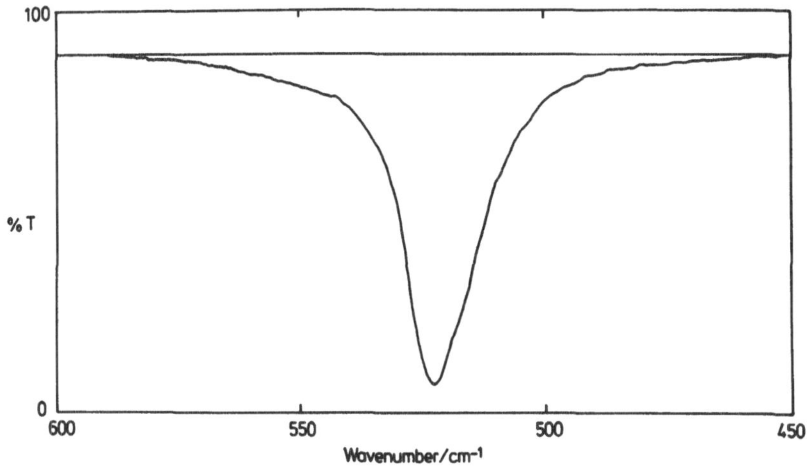

Fig. 2. Infrared absorption spectrum $I_{IR}(\omega)$ for ν_3 of liquid CH_3I.

- the ν(C-I) stretching mode. Within the limits of the experimental error (see below) we can derive three band profiles $I_{VV}(\bar{\nu})$, $I_{VH}(\bar{\nu})$ and $I_{IR}(\bar{\nu})$ from spectra such as these, between frequencies (determined by the S/N ratio and any other overlapping bands) ω_1 and ω_2 which defined the 'band' centred at ω_o. From these spectra it is usual (by Fourier Transformation) to compute three correlation functions. The three spectral profiles are thus written as [1 - 4].

$$I_{iso}(\omega) = I_{VV}(\omega) - \frac{4}{3}I_{VH}(\omega) = \frac{1}{2\pi}\left\{\frac{\omega-\omega_o}{\omega+\omega_o}\right\}^4 \left|\frac{\partial\alpha}{\partial q_v}\right|^2 \int_o^{t_m} \langle q_v(o).q_v(t)\rangle_{iso} \times \exp(i\omega t)dt \qquad (2.1)$$

$$I_{aniso}(\omega) = I_{VH}(\omega) = \frac{1}{2\pi}\left\{\frac{\omega-\omega_o}{\omega+\omega_o}\right\}^4 \int_o^{t_m} \langle Tr\beta'(o).\beta'(t)\rangle\langle q_v(o).q_v(t)\rangle_{aniso} \times \exp(i\omega t)dt \qquad (2.2)$$

and

$$I_{IR}(\omega) = \frac{B}{2\pi}\int_{o_1}^{t_m} \langle \underline{u}_i(o).\underline{u}_j(t)\rangle\langle q_v(o).q_v(t)\rangle_{IR} \exp(i\omega t)dt \qquad (2.3)$$

[$\omega = 2\pi\bar{\nu}c$ and B is a constant].

In these expressions $\underline{u}$ is a unit vector along the direction of the transition dipole moment $(\partial\vec{\mu}/\partial q_v)$ for the vibrational mode concerned and β' is the anistropic part of the transition polarisability tensor $\alpha'_{jk}(t)(\equiv|\partial\alpha_{jk}(t)/\partial q_v(t)|q_v{=}o)$ which can be conveniently divided into an isotropic part

$\alpha'(t) = \frac{1}{3}\sum \alpha'_{jj}(t)$ and an anisotropic part $\beta'_{jk}(t) = \alpha'_{jk}(t)-\alpha'(t)\delta_{jk}$.

It is usual to normalise [1-4,6] the correlation functions using the integrated band areas in such a way that we have,

$$\frac{\langle q_v(o).q_v(t)\rangle_{iso}}{\langle q_v(o)q_v(o)\rangle} \equiv C_{vib}^{iso}(t) = \frac{\int_{\omega_1}^{\omega_2} I_{iso}(\omega)\exp(-i\omega t)d\omega}{\int_{\omega_1}^{\omega_2} I_{iso}(\omega)d\omega} \qquad (2.4)$$

$$\frac{\langle q_v(o).q_v(t)\rangle_{aniso}}{\langle q_v(o).q_v(o)\rangle}\;\frac{\langle Tr\beta'(o).\beta'(t)\rangle}{\langle Tr\beta'^2(o)\rangle} \equiv C_{vib}^{aniso}(t)C_{2R}^{T}(t) =$$

$$= \frac{\int_{\omega_1}^{\omega_2} I_{aniso}(\omega)\exp(-i\omega t)d\omega}{\int_{\omega_1}^{\omega_2} I_{aniso}(\omega)d\omega} \qquad (2.5)$$

and

$$\frac{\langle q_v(o)q_v(t)\rangle_{IR}}{\langle q_v(o)q_v(o)\rangle}\;\frac{\langle \underline{u}(o).u(t)\rangle}{\langle \underline{u}(o).\underline{u}(o)\rangle} = C_{Vib}^{IR}(t)C_{1R}^{T}(t) =$$

$$= \frac{\int_{\omega_1}^{\omega_2} I_{IR}(\omega)\exp(-i\omega t)d\omega}{\int_{\omega_1}^{\omega_2} I_{IR}(\omega)d\omega} \qquad (2.6)$$

ensuring that C(t) = 1 at t = 0.

It is important to notice that the simplification implied in equations (1.2) and (1.3), by, for example

$$\langle \underline{u}_i(o)q_{vi}(o).u_j(t)q_{vj}(t)\rangle \cong \langle u_i(o).u_j(t)\rangle\langle q_{vi}(o).q_{vj}(t)\rangle \quad (2.7)$$

is only valid if the vibrational and reorientation relaxation processes are statistically independent. For polyatomic molecules, where the two processes have similar time constants, this is almost certainly not the case. And, indeed, there is experimental evidence (for CH_3I [7] - see later) that vibration rotation coupling is important for some bands. Further, it should be noticed that, in general, all the correlation functions contain both 'self' and 'distinct' parts [1-4]. For example, the dipole reorientation function may be written as,

$$\langle \underline{u}(o).\underline{u}(t)\rangle = \underbrace{\langle \underline{u}_i(o).\underline{u}_i(t)\rangle}_{\text{'self'}} + \underbrace{\langle \sum_{i\neq j} u_i(o).u_j(t)\rangle}_{\text{'distinct'}} \quad (2.8)$$

where i and j label different molecules. For infrared and Raman spectra it is usually assumed [2-4] that the reorientational correlation functions obtained (eqns. 2.9 and 2.10)

$$\langle \sum_{i,j} P_1(\cos\theta_{ij}(t))\rangle \equiv \langle \underline{u}(o).\underline{u}(t)\rangle \quad (2.9)$$

$$\langle \sum_{i,j} P_2(\cos\theta_{ij}(t))\rangle \equiv \langle \mathrm{Tr}\beta'(o).\beta'(t)\rangle \quad (2.10)$$

(where P_1 and P_2 are the 1st and 2nd order Legendre Polymomials; $\cos\theta$ and $\frac{1}{2}(3\cos^2\theta-1)$ respectively) contain only the 'self' (single particle, sp) terms. This means that in each case the distinct terms (eqn.2.8) are assumed to be zero. There is a good deal of experimental evidence, [4,8,9] however, that intermolecular coupling of the vibrational motions of molecules i and j does occur and a 'distinct' term in the expansion of $\langle q_v(o).q_v(t)\rangle$

$$\langle q_v(o).q_v(t)\rangle = \underbrace{\langle q_{vi}(o).q_{vi}(t)\rangle}_{\text{'self'}} + \underbrace{\langle \sum_{i\neq j} q_{vi}(o).q_{vj}(t)\rangle}_{\text{'distinct'}} \quad (2.11)$$

may be significant. Equation 2.7 then tells us that, unless vibration and rotational motions are totally uncorrelated, such coupling necessarily has an effect on the reorientational C.F.'s. Lynden-Bell [10] has shown that coupling of the transition dipoles leads to distinct differences between the $C_{vib}(t)$ functions obtained from infrared and Raman (isotropic and anisotropic) band profiles.

The treatment given so far takes account of neither the effects of local electrical fields [11-13] on the spectral intensity or dynamics nor the contribution to the spectra of interaction-induced intensity [14-16]. The first effect, caused by the fact that (due to interactions) the molecules in the fluid do not 'see' the same electric field as that propogated into the medium, is partially eliminated [6] by normalisation of C(t) - eqns. 2.4 to 2.6. The second effect, which usually distorts the spectral 'wings' (at high frequencies) is largely unknown for most liquids and certainly affects [16,17] the short time part of C(t).

Provided that the complications mentioned above are not too severe, then one can obtain expressions for the desired correlation functions by manipulation of equations 2.4 to 2.6, viz:

$$C_v^{iso}(t) = FT[I_{iso}(\omega)]/\phi_s(t) \tag{2.12}$$

$$C_{1R}(t) = FT[I_{IR}(\omega)]/FT[I_{iso}(\omega)] \tag{2.13}$$

$$C_{2R}(t) = FT[I_{aniso}(\omega)]/FT[I_{iso}(\omega)] \tag{2.14}$$

where $\phi_s(t)$ is a time domain representation of the spectral slit profile (which is assumed to remain the same in all three experiments - see below) and where $C_v^{iso}(t) = C_v^{aniso}(t) = C_v^{IR}(t)$.

III. EXPERIMENTAL PROBLEMS

There are a number of purely experimental problems which have to be overcome before one can be certain that the digitised band intensity is 'correct'.

In Raman spectroscopy it is essential to ensure that the incoming laser beam is fully vertically polarised. To this end, the laser light is usually passed through a Glan-Thompson prism [6] or a polariser placed in front of the sample. For highly polarised bands there may also be a problem [6] with leakage from the $I_{VV}(\omega)$ spectrum into that of $I_{VH}(\omega)$. The measured $I_{VH}^{obs}(\omega)$ may be different from the true value according to,

$$I^{obs}_{VH}(\omega) = I^{t}_{VH}(\omega) + CI_{VV}(\omega) \qquad (3.1)$$

Bartoli and Litovitz [6] found values of C of about 0.2% for some typical polarised (A) bands - by making measurements on the 459 cm^{-1} band of CCl_4 which has a well-known [18] depolarisation ratio. Calculation of $I^{t}_{VH}(\omega)$ is then straightforward. Polarisation scrambling behind the main entrance slit ensures that the monochromator is equally sensitive to transmission of I_{VH} and I_{VV} scattered light. There have been reports of local heating effects caused by a relatively high powered laser beam. However, we have never found this to be a problem although, of course, it is not possible to physically monitor the microscopic temperature. Stokes/Antistokes intensity ratios which measure the Boltzman population factors (and hence the microscopic temperature) have always corresponded well to the laboratory (bath) temperature even for input powers up to 2w.

Probably the most serious difficulty arises with achieving an accurate base line and good S/N ratio (especially in the wings of the band). Both factors seriously affect the derived C.F.

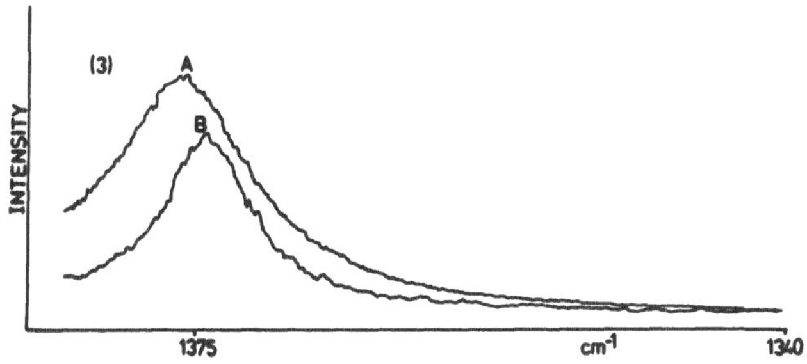

Fig. 3. Raman $I_{VV}(\omega)$ spectra of ν_3 of CH_3CN (A) liquid (B) in dilute solution in CCl_4((illustrating the long 'wing' to low frequency and relatively poor S/N ratio).
(Reproduced by permission from Molecular Association, Vol.2, Heyden, 1978, p.318)

Figure 3 illustrates this problem for ν_3 of acetonitrile which is only a weak Raman band with a long wing to low frequency. The resulting reorientational correlation functions (fig. 4) are very noisy and τ values are imprecise. With photon counting or multichannel detection (Raman) or fast scanning (F.T. infrared) these problems can be overcome, but only at the expense of a longer experimental time. Base line determination is an even more

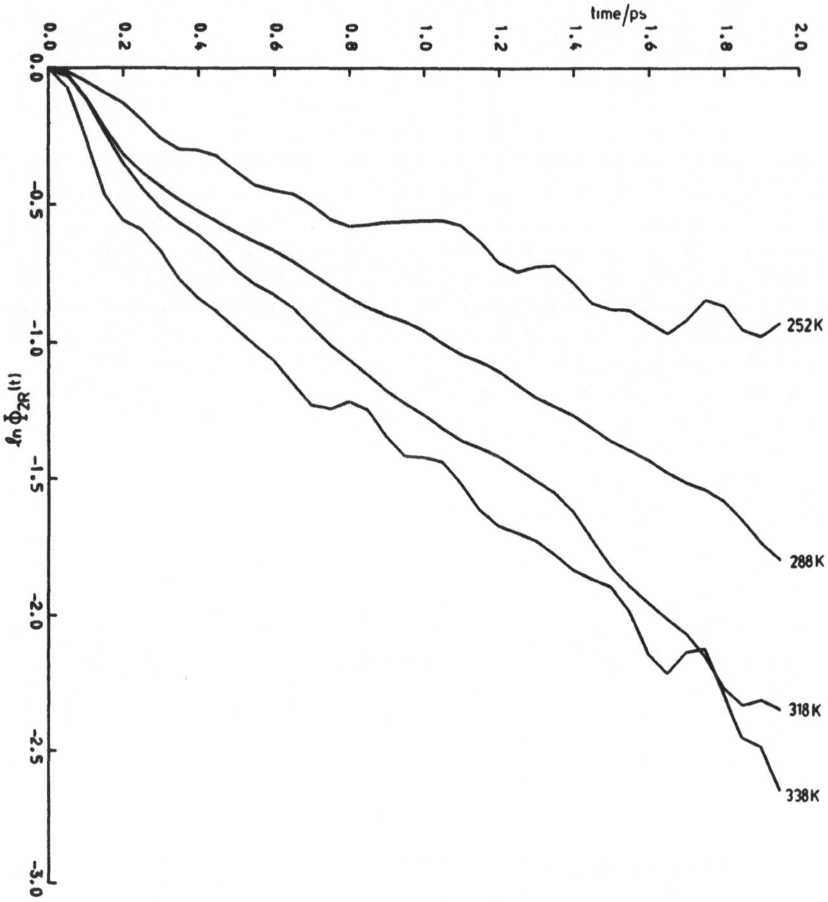

Fig. 4. The $C_{2R}(t)$ functions for ν_3 of CH_3CN at different temperatures. (Reproduced by permission from Chemical Physics, 1977, Vol. 25, p. 395.)

serious problem, mainly because of the difficulty of recording a meaningful 'background' (especially for a pure liquid). This problem is illustrated by figure 5 which shows the infrared band recorded for ν_8 of liquid CH_3CN. In this case there are interference fringes (arising from constructive and destructive interference of light rays suffering multiple reflections inside the cell). Computational procedures have been devised [19] to circumvent the problems of calculating the optical properties of thin liquid films from infrared data. These are referred to again below.

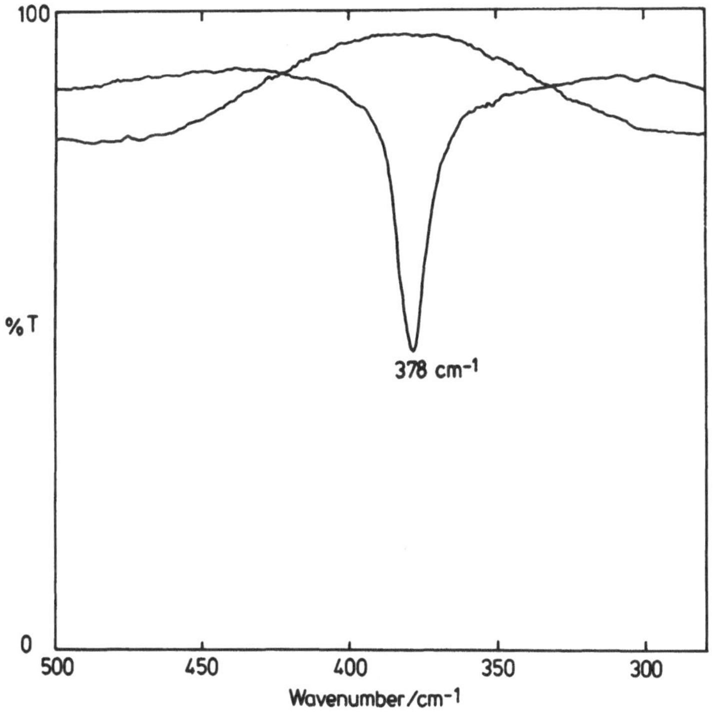

Fig. 5. Infrared spectrum of ν_8 of liquid CH_3CN showing 'fringes' due to interference effects in the thin CsI cell.

IV. COMPUTATIONAL PROBLEMS

There are also a number of computational problems which may arise in the calculation of a reliable correlation function from the experimental band profile.

(i) Determination of the band centre

In equations 1.1 to 1.6 it is assumed that the Fourier transform is performed with respect to the true band centre ω_0 (each ω should, strictly, be replaced by $\omega-\omega_0$). If attempts are made to compute the Fourier transform with respect to a false band centre then wild oscillations occur in the C(t) function. This is illustrated in figure 6 for a pair of simulated bands. Such possible errors are of fundamental importance when attempting

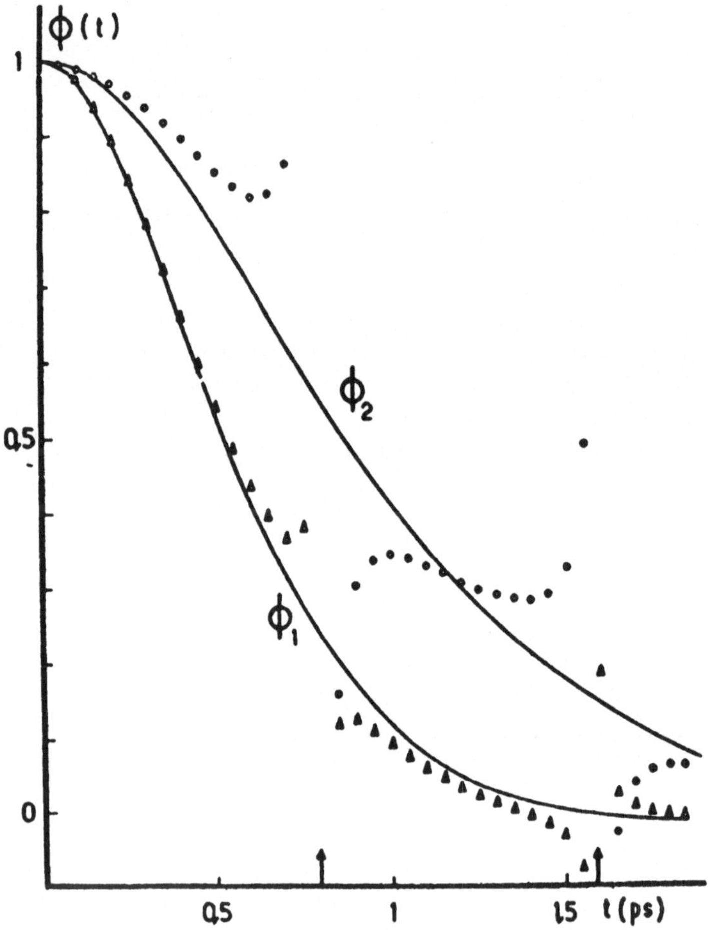

Fig. 6. Fourier transforms ϕ and ϕ_2 of a pair of bands centred at 1000 and 1020 cm^{-1} using equation 4.2. Solid lines are for correct band centres while the Δ and O points show the effects of transforming with the second band at 1021 cm^{-1}. (Reproduced by permission from Mol. Phys., 1977, 34, 145).

to deal with the problems of overlapping bands (section 2(ii)). It is therefore essential that one finds the band centre accurately, and this may justify collecting data at smaller frequency intervals than is dictated by the instrument slit width. Such determinations are, of course, also affected by the band width.

(ii) Corrections for the effects of overlapping bands

The fundamental vibrational bands (ν_i) of small polyatomic molecules are often disturbed by the presence of 'hot' bands (often of the type $\nu_i + \nu_j - \nu_j$ - see figure 7) and also by the corresponding fundamental bands of isotopically modified molecules (which are present in my sample in natural abundance). The most important isotopes in this respect are $^{37}Cl/^{35}Cl$, $^{79}Br/^{81}Br$ (^2H

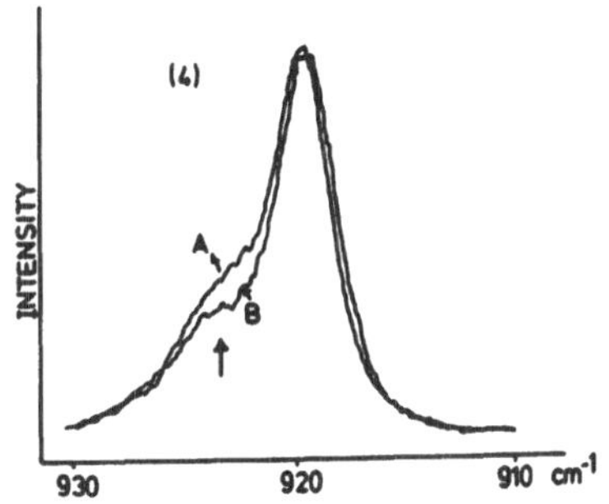

Fig. 7. Raman $I_{VV}(\omega)$ spectrum of ν_4 of CH_3CN showing 'hot' band in the high frequency wing in both liquid (A) and dilute solution (B). (Reproduced by permission from Molecular Association, Vol. 2, Heyden, 1978, p.318).

substitution usually results in bands without serious overlap, at least for those vibrations involving (largely) the hydrogen atoms). Van Konynenburg and Steele [20] some years ago showed how to eliminate the effects of such band 'structure' on the correlation function. The total intensity (at frequency ω) due to bands at frequency ω_i with fractional intensities x_i is

$$I(\omega) = \sum_i x_i I(\omega-\omega_i) \equiv \sum_i x_i I(\omega-\omega_o-\Delta\omega_i) \quad (\omega_i = \omega_o-\Delta\omega_i) \qquad (4.1)$$

where ω_o is the point about which Fourier transformation is performed and where $\Delta\omega_i$ is the frequency shift between ω_o and ω_i. The 'true' and 'apparent' correlation functions are related by,

$$C^t(t) = C^a(t)/\sum_i x_i \exp(-i\Delta\omega_i t) \qquad (4.2)$$

The obvious requirements are that (a) one knows the frequency shifts and fractional (relative) intensities of the various species present, (b) one may assume that the correlation functions derived from each individual band have the same numerical form (and hence that the relaxation processes affect each individual species of band in exactly the same way). In the case of the various chlorine-containing species (for example) this is reasonable. However, it is known [21] that the $\nu(CH_3)$ and $\nu(CD_3)$ vibrations of CH_3CN and CD_3CN are affected differently by the intermolecular potential - presumably because of the difference in amplitude between the two modes (i.e. the normal coordinates $q_v(t)$ are different). Furthermore, it is clear from figure 7 that the hot band $\nu_4+\nu_8-\nu_8$ for CH_3CN is obviously not the same shape as that of the ν_4 fundamental band. Figure 8 illustrates that the technique works well for the case of Cl_2 in solutions. The band profile has at least four components due to isotopic modifications of Cl_2.

(iii) Changes in refractive index through the band profile

For infrared bands Gordon [1] showed that the profile $I(\omega)$ referred to equations 1.3 and 1.6 is, in fact, the imaginary part of the complex permitivity, $\epsilon''(\omega)$

$$\hat{\epsilon}(\omega) = \epsilon(\omega) - i\epsilon''(\omega)$$

In fact, $I(\omega) = \epsilon''(\omega)/[1-\exp(-h\omega/kt)]$ (4.3)

$$\text{where } \epsilon''(\omega) = \frac{\alpha(\omega)n(\omega)c}{\omega} \tag{4.4}$$

and $\alpha(\omega) = \frac{1}{\ell} \log e(T_o/T)_\omega$ is the absorption coefficient. $\epsilon''(\omega)$ is also related to the absorption index $k(\omega)$ which is the imaginary part of the complex refractive index,

$$\hat{n}(\omega) = n(\omega) - ik(\omega)$$

where $\alpha(\omega)$ and $k(\omega)$ are related by

$$\alpha(\omega) = \frac{2\omega k(\omega)}{c} \tag{4.5}$$

$$\text{So } \epsilon''(\omega) = \frac{\alpha(\omega)n(\omega)c}{\omega} = 2n(\omega)k(\omega) \tag{4.6}$$

If therefore $n(\omega)$ and either $\alpha(\omega)$ or $k(\omega)$ have been measured separately [22-24] then

$\frac{\epsilon''(\omega)}{\omega}$ or $\frac{n(\omega)k(\omega)}{\omega}$ can easily be formulated. Now, it has been shown [22-24] that band profile is seriously distorted by

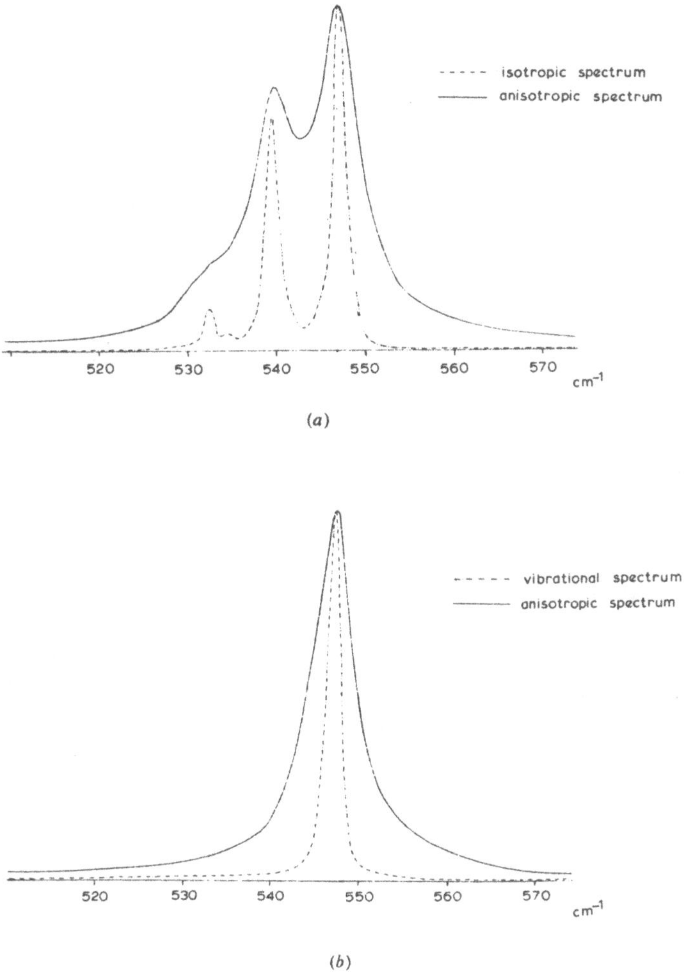

Fig. 8. Raman spectra of liquid chlorine (a) before correction for scattering by isotopic species, (b) after correction using equation 4.2. (Reproduced by permission from Mol. Phys., 1977, 34, 237.

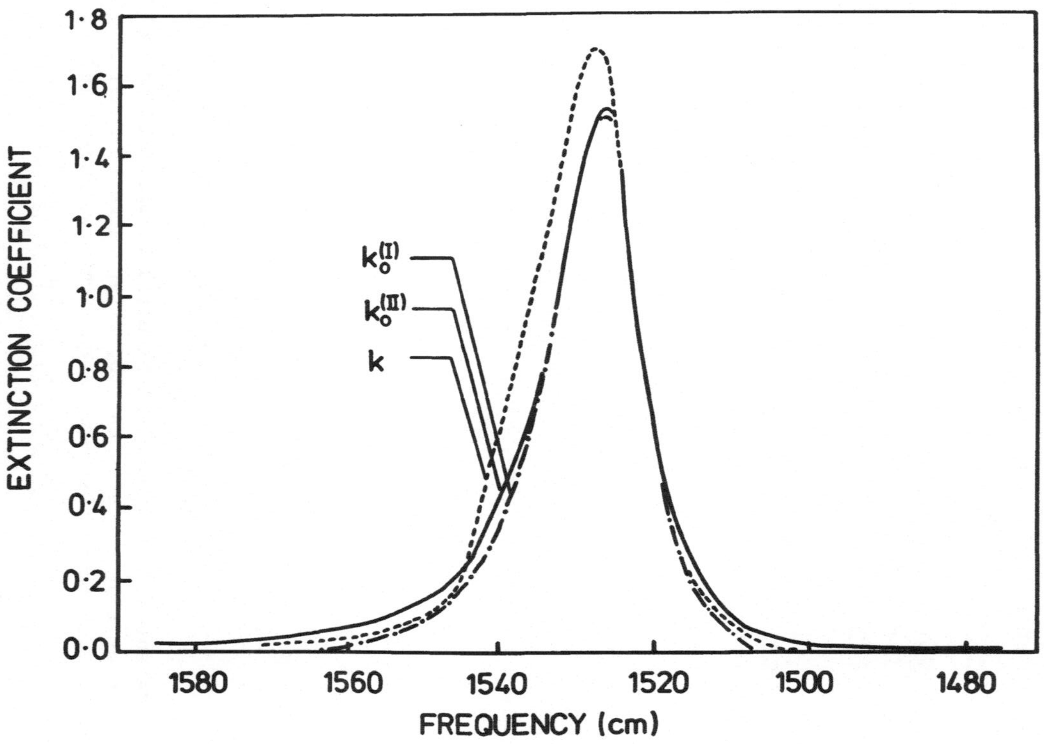

Fig. 9. The (very intense) infrared 1530 cm^{-1} band of liquid C_6F_6 showing the effects of spectral distortion due to refractive index changes and choice of base line. The distortion due to refractive index changes is greatest at the high frequency side of the band where n falls to 0.5. (Reproduced by permission from Appl. Spectroscopy, 1970, 24, 12).

the rapid variation in the refractive index $n(\omega)$ through the band (fig. 9) when the values of $k(\omega)$ are large near the band centre (i.e. for 'strong' bands). This is because the reflection of radiation from the liquid/cell interface varies with $n(\omega)$ through the band leading to false values of $\alpha(\omega)$. The 'corrected' band shape is given by $n(\omega)\,k(\omega)$ or $\epsilon''(\omega)$ which is easy to achieve if both $n(\omega)$ and $k(\omega)$ may be measured separately [22-24] or if $n(\omega)$ can be calculated [22,25] using the Kramers/Kronig relationships.

$$n(\omega_i) - 1 = \frac{2}{\pi}\int_0^\infty \frac{K(\omega)\omega - K(\omega_i)\omega_i \, d\omega}{\omega^2 - \omega_i^2} \tag{4.7}$$

$$\text{and } n(\omega_i)k(\omega_i) = \frac{-2\omega_i}{\pi}\int_0^\infty \frac{(n(\omega)-1) - (n(\omega_i)-1)d\omega}{\omega^2 - \omega_i^2} \tag{4.8}$$

Such a technique has been shown [26] to work well, for example, for the computation of the refractive index of CH_3CN in the microwave and far-infrared. Crawford and co-workers [22-24] have also advocated using a correlation function based on the complex part of the local susceptibility, (introduced to correct for 'dielectric' effects on the optical constants). In that case,

$$\hat{C}(\omega) = C'(\omega) = iC''(\omega) \tag{4.9}$$

$$\text{with } C'(\omega) = \tfrac{3}{4\pi}\left[1 - \frac{3(\epsilon'+2)}{(\epsilon'+2)^2 + \epsilon''^2}\right] \tag{4.10}$$

$$\text{and } C''(\omega) = \frac{9\epsilon''}{4\pi}\left[\frac{1}{(\epsilon'+2)^2 + \epsilon''^2}\right] \tag{4.11}$$

which are based on a Lorentz-Lorenz 'internal' field. The shape of $C''(\omega)$ is different again from that of $n(\omega)$ especially for intense bands in pure liquids (figure 10). In those cases there would clearly be an error in the computed correlation function if the values of $\alpha(\omega)$ were to be used instead of $\epsilon''(\omega)$ or $C''(\omega)$.

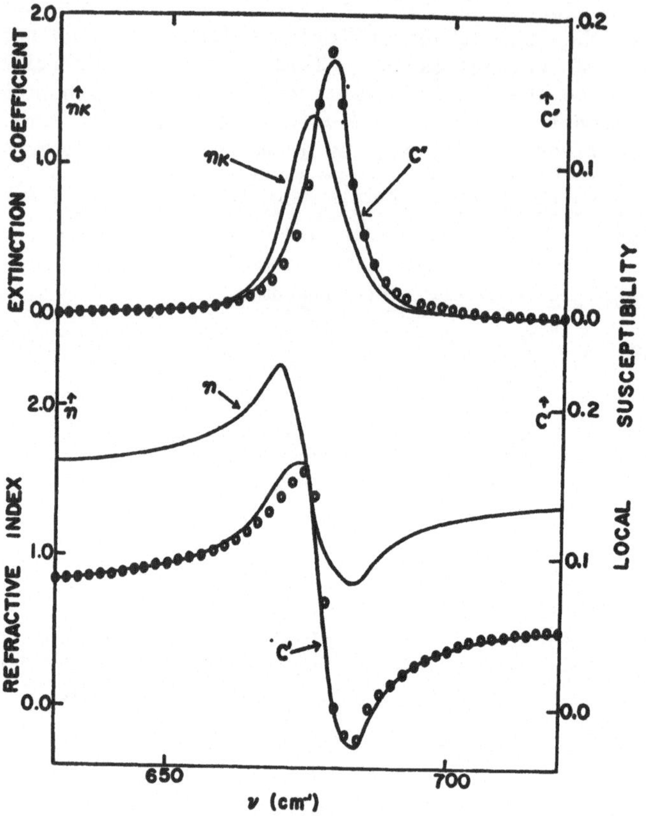

Fig. 10. Comparison of the functions n(ω) and n(ω)k(ω) with the real and imaginary parts of the complex susceptibility. The lines give experimental data while the prints give values calculated using a Van Vleck/Weisskopf model. (Reproduced by permission from J. Phys. Chem., 70, 1536, (1966)).

(iv) Effects of thermal population of rotational states

Mention has already been made of the effects of vibrational hot bands on the vibrational/rotational (infrared or Raman) spectrum. However, there are significant effects of the thermal distribution of molecules among rotational states. If one considers the Raman (I_{VH}) intensity of absorption (due to rotational transition) in the wings of a vibrational fundamental band then figure 11 and equation 1.2 show that the only difference between the $\Delta J = +2$ and $\Delta J = -2$ sides of the band is a factor $\left\{\frac{\omega - \omega_0}{\omega + \omega_0}\right\}^4 \exp(-h\omega_J/kT)$. For most liquids at room temperature $h\omega_J \ll kT$ and $\left\{\frac{\omega - \omega_0}{\omega + \omega_0}\right\}$ does not vary much over the vibrational

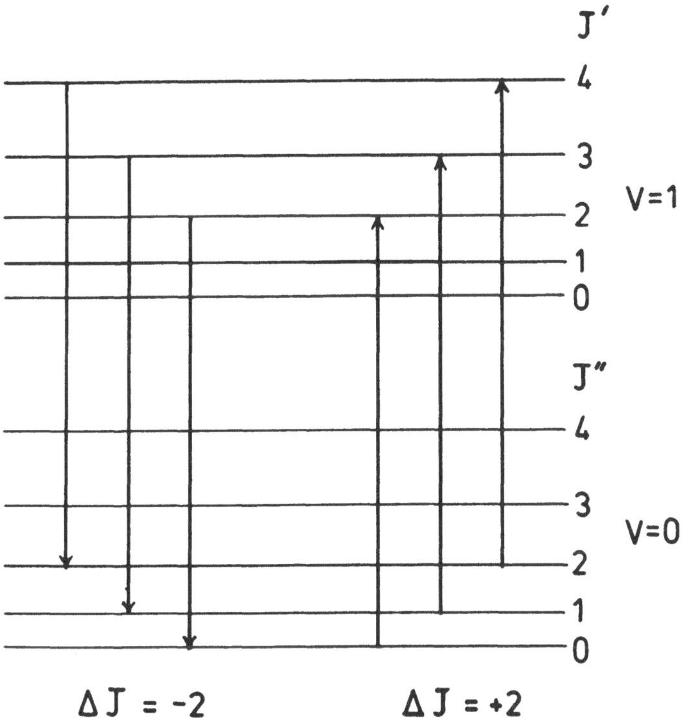

Fig. 11. Comparison of initial state population distributions for $\Delta J = \pm 2$ transitions of a Raman rotational/vibrational band.

band. The factor is thus not much different from unity. However, for small molecules (with small moments of inertia and large rotational energies) at low temperatures the factor may differ significantly from one and the high frequency wing (ΔJ, + ve) may differ in intensity from that of the low frequency wing. This situation is demonstrated by figure 12 for the ν_3 of liquid ethylene (160K). In terms of the correlation function $\langle P_2[\cos\theta_{ij}(t)]\rangle$ this means that the spectrum is asymmetric and the time domain function is uneven [25]. Under these circumstances, a quantum mechanical correction to the classical C.F. (usually referred to as 'detailed balance') [27] is required. It has been shown [28] that the correct recipe is to take

$$I_{VH}(\omega) = I_{VH}{}^{a}(\omega)\ \exp(h\omega/2kT) \tag{4.12}$$

for both branches (with the appropriate change of sign). This type of correction is usually only needed when working at low temperatures with small molecules. It should be noted that the recipe given by equation 4.12 accounts for only part of the quantum mechanical correction [29]. If moments higher than the 2nd moment are required then further terms are needed.

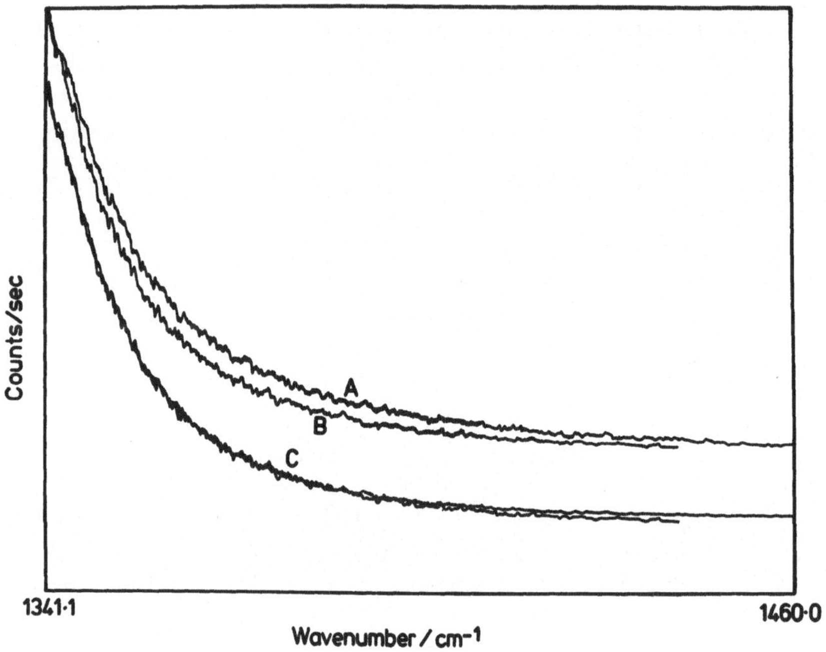

Fig. 12. Comparison of the intensity of the Raman $I_{VH}(\omega)$ band of ν_2 of liquid ethylene at 190K. A and B are the high and low frequency branches (respectively) while C shows the effect of using equation 4.12.

(v) Spectral Slit Profile

The simplest way of dealing with the removal of the instrument slit width before transformation is to perform a Fourier transform on the appropriate slit profile (which may be Gaussian or triangular or which may be measured directly in Raman spectroscopy by observation of a scattering profile from spheres). The method is illustrated by equation 2.12, and it should be emphasised that the procedure works because the observed band is, in effect, the result of a spectral convolution of the true band profile with the slit profile.

$$I^a(\omega) = I^t(\omega) * I_{slit}(\omega) \quad (4.13)$$

Since $FT[I^t(\omega)*I_{slit}(\omega)] = FT[I^t(\omega)] \times FT[I_{slit}(\omega)]$ (4.14)

it follows that division (or subtractions if log. ϕ is used) will suffice to correct for the effects of slit distortion. There are, however, a number of methods [30] available for doing slit profile corrections in the spectral domain, Jones et al [31] have described a method based on slit pseudo-deconvolution. If the slit function is $I_{sl}(\omega)$ and if the observed spectrum is $I_{obs}(\omega)$ then the true spectrum may be obtained from

$$I_{obs}(\omega) - \int I_{obs}(\omega').I_{sl}(\omega'-\omega)d\omega' \cong I^t(\omega) - I_{obs}(\omega) \quad (4.15)$$

In practice, the observed spectrum is convolved with the slit function and an approximation is made to the true spectrum by doing

$$I^t(\omega)(\text{first approx}) = I_{obs}(\omega).\frac{I_{obs}(\omega)}{I_{obs}(\omega)*I_{slit}(\omega)} \quad (4.16)$$

This procedure is illustrated in figure 13 where it is also shown how one may proceed to a 2nd approximated $I^t(\omega)$ by reconvolution of $I^t(\omega)$ profile. This iterative procedure is shown [31] to converge at the same rate as the subtraction method of equation 3, but has the advantage of avoiding possible negative ordinates. Figure 13 shows that the corrected band smaller width and a higher peak intensity.

Another technique is that originally considered by Strutt [32], subsequently extended by Runge [33] and Steele [34]. This method depends on the expansion of the Fourier transform of the slit profile in such a way that the true spectrum is given by

$$I^t(\omega) = I_{obs}(\omega) - \frac{s^2}{12}.\frac{d^2[I_{obs}(\omega)]}{d\omega^2} + \frac{s^4}{240}.\frac{d^4[I_{obs}(\omega)]}{d\omega^4} --- (4.17)$$

where s is the slit in the same units as ω (usually in cm^{-1}). If a Gaussian slit profile is used this becomes [34],

$$I^t(\omega) = I_{obs}(\omega) - \frac{s^2}{11.09}\frac{d^2[I_{obs}(\omega)]}{d\omega^2} - \frac{s^4}{240}\frac{d^4[I_{obs}(\omega)]}{d\omega^4} \quad (4.18)$$

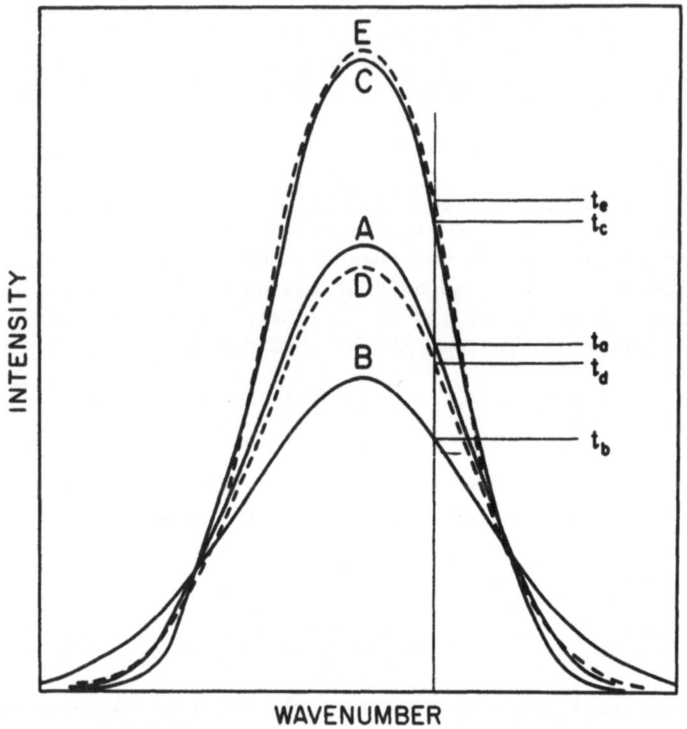

Fig. 13. Diagram illustrating the principal of spectral pseudo-deconvolution.
Curve A. The observed spectrum.
Curve B. Ordinates of Curve A convoluted with the slit function ordinates.
Curve C. Ratioed ordinates ($t_c = t_a.t_a/t_b$) to give the first approximation to the deconvoluted curves.
Curve D. Ordinates of Curve C convoluted with the slit function ordinates.
Curve E. Ratioed ordinates ($t_e = t_c.t_a/t_d$) to give the first iterative improvement of the deconvoluted curve.
(Reproduced, by permission, from Spectrochimica Acta 23A, 925 (1967).

This method has the disadvantage that noise tends to limit the precision with which derivative spectra may be obtained. Fortunately, the 2nd order term has been shown [34] to account for 84% of the necessary correction in a typical case.

V. CORRELATION FUNCTION RELIABILITY

Having computed the 'best' correlation function from the

available experimental data it is important to realise that it will be subject to uncertainties introduced during the measurement procedure. These uncertainties are a result of the following effects.

(i) Finite wavelength of radiation

Since C(t) measures an ensemble average it should measure a correlation between individual molecules of the ensemble. However, the radiation is of a wavelength much longer than the molecular size and so the C(t) measured corresponds to domains of molecules. In these circumstances interference terms are needed [1] between the molecules of the domain.

(ii) Discrete sampling of spectrum

The requirement of sampling at interval $\Delta\omega$ leads [35] to a periodic $\phi(t)$ function with period $\tau = 2\pi/\Delta\omega$. The calculated correlation function is therefore only reliable up to a time given by $\pi/\Delta\omega$. The value of $\Delta\omega$ is usually controlled by the instrumental slit width (δs) in such a way that $\Delta\omega \geqslant 2\delta s$. The true range is thus effectively given by,

$$t = \pi/2\pi c\delta\bar{\nu} \simeq \frac{16 \times 10^{-12}}{\delta\bar{\nu}} \text{ sec}$$

where $\delta\bar{\nu}$ is the slit width in cm^{-1}. For a limiting resolution of say 1 cm^{-1} the correlation function is valid to 32p sec . Except for very narrow bands obtained with poor resolution this is not normally a limiting factor.

(iii) Limited spectral frequency range

In practice, because of overlapping bands, it is only possible to measure the spectrum between ω_i and ω_f so the maximum frequency range is usually $(\omega_o - \omega_i)$. This limits the time resolution which may be achieved [35] and is given by,

$$\Delta t > \pi/2\pi c(\bar{\nu}_o - \bar{\nu}_i). \simeq \frac{16 \times 10^{-12} \text{ sec}}{(\bar{\nu}_o - \bar{\nu}_i)}$$

where $\bar{\nu}_o$ and $\bar{\nu}_i$ are in cm^{-1}. This limitation can be quite serious. For example, if $\bar{\nu}_o - \bar{\nu}_i$ is 100 cm^{-1} (this would correspond to 10 half widths of a 10 cm^{-1} (wide) band, a range which should always be aimed for) then the corresponding time resolution is 0.2psec. The effect of such a time resolution is shown in figure 14. The individual points demonstrate that it might be difficult to properly define the short time part of C(t) if the time resolution is insufficient.

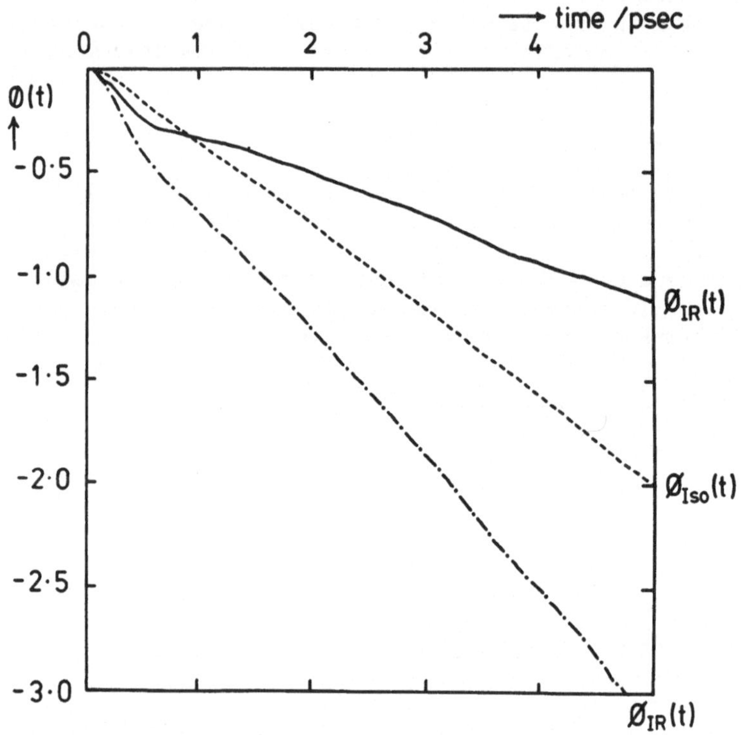

Fig. 14. Correlation functions for ν_3 of liquid CH_3I derived from data such as those given in figures 1 and 2. Illustrating the difficulty of accurately describing the short time (non-exponential) part of the CF with a time resolution of 0.2 psec

(iv) Noise in the spectrum

The effects of spectral noise on the correlation function have already been illustrated by figure 4. One of the few attempts to make a statistical assessment of such noise is that of Crawford and co-workers [22,24] The error bars in figure 15B are calculated by error propagation from the observed spectra, $n(\omega)$ and $k(\omega)$ (figure 15A) with due account being taken of the correlation between $n(\omega)$ and $k(\omega)$ at the same frequency It is clear that such an assessment is essential if an error on the final τ value (rarely better than ± 10%) is to be confidently stated.

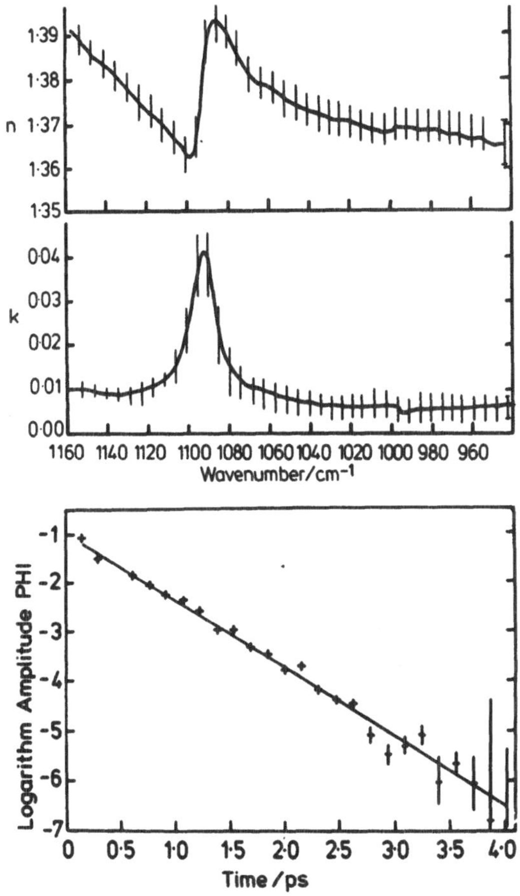

Fig. 15. Illustration of the random errors involved in $n(\omega)$, $k(\omega)$ and the resulting correlation function for the ν_{22} band of liquid acetone. (Reproduced by permission from Advances in Infrared and Raman Spectroscopy, Vol 4, Chapter 2, p.80-81.

SUMMARY

It is clear that, if the objectives mentioned in Section I are to be achieved, one needs to be very careful about the way in which measurements are made and the way in which the data are subsequently treated. There are many instances [36] in the literature of wild variations between C(t) functions and times obtained, ostensibly, from the same spectra, and it is not difficult to see why this may be the case. However, inaccurate data can lead to misleading or erroneous conclusions and they are absolutely no use to simulators or theoreticians who are trying to make meaningful comparisons.

REFERENCES

1. Gordon, R.G., 1968, Adv. Magn. Resonance, (Ed. Waugh, J.S.), 3, p.1.

2. Bailey, R.T. in 'Molecular Spectroscopy', Vol.2, 1974, Chem. Soc. Specialist Periodical Report, p.173.

3. Clarke, J.H.R. in 'Advances in Infrared and Raman Spectroscopy' Vol. 4, 1978, (Ed. Hester, R.E. and Clark, R.J.H.) Heyden, London, Chapter 4.

4. Yarwood, J. and Arndt, R. in 'Molecular Association' Vol.2, 1979, (Ed. Foster, R.), Wiley, New York, Chapter 4, pp.267-329.

5. Kubo, R. in 'Fluctuation, Relaxation and Resonance in Magnetic Systems', (Ed. D. Ter Haar), 1962, Oliver and Boyd, London, pp.23-68.

6. Bartoli, F.J. and Litovitz, T.A., 1972, J. Chem. Phys., 56, pp.404-425.

7. Doge, G. and Yarwood, J., 1984, Mol. Phys. (submitted).

8. Doge, G., Arndt, R. and Khuen, A., 1977, Chem. Phys., 21, p.53.

9. Kamoun, M. and Mirone, P., 1980, Chem. Phys. Lett., 75, p.287.

10. Lynden-Bell, R., 1977, Mol. Phys., 33, p.907, Faraday Symp. Chem. Soc., 11, p.167.

11. Brot, C, 1975, Dielectric and Related Molecular Processes, Vol. 2, Chem. Soc. Specialist Periodical Reports, pp.1-47; 1982, Mol. Phys., 45, p.543.

12. Kivelson, D. and Madden, P., 1975, Mol. Phys., 30, pp.1749-80.

13. Madden, P. and Kivelson, D., 1982, J. Phys. Chem., 86, pp.4244-4256.

14. Madden, P.A. and Cox, T.I., 1980, Mol. Phys., 39, p.1487; 1981, Mol. Phys., 43, p.287 and p.307.

15. Yarwood, J., 1983, Ann. Repts. Royal Soc. Chem., C, Chapter 7, pp.157-197, and references therein.

16. Kivelson, D. and Madden, P.A., 1980, Ann. Rev. Phys. Chem., 31, p.523; Madden, P.A., 1979, Phil. Trans. Roy. Soc., 293A, p.419.

17. Evans, M.W., Evans, G.J., Coffey, W.T., and Grigolini, P., 1982, Molecular Dynamics, Wiley-Interscience, New York, Chapter 11, pp.703-787.

18. Murphy, W.F., Evans, M.W., and Bender, P., 1967, J. Chem. Phys., 47, p.1836.

19. Fujiyama, T., Herrin, J. and Crawford, B.L., 1970, Appl. Spectrosc., 24, pp.9-15.

20. van Konynenburg, P. and Steele, W.A., 1972, J. Chem. Phys., 56, p.4776.

21. Doge, G., Arndt, R. and Yarwood, J. (unpublished data).

22. Crawford, B., Golpen, T.G., and Swanson, D., 1978, 'Advances in Infrared and Raman Spectroscopy' (Eds. Clark, R.J.H. and Hester, R.E.), Vol. 4, Chapter 2, pp.47-83 (and references therein).

23. Clifford, A.A. and Crawford, B.L., 1966, J. Phys. Chem., 70, pp.1536-1543.

24. Favelukes, C.E., Clifford, A.A. and Crawford, B.L., 1968, J. Phys. Chem., 72, pp.962-966.

25. McQuarrie, D.A., 1973, 'Statistical Mechanics', Harper & Row, New York, Chapter 21.

26. Arnold, K.E., 1981, Ph.D. Thesis (University of Durham).

27. Berne, B.J. and Harp, G.D., 1970, Adv. Chem. Phys., 17, p.63.

28. Steele, W.A., 1980, in 'Vibrational Spectroscopy of Molecular Liquids and Solids', (Eds. Bratos, S. and Pick, R.M.), Plenum, London, (also in personal communication to the author).

29. Sampoli, M. (personal communication).

30. Steele, D., 1982, in 'Vibrational Intensities in infrared and Raman spectroscopy', (Eds. Person, W.B. and Zerbi, Z.), Elsevier, New York, Chapter 19, pp.398-416.

31. Jones, R.N., Venkataraghavan, R. and Hopkins, J.W., 1967, Spectrochimica Acta, 23A, p.925, p.941.

32. Strutt, J.W., 1871, Phil. Mag., 42, p.441.

33. Runge, C., 1897, Z. Math., 42, 205.

34. Hill, I.R. and Steele, D., 1974, J. Chem. Soc., Faraday Trans., II, 70, p.1233.

35. Keller, B. and Kneubuhl, F., 1972, Hehr. Phys. Acta, 45, pp.1127-1164.

36. Evans, M.W. and Evans, G.J., 1983, J. Mol. Liquids, pp.149-260.

STRUCTURAL STUDIES OF MOLECULAR LIQUIDS BY NEUTRON AND X-RAY DIFFRACTION

J.C.Dore

Physics Laboratory, The University of Kent, Canterbury, CT2 7NR, England.

ABSTRACT

The principles of X-ray and neutron diffraction studies of molecular liquids are presented. The measured liquid structure factor $S_M(Q)$ provides detailed information on the conformation of the molecular unit which is interpreted through the molecular form-factor $f_1(Q)$. The structural characteristics of the liquid are contained in the term $D_M(Q)$ which may be transformed to give a real-space representation $d_L(r)$ which contains a weighted-sum of the partial pair-correlation functions, $g_{\alpha\beta}(r)$.

Examples of current work are given starting with relatively simple systems such as liquids composed of homo-nuclear diatomic molecules (N_2,O_2,$C\ell_2$,Br_2 etc). Other types of molecular species are presented with illustrative examples of the way in which the experimental observations may be interpreted. The use of isotopic substitution in neutron diffraction measurements is described and the special case of hydrogen/deuterium substitution is briefly treated. The particular case of liquids which exhibit strong hydrogen-bonding is discussed in relation to temperature variation studies of water.

1. INTRODUCTION

Previous chapters have emphasised the dynamic or time-dependent properties of the liquid state. In the present chapter the emphasis concerns the microscopic structural properties of the liquid as represented by the pair-correlation functions $g_{\alpha\beta}(r)$ at a separation, r. In certain respects, a liquid cannot have a

A. J. Barnes et al. (eds.), Molecular Liquids - Dynamics and Interactions, 383–409.

"structure" in the same sense as a crystallographic solid since the positions of the atoms and molecules are continuously changing and any local configuration will vary over a short time period with a small chance of re-forming an identical structure with the same atoms at any subsequent period of time. The term "structure" which applies to liquids is therefore essentially a time-averaged property which is firmly linked to probability concepts. Since atoms have size and molecules have shape there will be a spatial ordering around a single atom or molecule. This local ordering will be short-range (typically $\lesssim$ 20 Å) and will be influenced by the interatomic or inter-molecular forces. Experimental determinations of the structure by diffraction studies are therefore of fundamental importance in linking the basic interaction forces to the properties of the liquid and are an essential element in the explanation of dynamic processes. Unfortunately, the lack of long-range order and the complexity of inter-molecular interactions means that the required information cannot be easily obtained except for the simplest molecular systems consisting of up to five or six atoms. Even in this restricted area of interest, it is rarely possible to define the local structure with a high degree of precision. The reasons for this limitation and the ways in which progress can be made will be reviewed in the following sections using several different studies as illustrative examples of the current state of the art. No attempt will be made to present a comprehensive survey representative of the wide range of experimental studies that have been conducted on this topic.

2. DIFFRACTION FORMALISM

The most direct way of obtaining structural information is by diffraction of incident radiation having a wavelength comparable to the separation between scattering centres. The two probes that have been mainly used are X-rays and thermal neutrons although some results have also been obtained using electrons. The basic formalism has been presented in review articles by Powles [1] and Blum and Narten [2]. A brief outline will be given here to define the symbols, illustrate the principles of the diffraction method and to relate the functions of interest to experimental observables.

A schematic illustration of elastic neutron scattering from an isolated point scatterer (nuclei) is given in Figure 1a. An incidental plane wave with wave vector, $\underline{k}_o$ is incident from the left and produces an outgoing spherical wave of wave vector $\underline{k}_1$ ($k_1 = k_o$) and amplitude

$$a(\underline{Q}) = b\,\frac{e^{i\underline{Q}\cdot\underline{r}}}{r} \tag{2.1}$$

where b is the coherent neutron scattering length for the nucleus and $\underline{Q}$ is the scattering vector, $\underline{Q} = \underline{k}_1 - \underline{k}_0$. In the case of X-rays, the constant b is replaced by the atomic scattering factor f(Q) due to the spatial distribution of the electrons in the atom and is dependent on the magnitude of the scattering vector, Q. The scattered intensity, $|a(\underline{Q})|^2$ is isotropic for neutrons but decreases with increasing Q for X-rays. If the scattering is from a rigid molecular unit (as shown in Figure 1b), the scattered waves from the different nuclei or atoms will interfere and the scattering amplitude becomes a sum over all N atoms (nuclei) in the unit

$$a(\underline{Q}) = \sum_i^N b_i \, e^{i\underline{Q}.\underline{r}} \tag{2.2}$$

where $\underline{r}_i$ is the position co-ordinate. The intensity now becomes

$$|a(\underline{Q})|^2 = \langle \sum^N b_i \, b_j \, e^{i\underline{Q}.\underline{r}} \rangle \tag{2.3}$$

where $\underline{r}_{ij}$ is the distance between the i'th and j'th scatterers and the < > brackets denote an ensemble average; the sum is then over all pairs i, j of the N atoms. Since there is no preferential orientation of the molecule with respect to the direction $\underline{Q}$, an orientational average may be evaluated to give

$$|a(Q)|^2 = \sum_{ij}^N b_i \, b_j \, j_o(Qr_{ij}) \tag{2.4}$$

where

$$\langle e^{i\underline{Q}.\underline{r}_{ij}} \rangle_\omega = j_o(Qr_{ij}) = \frac{\sin Qr_{ij}}{Qr_{ij}}$$

It is therefore convenient to define a normalised molecular form-factor, $f_1(Q)$ which characterises the diffraction pattern for the isolated molecular unit and may be expressed as:-

$$f_1(Q) = \frac{1}{\left(\sum^N b_i\right)^2} \sum_{ij}^N b_i \, b_j \, j_o(Qr_{ij}) \, e^{-\gamma_{ij}Q^2} \tag{2.5}$$

The last term arises from variation in the atom positions due to thermal vibrations within the molecule and the γ_{ij} terms are related to the mean square amplitudes of vibration by $\gamma_{ij} = \frac{1}{2}\langle u_{ij}{}^2\rangle$ The form of $f_1(Q)$ is analogous to the expression for elastic electron diffraction by molecular gases [3].

For the condensed phase, such as a liquid or amorphous solid, there will be additional terms arising for interference contributions from neighbouring units (Figure 1c). The liquid structure factor, $S_m(Q)$ can therefore be written [1] as

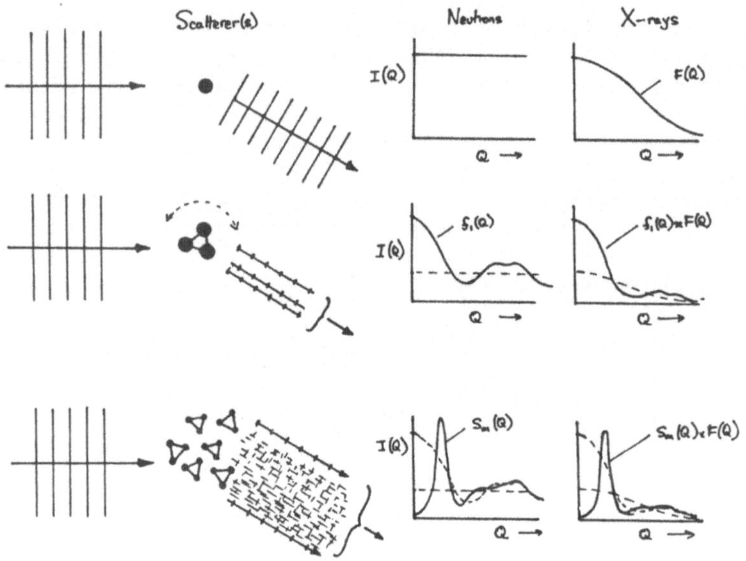

Figure 1. Diffraction from a) isolated atoms; b) isolated molecules and c) an assembly of molecules.

$$S_M(Q) = \frac{N_M}{(\Sigma b_i)^2} < \sum_{ij}^{N} b_i \, b_j \, e^{i\underline{Q}.\underline{r}_{ij}} > \qquad (2.6a)$$

where N_M now runs over all scatterers within the sample volume. It is convenient to divide this sum into intra- and inter-molecular contributions such that [4]:-

$$S_M(Q) = \frac{1}{(\sum^{N} b_n)^2} <|\Sigma \, b_n \exp(i\underline{Q}.\underline{r}_{cn})|^2>$$

$$+ \frac{1}{N_M(\Sigma b_n)^2} <|\sum_{\substack{i,j\\ i\neq j}} \exp(i\underline{Q}.\underline{r}_{cij}) \Sigma \, b_{n_i} b_{n_j}$$

$$\exp(i\underline{Q}(\underline{r}_{cn_i} - \underline{r}_{cn_j}))|> \qquad (2.6b)$$

where the summation n, N extends over atoms within a single molecular unit and the subscript c refers to the centre of the molecules such that $(\underline{r}_{c_{ij}} + (\underline{r}_{c_{ni}} - r_{c_{nj}}))$ represents the separation between atom i in one molecule and atom j in another molecule. The first term is simply the molecular form-factor $f_1(Q)$ and the second term, which is usually written as $D_M(Q)$, represents the interference terms due to neighbouring molecules which depend on the distribution of molecular centres and the relative orientation of the molecular axis. The total structure factor is therefore conveniently expressed as:-

$$S_M(Q) = f_1(Q) + D_M(Q) \qquad (2.7)$$

where $f_1(Q)$ characterises the molecular conformation and $D_M(Q)$ contains information about the arrangement of the molecules in the liquid. Strictly, $D_M(Q)$ is an ensemble average of the molecular configurations obtained in a snap-shot picture of the liquid but it is more convenient to visualize this as a time-average since each molecular unit effectively samples all possible configurations over an extended period of time. The information on the 'liquid structure', as opposed to that of the molecule, is therefore contained in the function, $D_M(Q)$. In the formalism used by Blum and Narten [2] this is equivalent to the function $h_{inter}(k)$ where k represents the scattering vector.

The real-space structure may be obtained from transform relations which convert from the Q-space representation to r-space. For a monatomic fluid the pair correlation function g(r) may be expressed as

$$4\pi r\rho[g(r) - 1] = \frac{2}{\pi}\int_0^\infty Q[S(Q) - 1]\sin Qr\, dQ \qquad (2.8)$$

where ρ is the atomic number density.

The analogue for a molecular system is

$$4\pi r\rho_M[g(r)-1] = \frac{2}{\pi}\int_0^\infty Q\, I(Q)\sin Qr\, dQ \qquad (2.9a)$$

where ρ_M is the molecular number density, g(r) is the total pair correlation function and

$$I(Q) = S_M(Q) - \frac{\sum^N b_i^2}{(\sum^N b_i)^2} \qquad (2.9b)$$

This expression for g(r) includes the intra-molecular terms and it is more usual to use the relation applying to the inter-molecular terms only:-

$$d_L(r) = 4\pi r\rho_M[g_L(r)-1] = \frac{2}{\pi}\int_0^\infty QD_M(Q)\sin Qr\, dQ \tag{2.10}$$

where $d_L(r)$ and $g_L(r)$ refer to the spatial correlations between different or distinct molecules.

In many cases the molecule will consist of several atom types α,β so that the total $g_L(r)$ consists of a number of independent partial correlation functions, $g_{\alpha\beta}(r)$. If the species α has a scattering length b_α and a concentration c_α, the total function may be expressed as a weighted sum

$$g_L(r) = \Sigma(2-\delta_{\alpha\beta})c_\alpha c_\beta b_\alpha b_\beta g_{\alpha\beta}(r); \quad \Sigma c_\alpha = 1 \tag{2.11}$$

where the summation runs over all pairs of atom types α,β. In the case of a two-component system such as liquid $HC\ell$, H_2O, $CC\ell_4$ or CS_2 there are therefore three partial correlation functions but the number rapidly increases for more complex molecules so that methanol (CH_3OH) or acetonitrile CH_3CN with four non-equivalent atoms have a total of ten different partial functions. Although it is sometimes possible to isolate the partial functions by use of isotopic substitution methods, the technique has a limited range of applicability and can only rarely be used to give a complete set of $g_{\alpha\beta}(r)$ functions with the required precision. This feature will be further discussed in Section 5d.

It is not necessary to represent the structure in terms of site-site correlation functions. An alternative form emphasises the orientational aspects of the correlation by reference to the distribution of molecular centres R and the alignment of molecular axes, Ω_i, Ω_j. The molecular pair correlation function can then be written in the form of a spherical harmonic expansion [5]:-

$$g(R,\Omega_1,\Omega_2) = \sum_{\ell_i\ell_j\ell_{ij}} \sum_{n_in_j} g_{n_in_j}^{(\ell_i\ell_j\ell_{ij})}(R)\Phi_{n_in_j}^{(\ell_i\ell_j\ell_{ij})}(\Omega_i\Omega_j\Omega_{ij}) \tag{2.12a}$$

The coefficients $g_{n_in_j}^{(\ell_i\ell_j\ell_{ij})}$, which define the structure in real space are related to the representation in momentum space, k, by the equation

$$g_{n_in_j}^{(\ell_i\ell_j\ell_{ij})}(R) = \frac{1}{\rho}\frac{1}{2\pi^2}\int_0^\infty h_{n_in_j}^{(\ell_i\ell_j\ell_{ij})}(\kappa)j_\ell(\kappa R)k^2dk \tag{2.12b}$$

The observed inter-molecular cross-section can therefore be used to define the various g(R) functions for chosen values of $\ell_i \ell_j n_i n_j$. This method has been used primarily by Zeidler and Bertagnolli for symmetric top molecules and will be discussed in Section 6a.

3. MOLECULAR SYSTEMS

It is convenient to divide the different molecular species into several categories as shown in Table 1. The degree of structural complexity increases as the number of atoms in the molecular unit is raised but some simplification can be achieved by symmetry. Diffraction studies have been made on most of the molecules listed in the examples column but there is little data for the bottom two categories.

4. EXPERIMENTAL METHODS

The scattering vector, Q, for elastic scattering is given by

$$Q = \frac{4\pi}{\lambda} \sin \theta/2 \tag{4.1}$$

where λ is the wavelength of the incident radiation and θ is the scattering angle. The conventional methods of measuring the diffraction pattern have utilised a monochromatic beam (fixed λ) with a detector which may be positioned at various scattering angles (variable θ). The X-ray technique is normally based on the utilization of characteristic lines (Cu K_α, Mo) from the spectrum of an X-ray generator and has been a standard routine for many years. Neutron diffraction methods follow a similar principle in which the white beam from a nuclear reactor is incident on a crystal monochromator which is used to select the required wavelength. The technical developments over recent years have led to the introduction of improved detection systems in which position-sensitive methods enable the diffraction pattern to be measured simultaneously over a wide angular range. This use of multidetectors has significantly improved data-collection rates but the accurate determination of the intensity profile still requires a period of about 10 hours, even on the most powerful sources [4].

An alternative method in which θ is fixed and a 'white' beam of variable wavelength is used, has more recently been employed in neutron diffraction studies. This technique is particularly advantageous for pulsed neutron beams available from accelerator-driven sources since time-of-flight detection systems may be used to define the neutron velocity and hence the incident wavelength. The initial work has been conducted on instruments such as the total-scattering spectrometer (TSS) at the Harwell Electron Linac

Molecule	Form	Type	Examples	Features/Interactions
Monatomic	rare gas metal	A	Ar, Kr, Ne Pb, Na etc.	spherical symmetry
Diatomic	homonuclear heteronuclear	A_2 AB	N_2, O_2, Br_2 etc. $HC\ell$, DBr , (CO) etc.	polar, orientation correlation
Triatomic	linear } bent }	AB_2	CS_2, (CO_2) H_2O, D_2O, (SO_2) etc.	orientation correlation hydrogen-bonding
Tetrahedral (spherical top)	homonuclear heteronuclear	A_4 AB_4	P_4, As_4 CD_4, $CC\ell_4$, $VC\ell_4$, $GeBr_4$ etc.	interlocking
Substituted Tetrahedral (symmetric top C_{3V})		AB_3C	$CHC\ell_3$, $VOC\ell_3$ etc.	interlocking
(broken C_{3V} symmetry)		AB_3R AB_3R	CH_3CN, CD_3OD	polar hydrogen-bonding
Hydrocarbons alkanes, polymers	symmetric chain		C_6D_6, C_6D_{12} (C_3H_8, C_nH_{2n}) etc.	six-fold symmetry inter-twined chains
Other organic systems	variable			complex

Table 1: Characterisation of different molecular systems and the particular features of the interaction between molecules in the liquid phase.

and an early report [6] outlined the main features required for investigation of both liquids and amorphous solids. The essential elements of the reactor (steady-state) method and the pulsed neutron method are illustrated in Figure 2 for measurements on a

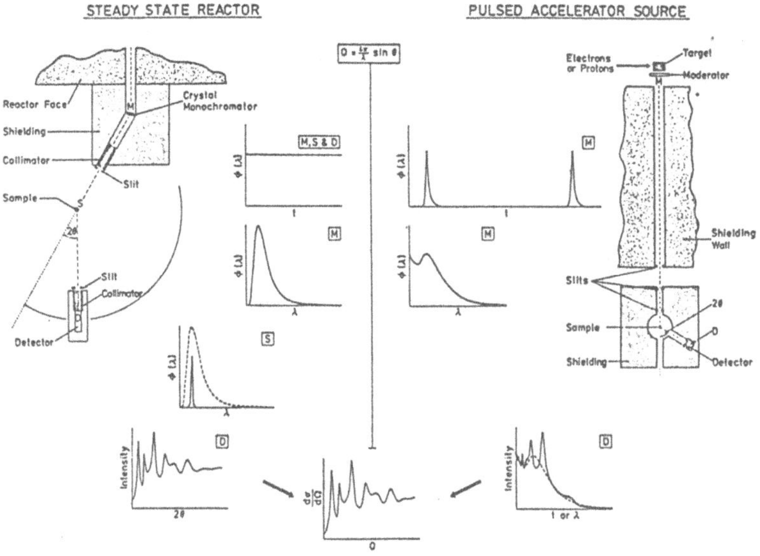

Figure 2. A comparison of steady-state (reactor) and pulsed (accelerator) techniques for neutron studies of liquids and amorphous solids. [7]

vitreous silica sample [7]. In the case of pulsed neutron diffraction it is necessary to define the shape of the incident spectrum, $\phi(\lambda)$ by scattering from a vanadium sample which is an almost completely incoherent scatterer. This measurement serves as a normalisation criterion for the measured intensity I(t) as a function of the time-of-flight. Since the neutron spectrum does not arise from equilibrium conditions in the moderator assembly there is an enhanced flux of epithermal neutrons ($\lambda \sim 0.05$-0.5Å) compared with the steady-state situation for the reactor case. As a result of this feature it is possible to make measurements for much higher Q-values than are normally available with reactor neutrons. A detailed comparison of the two methods has been reported by Dore and Clarke [8] and further discussion of new developments is deferred until the following Chapter (Section 6).
The X-ray analogue of this method utilizes energy-dispersive detection to measure the energy of the scattered radiation. Suit-

able germanium counters are already available for this purpose and the advent of high flux beams from synchrotron radiation sources [9] suggests that this could become an important area of future investigation. At the present time, little work has been conducted on disordered samples and the main emphasis has centered on crystallographic measurements [10].

The conversion of the raw data from the diffraction measurement to the final structure factor, $S_M(Q)$ requires various correction procedures. These can conveniently be divided into two types due to experimental and analytic influences. The experimental corrections depend on sample geometry and include absorption, container scattering and multiple scattering effects. Specific computer programs have been written to take account of these corrections and the data analysis procedure is fairly routine [11]. The analytic corrections are dependent on the physical properties of scattering mechanism which only approximates to the idealised form of elastic scattering. In the case of X-ray diffraction, the atomic form-factors are evaluated and combined on the assumption of the independent-atom approximation. Inelasticity effects in the form of Compton scattering and polarization corrections must also be applied. For neutron scattering allowance must be made for incoherent scattering contributions and in certain cases there may be large corrections due to inelasticity effects. These are particularly important for scattering nuclei of low effective mass such as hydrogen and deuterium. Methods of treating the inelasticity were originally presented by Placzek [12] for simple liquids and have been extended to molecular systems by Powles [13] Blum and Narten [2] and Egelstaff and Soper [14]. Satisfactory procedures have now been established for handling the measured data but the individual treatment may have slight variations depending on the methods that have been adopted. Fortunately the main effects apply to the intra-molecular terms, $f_1(Q)$, so that the systematic errors produced in the extracted $D_M(Q)$ results are expected to be small and hopefully, negligible for most cases. Various consistency checks can also be made to ensure that any errors in the final data are minimised. Details of the procedure adopted are usually described in the publications referring to the particular samples under investigation.

5. STRUCTURAL INFORMATION (SIMPLE MOLECULES)

The information that can be extracted from the observations is strongly dependent on the complexity of the molecular species being studied. The following sub-sections provide a sequence of examples which illustrate how the diffraction information may be related to the liquid structure for several simple molecular systems.

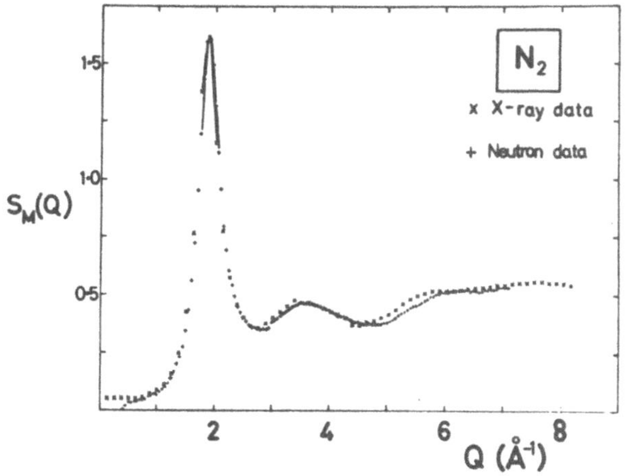

Figure 3. The structure factor for liquid N_2 from X-ray and neutron diffraction.

a) Diatomic Molecules

The simplest form of condensed molecular system is given by a homonuclear diatomic molecule such as nitrogen, N_2. In this case, there is only one pair correlation function $g_{NN}(r)$ which represents the site-site distribution in the liquid and it follows that both X-ray and neutron measurements should give the same or similar results. A comparison of two available datasets [15] is given in Figure 3 and shows good agreement in the general shape of the curve except at high Q-values where the oscillatory part of the curve due to the intra-molecular form-factor shows an apparent shift. This discrepancy was later resolved [16] when a full treatment of the inelasticity corrections was applied to the neutron data. The variation in the observed periodicity is explained primarily as a molecular recoil effect and the quantitative behaviour confirmed by several related experiments [17]. It has also been pointed out by Egelstaff [18] that the neutron and X-ray data are not strictly identical unless the centres of the electronic charge distribution co-incides with the nuclear positions. Using additional electron diffraction data for the gaseous phase, he has developed a set of related equations which suggest that there is a small variation in the intra-molecular distance, r_{NN}, in changing from the vapour phase to the liquid. In view of the possible systematic uncertainties involved in the treatment of diffraction data for different radiation probes it is perhaps premature to claim that this change has been positively identified but it does represent an interesting comparison which may be exploited more fully in the future, when greater precision can be achieved. Although nitrogen has served as a valuable reference material, the structural information is not of particular value in defining parameters for a model interaction potential. This is because the short bond-length ensures that the anisotropy is relatively weak and therefore the structure is not very sensitive to the details of the potential form. Various molecular dynamics computations are able to give satisfactory agreement with the diffraction data and do not distinguish amongst the fifty different forms [19] proposed for the nitrogen-nitrogen interaction.

Liquid oxygen is a similar homonuclear diatomic liquid to nitrogen but possesses paramagnetic properties so that the neutron diffraction contains an additional magnetic component (see following chapter). The halogens ($C\ell_2$, Br_2, I_2) have also been studied with both X-rays and neutrons. Results for liquid bromine [20] give a substantially different shape for the $S_M(Q)$ curve and the reason for this is not fully understood. Further studies [21] of these materials are in progress and it seems that the increased anisotropy of the molecule has an important effect on the structure. In contrast to nitrogen, the molecular dynamics simulations do not give a good representation of the experimental data. The hetero-nuclear diatomic systems ($HC\ell$, HBr) represent an even more complicated situation [22,23] and it is clear that more work is required. It would seem that even in the simplest molecular systems the characterization of liquid structure is far from complete.

b) Tetrahedral Molecules

Another frequently studied liquid is $CC\ell_4$ and is representative of a large class of molecules with T_d symmetry. A simpler homonuclear molecule with tetrahedral geometry is yellow phosphorus, P_4 and gives rise to a single oscillatory term in the form-factor which may be expressed as

$$f_1(Q) = \frac{1}{4}\left[1 + 3\,\frac{\sin Qr_{PP}}{Qr_{PP}}\; e^{-\gamma_{PP}Q^2}\right]$$

where r_{PP} is the PP bond-length and γ_{PP} is proportional to the mean square amplitude of vibration. A series of neutron diffraction measurements [24] have been made on the normal liquid, the super-cooled liquid and the plastic crystal phase. The structure factor and the fitted form factor are shown in Figure 4. The resulting $D_M(Q)$ curve has a more complicated form than for the diatomic liquids with a prominent sharp peak which signifies that correlations extend over a much longer range. The spatial distribution $d_L(r)$ is shown in Figure 5 for a number of different temperatures. The structural arrangement is relatively insensitive to temperature variation and the correlations are primarily dependent on the molecular shape. It can be shown that the short range order is influenced by the interlocking of the molecules so that the atom of one is situated in the hollow formed by the three atoms of the neighbouring molecule (Figure 6), but it is surprising

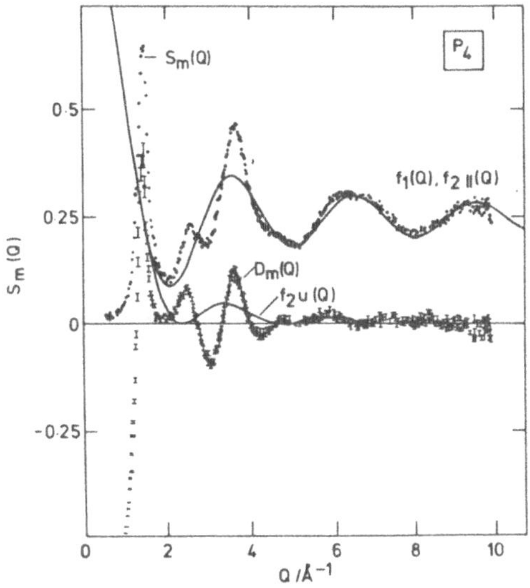

Figure 4. The structure factor and fitted molecular form-factor for neutron scattering by liquid P_4.

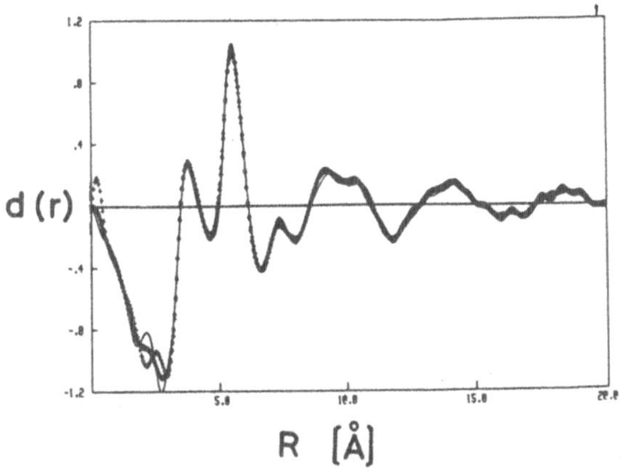

Figure 5. The $d_L(r)$ function for liquid P_4 at various temperatures.

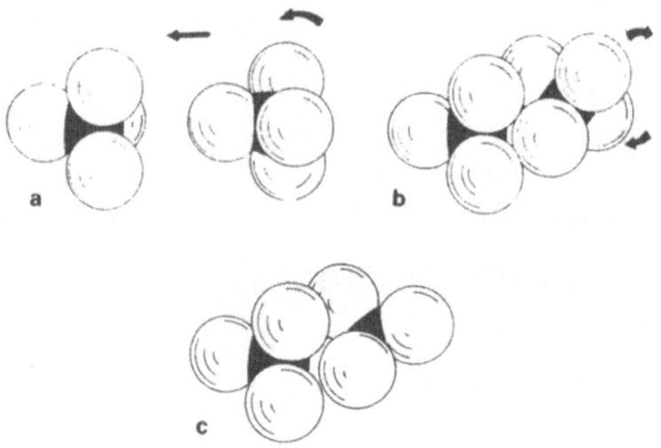

Figure 6. Interlocking effects in tetrahedral molecules.

that phosphorus exhibits this behaviour so strongly as the anisotropy is not large [25].

It is interesting to note that the diffraction measurement can only yield one piece of information and it is impossible to evaluate the centres structure factor $S_c(Q)$ directly from the observations since there is no scattering centre in this position. The complex relationship involving orientational correlations between molecules can only be extracted by comparing the data with model simulations; no molecular dynamics calculations have been reported for this liquid and therefore the details are not known.

The tetrachloride liquids $XC\ell_4$ (X = C, Si, Ti, V, Ge, Sn etc) make an interesting series for studying aspects of orientational correlation in relation to molecular contours. The various diffraction studies have been surveyed in a paper [25] which uses the reference interaction site model (RISM) to evaluate the partial distribution functions. Data for carbon tetrachloride are particularly instructive as they illustrate the complementary nature of X-ray and neutron scattering. There are three partial functions $g_{CC}(r)$, $g_{CC\ell}(r)$ and $g_{C\ell C\ell}(r)$ and the weighting factors for the diffraction measurements (eqn.2.11) are

$$g^x(r) = 0.01\, g_{CC}(r) + 0.15\, g_{CC\ell}(r) + 0.34\, g_{C\ell C\ell}(r)$$

for X-ray scattering and

$$g^N(r) = 0.02\, g_{CC}(r) + 0.25\, g_{CC\ell}(r) + 0.73\, g_{C\ell C\ell}(r)$$

for neutron scattering.

Since $g_{CC}(r)$ is a small contribution in both cases it can be neglected and the two datasets used to solve for $g_{CC\ell}(r)$ and $g_{C\ell C\ell}(r)$. This method was first used by Narten [26] and confirmed the inter-locking characteristics expected from the RISM predictions [27]. Further measurements by van Tricht [28] have given somewhat different conclusions for the $g_{CC\ell}(r)$ distribution but the basic behaviour of the experimental data appears to be similar. The third partial term, $g_{CC}(r)$ which represents the distribution of molecular centres cannot be measured easily because it makes such a small contribution to the total interference function. One possibility, investigated by Neilson [29], is to use a first order difference technique based on substitution of ^{13}C for ^{12}C in a neutron measurement. The use of isotopic substitution to give changed b-values in order to separate the partial terms will be discussed in Section 6.

Neutron diffraction studies of amorphous $CC\ell_4$ prepared by vapour-deposition [30] have also emphasised the inter-locked nature of the molecular units. The $D_M(Q)$ function has a different shape to that of the liquid, Figure 7, because of the higher packing density and loss of translational motion but it can be readily shown that the data are consistent with the inter-locking of the molecular contours as shown in Figure 6. It seems likely that the enhanced correlations of the amorphous solid can give useful information on the probable structure for the liquid phase but little work has yet been conducted along these lines for other molecular systems. Liquid carbon tetrachloride has also been used as an important reference molecule for pulsed neutron studies at high Q-values and this is reported separately [31,32].

The rest of the tetrachloride liquids have not been studied so extensively but $VC\ell_4$ is of some interest since neutron scattering is effectively limited to the chlorine atoms and the $g_{C\ell C\ell}(r)$ partial distribution is extracted from a single measurement. The results [33] are given in Figure 8 and are quite well represented by the MD computations of Gubbins and Murad [34].

c) Triatomic

Carbon disulphide has been studied by X-ray [35] and neutron scattering [36]. The resulting $d_L(r)$ functions are shown super-imposed in Figure 9. This is another good example of the complementary nature of the two measurements since sulphur has a small b-value for neutron scattering and the weight factors

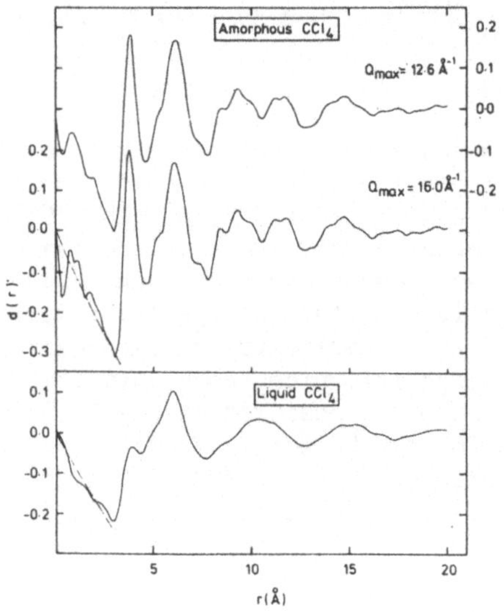

Figure 7. A comparison of data for liquid and amorphous CCl_4.

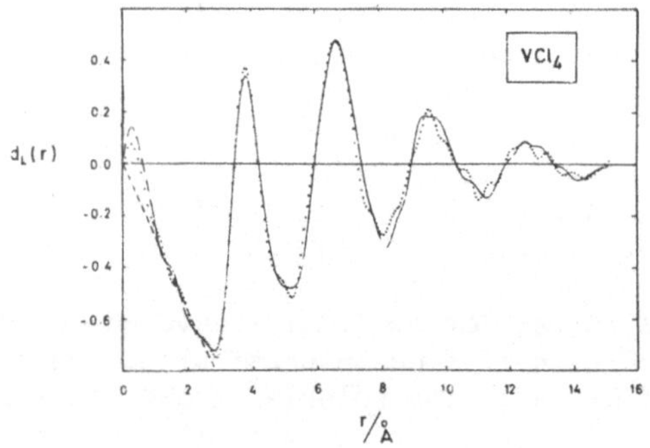

Figure 8. The $d_L(r)$ function representing $ClCl$ partial correlations in liquid VCl_4.

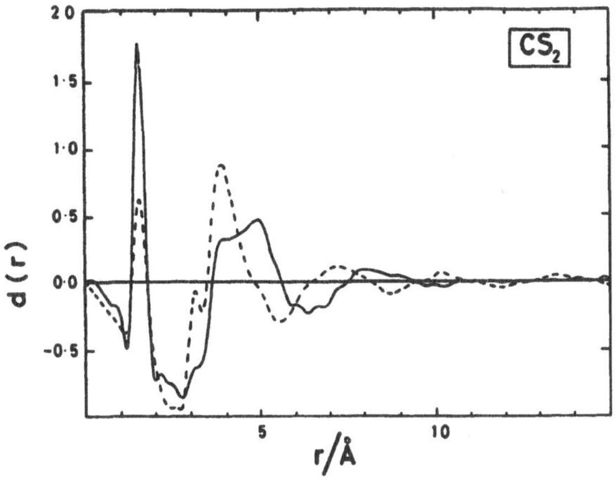

Figure 9. Combined X-ray and neutron data showing $d_L(r)$ for liquid CS_2 (... = X; —— = n).

become;-

$$g^X(r) = 0.03\ g_{CC}(r) + 0.26\ g_{CS}(r) + 0.71\ g_{SS}(r)$$

and

$$g^N(r) = 0.29\ g_{CC}(r) + 0.50\ g_{CS}(r) + 0.21\ g_{SS}(r)$$

It is not possible to separate the three partial functions from the two datasets but MD simulation [37] gives excellent agreement with the observations and again indicates that the structure is relatively insensitive to the details of the interaction potential. It is possible to use isotopic substitution for either the carbon or the sulphur atoms in order to obtain an independent third measurement but the structure is sufficiently well understood for this to be unnecessary in this case.

Water, H_2O, constitutes another triatomic molecule which is of major importance for liquid state studies but there are specific technical problems in obtaining the required information from diffraction experiments so that this discussion is deferred until Section 6.

d) General aspects for small molecules

The previous sections have illustrated how structural information in the form of g(r) functions can be extracted from X-ray and neutron diffraction measurements on liquids composed of relatively simple molecules. For homonuclear systems the data analysis is straightforward but for two-component liquids, three independent measurements are required to fully define the partial pair-correlation functions. A third independent measurement can sometimes be made for neutron diffraction with isotopic substitution to change the b-value of one of the components. Table 2 gives a list of isotopes relevant to simple molecular liquids for which this technique is feasible. The large difference between H and D is complicated by the large incoherent scattering from hydrogen and the extreme form of the inelasticity corrections. (Section 6c).

Z	Symbol	Natural Abundance (%)	b_{coh} (fm)	σ_{coh} (barn)	σ_{incoh} (barn)
1	H	99.98	-3.74	1.8	~80
	D	0.02	6.67	5.6	2.0
	T		4.7	2.8	?
6	C	(100)	6.65	5.6	0
	^{13}C	1.1	6.0	5.5	1.0
7	N	(100)	9.40	11.1	0.3
	^{15}N	0.4	6.5	5.3	?
8	O	(100)	5.83	4.2	0
	^{17}O	0.04	5.78	4.2	?
	^{18}O	0.2	6.00	4.5	?
16	S	(100)	2.85	1.02	0.2
	^{33}S	0.76	4.7	3.00	?
	^{34}S	4.22	3.48	1.54	?
17	Cℓ	(100)	9.58	11.5	5.9
	$^{35}Cℓ$	75.5	11.8	17.5	?
	$^{37}Cℓ$	24.5	2.6	0.8	?

Table 2. A compilation of coherent neutron scattering lengths for various isotopes relevant to studies of molecular liquids

The differences for Cℓ and N isotopes are sufficiently large for useful work to be conducted but C is marginal and the oxygen isotopes have insufficient variation. The isotopic difference method has proved immensely valuable for the study of aqueous solutions [38] and liquid alloy systems [39] but has so far had a limited application to molecular liquids. This is partly because there are less convenient isotopes but also because the additional information is not always required and a knowledge of the structure of the molecular unit often constitutes an important restriction on the possible modelling of the experimental results. It is likely that this situation will change when higher neutron fluxes are available (see next chapter) or when there is greater confidence in analysing data obtained from H/D substitution.

6. SPECIAL METHODS

a) Organic molecules ($CHCl_3$, CH_3CN, CH_3OH etc.)

Many organic liquids are of considerable importance as solvents and often consist of three or more atomic species. Under these circumstances a full description in terms of the individual partial contributions becomes extremely difficult; a three-component liquid would require six independent measurements to give a complete description. The most extensive work in this area has been reported by Zeidler, Bertagnolli and co-workers [40] who have used X-ray diffraction measurements with several sets of neutron data obtained from isotopic substitution. Chloroform may be characterised as a symmetric top molecule which has been extensively studied by spectroscopic methods and it was therefore of some interest to represent the data in terms of the spherical harmonic expansion (eqn.2.12). The diffraction data were therefore combined to give the various $g_{\ell\ell' m}(r)$ coefficients in the expansion and the results are shown in Figure 10. Since there are both statistical and systematic uncertainties in the treatment of the combined datasets, the final results have a band of error bars as shown in the graphs. This illustrates some of the difficulties associated with the extraction of orientational relationships from a set of diffraction measurements. Although the representation in terms of $Y_\ell^m(\theta,\phi)$ is useful for the discussion of molecular re-orientational motion, the series is slowly convergent and many terms may be required to give an accurate description of the structural data. The diffraction profile is effectively due to interference in the scattering from different atomic (nuclear) sites so the description in terms of pair-correlation functions is more directly linked to the observations. For simple molecules the g(r) formalism is clearly advantageous but as the molecular size and complexity increases, the set of individual partial terms become less informative and it is the behaviour of the total molecular unit that essentially characterises the properties of the liquid. Little work has yet been done on organic liquids of this type but it seems clear that MD simulation will be required to provide a physical interpretation of the experimental data in whatever form is most appropriate.

The work on chloroform has been used as an illustrative example and similar principles apply to the diffraction data for acetonitrile which have also been obtained from the isotopic substitution experiments [41]. A recent MD simulation due to

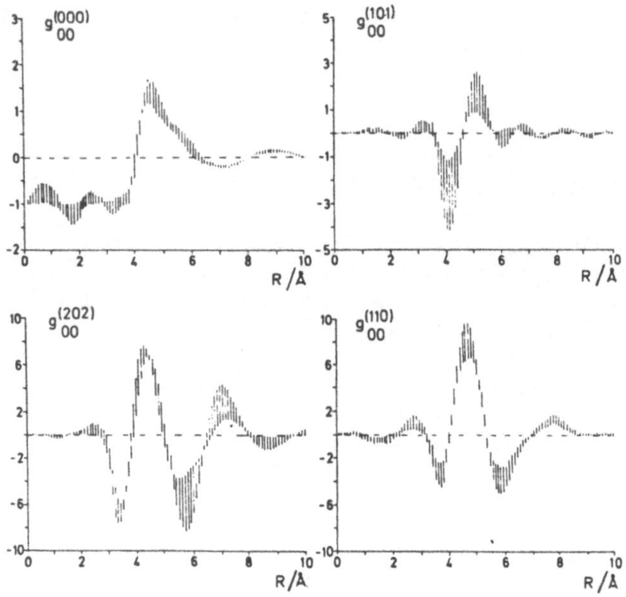

Figure 10. The coefficients g(R) in a spherical harmonics representation of orientation correlation in liquid chloroform $CHCl_3$

Madden [42] gives satisfactory agreement with the diffraction data and indicates quite well the way in which this field is likely to develop. Another topic of current interest centres on methyl alcohol in which H/D substitution methods have been used [43]. In this case it is the hydrogen-bonding interaction which is of major importance as this gives a strong orientational dependence to the interaction potential. The initial findings tend to confirm the predictions of a hydrogen-bonded chain-like structure predicted by Monte Carlo computations of Jorgensen [44]. Further work on methylene chloride, CH_2Cl_2 has been reported [45] and it seems likely that the experience gained from these studies will eventually define the limitations of the technique. However, it is clear that a series of very precise measurements must be made in order to define the structure for just one state point (fixed temperature and pressure) so that progress in this area will not be very rapid.

b) Temperature-difference measurements

The temperature variation experiments for liquid phosphorus showed little change over a range of 50°C but it has been known for some time that there is a significant structural re-arrangement in water with change in temperature. A series of X-ray measurements were reported by Narten [46] as early as 1970 and the first neutron measurements for D_2O [47] were made in 1977. The data are most conveniently analysed using a first-order difference procedure. This involves taking the difference between the structure factor $S_M(Q,T)$ at a temperature T and at a reference temperature, T_o, i.e.

$$\Delta S_M(Q,T) = S_M(Q,T) - S_M(Q,T_o) = \Delta D_M(Q,T)$$

assuming that the molecular form-factor $f_1(Q)$ is independent of small temperature changes. The function $\Delta D_M(Q)$ represents the structural change in the liquid and can be converted to a real-space representation by the transform relation

$$\Delta d_L(r,T) = \frac{2}{\pi}\int_0^{Q_m} Q\Delta D_M(Q,T)\sin Qr\, dQ$$

The advantage of this analysis method is that it removes many of the systematic uncertainties in the treatment of the data and in particular minimises the inelasticity corrections. Some of the early measurements on D_2O in the normal liquid phase [48] are shown in Figures 11 and 12; further results have been obtained in the super-cooled region [49] which illustrate a similar behaviour below the normal freezing point. Several groups have now made both X-ray and neutron measurements [50-52] which appear to show general agreement although the details of the interpretation tend to emphasise different features. The general conclusions may be briefly stated as supporting the importance of collective phenomena in the basic interaction potential which is strongly influenced by hydrogen-bonding. As the temperature is reduced the orientational correlations become stronger and deviations from a linear H-bond are more restricted. The structure appears to be evolving towards the continuous random H-bonded network of amorphous ice [53]. Additional data have now been taken for amorphous methyl alcohol [54] and there is much current activity on this topic which extends beyond the range of this brief review.

c) Isotopic substitution for water

No discussion of molecular liquids could be complete without reference to the *structure* of water. The determination of an accurate set of partial correlation functions $g_{OO}(r)$, $g_{OH}(r)$ and $g_{HH}(r)$ is probably the most important task in diffraction studies

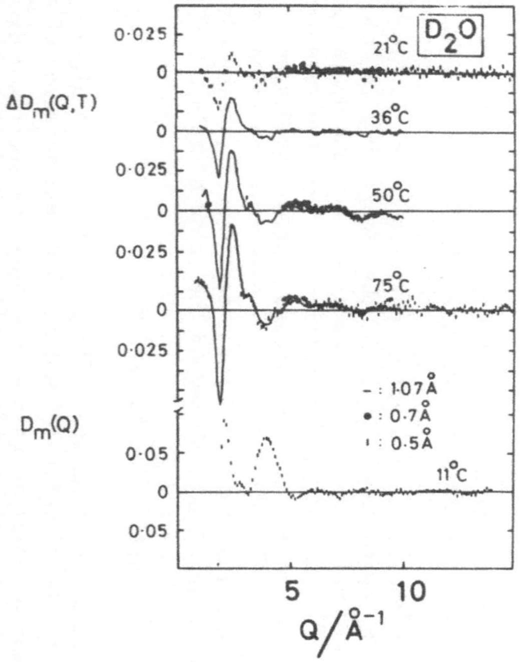

Figure 11. The difference function ΔD_M (Q,T) for neutron diffraction by D_2O water.

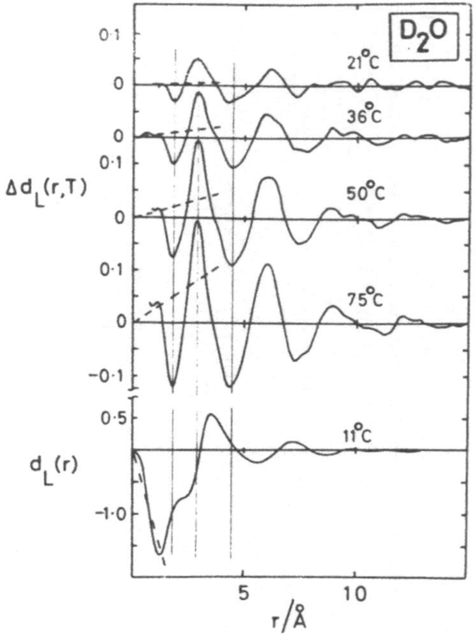

Figure 12. The spatial representation, $d_L(Q,T)$ of structural changes corresponding to Figure 11.

of the liquid state. Unfortunately, there are considerable problems to be overcome and a definitive dataset has not yet been obtained. In this short review it is only possible to give a brief indication of the complexities and how they may be resolved.

The use of H/D isotopic substitution is, in principle, very simple but the presence of a large incoherent scattering from hydrogen (Table 2) makes the accurate determination of the coherent differntial scattering cross-section very difficult to measure. The scattering from D_2O, H_2O and two special mixtures of H_2O/D_2O is shown in Figure 13 [55]. The general fall-off in intensity at high Q-values is due to inelasticity effects. The computed scattering from individual molecular units (intra-molecular scattering) is also shown and this must be subtracted to give the intermolecular contribution ($D_M(Q)$). Four datasets have been measured to overdetermine the three partial functions and to provide the necessary consistency checks. The experimental corrections, particularly multiple-scattering, need to be carefully evaluated to avoid systematic errors in the solution of the simultaneous equations. Similar experiments have been reported by Narten and Thiessen [56] and pulsed neutrons techniques used for different combinations of H_2O/D_2O mixtures by Soper and Silver [57]. Unfortunately the final conclusions by the different groups are not in good agreement although the general shape of the partial g(r) functions exhibit similar features at short distance (<3 Å). The major discrepancy is at longer distances (>4 Å) where the data of Reed and Dore [58] show oscillatory structure similar to that presented earlier by Palinkas et al [59] from a combination of X-ray, neutron and electron diffraction data. These problems remain unresolved and the form of the partial pair-correlation functions is still a controversial research topic. An additional factor in the analysis is the assumption that H and D substitution corresponds to isomorphous replacement. Recent results by Montague and Dore [43] on CD_3OH/D mixtures seem to confirm the early findings for water that H and D are not equivalent. This possibility is not really surprising in view of the extreme sensitivity of the structure to temperature variation and the knowledge that an isotopic change from H and D is equivalent to a temperature shift of about 6°C. Water is a complex liquid and the interpretation of the existing diffraction results appears to be rather contentious. The related work on temperature variation studies for liquid D_2O [48] and the established structure of amorphous D_2O ice are consistent with the existence of correlations in the 5 - 8 Å range. This most-common liquid seems well set to retain the mysteries of its complex hydrogen-bonded properties until some further date in the future. There is an increasing body of opinion that believes the key to a more complete understanding rests in the study of the deeply-supercooled state where the anomalous behaviour is more pronounced [60] and co-operative effects [61] are more evident.

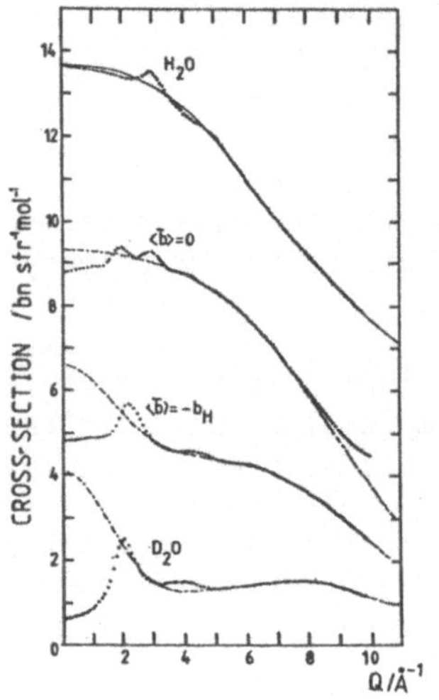

Figure 13. The scattering cross-section for various H_2O/D_2O mixtures of liquid water at 20°C.

7. CONCLUSIONS

The previous sections have illustrated the wide range of topics which are currently being investigated by diffraction methods. Although considerable developments have taken place during the last decade the structural study of liquids is still an experimental field in its infancy. The number of active groups is certainly small when compared with the large community of crystallographers but the techniques are now well established and the investigation of new inorganic liquids could be regarded as routine procedure. The main complexity and interest now seems firmly directed towards hydrogen (deuterium) -containing liquids and particularly the effects of hydrogen-bonding.

References

(1) Powles, J.G., 1973, Advances in Phys., 22, 1.
(2) Blum, L. and Narten, A.H., 1973, Adv.in Chem.Phys., 34, 203.
(3) Seip, H.M., in 'Molecular Structure by Diffraction Methods" G.A.Sim and L.E.Sutton (eds), Chem.Soc.,London, 1973 Vol.I.
(4) Forsyth, J.B. and Convert, P., 'Position Sensitive Detectors', to be published.
(5) Bertagnolli, H., and Zeidler, M.D., 1978, Mol.Phys., 35, 177.
(6) Sinclair, R.N., Johnson, D.A.G., Dore, J.C., Clarke, J.H. and Wright, A.C., 1974, Nuc.Inst. & Meth., 117, 445.
(7) Sinclair, R.N., Desa, J.A.E., Etherington, G., Johnson, P.A.V. and Wright, A.C., 1980, J.Non-cryst Solids, 42, 107.
(8) Dore, J.C. and Clarke J.H., 1976, Nuc.Inst. & Meth., 136, 79.
(9) Winnick, H. and Doniach, S. (Eds), 1980, 'Synchrotron Radiation Research', Plenum Press.
(10) Buras, B., Olsen, J.S. and Gerward, L., 1978, Nuc.Inst. & Meth., 152, 293.
Glazer, M. and Thompson, P., private communication
(11) Details of data analysis incorporate corrections such as given by Paalman, H.H. and Pings, C.J., 1962, J.Appl.Phys., 33, 2635 and Meardon, B.H., 1973, AERE Report R8121.
(12) Placzek, G., 1952, Phys.Rev., 86, 377.
(13) Powles, J.G., 1979, Mol.Phys., 37, 623, and references to earlier papers.
(14) Egelstaff, P.A., Adv. in Chem.Phys., to be published.
(15) Dore, J.C., Walford, G. and Page, D.I., 1975, Mol.Phys., 29, 565.
(16) Page, D.I. and Powles, J.G., 1975, Mol.Phys., 29, 1287.
Powles, J.G. and Rickayzen G., 1976, Mol.Phys., 32, 323.
Clarke, J.H., Dore, J.C. and Egger, H., 1979, Mol.Phys., 39, 533.
(17) Powles, J.G., Dore, J.C. and Osae, E.K., 1980, Mol.Phys., 40, 193 and 41, 475.
Deraman, M.D., Dore, J.C. and Powles, J.G., 1984, Mol.Phys., in press.
(18) Egelstaff, P.A., 1982, Phys.Chem.Liq., 11, 353.
(19) Powles, J.G., Private communication.
A recent survey of current models has been given by S.Romano to be published.
(20) Clarke, J.H., Dore, J.C., Walford, G. and Sinclair, R.N., 1976, Mol.Phys., 31, 883.
(21) Ricci, F.P. and Andreani, C., poster at this meeting and private communication.
(22) Soper, A.K. and Egelstaff, P.A., 1980, Mol.Phys., 39, 1201.
Powles, J.G., Osae, E.K., Dore, J.C. and Chieux, P., 1981, Mol.Phys., 43, 1051.
(23) Powles, J.G., Dore, J.C., Osae, E.K., Clarke, J.H., Chieux, P. and Cummings, S., 1981, Mol.Phys., 43.
(24) Granada, J.R. and Dore, J.C., 1982, Mol.Phys., 46, 757.

(25) Montague, D.G., Chowdhury, M.R., Dore, J.C. and Reed J., 1983, Mol.Phys., 50, 1.
(26) Narten, A.H., 1976, J.Chem.Phys., 65, 573.
(27) Lowden, L.J. and Chandler, D., 1973, J.Chem.Phys., 59, 6587; 1974, Ibid, 61, 5228.
(28) van Tricht, J.B., 1977, J.Chem.Phys., 66, 85.
van Tricht, J.B., 1977, Thesis, University of Delft.
(29) Neilson, G.W., private communication.
(30) Chowdhury, M.R. and Dore, J.C., 1981, J.Non.Cryst., 43, 267.
(31) Clarke, J.H., Granada, J.R. and Dore, J.C., 1979, Mol.Phys., 37, 1263.
Granada, J.R., Stanton, G.W., Clarke, J.H. and Dore, J.C., 1979, Mol.Phys. 37, 1297.
(32) Dore, J.C. and Bermėjo, F., in preparation.
Dore, J.C., 1984, J.Mol.Liquids, in press.
(33) Gibson, I.P. and Dore, J.C., 1979, Mol.Phys., 37, 1281.
(34) Murad, S. and Gubbins, K.E., 1980, Mol.Phys., 39, 271.
(35) Sandler, S.I. and Narten, A.H., 1975, Mol.Phys., 32, 1543.
(36) Gibson, I.P. and Dore, J.C., 1981, Mol.Phys., 42, 83.
(37) Tildesley, D.J., 1978, Faraday Disc., 66, 27 and private communication.
(38) Enderby, J.E. and Neilson, G.W., Vol.6 in 'Water: a comprehensive treatise', (ed) F.Franks, Plenum Press.
(39) Enderby, J.E., Egelstaff, P.A. and North, D.M., 1966, Phil. Mag., 14, 961.
A more recent review is Waseda, Y., 1980, 'The Structure of Non-crystalline Materials', (pub) McGraw-Hill.
(40) Bertagnolli, H., Leicht, D.O. and Zeidler, M.D., 1978, Mol. Phys., 35, 193.
(41) Bertagnolli, H., Leicht, D.O. and Zeidler, M.D., 1978, Mol. Phys., 36, 1769.
(42) Bohm, H.J., Madden, P.A. and McDonald, I.R., to be published; Madden P.A., private communication.
(43) Montague, D.G., Dore, J.C. and Cumming, S., to be published.
(44) Jorgensen, W.L., 1980, J.Am.Chem.Soc., 102, 53; Ibid 103, 341.
(45) Zeidler, M. and Jung, W., poster at this meeting and private communication.
(46) Narten, A.H., Danford, M.D. and Levy, H.A., 1967, Discuss. Faraday Soc., 43, 97.
(47) Walford, G. and Dore, J.C., 1977, Mol.Phys., 34, 22.
(48) Gibson, I.P. and Dore, J.C., 1983, Mol.Phys., 48, 1019.
(49) Bosio, L., Teixeira, J., Dore, J.C., Steytler, D.C. and Chieux, P., 1983, Mol.Phys., 50, 733.
(50) Bosio, L., Chen, S-H and Teixeira, J., 1983, Phys.Rev. A, 27, 1468.
(51) Egelstaff, P.A., Polo, J.A., Root, J.H., Hohn, L.J. and Chen, S.-H, 1981, Phys.Rev.Lett., 47, 1733.
Egelstaff, P.A. and Root, J.H., 1983, Chem.Phys., 76, 405.
(52) Ohtomo, N., Tokiwano, K. and Arakawa, K., 1982, Bull. Chem. Soc.Japan, 55, 2788.

(53) Chowdhury, M.R., Dore, J.C. and Wenzel. J.T., 1982, J.Non. Cryst.Sol., 53, 247.
(54) Montague, D.G., Gibson, I.P. and Dore, J.C., 1981, Mol.Phys., 44, 1355.
(55) Reed, J., 1981, Thesis, University of Kent.
(56) Thiessen W.H. and Narten, A.H., 1982.
Narten, A.H., Thiessen, W.H. and Blum, L., 1982, Science, 217, 1033.
(57) Soper, A.K. and Silver, R.N., to be published.
(58) Reed, J., Dore, J.C. in preparation.
(59) Palinkas, G., Kalman, E. and Kovacs, P., 1977, Mol.Phys., 34, 525.
(60) Angell, A., Chapter 1 in Vol.7, 'Water: A comprehensive treatise', F.Franks (ed), 1982, Plenum Press.
(61) Stanley, H.E., 1979, J.Phys A, 12, L329.
Stanley, H.E. and Teixeira, J., 1980, J.Chem.Phys., 73, 2404.

TECHNIQUES IN NEUTRON SCATTERING STUDIES OF MOLECULAR SYSTEMS

J.C.Dore

The Physics Laboratory, University of Kent, Canterbury, England. CT2 7NR

ABSTRACT

Thermal neutron diffraction by molecular liquids (previous chapter) does not cover the whole area of neutron research topics which have a bearing on molecular systems. Some of the special neutron methods which can be used are outlined in the present chapter, incorporating work on molecular gases which relates directly to the orientationally-averaged inter-molecular potential, polarized neutrons for the study of magnetic correlations and small-angle neutron scattering (SANS) for enhanced density fluctuations and hetero-phase systems such as microemulsions. Inelastic and quasi-elastic neutron scattering can also be used to study either collective or independent-particle motion. Finally, a brief survey of new prospects with the high-intensity sources (neutron and X-ray) based on accelerator-driven facilities currently under construction, is presented.

1. INTRODUCTION

The previous chapter has emphasised the use of neutron scattering in conjunction with X-ray diffraction for studying the structural characteristics of different molecular liquids. In some respects, the structure is only an intermediate piece of information which helps to establish the parameters needed to describe an effective interaction potential. It is therefore of some interest to study molecular systems of low density where pair interactions predominate and the potential can be studied more directly. The first section therefore deals with neutron diffraction by molecular gases and the interpretation in terms of the structural second virial

A. J. Barnes et al. (eds.), Molecular Liquids - Dynamics and Interactions, 411–429.

coefficient. Gaseous oxygen is of particular interest in this context as there is also a magnetic scattering component. Techniques of polarization analysis can be used to separate nuclear and magnetic contributions and these are found to be of particular relevance for the study of magnetic correlations in liquid oxygen.

The low Q-value region (< 0.5 Å^{-1}) can be investigated by small-angle neutron scattering (SANS) and corresponds to long-range correlations in the liquid. In certain cases, the behaviour in this region may show the effects of critical behaviour through density fluctuations in the liquid but a more direct use of the technique is for the determination of particle sizes in hetero-phase systems such as micellar solutions and microemulsions. Another major use of neutron facilities is by inelastic and quasi-elastic scattering where specific molecular motions may be investigated. Only a small number of coherent inelastic scattering experiments have been carried out for molecular liquids in order to study the dynamic structure factor, $S(Q,\omega)$. The main effort has been directed towards the study of proton motion through incoherent inelastic scattering which utilizes the high scattering cross-section of the H atom. This work is complementary to other forms of spectroscopic investigation and can be used to give detailed information on particle trajectories through the elastic incoherent structure factor (EISF). In the final section, some consideration is given to future prospects resulting from the construction of new improved central facilities with higher fluxes of both neutrons and X-rays.

2. DIFFRACTION STUDIES IN THE Q-RANGE UP TO 2 Å^{-1}.

a) Molecular correlations

As shown in the previous chapter, the oscillatory features of the structure factor are confined to Q-values larger than 2 Å^{-1} and there is usually a well-defined peak in the 1-2 Å^{-1} range. Below this main diffraction peak there is a smoothly varying curve which extrapolates to give a small intercept at Q = 0. Due to the reciprocity in the transform relation, the behaviour as $Q \to 0$ is due to the bulk properties of the liquid and the intercept is related to the isothermal compressibility, χ_T at a temperature T by the expression [1]:-

$$[S_M(Q)]_{Q\to 0} = \rho\, k_B\, T\, \chi_T$$

It is impracticable to measure χ_T by this means due to contributions from incoherent and multiple scattering effects. If the density is reduced the fluid becomes much more compressible and the intercept increases such that the scattering is equivalent to the molecular form-factor in the limit of zero density. This provides an interesting possibility for study in the region of finite low-density systems and several experiments have now been made for gas phase diffraction.

A theoretical formalism for neutron scattering from molecular gases is given by Powles, Dore and Osae [2]. A similar treatment has also been given by Egelstaff [3] and both groups have concentrated their experimental work on N_2 gas although Egelstaff and Teitsma [4] have also made an extended study of Kr gas.

For homonuclear diatomic molecules interacting via an 'almost-isotropic' potential, the structural second virial coefficient $B_2(Q,T)$ may be written in the form [2]:-

$$B_2(Q,T) = -2\pi \int_0^\infty R^2 \left(\frac{\sin QR}{QR}\right)\left\{\exp(-\beta\phi(R))-1\right\} dR$$

where R is the separation of molecular centres and $\phi(R)$ is an effective average over all molecular orientations. It can then be shown that the inter-molecular cross-section is given by:-

$$\left[\frac{d\sigma}{d\Omega}(Q)\right]_{inter} = \overline{b}^2\left[\frac{\sin(Qd/2)}{(Qd/2)}\right]^2 B_2(Q,T)\rho$$

where d is the intra-nuclear bond length and ρ is the density. A more recent measurement for N_2 gas at 100 bar[5] is shown in Figure 1 with a fit to the intra-molecular scattering superimposed. Figure 2 shows the inter-molecular cross-section with a fit based on the Kihara potential [6]. Since the N_2 molecule has a small anisotropy, very precise measurements are required to discriminate between different potential forms and this situation is similar to that reflected in the liquid studies presented in the previous chapter. A similar treatment of neutron measurements [7] on SF_6 gas, however, does reveal some interesting facts about the parameterization of the potential and discriminates between different models. It therefore seems that neutron diffraction studies of the gas phase can yield useful supplementary information about angle-averaged interaction potentials. In this sense $B_2(Q,T)$ can be seen as an extension of the conventional second virial coefficient $B_2(T)$. At present, these two molecules (N_2 and SF_6) are the only ones that have been studied in detail but the method clearly has scope for wider applications.

b) Magnetic correlations

Liquid oxygen appears initially to have very similar proper-

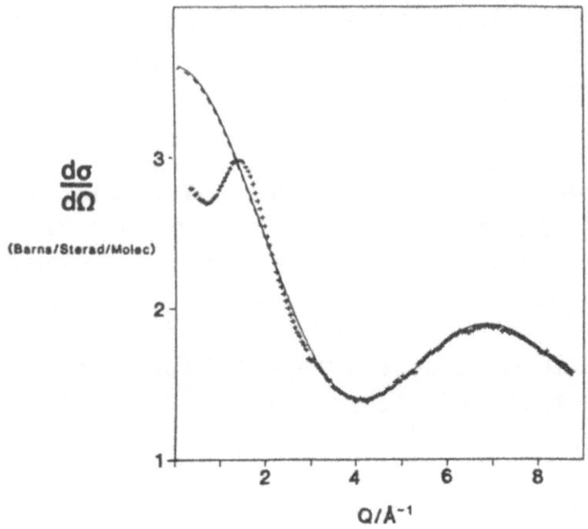

Figure 1.
The cross-section and form-factor fit for neutron scattering from N_2 gas at 100 bar and 293 K.

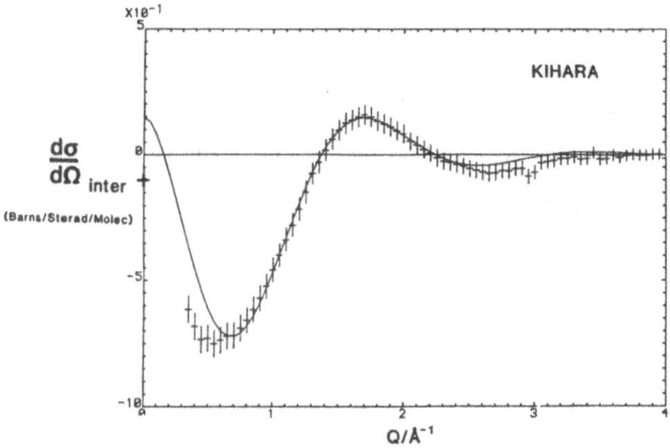

Figure 2.
The inter-molecular scattering contribution from Figure 1, compared with a prediction [5] based on the Kihara potential [6].

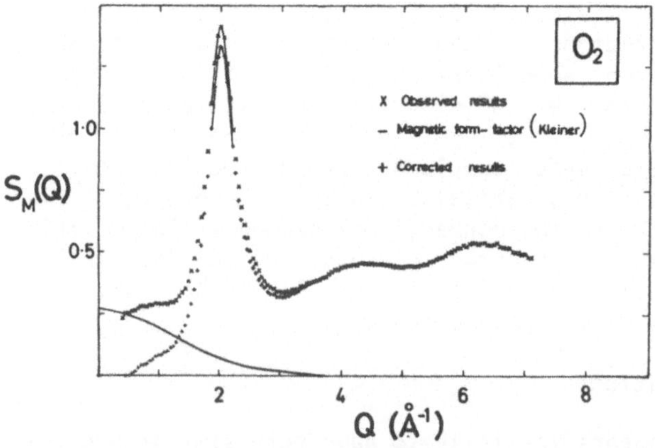

Figure 3.
Neutron scattering oxygen showing the Kleiner form-factor for magnetic scattering by the O_2 molecule.

ties to liquid nitrogen but due to antibonding orbitals the electron spins are paired to give a Σ^3 ground-state with spin S = 1. As a result of this, liquid oxygen displays paramagnetic properties. Since the neutron possesses an anomalous magnetic moment, there is an additional magnetic scattering contribution to the differential cross-section. This feature is clearly seen in the neutron diffraction pattern for liquid O_2 [8] as shown in Figure 3; the magnetic form-.actor prediction due to Kleiner [9] is also shown. A close examination revealed the shape of the curve in the low-Q region did not correspond to the expected behaviour. In order to investigate this discrepancy in more detail it is necessary to separate the nuclear and magnetic scattering components in a development of techniques first introduced by Moon, Riste and Koehler [10]. Figure 4 shows the polarization triple-axis spectrometer, D5, at the ILL. The white beam from the reactor is incident on a monochromator which gives an almost completely polarized monochromatic beam. After scattering from the sample, the spin-state of the scattered beam is analysed by another magnetic crystal. Separation of coherent and spin-incoherent scattering for liquid D_2O has been demonstrated [11] and a further development utilizing spin-rotation can be used to separate magnetic and nuclear scattering. The results of measurements on liquid O_2 [12] are presented in Figure 5. Diffraction studies of the gas phase [13] show satisfactory agreement with the Kleiner form-factor so that the magnetic scattering in the liquid phase exhibits an effective structure factor involving magnetic correlations. The results are consistent with antiferromagnetic correlations in which adjacent molecules tend to have opposed spin directions. These features have yet to be evaluated in detail but liquid oxygen appears to represent an unusual system of a liquid behaving as a disordered antiferromagnet. It is likely that other molecules,such as NO, will exhibit similar correlations but little theoretical work has yet been done. The topic of 'magnetic correlations in molecular liquids' is therefore a rather restricted one but will undoubtedly make interesting contributions to the study of magnetic interactions in general.

3. SMALL-ANGLE SCATTERING ($Q < 0.5$ Å^{-1})

a) Density fluctuations

At very low Q-values the liquid structure factor normally extrapolates smoothly to the isothermal compressibility limit at $Q = 0$ as expected. However, water has some anomalous properties and does not exhibit a monotonic change of slope as Q is decreased. The experimental studies of Bosio, Teixeira and Stanley [14] by small-angle X-ray scattering (SAXS) show that the scattered intensity increased at low Q and as the temperature is reduced. This was interpreted as enhanced density fluctuations resulting from

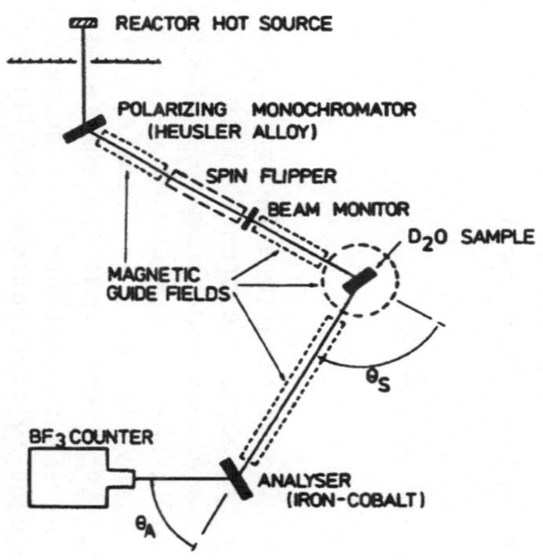

Figure 4. the polarization triple-axes-spectrometer, D5 at the ILL.

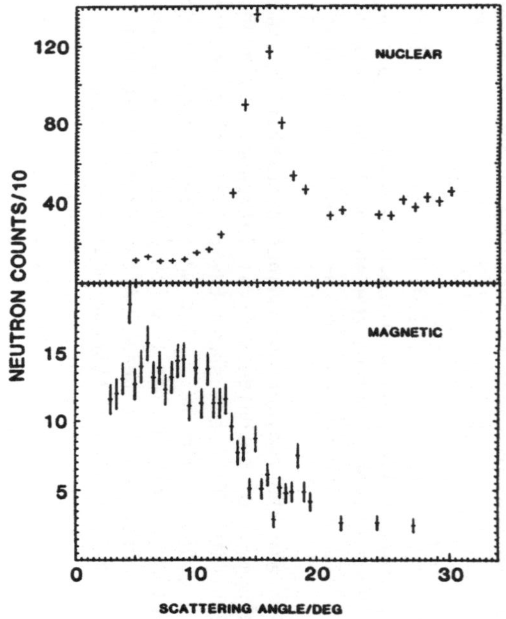

Figure 5. Separation of nuclear and magnetic scattering from liquid oxygen using polarized neutrons.

the formation of local 'patches' of low-density hydrogen-bonded clusters [15]. The addition of small quantities of a hydrogen-bonding impurity such as ethanol reduced the effect. Similar measurements have been made on liquid D_2O with neutrons [16] but the effects were less easy to identify due to the incoherent scattering contribution. Further work is needed to determine the shape of the SAS intensity profile, particularly for the super-cooled region.

b) Heterophase liquid systems

The main use of SAS techniques for topics concerning molecular liquids is in the study of heterophase systems such as micellar solutions or microemulsions. These experiments are relatively new and represent only a small part of a much wider field of investigation as shown in the general reviews of SAXS [17] and SANS [18]. The principles of small-angle neutron scattering can be readily illustrated by reference to the water-in-oil microemulsions formed with the surfactant aerosol-OT. The structure of a typical water droplet with a surfactant coat, dispersed in an oil medium is shown in Figure 6. The size of the water core may vary from 20 Å to 100 Å depending on the relative proportions of water and surfactant. The microemulsion appears visually clear since the droplets are too small to give substantial light scattering and the phase has well-defined stability limits. The properties have been surveyed in a conference compilation [19] from 1981 but the field is changing rapidly as new surfactant systems are studied.

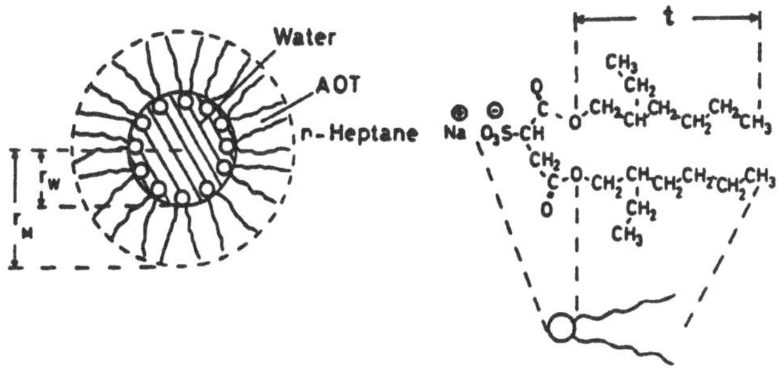

Figure 6. A schematic diagram of a microemulsion droplet formed by the surfactant AOT.

The principle of small angle-scattering is to create regions of different mean coherent scattering density in the heterophase liquid. Since the measurements are at low Q-values, the scattered intensity is not influenced by the detailed molecular structure but depends on the average scattering length density, ρ which may be written as

$$\rho \mp \Sigma n_i b_i / V \qquad (3.1)$$

where n_i represents the number of nuclei of type i with coherent scattering length b_i in the volume V. The scattering amplitude A(Q) for an isolated sphere of radius R and volume V, can then be written as

$$A(Q) = 3\Delta\rho\, V \left[\frac{\sin QR - QR \cos QR}{Q^3 R^3} \right] \qquad (3.2)$$

where Δρ is the difference or "contrast" between the ρ-values of the two media. The SANS intensity profile is given by

$$I_{obs}(Q) = |A(Q)|^2 = F(Q) \qquad (3.3)$$

where F(Q) is a form-factor for an individual droplet which can be experimentally determined using instruments such as D11 and D17 at the ILL [20].

The behaviour of a real microemulsion is much more complex than the idealised system described above and the SANS studies are capable of revealing much more detailed information on the structure. A few illustrative examples are provided below. Since the microemulsion consists of three different regions the use of H/D isotope substitution (see Table 2 in previous chapter) can be used to vary the ρ-value of each region independently. In practice, it is much easier to use D_2O in the central core and hydrogenated surfactant and oil components as shown in Figure 7a, but some experiments have also been done with the inverted system shown in Figure 7b or for various H/D combinations (Figure 7c). The examples given in Figure 8 are for measurements with a profile equivalent to that of Figure 7a so that the contrast region corresponds to the water droplet and the intensity profile is characterised by the inner radius r_w. In a dilute solution the inter-particle interference effects are small and the profile has a form given by equation 3.2. The radius parameter r_w is readily defined to an accuracy of better than ±1 Å. The detailed shape of the curve is not accurately predicted in the tail region of the intensity distribution but an excellent fit may be obtained if polydispersity is introduced [21], to take account of the

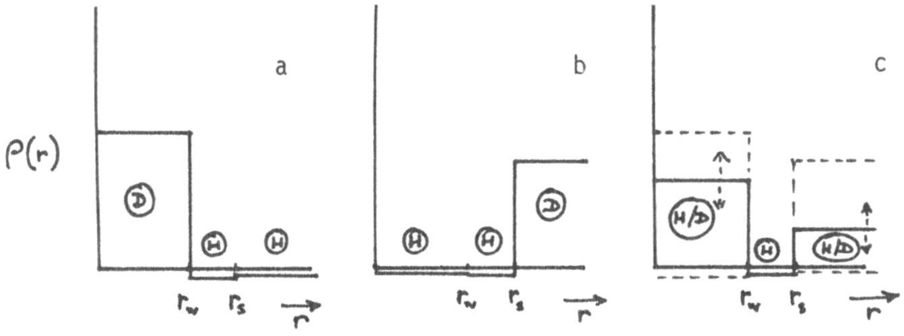

Figure 7. The contrast profile for a microemulsion droplet composed of a) D_2O/AOT/oil(H), b) H_2O/AOT/oil(D) and c) mixtures of H/D components.

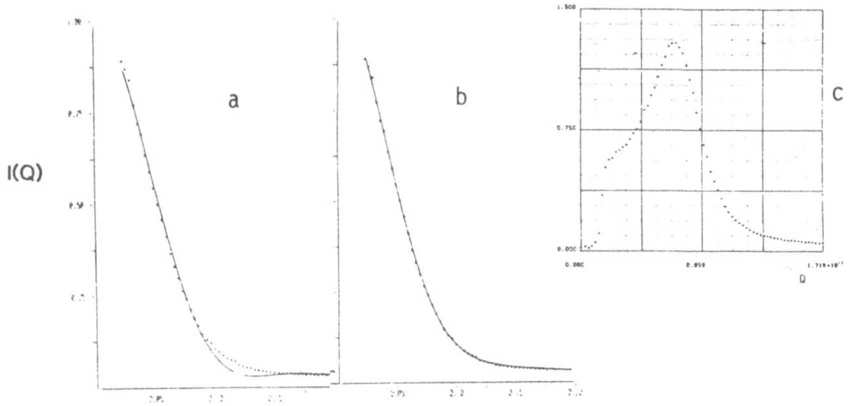

Figure 8. The SANS profile for AOT microemulsions:-
a) dilute with monodisperse form-factor fit
b) as a) but with a polydispersity function and
c) a more concentrated solution.

variation in droplet sizes about the mean value; Figures 8a and 8b are for a D_2O/AOT/heptane microemulsion and show the improved fit (b) when polydispersity is included. For more concentrated solutions there are inter-particle contributions which may be represented by a structure factor S(Q) in much the same way as for a monatomic fluid. The observed intensity now becomes

$$I_{obs}(Q) = S(Q)F(Q)$$

so that S(Q) can be experimentally determined; Figure 8c shows the SANS intensity for this situation. Another phenomenon occurs as the temperature is increased towards the upper phase stability limit. This is illustrated in Figure 9 for a dilute D_2O/AOT/n-undecane microemulsion where there is a large increase in intensity at low Q-values due to critical scattering [22]. This additional contribution may be expressed in an Ornstein-Zernike form as

$$I_{obs}(Q) = F(Q)\left[1 + \frac{\kappa}{1 + \xi^2 Q^2}\right]$$

where ξ is a correlation length which obeys a scaling relation

$$\xi(T) = \xi_o\left(\frac{T - T_c}{T_c}\right)^{-\nu}$$

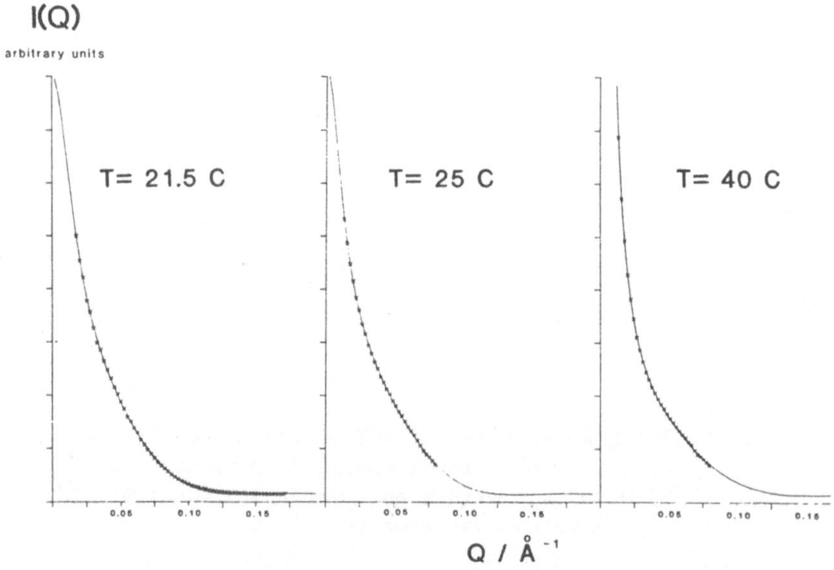

Figure 9. Temperature variation of the SANS intensity for a D_2O/AOT/undecane microemulsion near the phase transition region.

for a critical temperature and ν is a critical constant. Existing data [22,23] suggests that ν is significantly larger than the expected classical value of 0.5. Much of the current SANS work has switched from the study of droplet sizes to a more fundamental investigation of the stability criteria in the phase transition region. The critical scattering component in the intensity profile can be regarded as due to interference effects from incipient clustering of droplets prior to aggregation and phase separation. This is a relatively new field and seems likely to make a major contribution to the understanding of stability criteria in microemulsion systems. There is a growing theoretical involvement and much current interest [24] in this rapidly-changing field of science.

4. INELASTIC AND QUASIELASTIC SCATTERING

a) Formalism

The previous discussion has concentrated primarily on the observation of the total scattering intensity for a particular angle θ and incident wavelength, λ to give the differential scattering cross-section $d\sigma/d\Omega$. It is also possible using a range of different types of instrument [25] to analyse the energy of the scattered beam and to determine the double differential cross-section, $d^2\sigma/d\Omega d\omega$. This quantity is related to the dynamic structural factor or scattering law, $S(Q,\omega)$ by the general expression:-

$$\frac{d^2\sigma}{d\Omega d\Omega} = \frac{N}{4\pi}\frac{k_1}{k_o}\left[\sigma_{coh}\, S(\underline{Q},\omega) + \sigma_{inc}\, S_s(\underline{Q},\omega)\right]$$

where N is the number of scattering centres, $\underline{k}_o$ and $\underline{k}_1$ are the incident and scattering wave vectors and σ_{coh} and σ_{inc} are the coherent and incoherent total cross-sections. In this form, $S_s(\underline{Q},\omega)$ contains information on the motion of individual atoms, while $S(\underline{Q},\omega)$ includes collective motion behaviour and may be written as

$$S(Q,\omega) \doteqdot S_D(Q,\omega) + S_s(Q,\omega)$$

where $S_s(Q,\omega)$ arises from self-terms ($i=j$) in the expansion and $S_D(Q,\omega)$ arises from distinct-terms ($i\neq j$) in a similar manner to that used for the diffraction formalism. The scattering law is a representation in Q-ω space of the van Hove correlation functions $G_{ij}(\underline{r},t)$ which may be formally written [26] as a double transform:-

$$S_s(Q,\omega) = \frac{\sigma_i}{2N}\int G_{ij}(\underline{r},t)\exp i(\underline{Q}.\underline{r}-\omega t)d\underline{r}\, dt$$

and

$$S_D(Q,\omega) = \frac{1}{2\pi N}\sum b_i b_j \int G_{ij}(\underline{r},t)\exp i(\underline{Q}.r-\omega t)d\underline{r}\, dt$$

The structure factor, $S(Q)$ is simply the integral over all energy transfer values, i.e.

$$S(Q) = \int_{-\infty}^{\infty} S(Q,\omega)d\omega$$

and the pair-correlation function $g_{ij}(r)$ an ensemble average at time, t=0 corresponding to a snap-shot picture of the instantaneous configuration i.e.

$$g_{ij}(r) = G_{ij}(r,t).\delta(t) = G_{ij}(r,o)$$

The $S(\underline{Q},\omega)$ or $G_{ij}(\underline{r},t)$ functions contain all the available information concerning the motion of the scattering centres. In the case of monatomic fluids the $S_s(Q,\omega)$ will characterise the diffusive motion of individual atoms and these can conveniently be divided into three time domains corresponding to cage-like vibrations, short-time collective motion and long time diffusion. Considerable experimental effort has been expended on the accurate determination of this information for rare gas fluids (e.g. Ar) and for liquid metals (e.g. Rb) using neutron methods; this is reviewed in a comprehensive article by Copley and Lovesey [26]. The situation is even more complicated for molecular systems but it is again convenient to distinguish different types of motion. The intra-molecular motion corresponding to normal mode vibrations will usually involve fairly large energy exchange (> 5 meV, 40 cm^{-1}) and this can be studied by inelastic neutron scattering whereas the translational and diffusive motion will be on a much longer time-scale and may be studied by quasi-elastic scattering. For neutron scattering by liquids there can, strictly, be no elastic scattering as for crystalline solids since all the scattering centres are in motion and therefore the peak in $S_s(Q,\omega)$ in the region of ω=0 becomes broadened by diffusive motion. The shape of the quasi-elastic peak varies with Q-value and for simple translational diffusion at intermediate values of Q, takes the form of a Lorentzian curve

$$S_s(Q,\omega) = \frac{1}{\pi}\left[\frac{DQ^2}{(DQ)^2+\omega^2}\right]$$

with a width $2DQ^2$ where D is the self-diffusion coefficient. It is possible to introduce different models for varying kinds of motion and a more detailed discussion has been given by Volino and Dianoux [27] with a specific emphasis on organic liquids.

b) Experimental studies: coherent scattering

Although much work has been conducted on both classical and quantum fluids composed of atoms, very few experimental measurements have been reported for molecular systems. Once again, it is liquid nitrogen that has been used as the main system for initial studies. The work of Carneiro and collaborators [28] in the low Q-value region is typical and used a triple-axis spectrometer to extract $S(Q,\omega)$ from the observations. The results showed good agreement with the consistency check for detailed balance:-

$$S(Q,\omega) = \exp\left(\frac{h\omega}{k_B T}\right) S(-Q,-\omega)$$

and enabled the symmetrized dynamic structure factor $\tilde{S}(Q,\omega)$ to be evaluated where

$$\tilde{S}(Q,\omega) = \exp\left(-\frac{h\omega}{2k_B T}\right) S(Q,\omega)$$

The results were compared with MD computations and were in reasonable agreement with the predictions but suggested that the chosen interaction potential underestimates the compressibility. Measurements of $S(Q,\omega)$ for liquid nitrogen have also been reported [29] at much higher Q-values where the main interest was centered on the study of molecular recoil effects.

These two experiments demonstrate the versatility of neutrons in probing the form of $S(Q,\omega)$ over a wide range of Q-ω space but the technique has not been more fully utilised due to the large experimental correction factors for multiple-scattering. Compared with other forms of spectroscopy, neutron studies with a triple axis spectrometer give very low data-collection rates. For this reason large samples are required and some compromise with respect of energy resolution is also necessary. It would seem that this technique is restricted due to the limitations in the incident neutron flux.

c) Experimental studies: incoherent scattering

The main use of quasi-elastic neutron scattering is in the study of individual proton motion. This is possible because of the large incoherent scattering cross-section of H (Table 1; previous chapter) which is an order of magnitude higher than that of most other atoms. In addition it is often convenient to use a first-order difference method by selective H/D substitution in order to isolate the individual H contributions[30]. The type of instrument used for the measurement is influenced by the characteristic time scale of the motion. The principles of the measurements have been reviewed by Volino and Dianoux [27] who also discuss the way the data are analysed in terms of the elastic incoherent structure factor, EISF, which is representative of the proton trajectory and can be used to distinguish between different types of proton motion.

For certain liquids, two relaxation mechanisms with similar time scales can complicate the interpretation of the results. This appears to be the case for water where translational and rotational motions are comparable at room temperature but can be more clearly distinguished in the super-cooled regime. Recent measurements by Chen, Teixeira and Nicklow [31] provide an interesting illustration of how high and low-resolution measurements over a range of temperatures can give important information to aid the interpretation of results obtained by other spectroscopic techniques. In this case a sharp-line in $S_s(Q,\omega)$ is attributed to translational diffusion and a broad-line to motions associated with rotational effects from hydrogen-bond breaking. It is clear that neutron quasi-elastic measurements can play an important role in the general study of molecular motion but the detailed discussion of the status of neutron scattering in relation to other spectroscopic techniques goes beyond the scope of this present report [32].

5. FUTURE DEVELOPMENTS

The previous sections have outlined several of the wide-ranging applications of X-ray and neutron facilities to the investigation of molecular liquids and closely related systems. Many of the methods are limited by the available beam intensity and there are already well-advanced plans to build new facilities with increased power in several different countries. The development in the U.K. is typical as two major projects are in progress and these will be briefly described in order to set the context for the type of experimental work which is likely to develop over the next decade.

a) Synchrotron Radiation

Most X-ray diffraction measurements have utilised laboratory-based equipment but the advent of purpose-built synchrotron radiation sources [33] to provide high intensity beams of electromagnetic radiation has now opened up new possibilities. The Synchrotron Radiation Source (SRS) at the Daresbury Laboratory in Cheshire, England, is illustrated in Figure 10. It is based on an electron synchrotron which feeds a storage ring. Due to relativistic effects, the circulating beam of electrons produces radiation in the direction tangential to the orbit and this is channelled down a number of different beam lines. The incorporation of a 'wiggler' magnet to give a greater curvature to the orbit produces a shift in the spectrum to shorter wavelengths. The intensity in the X-ray region (0.2 - 1.5 Å) is several orders of magnitude higher than that of a conventional laboratory X-ray generator. This means that a diffraction pattern can, in principle, be measured in a few minutes if the high count-rates can be satisfactorily handled. Furthermore it is possible with an incident white beam to use energy-dispersive methods with an energy sensitive detector. This method

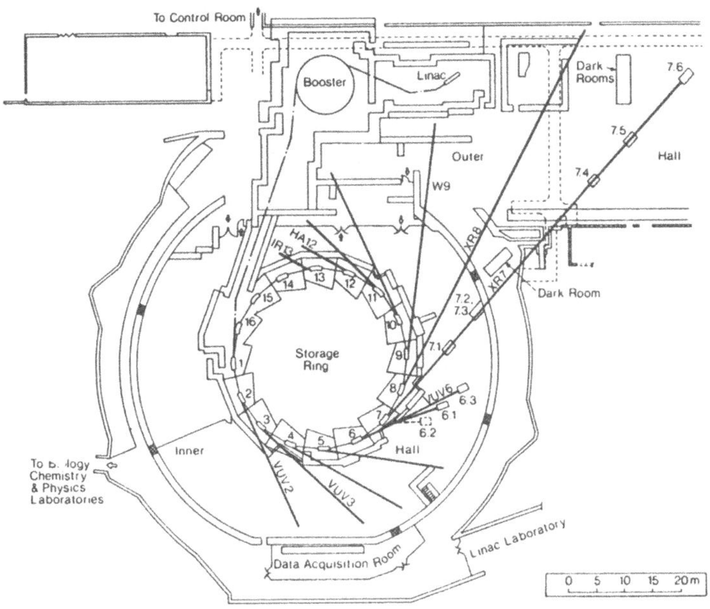

Figure 10. Layout of the Synchrotron Radiation Source at the Daresbury Laboratory, England.

employs a fixed-angle counter and is the analogue of the time-of-flight method for neutrons. Other developments based on anomalous dispersion to change the effective scattering length may also be utilized. No measurements on liquids have yet been reported but the wiggler line (W9) on the SRS has been commissioned and initial experiments should be possible in 1984. Corresponding developments in the use of SAXS techniques also offer considerable advantages over more conventional methods.

b) Pulsed Neutrons

The maximum neutron flux from research reactors is limited by heat dissipation in the moderator. The optimum performance of a moderator assembly can therefore be improved by pulsed neutron methods. The first phase in the development of pulsed neutron sources was based on electron linear accelerators which produce fast neutrons in a heavy metal target by (γ,n) and (γ,f) processes [34] . The problems of heat dissipation again limit the flux but the use of an incident proton beam overcomes this difficulty. The Spallation Neutron Source (SNS) [35] is based on a 800 MeV proton

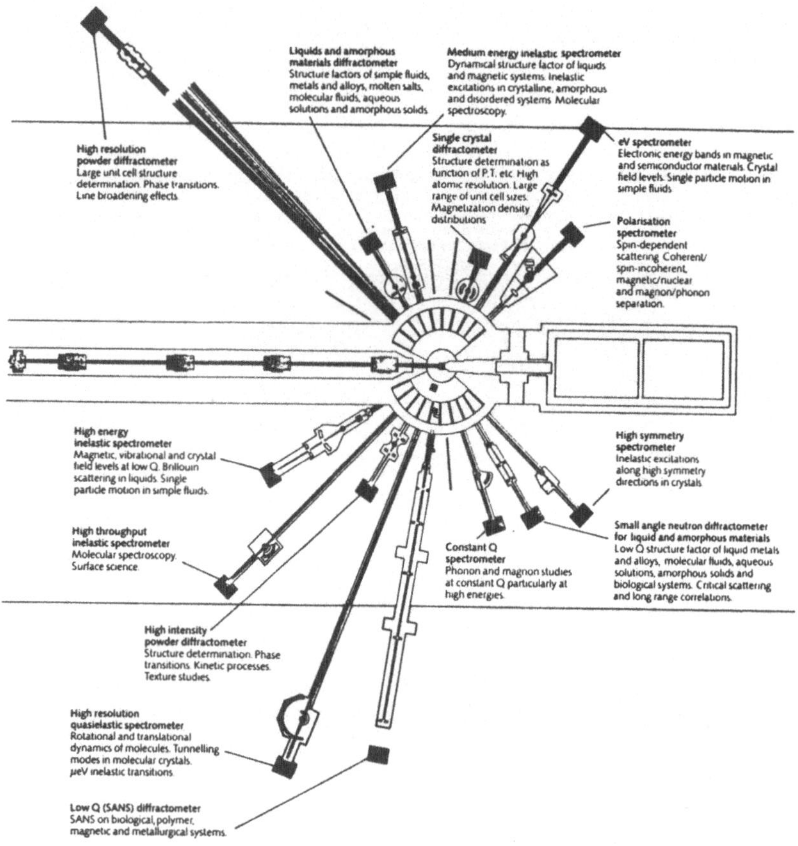

Figure 11. Layout of the Spallation Neutron Source at the Rutherford-Appleton Laboratory, England, showing the proposed location of various neutron instruments.

synchrotron with a uranium target station and is currently under constructionat the Rutherford-Appleton Laboratory, England. Fast neutrons are produced by spallation processes and are slowed down in the surrounding moderator assembly. Various beam lines are arranged around the target assembly as shown in Figure 11. The characteristics for each line are dependent on the type of moderator viewed and are optimised for several different kinds of instrument. The SNS is scheduled to be operational in late 1984. and will run up to full intensity over a two-year period. Several instruments related to work on molecular systems are planned or under construction and are shown in the Figure. The design figure represents an increase of > 10^3 over the flux available for the initial studies on the Harwell Linac [36]. This factor also implies a time of a few minutes for the measurement of a typical diffraction pattern if suitable high-speed detection and counting systems can be developed.

c) Prospects

It is clear that the SRS, SNS and other similar machines in U.S.A., France, Germany, Japan, etc., will radically alter the state of present-day research. A much higher statistical accuracy will be obtained in the measured data and this will open up new possibilities for detailed study. Uncertainties in the treatment and analysis of raw diffraction data can be systematically investigated and hopefully eliminated. The careful combination of X-ray and neutron measurements is capable of providing a much more precise tool for the investigation of molecular interactions, by exploiting the basic formalism and experimental principles presented in the previous sections. It will inevitably take several years for these facilities to develop their full potential but the stage is set for a dramatic increase in information concerning the liquid state. This prospect does not guarantee an automatic break-through in the understanding of this highly complex state but it does give encouragement that progress can be made. It is therefore to be hoped that the new central facilities will lead to the establishment of a closer network of communications between groups and that this collaborative approach will define the most profitable lines for future research.

6. ACKNOWLEDGEMENTS

The author is grateful to many colleagues who have provided information for this brief survey and wishes to acknowledge with thanks, permission to reproduce figures that are used in this and the preceding chapter.

References

(1) Powles, J.G., 1973, Advances in Phys., 22, 1.
(2) Powles, J.G., Dore, J.C. and Osae, E.K., 1980, Mol.Phys., 40, 193.
(3) Sullivan J.D. and Egelstaff, P.A., Mol.Phys., 39, 329, and 44, 287.
(4) Teitsma, A. and Egelstaff, P.A., 1980, Phys.Rev., 21A, 367. Egelstaff, P.A., Teitsma, A. and Wang, S.S., 1980, Phys. Rev., 22A, 1702.
(5) Deraman, M., Dore, J.C. and Powles, J.G., Mol.Phys., to be published.
(6) Kihara, T., Chapter 7, in 'Intermolecular Forces', 1976, (pub) John Wiley and Son. Kihara, T., 1982, Chem.Phys.Lett., 92, 175.
(7) Powles, J.G., Dore, J.C., Deraman, M.B. and Osae, E.K., 1983, Mol.Phys., 50, 1039.
(8) Dore, J.C., Walford, G. and Page, D.I., 1975, Mol.Phys., 29,565.
(9) Kleiner, W., 1955, Phys.Rev., 97, 411.
(10) Moon, R.M., Riste, T. and Koehler, W.C., 1969, Phys.Rev., 181, 920.
(11) Dore, J.C., Clarke, J.H. and Wenzel, J.T., 1976, Nuc.Inst. & Meth., 138, 317.
(12) Deraman, M., Dore, J.C. and Schweizer, J., in preparation for J.Mag & Mag.Materials.
(13) Derman, M., Dore, J.C. and Schweizer, J., in preparation for Mol.Phys.
(14) Bosio, L., Teixeira, J. and Stanley, H.E., 1981, Phys.Rev.Lett., 46, 597.
(15) Stanley, H.E., Teixeira, J., Geiger, A. and Blumberg, R., 1981, Physica, 106A, 260.
(16) Dore, J.C., Bosio, L., Teixeira, J. and Stanley, H.E., unpublished results.
(17) Glatter, O. and Kratky, O., 1982, 'Small Angle X-ray Scattering' Academic Press.
(18) Jacrot, B., 1976, Rep.Prog.Phys., 39, 911. Kostorz, G., in Treatise on Materials Science & Technology Vol.15 Neutron Scattering, 1979, Academic Press.
(19) Robb, I. (ed), 1982, 'Microemulsions' (pub) Plenum Press.
(20) Neutron Beam Facilities at the HFR Institut Laue-Langevin, Grenoble, 1977.
(21) Robinson, B.H., Toprakioglu, C., Dore, J.C. and Chieux, P., Trans.Chem.Soc.(Faraday), to be published.
(22) Toprakcioglu, Dore, J.C., Robinson, B.H., Howe, A. and Chieux, P, Trans.Chem.Soc.(Faraday), to be published.
(23) Toprakcioglu, C., Dore, J.C., and Robinson, B.H., in preparation and more recent unpublished results.
(24) Faraday Discussion, 1983, 'Concentrated Colloidal Dispersions'.
(25) Montague, D.G., Chowdhury, M.R., Dore, J.C. and Reed, J., 1983, Mol.Phys., 50, 1. (ILL facilities)
(26) Copley, J.R.D. and Lovesey, S.W., 1975, Rep.Prog.Phys., 38, 1.

(27) Volino, F. and Dianoux, A.J., Chapter 2 in 'Organic Liquids' (ed) A.D.Buckingham, E.Lippert and S.Bratos, 1978, (pub) Wiley & Sons Ltd.
(28) Pedersen, K.S., Carneiro, K. and Hansen, F.Y., 1981, University of Copenhagen, preprint SP81-41.
(29) Sinclair, R.N., Clarke, J.H. and Dore, J.C., 1975, J.Phys.C. 8, L41.
(30) White, J.W. in 'Chemical Applications of Neutron Scattering' 1972, (ed) Willis, B.J.M., (pub) Oxford Press.
(31) Chen, H-S., Teixeira, J. and Nicklow, R., 1982, Phys.Rev.A, 26, 3477.
(32) Volino, F., in 'Microscopic Structure and Dynamics of liquids' (eds) J.Dupuy and A.J.Dianoux, 1978, Plenum Press.
(33) A description of the general facilities at the SRS is contained in the Appendix to the Daresbury Annual Report 1982/3, 'Synchrotron Radiation', A.Jackson and K.R.Lea (eds), SERC(pub).
(34) Lynn, J.E., 1980, Contemp.Phys., 21, 483.
(35) Manning, G., 1978, Contemp.Phys.19, 505.
(36) Dore, J.C. and Clarke, J.H., 1976, Nuc.Inst. & Meth., 136, 79.

INTERACTION-INDUCED PHENOMENA

P A Madden

Royal Signals and Radar Establishment
St Andrews Road, Great Malvern,
Worcs WR14 3PS

INTRODUCTION

As earlier lectures have shown, an absorption or light scattering lineshape ($I(\omega)$) is related to molecular properties via a correlation function

$$I(\omega) \equiv I\left[\int_0^\infty dt\; e^{i\omega t} \langle \underset{\sim}{M}(t) . \underset{\sim}{M}(0) \rangle\right], \tag{1.1}$$

where $I[\ldots]$ indicates that $I(\omega)$ is some functional of the correlation function, $\underset{\sim}{M}$ is the total dipole moment of the sample. In conventional lineshape analyses, $\underset{\sim}{M}$ is regarded as a sum of molecular moments ($\underset{\sim}{m}^i$) which depend only on the position and orientation of one molecule,

$$\underset{\sim}{M} = \sum_i \underset{\sim}{m}^i \tag{1.2}$$

In microwave and far infra-red spectroscopy, $\underset{\sim}{m}^i$ is the molecular dipole moment

$$\underset{\sim}{m}^i = \underset{\sim}{\mu}^i \tag{1.3}$$

In vibrational infra-red it is the part of the molecular dipole which is modulated by the vibrational normal coordinate q^i

$$\underset{\sim}{m}^i = q^i \left(\frac{\partial \underset{\sim}{\mu}^i}{\partial q^i}\right)_o \equiv \underset{\sim}{\tilde{\mu}}^i \tag{1.4}$$

A. J. Barnes et al. (eds.), Molecular Liquids - Dynamics and Interactions, 431–474.

In Rayleigh scattering $\underset{\sim}{m}^i$ is the laser-induced dipole

$$\underset{\sim}{m}^i = \underset{\approx}{\alpha}^i . \underset{\sim}{E}^L(\underset{\sim}{r}^i) \tag{1.5}$$

where $\underset{\approx}{\alpha}$ is the molecular polarisability and $E^L(r^i)$ the laser field at molecule i. Finally, in Raman scattering $\underset{\sim}{m}^i$ is the vibrationally modulated part of the laser-induced dipole,

$$\underset{\sim}{m}^i = q^i \left(\frac{\partial \underset{\approx}{\alpha}^i}{\partial q^i} \right)_o \underset{\sim}{E}^L(\underset{\sim}{r}^i) \equiv \underset{\approx}{\tilde{\alpha}}^i . \underset{\sim}{E}^L(\underset{\sim}{r}^i) . \tag{1.6}$$

This conventional viewpoint is <u>not</u> an adequate basis for detailed studies of spectroscopic lineshapes in dense systems(1), one must also account for the influence on the total moment of the interaction between pairs, triplets etc of molecules and write

$$\underset{\sim}{M} = \sum_i \underset{\sim}{m}^i + \sum_{ij}{}' \underset{\sim}{m}^{ij} + \sum_{ijk}{}' \underset{\sim}{m}^{ijk} + .. \tag{1.7}$$

The effect of these additional <u>interaction-induced</u> terms is most readily appreciated when considering spectra which are forbidden in the conventional viewpoint, because of some symmetry property of the isolated molecule which implies

$$\underset{\sim}{m}^i \equiv \underset{\sim}{0} . \tag{1.8}$$

Examples of forbidden processes are thus: the far infra-red and vibrational infra-red spectra of N_2, the depolarised Rayleigh spectrum of Argon or the Raman spectrum of the bending vibration of CO_2. These spectra are observed in condensed phases; they have been studied extensively over the last thirty years or so, though interest has shifted to liquid phase studies only comparatively recently (see refs. 1-3, for reviews). Whilst the interaction-induced phenomena are readily recognised in forbidden spectra, the fact that they <u>invariably</u> contribute to allowed spectra (where $\underset{\sim}{m}^i \neq 0$) is frequently ignored. It is conveniently assumed that interaction-induced contributions to allowed bands may be dealt with by the use of dismissive euphemisms such as "local-field effect" or "background", and then forgotten! In fact the interaction-induced component can have an important influence on the shape and intensity of the band.

These lectures then address two aspects of the study of interaction-induced phenomena. On the one hand we may study the limitations imposed on the interpretation of allowed band data by the presence of an interaction-induced component. On the other hand we may study the induced processes themselves, this is clearly best effected on the forbidden bands. The intrinsic interest of induced spectra is that they reflect <u>directly</u> molecular encounters (rather than indirectly, in the way that the intermolecular torque determines the molecular reorientation time, for example). Induced spectra markedly reflect phase changes, eg such important processes as the superionic or glass-crystal transition. They may also acquire relevance through interest in forbidden transitions for possible laser systems.

2. MECHANISMS FOR THE INDUCED-MOMENTS

To date induced phenomena have, almost invariably, been discussed in terms of the pair contribution, $\underset{\sim}{m}^{ij}$ alone. Information about $\underset{\sim}{m}^{ij}$ comes largely from Rayleigh and far i.r studies on low density systems[3,4] relatively recently <u>ab initio</u> electronic structure calculations have begun to provide objective results on simple systems. For the spectroscopic studies of molecular systems, of primary interest in these lectures, I believe that this work may be summarised in the following way:- the dominant part of the induced moment may be regarded as a property of the coulomb interactions between a pair of non-overlapping molecules. By dominant I imply that 90%, or better, of an observation can be described whilst neglecting overlap of charge clouds and dispersion effects. Not all induced phenomena can be described in this simple framework, but the exceptions (eg far i.r absorption in rare gas mixtures or isotropic Rayleigh scattering by rare gases) can be recognised within the framework, they are usually associated with atomic fluids.

For non-overlapping charge clouds the total moment of the system may be written as[5]

$$
\underset{\sim}{M} = \sum_i \left[\underset{\sim}{\mu}^i + \underset{\approx}{\alpha}^i \cdot \underset{\sim}{F}(\underset{\sim}{r}^i) + \tfrac{1}{2}\underset{\approx}{\beta}^i : \underset{\sim}{F}(\underset{\sim}{r}^i)\underset{\sim}{F}(\underset{\sim}{r}^i) + \right.
$$

$$
\left. + (1/3)\underset{\approx}{A}^i : \underset{\approx}{F}'(\underset{\sim}{r}^i) + \ldots \right] \tag{2.1}
$$

where $F(\underset{\sim}{r}^i)$ is the field at molecule i and F' is its gradient. β^i is the hyperpolarizability and A^i is the tensor which gives the dipole induced by a field gradient (or the quadrupole induced by a field c.f eqn 2.2 below). Equation 2.1 is the result of a double Taylor expansion of the effect of the distortion of the electron density of a molecule by the field acting on it, in powers of the field and in the spatial gradients of the field at the molecular centre. In normal fluids we expect the intermolecular fields to be sufficiently weak that the electron density resembles that of the isolated molecule and the first expansion converges rapidly (an exception would be an ion in a molten salt[6]). The convergence of the second expansion is likely to be more finely balanced; it may be regarded as an expansion in the ratio of an intra- to an inter-molecular length (r_{intra}/r_{inter}) thus $A \cong r_{intra}\,\alpha$ and $F' \cong F/r_{inter}$. Neglect of the gradient terms is tantamount to treating the molecular polarizability as concentrated at a point. These issues have been discussed in the context of the inter-molecular potential by Stone(7).

The symmetry properties of the molecular properties are of great importance, they are discussed in detail in ref. 5. For a molecule with a centre of symmetry $\underset{\sim}{\mu}, \underset{\approx}{\beta}$ and $\underset{\approx}{A}$ are all "u-" symmetry properties and vanish for the isolated molecule and are not modulated by g-symmetry vibrations. $\underset{\approx}{\alpha}$ is a g-symmetry property, it is always non-zero and is not modulated by u-symmetry vibrations. For molecules with tetrahedral, octahedral or spherical symmety, $\underset{\approx}{\alpha}$ is a scalar and the anisotropic part of the tensor is not modulated by totally symmetric vibrations.

We also need an expression for the field $\underset{\sim}{F}$. To deal with light scattering we need the total field at a molecule in the presence of a laser field:

$$\underset{\sim}{F}(\underset{\sim}{r}^i) = \underset{\sim}{E}^L + \sum_{j\neq i} \underset{\approx}{T}^{(2)}(\underset{\sim}{r}^{ij}).(\underset{\sim}{\mu}^j + \underset{\approx}{\alpha}^j.\underset{\sim}{F} + ..)$$
$$- 1/3 \sum_j \underset{\approx}{T}^{(3)}(r^{ij}):(\underset{\sim}{\theta}^j + \underset{\approx}{A}^j.\underset{\sim}{F} + ...) \qquad (2.2)$$

This expression is governed by the same convergence provisos as apply to equation 2.1. It expresses the total field at a molecule as the sum of the laser field and the fields from the permanent ($\underset{\sim}{\mu}$, $\underset{\sim}{\theta}$...) and field- induced (αF, AF, ...) dipoles, quadrupoles etc. on other molecules. $\underset{\sim}{T}^{(2)}(\underset{\sim}{r}^{ij})$ and $\underset{\sim}{T}^{(3)}(\underset{\sim}{r}^{ij})$ give the field at i due to a dipole and quadrupole on j, respectively; they are given by

$$\underset{\sim}{T}^{(2)}(\underset{\sim}{r}^{ij}) = \nabla\nabla(r^{ij})^{-1} \qquad (2.3)$$

$$\underset{\sim}{T}^{(3)}(\underset{\sim}{r}^{ij}) = \nabla\underset{\sim}{T}^{(2)}(\underset{\sim}{r}^{ij}), \qquad (2.4)$$

note that

$$\underset{\sim}{T}^{(n)}(\underset{\sim}{r}^{ij}) = (-1)^n \underset{\sim}{T}^{(n)}(\underset{\sim}{r}^{ij}) . \qquad (2.5)$$

In almost all applications the expression for $\underset{\sim}{F}$ is simplified by the replacement of all $\underset{\sim}{F}$ factors on the right hand side by the bare laser field $\underset{\sim}{E}^L$. In any real system additional terms such as

$$\sum_{ijk} \underset{\approx}{\alpha}^i.\underset{\approx}{T}^{(2)}(\underset{\sim}{r}^{ij}).\ \underset{\approx}{\alpha}^j.\underset{\approx}{T}^{(2)}(\underset{\sim}{r}^{jk}).(\underset{\sim}{\mu}^k + \underset{\approx}{\alpha}^k \underset{\sim}{E}^L) \qquad (2.2a)$$

are present, these contribute higher order terms to $\underset{\sim}{m}^{ij}$ as well as giving the leading contributions to m^{ijk}. Roughly speaking we might expect this term to be of order $(\tilde{r}_{intra}/r_{inter})^3$ smaller than the corresponding first order term which appears in equation 2.2.

In most of what follows we shall focus on the leading term which emerges when equations 2.1 and 2.2 are combined, ie the term which is lowest order in the field, field gradient and polarizability and comes from the lowest order non-vanishing multipole. If, for some reason, this leading term is "small" (as for example in the isotropic Rayleigh spectrum below) then it should be clear from the development that a whole slew of terms will start to contribute in next order; in addition the overlap and dispersion contributions which were neglected at the outset may become comparable to the residual terms which occur within the model.

We are now in a position to obtain expressions for $\underset{\sim}{m}^{ij}$ for various spectra[8]. For the microwave and far-infra-red absorption spectrum we need to find a contribution to $\underset{\sim}{M}$ which occurs when $\underset{\sim}{E}^{L}$ is not present. In the case of a molecule with no symmetry equations 2.1 and 2.2 give

$$\underset{\sim}{M} = \sum_i \underset{\sim}{\mu}^i + \sum_{ij}{}' \underset{\approx}{\alpha}^i \left\{ \underset{\approx}{T}^{(2)}(\underset{\sim}{r}^{ij}).\underset{\sim}{\mu}^j + \underset{\approx}{T}^{(3)}(\underset{\sim}{r}^{ij}):\underset{\sim}{\theta}^j + \ldots \right\}$$

$$+ \sum_{ij}{}' \underset{\approx}{A}^i : \left\{ \underset{\approx}{T}^{(3)}(\underset{\sim}{r}^{ij}).\underset{\sim}{\mu}^j + \ldots \right\}$$

$$+ \sum_i \underset{\sim}{\beta}^i : \left(\sum_{j\neq i} \underset{\approx}{T}^{(2)}(\underset{\sim}{r}^{ij}).\underset{\sim}{\mu}^j \right) \left(\sum_{k\neq i} \underset{\approx}{T}^{(2)}(\underset{\sim}{r}^{ik}).\underset{\sim}{\mu}^k \right) \qquad (2.6)$$

As discussed above, we might expect to simplify this model by the assumption that the intermolecular fields are weak - so that the last term may be neglected - and that we may neglect gradient-induced dipoles with respect to field-induced ones (ie neglect the third term). For molecules with some symmetry the equation simplifies considerably and these assumptions become better founded. In particular for atoms none of the terms survive! Thus to account for the far i.r absorption of rare gas mixtures[4] it is necessary to go outside the no-overlap model. For the important case of linear centro-symmetric molecules $\underset{\sim}{\mu}^i$, $\underset{\sim}{A}^i$ and $\underset{\sim}{\beta}^i$ vanish[5] so that

$$\underset{\sim}{M} = \sum_{ij} \underset{\approx}{\alpha}^i . T^{(3)}(r^{ij}).\underset{\approx}{\theta}^j + \ldots \qquad (2.7)$$

and the next non-vanishing term is of order $(r_{intra}/r_{inter})^2$ smaller (eg the hexadecapole-induced dipole).

To obtain the moment responsible for a vibrational infra-red spectrum we need to consider the derivative of M with respect to the normal coordinate on a molecule $\left(\text{ie} \sum_i q^i \frac{\partial \tilde{M}}{\partial q^i}\right)$. The result involves products of normal coordinate derivatives of molecular properties - denoted by tildes - and properties of undistorted molecules. Again symmetry leads to great simplifications. For a linear centrosymmetric molecule $\tilde{\mu}^i$, $\tilde{A}^i$ and $\tilde{\beta}^i$ vanish for the symmetric stretching vibration as do $\underset{\sim}{\mu}^i$, $\underset{\sim}{A}^i$, $\underset{\sim}{\beta}^i$. Therefore we obtain

$$\sum_i q^i \frac{\partial M}{\partial q^i} = \sum_{ij} \left[\tilde{\underset{\approx}{\alpha}}^i . \underset{\approx}{T}^{(3)} (r^{ij}) : \underset{\approx}{\theta}^j + \underset{\approx}{\alpha}^j . \underset{\approx}{T}^{(3)} (\underset{\sim}{r}^{ji}) : \tilde{\underset{\sim}{\theta}}^i + .. \right] \tag{2.8}$$

The effect of the vibrationally modulated quadrupole term has often been ignored in analysing forbidden infra-red bands (ie the second term in this equation); studies of CS_2 have shown that it is responsible for about 60% of the absorption intensity[8-10].

In Rayleigh scattering we see the radiation caused by a moment which oscillates at the laser frequency; we therefore need to pick out of equations 2.1 and 2.2 terms which involve at least one factor of $\underset{\sim}{E}^L$. In normal Rayleigh scattering only the linear terms are seen. For molecules with an inversion centre, $\underset{\sim}{A}^i$ and $\underset{\sim}{\beta}^i$ vanish and we obtain

$$\underset{\sim}{M} = \sum_i \underset{\approx}{\alpha}^i . \underset{\sim}{E}^L + \sum_{ij}{}' \underset{\approx}{\alpha}^i . \underset{\approx}{T}^{(2)} (\underset{\sim}{r}^{ij}) . \underset{\approx}{\alpha}^j . \underset{\sim}{E}^L + \ldots \tag{2.9}$$

where the correction terms are at least of order (r_{intra}/r_{inter}) smaller. An important special case is that of atoms, for which

$$\alpha^i_{\alpha\beta} = \alpha \delta_{\alpha\beta} \tag{2.10}$$

For depolarised scattering, where $\underset{\sim}{E}^L$ defines the z direction and we observe M_x, the first term does not contribute and

$$M_x = \alpha^2 \sum_{ij}{}' T^{(2)}_{xz} (\underset{\sim}{r}^{ij}) \, E^L \tag{2.11}$$

this is the dipole-induced dipole mechanism for the (forbidden) depolarised Rayleigh scattering in atomic fluids. There is a somewhat more subtle symmetry in the atomic systems; we may write equation 2.9 as

$$\underset{\sim}{M} = \underset{\approx}{\Pi}.\underset{\sim}{E}^{L} \tag{2.12}$$

where $\underset{\approx}{\Pi}$ is the total polarizability of the system. The depolarised scattering arises from the off-diagonal elements of $\underset{\approx}{\Pi}$, ie the anisotropic part of the tensor. The polarised scattering contains an isotropic and an anisotropic part, but the latter can be removed to give a pure isotropic spectrum associated with the trace of $\underset{\approx}{\Pi}$[11]. Now $T^{(2)}$ is a traceless tensor[5], so that

$$\text{tr}\ \underset{\approx}{\Pi} = 3\alpha + 0 + \ldots \tag{2.13}$$

ie the lowest order DID contribution to the interaction-induced isotropic polarizability vanishes for atoms (and also tetrahedral and octahedral molecules). Therefore, as discussed above, we would expect the isotropic Rayleigh wing to be weaker and to be more sensitive to overlap (and the whole "slew" of higher order terms neglected in the model,) than the depolarised spectrum for which equation 2.11 should be satisfactory. This does seem to be the case[12], although simple DID does seem to fall below the requisite 90% accuracy in dealing with depolarized scattering in dense systems (see below).

In Raman scattering we need the vibrationally modulated part of $\underset{\sim}{M}$ which is linear in $\underset{\sim}{E}^L$. Let us consider the case of a centro-symmetric linear molecule again - then, for a Raman allowed ($\underset{\approx}{\tilde{\alpha}} \neq 0$, $\underset{\sim}{\tilde{A}} = 0$, $\underset{\sim}{\tilde{\beta}} = 0$) vibration of g-symmetry we obtain

$$\sum_i q^i \frac{\partial M}{\partial q^i} = \sum_i \underset{\sim}{\tilde{\alpha}}^i.\underset{\sim}{E}^L + 2\sum_{ij} \underset{\sim}{\tilde{\alpha}}^i.\underset{\approx}{T}^{(2)}(\underset{\sim}{r}^{ij}).\underset{\sim}{\alpha}^j.\underset{\sim}{E}^L + \ldots \tag{2.14}$$

(the factor 2 arises because we must differentiate both the i and j terms in eqn 2.2 and make use of eqn 2.5). For a Raman forbidden u-vibration ($\underset{\sim}{\tilde{\alpha}} = 0$, $\underset{\sim}{\tilde{A}} \neq 0$, $\underset{\sim}{\tilde{\beta}} \neq 0$, $\underset{\sim}{\mu} = 0$)[8]

$$\sum_i q^i \frac{\partial M}{\partial q^i} = 2/3 \sum_{i,j}{}' \underset{\approx}{\tilde{A}}^i : \underset{\approx}{T}^{(3)}(\underset{\sim}{r}^{ij}).\underset{\approx}{\alpha}^j.\underset{\sim}{E}^L$$

$$+ \sum_{ij} \underset{\sim}{\tilde{\beta}}^i : \left[\underset{\approx}{T}^{(3)}(\underset{\sim}{r}^{ij}) : \underset{\sim}{\theta}^j\right] \underset{\sim}{E}^L. \tag{2.15}$$

This term offers an interesting prospect as it suggests a way of checking upon our ansatz that it is more important to describe the distribution of polarizable matter in a molecule (by retention of the gradient terms in eqn 2.1) than to account for the non-linear distortion of the electron density by the intermolecular fields (by keeping the F^2 terms in eqn 2.1). The relative importance of the two terms has been investigated for the asymmetric stretch and bending vibrations of CS_2, by observing the effect on the Raman intensity of dissolving CS_2 in solvents of differing polarity (which will change the β term) and of different polarizability (which should affect the A term)[13]. It was concluded that the A terms were overwhelmingly dominant. A similar conclusion was reached for the corresponding modes of CO_2 from values of the molecular properties calculated ab initio [14].

Induced phenomena may be studied by non-spectroscopic techniques. The refractive index is determined by the average value of the trace of the total polarizability[15] of the system (ie $\langle tr\, \underline{\underline{\Pi}} \rangle$, c.f. eqn 2.12); measurements of the refractive index virial coefficients therefore give information which is complementary to that obtained by isotropic light scattering. An even more direct relationship exists[16] between the static Kerr constant of a non-polar material (or the optical Kerr constant in the general case) and the depolarised light scattering intensity; apart from hyperpolarizability effects and the frequency dispersion of the molecular polarizability they are both given by the correlation function of the anisotropic part of $\underline{\underline{\Pi}}$.

3. GAS PHASE STUDIES OF INDUCED-MOMENTS

3.1 General Considerations

Until recently, quantitative studies on induced moments were made only on gases, where the result of isolated collision events between pairs of molecules can be separated. The dependence of the induced moment on the relative separation and orientation of the collision partners has an important influence on the lineshape as well as on the integrated intensity, the effect is a consequence of molecular rotation[4]. Consider, for example, the induced far i.r spectrum of N_2; in the model the moment responsible is the quadrupole-induced dipole (QID) given in eqn 2.7. At the start of the collision the two molecules are in well defined rotational states; the QID is rather long-ranged ($T^{(3)} \alpha\ r^{-4}$) so that the observations will be dominated by large impact parameter collisions which are rotationally elastic and we can imagine the rotational quantum numbers to be conserved throughout the collision. Now the quadrupole moment is a traceless second rank tensor so that the field it causes at another molecule will go through two cycles in every rotational period (see Figure 1). The rotationally modulated dipole which results will give rise to a rotational absorption spectrum. If the polarizability of a N_2 molecule were purely isotropic the selection rule would be $\Delta J = 0,2$ (c.f the usual discussion[17] of the Raman selection rule); the anisotropy in α gives an additional rotational

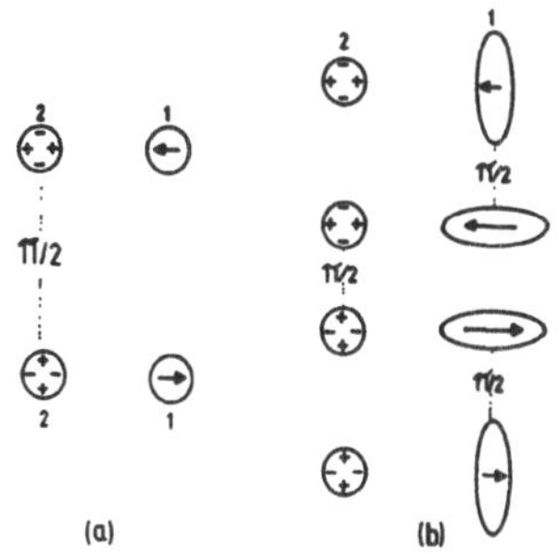

Fig 1: The modulation of the QID by molecular reorientation. (a) The dipole induced in 1 via its isotropic polarisability reverses twice in a single rotation of the quadrupole (2). (b) The anisotropic polarisability of 1 causes a further modulation of the dipole, at twice the rotation frequency of 1.

modulation and gives rise to a second branch with $\Delta J = 0,2,4$. Since the molecules are moving past each other, each rotational line is translationally broadened. Depending on the ratio of the mass to the moment of inertia, separated rotational lines or a broad featureless spectrum may be observed. A model such as QID therefore makes specific predictions about the relative intensity of different rotational lines (or regions of the spectrum), it is possible to check the model in a way which is relatively insensitive to the form of the intermolecular potential. The quadrupole-induced dipole gives rise to $\Delta J = 2$ transitions via the isotropic part of the polarizability of its collision partner; in general a 2^{ℓ}-pole-induced dipole gives rise to $\Delta J = \ell$ transitions (eg the hexadecapole (ℓ=4) -induced dipole has $\Delta J = 4$).[4]

3.2 The Far Infra-red Spectra of Molecular Gases

The far i.r spectra of simple molecular gases are now extremely well characterised. For H_2 and D_2 the individual rotational lines are easily resolved [18], for larger species such as N_2[19] or CO_2[20] the individual lines are merged by the translational broadening. As a representative example we may consider the work of Poll and Hunt[19] on N_2. They fit the spectroscopic lineshape[21,22] with the QID model and with a more sophisticated model in which the hexadecapole-induced dipole (HID) and overlap terms were included. The simple QID model gave a remarkably good description of the lineshape and intensity (see Fig 4 of Ref 19); the predicted absorption was too low at high frequency indicating that some higher order J transitions are present. This deficiency was remedied in the second model and an excellent fit obtained. It was found that the QID, HID and overlap terms gave respectively 93.7%, 5.8% and 0.5% of the total intensity at 300°K. The value of the quadrupole moment which came from the spectral fit was within 1% of the value obtained from birefringence measurements[23]; the hexadecapole was in reasonable agreement with *ab initio* results.

3.3 Rayleigh Spectra of Atomic and Molecular Gases

The Rayleigh spectra of the inert gases have been intensively studied over the last six years or so; the work has been reviewed by Frommhold[12]. The situation with regard to the *depolarised* spectra is extremely satisfactory. Three groups have obtained identical results for Argon (see Ref 24 for a discussion of experimental technique) and the simple DID model (eqn 2.11) gives 92% of the measured intensity[25]. The lineshape is determined solely by translational motion, it may be calculated directly, by trajectory methods[12], or analysed through the spectral moments[26]. For Argon the lineshape out to 120 cm^{-1} is satisfactorily accounted for by the DID model[25]. Small dispersion[27], higher order DID (as in eqn 2.2a), and overlap terms have been introduced into the model and lead to good representations of the whole lineshape over a wide range of density[28]. For He and Ne these additional terms become slightly more important but they are negligibly small for the heavier cases of Kr and Xe[28].

For the *isotropic* inert gas spectra[12] the situation is rather less well settled. As anticipated above, from the vanishing first order DID term, the spectra are very weak and difficult to observe, and a reliable theoretical basis for the interpretation of the data is lacking. As yet no model for the isotropic part of the polarizability of an atomic pair has emerged which can simultaneously account for the light scattering spectra and the refractive index (or dielectric) virial data and which is consistent with *ab initio* calculations. However there has been a good deal of recent activity on this topic[29,30].

The comparison between the calculated lineshape for the DID model and the *depolarised* spectra of tetrahedral molecular gases is much less satisfactory than for the atomic case; CH_4, CF_4[31] and $C(CH_3)_4$ [32] have been studied. The molecular spectra show a marked excess intensity at high frequency; however, this discrepancy can be accounted for within the no-overlap model[33]. For a tetrahedral molecule $\underset{\approx}{\alpha}$ is a scalar, as in the atomic case, but the $\underset{\approx}{A}$-tensor is non-zero. Thus there are corrections of the distribution of polarizable matter type, to eqn 2.9, which are only of order $(r_{intra}/r_{inter})^1$ smaller than the first order DID term which accounts for the atomic spectra. Equation 2.11 becomes

$$M_x = \alpha^2 \sum_{ij}{}' T^{(2)}_{xz}(\underset{\sim}{r}^{ij}) E^L$$

$$+ \left(\sum_{ij}{}' (\underset{\approx}{A}^i : \underset{\approx}{T}^{(3)}(r^{ij}) . \underset{\approx}{\alpha}^j - \underset{\approx}{\alpha}^i . \underset{\approx}{T}^{(3)}(\underset{\sim}{r}^{ij}) : \underset{\approx}{A}^j) . \underset{\sim}{E}^L \right) \quad (3.1)$$

and the extra term also gives a contribution to the isotropic spectrum. The importance of the new term is that the $\underset{\sim}{A}$-tensor is modulated by molecular reorientation so that the extra contribution appears at rotational transition frequencies (as in the far i.r spectra discussed above) out in the wings of the spectrum, where the DID scattering is weak. For this reason the effect is known as collision-induced rotational scattering (CIRS). Whilst the effect on the lineshape is pronounced, the contribution of the CIRS to the total scattering is small. For the depolarised spectrum of CH_4 and CF_4 it is 1.4% and 2.8% of the DID terms(25); in the isotropic intensity it is much more important, due to the vanishing DID effect.

By way of a conclusion to this gas phase section we re-iterate that: unless it vanishes (as in the isotropic, atomic Rayleigh spectrum), the leading term from the no-overlap model provides the dominant contribution to the interaction-induced moment. The influence of the succeeding terms in the expansion can sometimes be distinguished in gases by their effect on the lineshape (as in the HID contribution to the far i.r or the CIRS in CH_4). They are found to give effects on the total intensity which are considerably smaller than the leading term and comparable to the neglected overlap and dispersion terms.

4. LIQUID PHASE STUDIES OF INDUCED MOMENTS

4.1 General Considerations

Of central importance to an appreciation of intensity and lineshape data on liquids is an appreciation of the cancellation problem. The density of a liquid at its triple point is usually only about 10% smaller than that of the associated crystal - this implies a high degree of local positional order which may cause a tremendous reduction in the intensity of an induced spectrum from expectations based on uncorrelated binary collisions. The best documented example is the atomic depolarised Rayleigh spectrum. In a dense system the neighbours of a given atom are arranged in shells, consider the net contribution

$$\alpha^2 E^L \sum_{j\epsilon shell} T_{zz}(\underset{\sim}{r}^{ij}) = \alpha^2 E^L \sum_{j\epsilon shell} \frac{3\cos^2\theta^{ij}-1}{(r^{ij})^3}, \qquad (4.1)$$

to the induced moment of an atom from a particular shell, in the DID model (θ^{ij} is the angle subtended by r^{ij} in the laboratory frame). For perfectly formed shells (ie perfect positional correlation) the contribution from any shell vanishes - as illustrated in Figure 2. Consequently the total moment $\sum_{j\neq i} m^{ij}$ on any atom vanishes as does the total intensity

$$I = \langle \sum_{ij}{}' m^{ij} \sum_{k\ell}{}' m^{k\ell} \rangle = 0. \qquad (4.2)$$

I may be written as a sum of 2-, 3- and 4- particle contributions

$$I = N^2\langle (m^{12})^2\rangle + 2N^3\langle m^{12}m^{13}\rangle + N^4\langle m^{12}m^{34}\rangle, = 0, \qquad (4.3)$$

but the first term is certainly non-zero, as it is a squared term.

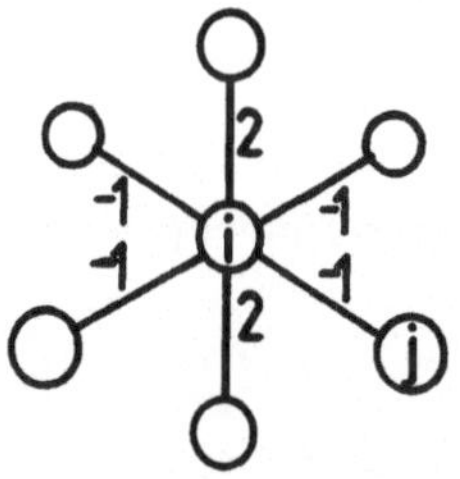

Fig 2: The cancellation of the DID contributions when i is surrounded by a perfect (octahedral) shell of neighbours. The numbers show the relative size of the m_z^{ij} contributions when the applied field is along the z direction.

In fact in a perfect crystal this two-body term (conventionally called S_2) is equal to the four-body term (S_4) and to minus one-half the three body term ($2S_3$). Whilst the dense fluid does not have this perfect positional ocrrelation a high degree of cancellation is found, I is only about 1/100 of the two-body term for liquid argon at the triple point. I/ρ vs ρ is shown for argon in figure 3[34]. These considerations show that we should regard the scattering as a property of the local density fluctuations in a liquid, rather than a consequence of collisions between pairs of molecules.

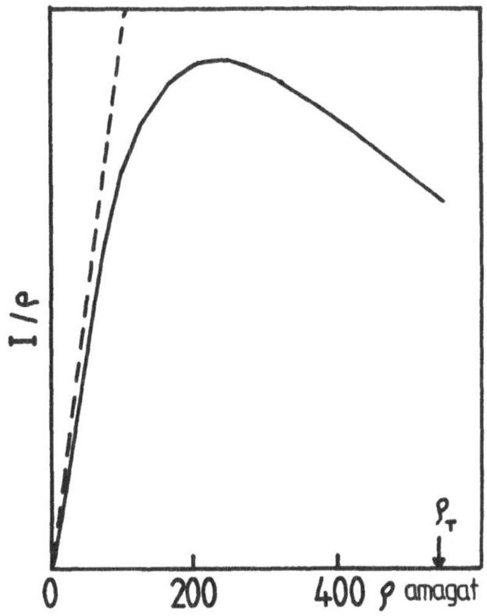

Fig.3: The intensity of the depolarised Rayleigh scattering from argon as a function of density at room temperature (ref 34). I/density is plotted; the dashed line shows the 2-body intensity.

The cancellation effects are not so pronounced in fluids of molecules with anisotropic shapes (ie lower than tetrahedral symmetry). Let us consider the induced Rayleigh scattering in a liquid of linear, centrosymmetric molecules[8]. Then the polarizability tensor is characterised by its scalar ($\alpha=(\alpha_{xx}+\alpha_{yy}+\alpha_{zz})/3$) and anisotropic ($\gamma=(\alpha_{zz}-\alpha_{xx})$) parts,

$$\alpha^{i}_{\alpha\beta} = \alpha\delta_{\alpha\beta} + (2/3)\gamma\left[1/2(3\hat{e}^{i}_{\alpha}\hat{e}^{1}_{\beta} - \delta_{\alpha\beta})\right]$$

$$\equiv \alpha\delta_{\alpha\beta} + (2/3)\gamma Q^{i}_{\alpha\beta} \qquad (4.4)$$

were $\hat{e}^i$ is a component of a unit vector along the molecular axis and Q^i is a traceless second-rank tensor function of the angles subtended by $\hat{e}^i$. $Q^i_{\alpha\beta}$ may be regarded as a combination of second rank spherical harmonics, if there is no orientational correlation between different molecules in the fluid

$$\langle Q^i_{\alpha\beta}\, Q^j_{\gamma\delta}\rangle \;\propto\; \delta_{ij}(\delta_{\alpha\gamma}\delta_{\beta\delta} + \delta_{\alpha\delta}\delta_{\beta\gamma}) \tag{4.5}$$

The induced moment (DID) may now be written as a sum of four terms (let $\underset{\approx}{T}^{ij} = \underset{\approx}{T}^{(2)}(\underset{\sim}{r}^{ij})$)

$$\begin{aligned} m^{ij}_\alpha = \Big\{ & \alpha^2 T^{ij}_{\alpha z} + (2/3)\alpha\gamma T^{ij}_{\alpha\gamma}\, Q^j_{\gamma z} \\ & + (2/3)\alpha\gamma Q^i_{\alpha z}\, T^{ij}_{\gamma z} + (4/9)\gamma^2 Q^i_{\alpha\gamma}\, T^{ij}_{\gamma\delta} Q^i_{\delta z} \Big\} E^L_z \,. \end{aligned} \tag{4.6}$$

The expression for I (eqn 4.3) is very complicated, it involves 48 terms! The first contribution to m^{ij}_α gives rise to a set of purely positional averages

$$\begin{aligned} &(\alpha^2 E^L)^2 \Big[N^2 \langle T^{12}_{\alpha z} T^{12}_{\alpha z}\rangle + 2N^3 \langle T^{12}_{\alpha z} T^{13}_{\alpha z}\rangle + N^4 \langle T^{12}_{\alpha z} T^{34}_{\alpha z}\rangle \Big] \\ &\qquad \equiv (\alpha^2 E^L)^2 \Big[S_2 + 2S_3 + S_4 \Big] \end{aligned} \tag{4.7}$$

which we expect to be subject to the same extensive cancellation as found in atomic fluids. But consider the correlation functions which arise from the last term in eqn 4.6.

$$\begin{aligned} &((4/9)\gamma^2 E^L)^2 \Big\{ N^2 \langle (Q^1_{\alpha\gamma} T^{12}_{\gamma\delta} Q^2_{\delta z})(Q^1_{\alpha\gamma'} T^{12}_{\gamma\delta'} Q^2_{\delta' z})\rangle \\ &+ 2N^3 \langle (Q^1_{\alpha\gamma} T^{12}_{\gamma\delta} Q^2_{\delta z})(Q^1_{\alpha\gamma'} T^{13}_{\gamma'\delta'} Q^3_{\delta' z})\rangle \\ &+ N^4 \langle (Q^1_{\alpha\gamma} T^{12}_{\gamma\delta} Q^2_{\delta z})(Q^3_{\alpha\gamma'} T^{34}_{\gamma'\delta'} Q^4_{\delta' z})\rangle \Big\} \end{aligned} \tag{4.8}$$

These averages involve orientational functions on different moleclues. Whilst the positional correlations between molecules in a dense fluid are well developed, the orientational correlations are relatively weak. If we take the extreme case in which the position and orientation of one molecule are completely uncorrelated with the orientation of another then this group of terms would reduce to

$$((4/9)\gamma^2 E^L)^2 \left\{ S_2 + 0 + 0 \right\} \quad (4.9)$$

ie there would be an un-cancelled two-body contribution! Whilst this model is clearly naive, it semi-quantitatively accounts for the different density dependence of the depolarised Rayleigh intensity in CS_2 from the intensity of the isotropic Rayleigh and the forbidden bands[8]. The concept of uncancelled two-body contributions also has far reaching implications for the interpretation of induced spectral lineshapes in molecular fluids[35]. The sudden switching on of the orientational correlations upon crystallization is signalled by the tremendous decrease in the intensity of the forbidden Raman lines in CS_2[35]; when the orientations become correlated, cancellation between groups of terms similar to those appearing in equation 4.8 is suddenly induced.

4.2 Experiment and Computer Simulation

Quantitative studies of liquids began with the advent of computer simulation, which permits the calculation of a spectrum for a given microscopic model, (see the lectures of Tildesley). In this section we shall be concerned only with the way in which studies of the intensity reflect upon the induced moment models described in section 2.

The most extensively studied system has been the Rayleigh scattering in atomic fluids, it was the subject of the pioneering work of Alder et al[36]. Paradoxically, this may be the system for which achieving agreement with experiment is most difficult. This is a consequence of the cancellation problem described above; as we showed, the contribution of the first-order DID term (which dominates in the gas-phase) almost completely disappears near the triple point. Thus the depolarised scattering in the dense fluid acquires many of the unfortunate attributes of the isotropic scattering in the gas! The lowest order contribution of the no-overlap model disappears (almost) and the effect becomes susceptible to overlap and the "slew" of higher order terms which were small in the gas.

For many years it was thought that there was a very large discrepancy between the results of simulation with the DID model and experiment, but recent experimental results have shown that the DID model gives only about 25% too much intensity[37,38] for Argon. The extent to which overlap contributions to the pair polarizability can account for this difference has been investigated, but the most recent work appears to show that the pair polarizability which best fits the gas-phase data does not give a good representation[39] of the liquid. Alder[36] has expressed the view that overlap in a dense system has a cooperative effect which is not attributable to the pair terms alone. An alternative explanation is that normally neglected terms from within the no-overlap model become significant by virtue of the cancellation of the leading term. In particular, it has been argued that higher order DID terms, like that which appears in eqn 2.2a, may reduce the effective polarizability of a pair, by analogy with the way in which local field factors (v.i) reduce the effective polarizability in a fluid of anisotropic molecules[40,41]. This proposal is rather difficult to test as the higher-order terms are also subject to cancellation![40]. A simulation study by Alder et al has suggested[42] that they are unimportant. In reality, it is likely that many extra-DID terms contribute simultaneously.

There have now been a number of detailed studies of molecular liquids. This work will be described in detail below; here we shall only summarise how the multipole concepts have been applied to forbidden spectra; because cancellation effects are not so important as in the inert gases these concepts should form a good basis. Probably the first work of this kind was the[43] examination of the QID mechanism in a computer simulation of the far i.r. spectrum of liquid N_2. More recently, the forbidden i.r spectra of Cl_2[44] and of the symmetric stretching vibration of CS_2[45,46], as predicted by eqn 2.8, the forbidden Raman spectra of CS_2[45], given by eqn 2.15, and the far i.r of CS_2[45] have been simulated. In general this work has found good agreement between calculated and experimental lineshapes but the calculated intensities have differed from experiment, typically by a factor of ~2. In part this may be attributable to uncertainty in the correct values to be used for molecular parameters such as $\tilde{A}$ and $\tilde{\theta}$ and to deficiencies in the potentials used in the simulations. However a systematic source of disagreement must be the inability to handle the type of higher order term appearing in eqn 2.2a. These give intensity contributions of order $\alpha\rho$ smaller than the lowest order terms which appear in eqns 2.7, 2.11, 2.8, and 2.14. For a fluid like Argon $\alpha\rho \simeq 0.04$ but for CS_2 it reaches ~0.1. Another way of describing the problem is to allude to the difficulties caused by "local field effects" for the interpretation of the intensities of <u>allowed</u> molecular spectra - the "effects" are a way of accounting for the influence of the many particle moments ($\tilde{m}^{ij}$, $\tilde{m}^{ijk}$, etc) on the intensities predicted in the conventional viewpoint (ie for $\tilde{m}^i$ alone). Local field factors can change intensities by 50% or so. Unfortuantely there is, as yet, no way of describing "local field effects" on the spectra predicted by the pair moments $\tilde{m}^{ij}$ for <u>induced</u> phenomena. In this view the level of disagreement between the simulation and experiment is about that which might be anticipated[45].

At the conceptual level, simulations have been used to examine the idea that un-cancelled two-body terms play an important role in determining the properties of induced molecular spectra[45]. Ideas used in the interpretation[8] of data on CS_2 were fully vindicated[45], the neglect of angular correlation model described above successfully identified the dominant contributions to the forbidden spectra.

5 THE INTERFERENCE OF ALLOWED AND INDUCED PROCESSES

5.1 General Considerations

In this section we shall examine the problems posed for the interpretation of "allowed" spectroscopic lineshapes by the simultaneous influence of permanent and induced moments. These problems are quite general; they arise in dielectric spectroscopy, infra-red absorption and light scattering.

The total moment $\underset{\sim}{M}$, which appears in eqn 1.1, may be written as the sum of the moment of the "conventional" picture (${}^{o}\underset{\sim}{M}$ - the sum of molecular moments eqn 1.2) and an induced contribution ${}^{I}\underset{\sim}{M}$, ie

$$\underset{\sim}{M} = {}^{o}\underset{\sim}{M} + {}^{I}\underset{\sim}{M} \tag{5.1}$$

${}^{o}\underset{\sim}{M}$ typically relaxes through molecular reorientation (or dephasing) but ${}^{I}\underset{\sim}{M}$ depends upon the <u>relative</u> positions and orientation of molecules in the fluid. The question which arises is:- what aspects of the total moment correlation function (or its associated spectrum) reflect on the properties of ${}^{o}\underset{\sim}{M}$, or, to what extent can the spectrum be used to study reorientation?

The relaxation times associated with the purely induced moments seen in the depolarised Rayleigh spectrum of an inert gas or the far i.r of a non-polar fluid are usually much shorter than reorientation times; furthermore such spectra are usually much weaker than allowed spectra. We might (naively) hope that we can write

$$\langle \underset{\sim}{M}(t).\underset{\sim}{M}\rangle = \langle {}^{o}\underset{\sim}{M}(t).{}^{o}\underset{\sim}{M}\rangle + \langle {}^{I}\underset{\sim}{M}(t).{}^{I}\underset{\sim}{M}\rangle + 2\langle {}^{I}\underset{\sim}{M}(t).{}^{o}\underset{\sim}{M}\rangle \tag{5.2}$$

and identify the first term with a low frequency spectral feature, the second with a broad, weak background and neglect the cross-term altogether. However, this argument neglects the fact that in the light scattering of anisotropic molecules or the dielectric spectra of polar ones there are new induced effects not present in the purely induced counterparts. Also there is no reason to suspect that ${}^{I}\underset{\sim}{M}$ is uncorrelated with ${}^{o}\underset{\sim}{M}$, so that neglect of the cross term is without foundation.

An alternative approach follows from an analysis originally described in a paper by Keyes, Kivelson and McTague[47]. The starting point is to project out of $^I\underset{\sim}{M}$ the part which depends on the orientation variables which determine $^o\underset{\sim}{M}$, that is to rewrite eqn 5.1 as[48]

$$\underset{\sim}{M} = (1 + G)\,{}^o\underset{\sim}{M} + \Delta\underset{\sim}{M}, \qquad (5.3)$$

where

$$G = \langle {}^I\underset{\sim}{M}\,.\,{}^o\underset{\sim}{M}\rangle / \langle |{}^o\underset{\sim}{M}|^2\rangle \,, \qquad (5.4)$$

and

$$\Delta\underset{\sim}{M} = {}^I\underset{\sim}{M} - G\,{}^o\underset{\sim}{M}\,. \qquad (5.5)$$

$G\,{}^o\underset{\sim}{M}$ may be interpreted as the mean value of the induced moment for a given value of $^o\underset{\sim}{M}$ and $\Delta\underset{\sim}{M}$ the fluctuating part of $^I\underset{\sim}{M}$

Consider, for illustration, an allowed vibrational i.r spectrum. Then

$$\underset{\sim}{M} = \sum_i \underset{\sim}{\tilde{\mu}}^i + \sum_{ij}{}' \underset{\sim}{\delta}^{ij} \qquad (5.6)$$

where $\underset{\sim}{\delta}^{ij}$ represents the QID term given in eqn 2.8, and

$$G = \langle \sum_i \underset{\sim}{\tilde{\mu}}^i \,.\, \sum_{k\ell}{}' \underset{\sim}{\delta}^{k\ell}\rangle / \langle |\sum_i \underset{\sim}{\tilde{\mu}}^i|^2\rangle$$

$$= \langle \underset{\sim}{\tilde{\mu}}^1 \,.\, \sum_{\ell} \underset{\sim}{\delta}^{1\ell}\rangle / \langle |\underset{\sim}{\mu}^1|^2\rangle, \qquad (5.7)$$

where in the second equality the absence of correlation between the vibrational phases of different molecules has been used (see lecture of Bratos). The numerator "reads" - hold the value of $\tilde{\mu}_\alpha^1$ at a given value (ie fix the orientation of molecule 1) and now determine the αth component of the <u>induced</u> dipole in molecule 1 averaged over the possible configurations of all other molecules in the sample. This mean is not trivial because the positions and orientations of the other molecules are not random when the orientation of molecule 1 is fixed. The first term in 5.3 is then just a modified molecular dipole density

$$(1 + G)\,{}^o\underset{\sim}{M} = \sum_i (1 + G)\, \underset{\sim}{\tilde{\mu}}^i \qquad (5.8)$$

in which the magnitude of each molecular moment has been changed
from the gas-phase value of to a new "effective" value which
includes the induced moment $(G\underset{\sim}{\mu}^i)$ from the neighbours of a molecule in their equilibrium average configuration. $\Delta\underset{\sim}{M}$ contains the
difference between the instantaneous induced moment on a molecule
and this mean induced value

$$\Delta\underset{\sim}{M} = \sum_i \left[\sum_{\ell\neq i} (\underset{\sim}{\delta}^{i\ell} - \underset{\sim}{\mu}^i <\underset{\sim}{\delta}^{i\ell}.\underset{\sim}{\mu}^i>/<|\underset{\sim}{\mu}^i|^2>) \right] \qquad (5.9)$$

ie it is the fluctuating part of the induced dipole.

In terms of the projected quantities the total moment correlation function becomes

$$<\underset{\sim}{M}(t).\underset{\sim}{M}> = (1 + G)^2 <{}^o\underset{\sim}{M}(t).{}^o\underset{\sim}{M}>$$

$$+ \left[2(1+G)<\Delta\underset{\sim}{M}(t).{}^o\underset{\sim}{M}> + <\Delta\underset{\sim}{M}(t).\Delta\underset{\sim}{M}> \right] \qquad (5.10)$$

this formally separates the correlation function into a reorientational and a fluctuation-or "collision-" induced part (ie both
terms in square brackets). The square bracketed terms depend upon
the reorientation variable ${}^o\underset{\sim}{M}$ only in so far as the fluctuations
responsible for $\Delta\underset{\sim}{M}(t)$ depend upon the value of ${}^o\underset{\sim}{M}$ at earlier
times, as the equal time correlation is zero

$$<\Delta\underset{\sim}{M}(t).{}^o\underset{\sim}{M}(t)> \equiv <\Delta\underset{\sim}{M}.{}^o\underset{\sim}{M}> \equiv 0, \qquad (5.11)$$

by construction. One might hope that the reorientational character
of the square bracketed term is small.

Because of eqn 5.11 it is always meaningful to speak of
dividing the intensity into a reorientational (the first term) and
a truly collision-induced part associated purely with $<|\Delta\underset{\sim}{M}|^2>$
However, this division is only useful if we can recognise distinctive features in the spectrum associated with each intensity term
and thereby separate them.

Consider what happens as the molecule rotates in the terms
of the i.r example. If the neighbours were to relax very rapidly
to their new equilibrium positions upon a change in the molecular
orientation then the mean induced moment will follow the molecular
orientation and behave as an extra contribution to the molecular
dipole, the fluctuating part of the induced moment (ΔM) will
rapidly forget the earlier molecular orientation. Under this
condition of timescale separation the cross-term will be small at
all times (it is zero at t=0, by 5.11) and the first term in
eqn 5.10 will contain all the reorientational time dependence.
At low frequencies the spectrum will comprise a reorientational

part whose relaxation is determined by $\langle {}^{o}\tilde{M}(t).{}^{o}\tilde{M} \rangle$ but whose amplitude is not $\langle |{}^{o}\tilde{M}|^2 \rangle$ - rather the "local field" modified reorientational amplitude $(1 + G)^2 \langle |{}^{o}\tilde{M}|^2 \rangle$. At high frequencies the spectrum will reflect the relaxation of $\langle \Delta\tilde{M}(t).\Delta\tilde{M} \rangle$, which is governed by fluctuations in the local fluid structure.

It is easy to convince oneself, by inspection, that this description applies to the spectra of slowly rotating molecules in dense fluids - as an example fig 4a shows a plausible separation of the Rayleigh spectrum of CS_2 (49). However the spectra of more inertial rotors - such as the Rayleigh spectrum of CO_2 at the triple point (fig 4b) (50) does not appear to separate. Without the guidance which computer simulation has provided it would not be possible to proceed objectively.

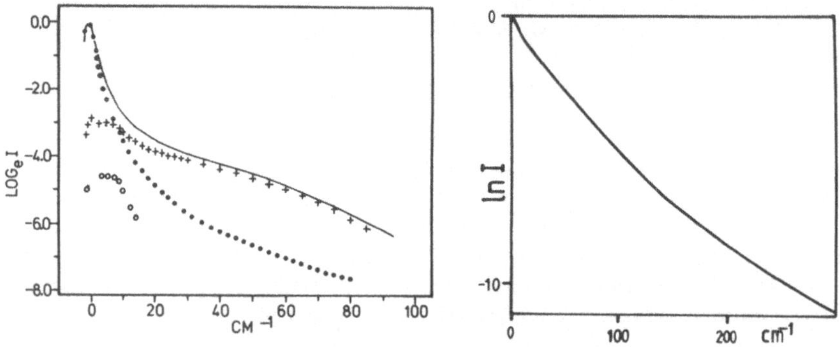

Fig 4: Separation of induced and reorientational Rayleigh spectra. (a) The spectrum of CS_2 at 232K, showing a plausible separation into a reorientation2 ..... and collision-induced +++ component. (b) The spectrum of CO_2 (ref 50) at the triple point.

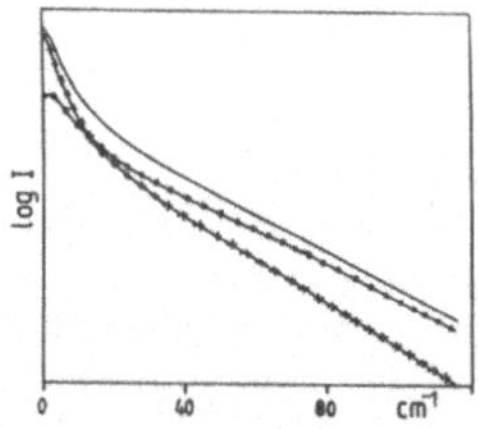

Fig 5: The Raman spectrum of CS_2 at 297K separated into reorientational +++++ and collision-induced ..... components.

5.2 Computer Simulation Studies - Light Scattering

There have been three[51,52,53] simulation studies of the issues raised above which address the Rayleigh and Raman spectra of small linear molecules. These studies will be described in more detail by Tildesley and so I will only précis those points which are necessary for continuity of my presentation.

Frenkel and McTague[51] studied two-site models of N_2, O_2, Cl_2 and CO_2 and used the DID model for the induced polarizability. They showed that for CO_2 the collision induced contribution to the depolarised Rayleigh intensity was 25% of the total intensity and that the second moment was increased by about 50% by induced effects. The timescale separation was examined in N_2 and CO_2. In their terminology $\langle \Delta \underset{\sim}{M}(t).\Delta \underset{\sim}{M} \rangle$ is the collision induced contribution; it was found to relax in a very similar way to the orientational function $\langle {}^{o}\underset{\sim}{M}(t).{}^{o}\underset{\sim}{M} \rangle$ and the spectra of the two terms were indistinguishable for practical purposes. Furthermore the cross-term $\langle \Delta \underset{\sim}{M}(t).{}^{o}\underset{\sim}{M} \rangle$ was quite large. The net effect of the non-orientational terms was to reduce the amplitude of the spectrum at low frequencies and to increase it in the wings. Frenkel and McTague's results on nitrogen have been carefully compared with experiment by Sampoli, de Santis and co-workers[54].

Recently Ladanyi[52] has extended the work on N_2 and CO_2 by examining the induced terms given by a site-site DID model. This model allows for the distribution of polarizable matter within the molecule (c.f the discussion of section 2) by representing the molecular polarizability by an isotropic point polarizability on each interaction site and taking all orders of intra-molecular DID interactions into account. For CO_2 her results are appreciably different from Frenkel and McTague's. In particular there is an enhanced projection of the induced terms along ${}^{o}\underset{\sim}{M}$ and a better timescale separation between the allowed and collision-induced processes so that the total spectrum resembles the reorientational spectrum more closely.

Madden and Tildesley[53] have studied the spectra of a model of CS_2; they used the normal centre-centre DID model. In CS_2 the molecular reorientation is slow and diffusional in contrast to the rapid inertial tumbling found in N_2 and CO_2; in this regard CS_2 may be regarded as a representative normal organic fluid. In figure 5 the simulated depolarised Raman spectrum (of the ν_1 mode is shown for a density and temperature appropriate to STP. In this example clear evidence of a timescale separation is found, the total spectrum is dominated by the reorientational spectrum at low frequencies. This separation of timescales improves even further as the fluid becomes denser and cooler. By comparing figures 4 and 5 it can be seen thatthe intuitive separation of the experimental spectrum is not far from the truth.

5.3 Computer Simulation Studies - Dielectric Spectra

Edwards and Madden[55] have recently completed a simulation of a three-site model of CH_3CN in which the induced dipole contributions to the dielectric permittivity were examined. A molecular theory of dielectric properties and local field factors has been developed and the simulation was designed as part of a systematic investigation of this work[56].

At the simplest level the calculation can be regarded as including the intermolecular field-induced dipole contribution to the total moment of the sample. The potential model used has point charges on the three interaction sites - these are chosen to reproduce the dipole and quadrupole moment of the molecule[57]. The intermolecular ($\underset{\sim}{F}$) field is calculated as the field at the central carbon atom due to these charges on all the other molecules in the simulation (500 in total). The induced dipole is given by

$$^{I}\underset{\sim}{M} = \sum_i \underset{\approx}{\alpha}^i . \underset{\sim}{F}(\underset{\sim}{r}^i), \tag{5.12}$$

in terms of the multipole expansion of §2 the dipole-induced, quadrupole-induced and all higher multipole-induced dipoles are included.

The calculation of dielectric properties in computer simulations is complicated by the long-ranged nature of the coulomb interactions and by the fact that the total moment correlation function is a "collective" function which requires long runs to properly average. These issues have been discussed by Tildesley in these lectures and in several reviews[36,58].

Contact is made between properties calculable in the simulation and the permittivity via[57,59].

$$(\varepsilon(\omega)-\varepsilon(\infty))/\varepsilon(0)-\varepsilon(\infty)) = 1 + i\omega \lim_{k\to 0}[\Phi_T(k,\omega)] \tag{5.13}$$

where

$$\Phi_T(k,\omega) = \int_0^\infty dt \exp i\omega t \frac{\langle \underset{\sim}{M}_T(\underset{\sim}{k},t) . \underset{\sim}{M}_T(\underset{\sim}{k}) \rangle}{\langle |\underset{\sim}{M}_T(\underset{\sim}{k})|^2 \rangle} \tag{5.14}$$

and $\underset{\sim}{M}_T(= {}^{o}\underset{\sim}{M}_T + {}^{I}\underset{\sim}{M}_T)$ is the transverse part of the total dipole density

$${}^{o}\underset{\sim}{M}_T(\underset{\sim}{k},t) = \sum_i (\underset{\approx}{1} - \hat{\underset{\sim}{k}}\hat{\underset{\sim}{k}}) . \underset{\sim}{\mu}^i \exp[i\underset{\sim}{k}.\underset{\sim}{r}^i] \tag{5.15}$$

$$^{I}\underset{\sim}{M}_T(\underset{\sim}{k},t) = \sum_i (\underset{\approx}{1} - \underset{\sim\sim}{\hat{k}\hat{k}}).\underset{\approx}{\alpha}^i.\underset{\sim}{F}(r^i)\exp[i\underset{\sim}{k}.\underset{\sim}{r}^i] . \quad (5.16)$$

The forces are evaluated by the Ewald method - this means that $\underset{\sim}{F}$, and hence the induced dipoles, may be calculated as a biproduct of the force loop with no extra work!

Figure 6a shows the correlation functions of the unprojected quantities $^{O}\underset{\sim}{M}_T$ and $^{I}\underset{\sim}{M}_T$. They are seen to relax on very similar timescales; note too that the interaction-induced moments constitute a large fraction of the total moment function, (~40%). When the projection is taken (fig 6b) a very large fraction of $^{I}M_T$ is found to be correlated with the orientation $^{O}\underset{\sim}{M}_T$. The total spectrum $\Phi_T(k,\omega)$ closely resembles the orientational spectrum. the spectrum associated with the autocorrelation function of $\Delta\underset{\sim}{M}_T$ is weak and much broader than the orientational one.

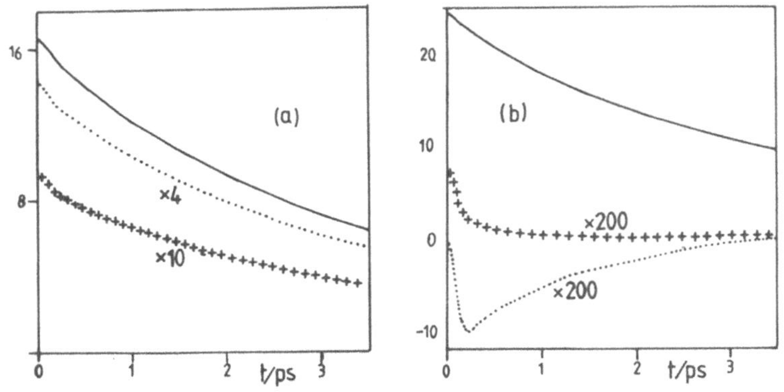

Fig 6: Two decompositions of the total moment correlation function for a model of methyl cyanide, from Ref. 55.
(a) The correlation functions of the rigid (^{O}M) ——— and interaction-induced (^{I}M) +++ moments and their cross term ...
(b) The reorientational ($(1+G)^{O}M$) ——— and collision-induced ΔM ++++ moments and their cross term ......

The results may be compared with the experimental static permittivity

$$\varepsilon(0) - \varepsilon(\infty) = \lim_{k\to 0} \langle|\underset{\sim}{M}_T(k)|^2\rangle/2k_BTV\varepsilon_o \quad . \qquad (5.17)$$

The experimental value of $\varepsilon(0)-\varepsilon(\infty)$ is 38 at 297K[60]. If only the rigid dipole contribution (ie $\langle|{}^oM_T|^2\rangle$ is taken the simulation gives 27; however, including the interaction induced term gives $\varepsilon(0)-\varepsilon(\infty) = 40$. The contribution to $\langle|M_T|^2\rangle$ (and hence to $\varepsilon(0)$) from $\langle|\Delta M_T|^2\rangle$ is very small; thus the static permittivity calculated in the simulation is well represented by the "local field" modified reorientational term alone

$$\varepsilon(0) - \varepsilon(\infty) = \lim_{k\to 0} (1 + G_T)^2\langle|{}^o\underset{\sim}{M}_T(k)|^2\rangle/2k_BTV\varepsilon_o \qquad (5.18)$$

We will discuss the value of the local field factor below.

Experiment and simulation may also be compared via the far i.r absorption coefficient $\alpha(\omega)$[60],

$$\alpha(\omega) = (\omega/c)\ \varepsilon''(\omega)/n(\omega) \qquad (5.19)$$

(where $n(\omega)$ is the refractive index - see the lectures of Cole). On the whole this comparison is extremely favourable; in particular, the spectral amplitudes are in excellent agreement. Fig 7 shows that the "collision-induced" contribution (ie the $\Delta M\Delta M$ term and the cross term as bracketed in eqn 5.10) have an appreciable affect on $\alpha(\omega)$. However, the reorientational absorption coefficient alone does show a Poley absorption - that is, the absorption rises well above the Debye Plateau. The spectrum associated with the $\Delta M\Delta M$ autocorrelation function is very similar in amplitude and shape to the absorption coefficient of non-polar materials[61] (in contrast to ${}^IM$). Note that whilst the "collision-induced" contribution to the peak height is not enormous the total interaction induced contribution is very large - the rigid dipole model absorption coefficient shows a peak of only about 400 neper cm^{-1}.

In summary, we note that the simulation results have shown a large difference between the role of induced phenomena in simple inertial liquids like N_2 and in "organic" fluids of relatively slowly reorienting molecules like CS_2 and CH_3CN. In the latter case the possibility of separating the spectrum into a "local-field" modified rotational term and a "collision-induced" term would seem to exist. All the simulations show that collision-induced effects have an important effect on lineshapes at high frequency, but it is somewhat surprising to find that the reorientational spectra at high frequency are very similar in shape to the induced ones; this presumably accounts for the very plausible looking "short-time orientational correlation functions" (or "angular velocity correlation functions") which are obtained by naively Fourier transforming experimental lineshapes, with no regard for induced effects.

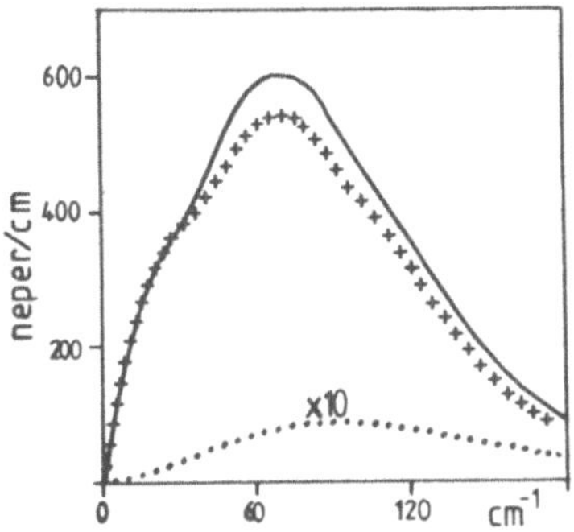

Fig 7: The resolution of the far infrared spectrum of methyl cyanide ——— into a reorientational part from the a.c.f. of $(1+G)\,{}^{o}M$ +++++ and a collision-induced (a.c.f. of ΔM) ••••• part.

6. LOCAL FIELD FACTORS AND ORIENTATIONAL CORRELATION PARAMETERS

6.1 General Considerations

As shown in the last section a spectral intensity (which for convenience will be taken to cover $\varepsilon(0) - \varepsilon(\infty)$) may always be formally divided into a reorientational part, I_R, and a collision-induced part. Under the favourable circumstance that a timescale separation exists (as in the CS_2 and CH_3CN examples of §5) it may be possible to separately determine the two parts and obtain a measured value for I_R, ie

$$I_R = (1 + G)^2 \langle |{}^{o}\underset{\sim}{M}|^2 \rangle \tag{6.1}$$

Indeed it was seen that the collision induced contribution to $\varepsilon(0) - \varepsilon(\infty)$ was very small in the simulation of CH_3CN so that "I_R" was equal to the total "intensity" in this case.

For the collective spectra (ie dielectric and Rayleigh) $<|{}^{o}M|^2>$ includes an orientational correlation parameter; for example, in the Rayleigh case

$$I_R = (1 + G)^2 < \sum_i \alpha^i_{xz} \sum_j \alpha^j_{xz} > (E^L)^2 \qquad (6.2)$$

or, using eqn 4.4,

$$I_R = (2/15)(1 + G)^2 \gamma^2 \rho g_2 (E^L)^2 \qquad (6.3)$$

where g_2 is the second rank orientational correlation parameter

$$g_2 = < \sum_i Q^i_{xz} Q^j_{xz} > (20/3) = 1 + \sum_{j \neq 1} <P_2(\cos\theta_{ij})> . \qquad (6.4)$$

An alternative way of writing this expression is to introduce an "effective" polarizability anisotropy $\gamma_{eff} = \gamma(1+G)$ (as discussed following eqn 5.8), viz

$$I_R = (2/15)\gamma^2_{eff} g_2 (E^L)^2 \qquad (6.5)$$

Similarly $\varepsilon(0) - \varepsilon(\infty)$ may be related to the Kirkwood g-factor (see the lectures of Cole)

$$g_1 = 1 + \sum_{j \varepsilon s_1} <P_1(\cos\theta_{ij})> \qquad (6.6)$$

where the sum runs over those molecules j within a small sphere around molecule 1 in an infinite sample.

To measure these orientational correlation parameters (or, more generally, to interpret spectral intensities in condensed phases) it is necessary to deal with the local field factor (1+G). The basis of a molecular theory of this factor has been formulated above as

$$G = <{}^{o}\underset{\sim}{M} . {}^{I}\underset{\sim}{M}> / <{}^{o}\underset{\sim}{M} . {}^{o}\underset{\sim}{M}> \qquad (6.7)$$

and we have given molecular expressions for the induced moments. The outstanding requirement of the theory is to evaluate the ensemble averages. It is desirable that in so doing the molecular theory should make contact with the classical expressions for the local field factors which were obtained on the basis of continuum electrostatic considerations many years ago (see the lectures of Cole and ref 62). Theoretical developments of this kind have been made in both the light scattering and dielectric fields.

6.2 Molecular Theories of Local Field Factors

Keyes and Ladanyi (KL) have developed a theory of local field factors in the depolarised Rayleigh experiment[63]. With the DID model for the induced polarizability equations 5.4 and 2.11 give

$$\gamma_{eff}/\gamma = 1 + G = 1 + \frac{\langle \sum_i Q^i_{xz} (\sum'_{jk} \underset{\approx}{\alpha}^j \cdot \underset{\approx}{T}^{jk} \underset{\approx}{\alpha}^k)_{xz} \rangle}{(2/3)\gamma \langle \sum_{ij} Q^i_{xz} Q^j_{xz} \rangle} \tag{6.8}$$

The 1-, 2-, and 3- particle contributions to the numerator may be separated giving

$$\gamma_{eff}/\gamma = 1 + \frac{2N\langle Q^1_{xz} (\underset{\approx}{\alpha}^1 \cdot \underset{\approx}{T}^{12} \cdot \underset{\approx}{\alpha}^2)_{xz} \rangle + N^2 \langle Q^1_{xz} (\underset{\approx}{\alpha}^2 \cdot \underset{\approx}{T}^{23} \cdot \alpha^3)_{xz} \rangle}{(2/3)\gamma g_2} \tag{6.9}$$

The 2-particle term may be evaluated in terms of the invariant spherical harmonic expansion of the pair distribution function[63]. The 3-molecule term presents a greater problem; KL assumed that certain "irreducible" terms which appear in the superposition approximation to the 3-particle function may be ignored ("the factorisation approximation") which leads to the result that the 3-particle term is just (g_2-1) times the 2-particle one and

$$\begin{aligned} \gamma_{eff}/\gamma &= 1 + (6\rho\tau_{20}/5\gamma)[\alpha^2 + \alpha\gamma/3 + 2\gamma^2/9] \\ &\quad - (\tau_{22}/15)[\alpha\gamma + \gamma^2/3] \end{aligned} \tag{6.10}$$

In this expression τ_{20} is determined by the way in which the shape of one molecule distorts the radial distribution function away from spherical symmetry whilst τ_{22} is determined by mutual orientational correlation between molecules. The distribution function of model diatomics obtained in computer simulations showed that the τ_{22} effects were much smaller than the orientation-position effects given by τ_{20}. The effect of the τ_{20} terms on γ_{eff} can be simply interpreted; for a cylindrical molecule the closest neighbours must lie around its waist. If the cylinder is aligned with an external field (ie if the field is parallel to the molecular z axis) then the dipole induced by the field in the near neighbours will tend to reduce the net field along the cylinder axis; as the dipole induced in the cylinder by a given external field is thereby reduced the effect is to reduce α_{zz}. Conversely if the cylinder is perpendicular to the field (ie $\underset{\sim}{F}$ parallel to x) the dipole induced in the near neighbours enhances the applied field and hence α_{xx}. Consequently the radial correlation effect reduces γ_{eff}, ie τ_{20} is negative. This feature is incorporated into the classical Onsager-Scholte(62) description of the local ("depolarising") field. KL showed that, for a molecule of ellipsoidal shape, if they approximated the pair distribution function by simply excluding molecular centres from the region occupied by a molecule they recovered the classical result through order $\alpha\rho$(64).

KL also considered the effective polarizability (α_{eff}) which determines the refractive index of a fluid(63) and the appropriate local field factors for the Kerr effect(16). Cox et al obtained analagous expressions for the local field factors in the Raman effect(49).

KL's numerical results for the effective polarizabilities of model diatomics were examined by Frenkel and McTague(51), these authors obtain G directly in the course of their simulation and without resorting to the factorisation approximation for the 3-particle distribution function. Their result for $(\gamma_{eff}-\gamma)/\gamma$ was about 50% smaller than KL's for Br_2. Perhaps significantly, the value of α_{eff} obtained by both groups agreed, and the factorisation approximation is not invoked by KL in obtaining this quantity. Notice that the factorisation approximation is implicit in the application of the classical local field factor to the Rayleigh experiment. A factorisation is not needed to evaluate the Raman factors(49).

A molecular theory of the local field factors appearing in the expression for the dielectric permittivity has been considered by Madden and Kivelson(56). They have shown how the introduction of an idealised distribution function leads to the Fröhlich formula for the permittivity, ie.

$$\frac{(\varepsilon(0) - \varepsilon(\infty))(2\varepsilon(0) + \varepsilon(\infty))}{\varepsilon(0)} = 4\pi\left(\frac{\varepsilon(\infty)+2}{3}\right)^2 \rho\mu^2 g_1/k_B T \quad (6.11)$$

from the general molecular result given in equations 5.4 and 5.13. Edwards and Madden(55) have compared their results for G_T, obtained in the simulation of CH_3CN, with the predictions of this theory. The simulation results are in good agreement with the values calculated with the idealised distribution function though significant short-ranged structural effects, not contained in the theory, are found.

6.3 Orientational Correlation Parameters

To obtain the orientational correlation parameters g_1 and g_2 from the reorientational parts of the permittivity or the depolarised

Rayleigh intensity one must determine the local field factors (c.f eqn 6.1). In the dielectric case it seems that one must calculate this quantity (normally an approximate formula of the Fröhlich type is used), but in the case of g_2 it is possible to avoid this by combining the results of several experiments in order to eliminate the local field effects (see also the lectures of Versmöld).

The key is the Cotton-Mouton, or magnetically induced birefringence, experiment. If the multipole analysis of induced effects, as outlined in §2 is justified then it may be shown[65] that the Cotton-Mouton constant (${}_mC$) is proportional to g_2, in fact

$$ {}_mC \;\alpha\; \gamma_{eff}\, \Delta\chi\, g_2/k_B T \tag{6.12} $$

where $\Delta\chi$ is the gas-phase magnetizability anisotropy. The important point is that because magnetic dipoles are so small (relative to electric dipoles) induced magnetic effects are insignificant. The collision-induced Cotton-Mouton effect is thus negligibly small and the effective magnetizability anisotropy is equal to the isolated molecule value. g_2 and γ_{eff} values for CS_2 and a number of benzenoids[65,66] have been determined by combining ${}_mC$ and I_R values with measurements of the gas-phase magnetizability.

7 LINESHAPES OF INDUCED SPECTRA IN LIQUIDS

7.1 General Considerations

As stressed in section 4, induced spectra in liquids arise from fluctuations in the local density around a particle. The interest in the study of induced spectra *per se* is that these fluctuations are the elementary events involved in such transport processes as viscous flow, diffusion or vibrational dephasing. Consider the diffusion coefficient, it is related to a friction coefficient by

$$ D = k_B T/\xi(0) \tag{7.1} $$

and

$$ \xi(\omega) = \mathrm{Re}\int_0^\infty dt\, \exp[i\omega t]\langle \underset{\sim}{f}(t).\underset{\sim}{f}\rangle \tag{7.2} $$

where $\underset{\sim}{f}$ is a "random force". The random force arises from the local density fluctuations and eqn 7.1 shows that the diffusion coefficient is determined by its spectrum at one point, ie $\omega = 0$. Since the induced spectra reflect the spectra of the density fluctuations at all frequencies they should provide a good testing ground for theories of intermolecular dynamics which might then be applied to the more technically important transport processes.

We will begin by showing how such a theory can be formulated, with reference to the induced Rayleigh spectra in atomic fluids. We will then describe how the concepts may be modified to deal with the induced spectra of molecular liquids. Finally we will illustrate, with the aid of simulation results, the extent to which the connection between transport processes and the theory of induced spectra can be made.

7.2 Atomic Liquids

In the DID model the atomic induced Rayleigh scattering is due to the moment (we consider the polarised spectrum for convenience)

$$M_z = \sum_{ij}{}' m_z^{ij} \tag{7.3}$$

where

$$m_z^{ij} = (\alpha^2 E^L)\, T_{zz}^{(2)}(\underset{\sim}{r}^{ij}) \tag{7.4}$$

and

$$\begin{aligned} T_{zz}^{(2)}(\underset{\sim}{r}^{ij}) &= 2P_2(\cos\theta^{ij})(r^{ij})^{-3} \quad ; r_{ij} > \sigma \\ &= 0 \quad ; r_{ij} < \sigma \end{aligned} \tag{7.5}$$

The interaction tensor is cut off at the distance of closest approach, σ.

It is tempting to write the spectrum ($I(\omega)$) as a sum of 2-, 3-, and 4- particle contributions, viz.

$$\begin{aligned} I(\omega) = \int_0^\infty dt\, \exp[i\omega t] \Big\{ & N^2 \langle m_z^{12}(t) m_z^{12} \rangle + 2N^3 \langle m_z^{12}(t) m_z^{13} \rangle \\ & + N^4 \langle m_z^{12}(t) m_z^{34} \rangle \Big\}; \end{aligned} \tag{7.6}$$

however, this is not a good way to approach the dense fluid case. This is because of the cancellation effects which we discussed in the context of the analagous decomposition of the intensity in section 4. As well as the intensity, distinctive spectral characteristics associated with each of the n-particle terms also cancel. This is beautifully illustrated by the simulation results obtained

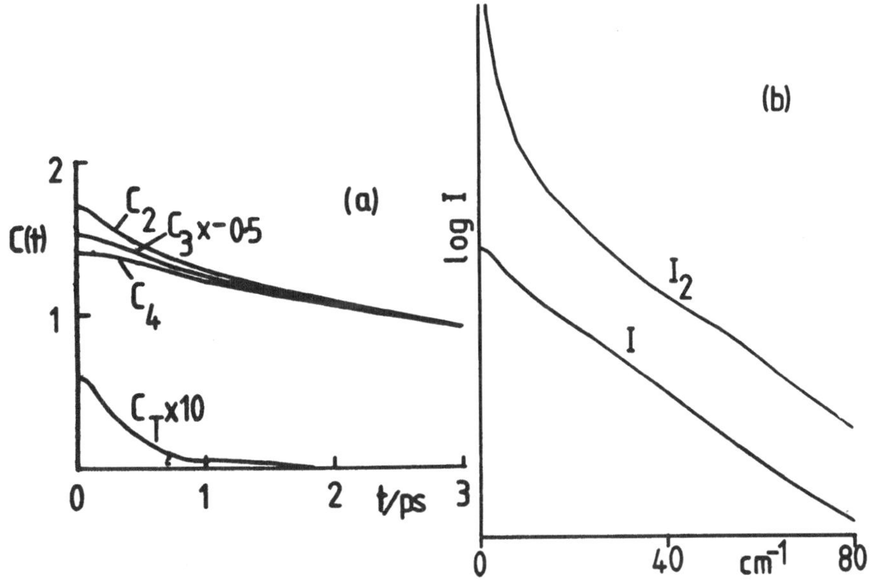

Fig 8: (a) The 2- (C_2), 3-(C_3) and 4-(C_4) particle correlation functions (eqn 7.6) for the depolarised Rayleigh spectrum of argon at its triple point (from ref. 67) and the total correlation function (C_T).(b) The 2-particle and total spectra (I_2&I).

by Ladd et al[67], which are schematically reproduced in Figure 8. The 2-body functions decays very slowly in time, due to the slow dependence of $T^{(2)}$ on rij and $\hat{r}^{ij}$; in fact it can be shown to decay as $t^{-3/2}$ at long times (v.i). If this slow decay were a feature of the total correlation function the observed spectrum would show an $\omega^{-1/2}$ cusp at $\omega = 0$ (see fig 8). However, the long time parts of the 3-, and 4- particle functions cancel this slow decay, the character of the total correlation function reflects only short time effects. Note that the shapes of the total and 2-body spectra agree at frequencies beyond about 40 cm^{-1}. The cancellation effect becomes less pronounced at lower densities and the spectra show an enhancement of intensity at low frequency as a result.

To model the spectra, then, the decomposition we seek is into the fluctuations in the local structure, ie the excitations which cause the departure from cancellation, rather than into the n-particle contributions. The total dipole on a given atom$^{(i)}$ may be written[68]

$$\sum_{j\neq i} m_z^{ij} = (\alpha^2 E^L) \int \underset{\sim}{dr}\, T_{zz}^{(2)}(\underset{\sim}{r}) \sum_{j\neq i} \delta(\underset{\sim}{r} - (\underset{\sim}{r}^j - \underset{\sim}{r}^i)) \qquad (7.7)$$

Notice that $\sum_{j\neq i} \delta(\underset{\sim}{r} - \underset{\sim}{r}^{ij})$ is the particle density at a position $\underset{\sim}{r}$ from atom i $(n^i(\underset{\sim}{r}))$ (ie the delta function counts one if a particle is found at the point $\underset{\sim}{r} = \underset{\sim}{r}^j - \underset{\sim}{r}^i$). The integral in eqn 7. is in the form of a convolution so that

$$\sum_{j\neq i} m_z^{ij} = (\alpha^2 E^L) \int d\underset{\sim}{k}\; T_{zz}(\underset{\sim}{k}) n^i(\underset{\sim}{k}) \tag{7.8}$$

The description suggested by this representation of the induced dipole is that it is made up of a sum (integral) of contributions from each fourier component of the local density around i $(n^i(\underset{\sim}{k})$ may be thought of as a fluctuation in the form of a plane wave in the local density), $T(\underset{\sim}{k})$ assigns a weight to each component. $n^i(\underset{\sim}{k})$ is just

$$n^i(\underset{\sim}{k}) = \sum_{j\neq i} \exp[ik.(\underset{\sim}{r}^i - \underset{\sim}{r}^j)] \tag{7.9}$$

and $T(\underset{\sim}{k})$ is

$$T_{zz}(\underset{\sim}{k}) = 2\, P_2\, (\hat{\underset{\sim}{k}}) [j_1(k\sigma)/k\sigma] \tag{7.10}$$

In the absence of the Bessel function factor (which arises from the cut-off in the dipole tensor) $T(\underset{\sim}{k})$ would not depend on $|k|$ and the dominant contributions to the induced moment would come from very large k (as $d\underset{\sim}{k} = k^2 dk d\hat{\underset{\sim}{k}}$). The cut-off diminishes these high k contributions (as density fluctuations cannot have arbitrarily small wavelengths in a fluid of particles of finite size). The spectrum may now be written:

$$I(\omega) = (\alpha^2 E^L)^2 \int d\underset{\sim}{k} d\underset{\sim}{k}^1 T_{zz}(\underset{\sim}{k}) T_{zz}(\underset{\sim}{k}^1) \int dt\, \exp[i\omega t]$$
$$\langle \sum_i n^i(\underset{\sim}{k},t) \sum_j n^j(\underset{\sim}{k}^1) \rangle\,, \tag{7.11}$$

as a weighted sum of subspectra associated with local density fluctuations of a given wavelength.

It is now useful to note the following: nothing in the above development, except for the specific form of $T(\underset{\sim}{k})$, depends upon the fact that we are trying to describe a light scattering spectrum - the form of the analysis would go through unchanged for any function of an intermolecular vector. In particular we would expect to be able to write an equation of identical form for the spectral density of the random force and similar properties of intermolecular interactions, only the detailed form of $T(\underset{\sim}{k})$ would differ. For example, such expressions have been used as the basis of molecular theories of the Stokes-Einstein relationships for translational and rotational diffusion coefficients[69] and of dielectric friction on a rotating molecule[70]; indeed the decomposion into local density subspectra may be recognised[69] as the underlying concept in "mode-coupling" theories of transport processes[71]. The local density subspectra can therefore be used to unify the description of diverse fluid properties which depend on intermolecular interactions, and they may be observed in induced spectra!

In the calculation of the spectrum[58] from 7.11 the contribution of the $\underset{\sim}{k} \neq \underset{\sim}{k}'$ terms have been neglected (in the manner suggested by mode-coupling theory). Unfortunately this assumption becomes unreliable for large k which means (v.i) that a quantitative theory has not yet been developed for high frequency[72]. Subspectra with $k \sim 2\pi/\sigma$ and larger are found to dominate the total spectrum (ie the important density fluctuations have a spatial extent of about an intermolecular length or shorter). It is known from neutron scattering[73] that to model the relaxation of density fluctuations with this kind of wavelength viscoelastic effects must be considered[73]. The theory uses input from the analysis of neutron experiments[68] and is found to predict the observed lineshape rather well[72], at least through low and intermediate frequencies. Rather than follow this development here an (oversimplified) physical picture will be given.

Despite the strictures given at the beginning of this section, let us consider the two-body part of a density subspectrum! (The limitations will be considered at the end). This is determined by the correlation function

$$<\exp\left[i\underset{\sim}{k}.(\underset{\sim}{r}^1(t)-\underset{\sim}{r}^2(t))\right]\exp\left[-i\underset{\sim}{k}.(\underset{\sim}{r}^1(0)-\underset{\sim}{r}^2(0))\right]$$
$$= j_o(k<(\Delta r^{12}(t))^2>^{\frac{1}{2}}) \tag{7.12}$$

where $j_o(x) = \sin x/x$ and $\Delta r^{12}(t) = \underset{\sim}{r}^{12}(t)-\underset{\sim}{r}^{12}(0)$. Now $j_o(0) = 1$ and $j_o(x)$ passes through zero when $\tilde{\Delta}r^{12}(t) \tilde{=} \pi/k$, a subspectrum with a given k thus reflects particle displacements over distances of order πk^{-1}. This means that the subspectra acquire distinctive relaxation characteristics which are determined by the value of k. Consider figure 9 which shows an impression of the behaviour of $\Delta r^{12}(t)$ in a gas, a liquid and a solid. We can see that low k subspectra will predominantly reflect the diffusional aspects of molecular motion,with translation over distances larger than an interatomic separation involved. Since this motion is slow such subspectra will dominate the observed spectrum at low frequency. Intermediate k will reflect the residual solidlike motion which persists into the liquid (most pronouncedly close to

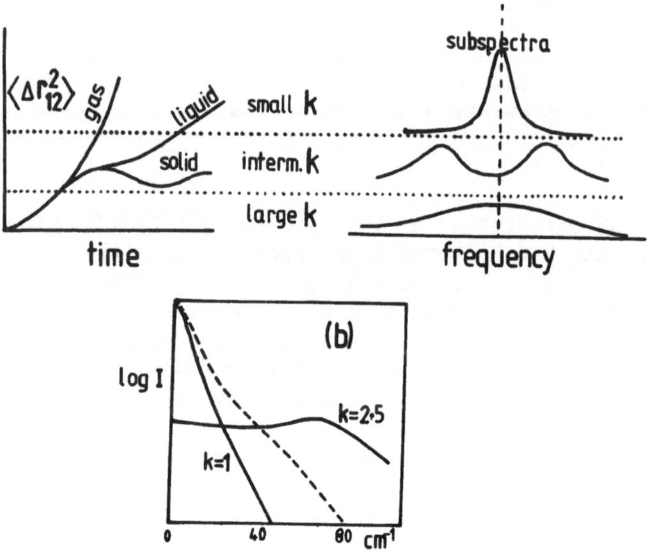

Fig 9: Mean interparticle displacement in a liquid vs. time is compared with the same quantity in a solid and a gas. The subspectra for large, intermediate and small wavevectors expected from this behaviour are shown to the right. (b) Subspectra calculated from theory (ref 68) at $k\sigma/2\pi=1$ (small k) and $k\sigma/2\pi=2.5$ (intermediate k) and the full spectrum ------.

the triple point). Whereas high k will reflect the very rapid motions over distances so small that the translation is gas-like or force free, ie the distances are so short that the potential felt by an atom barely changes. We therefore expect a gas-like character for the behaviour of the spectral wings. These qualitative considerations are supported by the predictions of the full theory (fig 9).

The principle deficiency of looking at the two-atom term alone is that the low k behaviour is overemphasised, it is the low k subspectra which are most affected by cancellation.

These ideas show why the shape of the spectrum is relatively insensitive to the detailed from of the induced moment[(68)] (ie to departures from the DID model), as observed in computer simulations[(36)]. The weighting of different k vectors (eg the dominance of $k \cong 2\pi/\sigma$) is largely due to the intrinsic structural and dynamical properties of the fluid, $T(\underset{\sim}{k})$ is fairly slowly varying over the important range of k and exerts comparatively little effect. Therefore reasonable changes in $T(\underset{\sim}{k})$ (ie compatible with the departures from the DID model found in gases) only change in a minor way the relative weighting of subspectra of different k. (The same is not true in gases where there is no lower cut-off in the range of k by cancellation effects.)

What happens upon crystallisation? The low k subspectra which reflect the interchange of molecules between coordination shells disappear and we are left with the shifted peaks and broad background arising from intermediate and high wavevectors[74,35].

Other descriptions of the Rayleigh lineshape have emerged which give rise to a function which may be fit to the spectra, these descriptions are conceptually simpler than that outlined above[75,76]. They address the whole induced moment $\sum_{ij}' m^{ij}$ as a single variable and do not develop the density fluctuation idea. They are thus not predictive and do not contain this feature of unifying intermolecular processes. However, they are undoubtedly easier to use in correlating data on different liquids or at different state points.

7.3 Molecular Liquids:

In molecular liquids an additional complication arises, the induced moment will normally depend upon the molecular orientation as well as the relative position of molecules in the fluid. Consider, for example, the QID expression for the moment responsible for far i.r absorption (eqn 2.7); making the dependence upon orientational variables explicit, we obtain

$$m^{ij}_{\alpha} = (\alpha\delta_{\alpha\beta} + (2/3)\gamma Q^{i}_{\alpha\beta})T^{(3)}_{\beta\gamma\delta}(\underset{\sim}{r}_{ij})\,\theta\, Q^{j}_{\gamma\delta} \tag{7.12}$$

There are two terms, both depend upon the orientation of j (through $\underset{\sim}{Q}^{j}$ c.f eqn 4.4) and one of which depends upon the orientation of i as well. This extra complexity precludes a predictive theory of the whole lineshape, at least for the present. However, when dealing with molecular spectra it is sometimes possible to compare several different induced spectra for the same fluid. It <u>is</u> within the range of theory to account for the common features of such spectra in a semi-quantitative fashion.

Consider, for example, the induced spectra of liquid CS_2 shown in figure 10[35]. It is noticeable, particularly in the high temperature spectra which are almost exponential, that the forbidden bands have a common shape and the DID wings of the

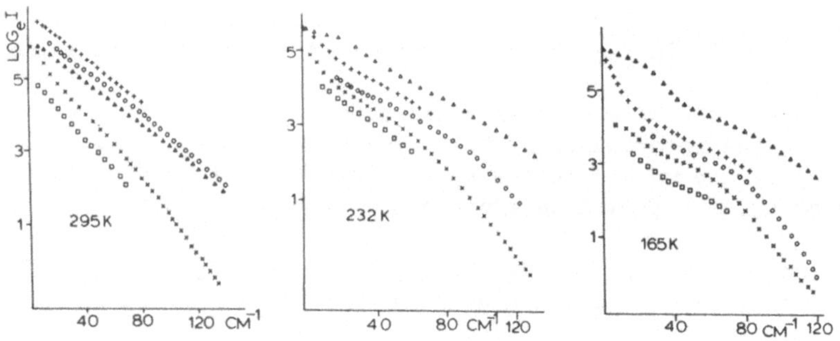

Fig 10: Forbidden spectra of CS_2: ν_2 Raman ++++, ν_3 Raman ΔΔΔ Rayleigh DID xxxxx, ν_1 Raman DID □□□ and far i.r ooooo (from ref. 35a).

allowed bands have a common shape, and that the forbidden bands are broader than the DID ones. At lower temperatures the band-shapes become more complex, but the relationship between the forbidden spectra and the DID spectra is preserved at frequencies above about 40 cm^{-1}. A prominent low frequency feature with a very temperature dependent width and unusual (non-lorentzian) shape is formed (the best low frequency data is for ν_2 +++++). At higher frequency a shoulder appears and at still higher frequencies the line becomes exponential again with a frequency exponent which is rather insensitive to temperature.

It is possible to account for these features by extending the atomic liquid theory described above[35]. This account has been checked in recent simulation work[45] which provides detailed corroboration for the concepts advanced here.

The formal decomposition of the total spectrum into sub-spectra associated with density fluctuations of a given wavelength (k^{-1}), as in eqn 7.11, may still be performed. However, these subspectra are now properties of an orientation density and may be relaxed by molecular reorientation as well as translation. We must examine how the presence of this additional relaxation channel affects the behaviour of the subspectra in each of the characteristic k regimes introduced above.

In the atomic fluid the high k subspectra were shown to dominate the spectral wings. The rotational degrees of freedom should have little effect on the relaxation of the high k subspectra as the translational relaxation requires motion only over distances of order k^{-1} (say $\sigma/10$ typically) which should be complete before any appreciable reorientation occurs (to relax $\underset{\sim}{Q}^1$ requires reorientation through the magic angle of $\sim 54^o$). The high frequency molecular spectra should thus reflect translational motion only and be similar in character to the atomic liquid spectra in the same frequency domain. In particular the temperature dependence of the exponent which characterises the wings should be slow ($\sim T^{1/2}$). Furthermore, since rotational effects are irrelevant at high frequency and since the dependence on translational variables is given by the same function for all forbidden and all DID spectra (ie $\underset{\sim}{T}^{(3)}$ and $\underset{\sim}{T}^{(2)}$ respectively c.f eqns 2.7, 2.8 and 2.15 and 2.9 and 2.14), the observed similarity of all forbidden and all DID spectra is explained (fig 10). As $\underset{\sim}{T}^{(3)}$ is a more rapidly varying function of the intermolecular vector than is $\underset{\sim}{T}^{(2)}$ ($T^{(3)}_{zzz}(r) \propto P_3(\cos\theta_r)/r^4$ vs. $T^{(2)}_{zz}(r) \propto P_2(\cos\theta_r)/r^3$) we can see why the forbidden spectra are broader [35].

At low frequency the atomic spectra reflect subspectra with k values of order $2\pi/\sigma$. From this we can expect rotational effects to be significant in the molecular spectra at low frequency, as the rate of translational relaxation ($\cong Dk^2$) becomes comparable to the rate of relaxation of the orientation variables. We might expect that by providing an extra relaxation channel the effect of the rotational degrees of freedom would be to make the molecular spectra broader than the atomic ones. In fact the rotational degrees of freedom play another role which makes their influence at low frequency dramatically different. We recall (c.f section 4.3) that due to imperfect orientational correlation in the liquid there are "un-cancelled" two-body contributions to the intensity in molecular liquids. In fact there are un-cancelled two-body contributions to the lineshape as well, for the same reason[45]. Thus (apropos fig 8) we should expect the slow, long-time relaxation of the two-body time dependent correlation function to make its presence felt at low frequency - as a consequence of rotational effects! In the subspectra language, the low k values give a larger contribution to the molecular spectra and enhance the diffusive peak at $\omega = 0$. The prominent, cusp-shaped low equency frequency peaks which are seen in the CS_2 spectra are accurately described by a diffusional theory of the un-cancelled two-body correlation functions alone[35].

7.4 Induced Lineshapes and Transport Processes

We now return to the questionof how much useful dynamical information, relating to processes not directly linked to spectroscopy *per se*, can be obtained from studies of induced lineshapes.

In materials such as superionic solids and ionic melts[77] induced spectroscopy promises to be an important means of probing events which are difficult to monitor in any other way; induced spectra of H_2 isotopes may be used to monitor the state of the materials in laser fusion[78]. In line with the material actually covered in the lectures though, the question of relevance should be confined to what may be learnt about weakly interacting (Van der Waals) fluids! The issue is how closely do induced spectra resemble the spectral densities of the "random forces" which drive transport processes. As we have shown, both types of spectra can be regarded as properties of the local density fluctuation subspectra, as discussed following eqn 7.11. Any differences may be said to arise from the different k dependences of the weight factors T(k). Short-ranged intermolecular effects, such as the force resulting from the repulsive branch of the potential, weight high k; whereas long-ranged effects, attributable to multipolar interactions, are relatively flat.

There have been limited attempts to correlate the spectrum of the random force responsible for vibrational dephasing with induced spectra. In the fast modulation limit[79] the dephasing time (τ) is determined by the amplitude $(\langle|\delta\nu|^2\rangle)^{\frac{1}{2}}$ and relaxation time τ_ν of the solvent-induced fluctuations in the oscillator frequency

$$\tau^{-1} = \langle|\delta\nu|^2\rangle \; \tau_\nu \tag{7.12}$$

Similarities between the values of τ_ν and the correlation times of induced moments[80] and between the temperature and density dependence of τ^{-1} and the zero-frequency amplitudes of induced spectra[49] have been noted.

It is hard to firmly base such comparisons without knowledge of the nature of the interaction responsible for the transport process (ie in the above cases it is not clear whether the dephasing is due to short or long-ranged interactions). In such circumstances one can again resort to computer simulation.

Figure 11 shows a comparison of the force correlation function of CS_2 with the correlation function of the induced dipole which determines the near i.r spectrum of ν_1. The force is due to the inter-site Lennard-Jones interactions, whereas the induced moment is the quadrupole-induced dipole. Because the force is short-range its correlation function has almost no contribution from the diffusive, low k modes which dominate the induced moment a.c.f. However the oscillation which occurs in the force correlation function occurs at the same frequency as the shoulder which appears in the spectrum of the induced moment (~ 80 cm^{-1}). As we argued above, this shoulder is the signature of the residual

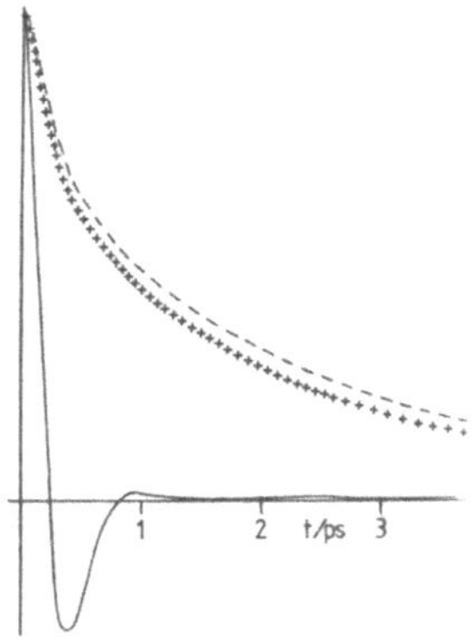

Fig 11: The autocorrelation function of the QID infra-red spectrum of CS_2 at 191K (ref 9) +++++ is compared with the force correlation function calculated in the same simulation ——. The dashed line shows the correlation function of the random frequency fluctuations which drive the dephasing of the ν_3 mode.

solidlike oscillations which enter the induced spectra through the intermediate k subspectra. The implication is that such subspectra have a much larger weight in the spectral densities of the force.

It should not be concluded from this poor comparison that there is no relationship between induced spectra and transport, rather that we must look for the relationship more selectively. If we compare a transport process with a long-range random force with a multipole-induced spectrum things become much better. Shown in figure 11 is the autocorrelation function of the dipole-quadrupole dephasing (which might contribute to the i.r linewidth of the asymmetric stretching mode ν_3), this resembles the induced moment a.c.f. closely. Long-ranged random forces also contribute to phenomena such as dielectric friction[70], and to electronic energy transport in liquids. On the other hand, if one wished to find an induced spectral analogue for a short-ranged transport process (such as the diffusional friction coefficient) one should pick a spectrum for which the moment is caused by overlap, eg the i.r spectrum of inert gas mixtures.

REFERENCES

1. D. Kivelson and P. A. Madden, Ann. Rev. Phys. Chem., 31 523 (1980).

2. J. Van Kranendonk (Editor), Intermolecular Spectroscopy and the Dynamical Properties of Dense Systems (North Holland, 1980).

3. G. C. Tabisz, Specialist Periodical Report on Molecular Spectroscopy, Vol. 6 Ed. J. Barrow (Chemical Society, London, 1979).

4. G. Birnbaum, B. Guillot and S. Bratos, Adv. Chem. Phys. 51, 49 (1982).

5. A. D. Buckingham, Intermolecular Interactions, Ed. B. Pullman (Wiley, 1978).

6. P. W. Fowler and P. A. Madden, Mol. Phys. 49, 913 (1983).

7. A. J. Stone - this volume.

8. T. I. Cox and P. A. Madden, Mol. Phys. 39, 1487 (1980).

9. D. A. Madden and D. J. Tildesley, Mol. Phys. 49 193 (1983).

10. R. D. Amos and J. H. Williams, Chem. Phys. Lett., 66, 370 (1979).

11. W. A. Steele - this volume.

12. L. Frommhold, Adv. Chem. Phys. 46, 1 (1980).

13. T. I. Cox and P. A. Madden, Mol. Phys. 43, 307 (1981).

14. R. D. Amos, A. D. Buckingham and J. H. Williams, Mol. Phys. 39, 1579 (1980).

15. S. Keilich, Specialist Periodical Report on Dielectric and Related Processes, Vol 1. Ed. M. Davies (Chemical Society, London, 1972).

16. T. F. Keyes and B. M. Ladanyi, Mol. Phys. 37, 1643 (1979).

17. D. A. Long, Raman Spectroscopy (McGraw-Hill, 1977).

18. G. Birnbaum, J. Quant. Spectrosc. Radiat. Trans. 19, 51 (1978).

19. J. D. Poll and J. L. Hunt, Can. J. Phys. 59, 1448 (1980).

20. G. Birnhaum, W. Ho, and A. Rosenberg, J. Chem. Phys. 55, 1028, 1039 (1971).

21. U. Buontempo, S. Cunsolo, G. Jacucci and J. J. Weis, J. Chem. Phys. 63, 2570 (1975).

22. D. R. Bosomworth and H. P. Gush, Can. J. Phys. 43, 751 (1965).

23. A. D. Buckingham, R. L. Disch and D. A. Dunmur, J. Am. Chem Soc. 90, 3104 (1968).

24. F. Barocchi, M. Zoppi, M. H. Profitt and L. Frommhold, Can. J. Phys. 59, 1418 (1980).

25. D. P. Shelton and G. C. Tabisz, Can. J. Phys., 59, 1430 (1980).

26. M. Zoppi, F. Barocchi, D. Varshneya, M. Neumann and T. A. Litovitz, Can. J. Phys. 59, 1475 (1980).

27. A. D. Buckingham and K. L. Clarke, Chem. Phys. Lett. 57, 321 (1978).

28. F. Barocchi, G. Spinelli and M. Zoppi, Chem. Phys. Lett. 90, 22 (1982).

29. P. D. Dacre, Mol. Phys. 45, 1 (1982) and 47, 2071 (1982).

30. D. E. Logan and P. A. Madden, Mol. Phys. 46, 1195 (1982).

31. D. P. Shelton and G. C. Tabisz, Mol. Phys. 40, 299 (1980).

32. H. A. Posch, Mol. Phys. 46, 1213 (1982).

33. A. D. Buckingham and G. C. Tabisz, Mol. Phys. 36, 583 (1978).

34. Ref. 26, Figure 1.

35. a) P. A. Madden and T. I. Cox, Mol. Phys. 43, 287 (1981); b) Chem. Phys. Lett. 77, 511 (1981).

36. B. J. Alder and E. L. Pollock, Ann. Rev. Phys. Chem. 32, 311 (1981).

37. D. Varshneya, S. F. Shirron, T. A. Litovitz, M. Zoppi, F. Barocchi, Phys. Rev. A23, 78 (1981).

38. J. H. R. Clarke and J. Bruining, Chem. Phys. Lett, 80, 42 (1980).

39. See ref. 28.

40. T. Keyes, B. M. Ladanyi, and P. A. Madden, Chem. Phys. Lett 64, 479 (1979).

41. T. Keyes, Chem. Phys. Lett., 70 194 (1980).

42. J. J. Weis and B. J. Alder, Chem. Phys. Lett., 81, 113 (1981).

43. Ref 21.

44. C. S. Murthy, K. Singer, D. Steele, J. J. Tindle, and R. Vallauri, Chem. Phys. Lett. 90, 95 (1982).

45. Ref. 9.

46. S. Weiss and U. Dinur, Chem. Phys. Lett. 99, 197 (1983).

47. T. F. Keyes, D. Kivelson and J. P. McTague, J. Chem. Phys. 55. 4096 (1971).

48. P. A. Madden, Phil. Trans. R. Soc. Lond. A293, 419 (1979).

49. T. I. Cox, M. R. Battaglia and P. A. Madden, Mol. Phys. 38, 1539 (1979).

50. Taken from Ref. 51, see also H. Versmöld, Mol. Phys. 43, 383 (1981)

51. D. Frenkel and J. P. McTague, 72, 2801 (1980).

52. B. Ladanyi, J. Chem. Phys. 78, 2189 (1983).

53. P. A. Madden and D. Tildesley, Mol. Phys. - to be published.

54. M. Sampoli and A. DeSantis, Mol. Phys. 51 - to be published, see also Mol. Phys. 46, 1271 (1982).

55. D. M. F Edwards and P. A. Madden, Mol. Phys. 51 (1984).

56. P. A. Madden and D. Kivelson, Adv. Chem. Phys., Vol. 56 (1984).

57. D. M. F. Edwards and P. A. Madden, Mol. Phys. 51, (1984).

58. G. Stell, G. N. Patey and J. S. Høye, Adv. Chem. Phys., 48, 183 (1981).

59. E. L. Pollock, and B. J. Alder, Phys. Rev. Lett. 46, 950/1981).

60. K. E. Arnold, J. Yarwood and A. H. Price, Mol. Phys. 48, 451 (1983).

61. See e.g M. Evans, G Evans and R. Davies, Adv. Chem. Phys. XLIV, 225 (1980), Figure 79, pg.468.

62. C. J. F. Böttcher, Theory of Electric Polarization, Vol. 1 (Elsevier, New York 1973).

63. T. Keyes and B. M. Ladanyi, Mol. Phys. 33, 1271 (1977).

64. T. Keyes and B. M. Ladanyi, Mol. Phys. 34, 765.

65. M. R. Battaglia, T. I. Cox and P. A. Madden, Mol. Phys. 37, 1413/1978.

66. P. A. Madden, M. R. Battaglia, T. I. Cox, R. K. Pierens and J. Champion, Chem. Phys. Lett., 76, 604 (1980).

67. A. J. C. Ladd, T. A. Litovitz and C. Montrose, J. Chem. Phys. 71, 4242 (1979).

68. P. A. Madden, Mol. Phys. 36, 365 (1978), Chem. Phys. Lett., 47, 174 (1977).

69. A. J. Masters and P. A. Madden, J. Chem. Phys., 74, 2450, 2460 (1981).

70. P. A. Madden and D. Kivelson, J. Phys. Chem., 86, 4244 (1982).

71. T. F. Keyes, Modern Theoretical Chemistry, Vol, 6. Ed. B. J. Berne (Plenum, 1977).

72. S. C. An, L. Fishman, T. A. Litovitz ,C. J. Montrose, H. A. Posch, J. Chem. Phys. 70, 4626 (1979).

73. J. R. D. Copley and S. W. Lovesey, Rept. Prog. Phys. 38, 461 (1975).

74. P. A. Fleury, J. M. Worlock and H. L. Carter, Phys. Rev. Lett. 30, 591 (1973).

75. B. Guillot, S. Bratos and G. Birnbaum, Phys. Rev. A22, 2230 (1980).

76. Ref. 61.

77. C. Raptis, R. A. J. Bunten and E. W. J. Mitchell, J. Phys. C 16, 5351 (1983).

78. P. C. Souers, E. M. Feason, R. L. Stark, R. T. Tsugawa, J. D. Poll and J. L. Hunt, Can. J. Phys. 59, 1408 (1980).

79. D. Oxtoby, Adv. Chem. Phys. 40, 1(1979).

80. M. R. Battaglia and P. A. Madden, Mol. Phys. 36, 1601 (1978).

MOLECULAR DYNAMICS SIMULATIONS: TECHNIQUES AND APPROACHES.

H.J.C. Berendsen and W.F. van Gunsteren
Laboratory of Physical Chemistry,
The University of Groningen
Nijenborgh 16
9747 AG GRONINGEN
The Netherlands

ABSTRACT

Methods and algorithms for molecular dynamics simulations of molecular fluids are described. Those techniques are emphasized that in the experience of the authors are reliable and easy to implement. This implies the use of cartesian coordinates throughout, the use of the SHAKE procedure to satisfy constraints and the incorporation of an adjustable coupling to an external bath with constant temperature and pressure.

CONTENTS

A. J. Barnes et al. (eds.), Molecular Liquids - Dynamics and Interactions, 475–500.

INTRODUCTION

The aim of the theory of molecular systems is to predict macroscopic properties on the basis of fundamental interactions between the constituent particles. These are ultimately nuclei and electrons with their Coulomb interaction, but we can divide the prediction process in two steps: 1 ab initio quantum mechanical calculations based on Schrödinger's equation to produce a simplified description of interaction potentials between pairs of molecules, 2 predict properties of large ensembles of molecules based on such simplified pair potentials. In dividing the prediction process this way, two sources of error are already introduced:

a The description of the interaction between a large number of particles as a sum of pair potentials neglects many-body interactions, as is discussed in other contributions to this volume. For condensed systems in a limited range of temperatures and densities the average effect of many-body potentials can be incorporated into effective pair potentials. For example, a polar molecular species will in the condensed phase be preferably oriented with its dipole moment along the field due to neighbouring molecules. The polarisability (giving rise to a many body interaction) will on the average produce an enhancement of the dipole moment. Thus a pair potential with an enhanced dipole moment could act as an effective potential that incorporates the average effect of the polarisability; it would never incorporate fluctuations in space and time of the many-body effects.

b The description of pair potentials in a simplified way is computationally necessary and conceptually desirable, but it represents an approximation. The accuracy of the approximation can be judged by the results. Some features of a model description are much more critical than others in determining bulk properties.

One difficulty with ab initio potentials is that it is hardly possible to obtain accuracies in the potential energies much better than kT, as is required to derive thermodynamic quantities of large aggregates. Therefore one often relies on empirical or semi-empirical potentials, in which parameters have been determined from experimental data. As long as data are used from solid or gas, with the aim to simulate the liquid, this is not objectionable. However, the necessity to use effective pair potentials will in many cases require the use of experimental data on the liquid itself. It is in such cases necessary to test the potentials on independent experimental properties of the liquid.

The second step, prediction of the properties of a large ensemble of molecules based on simplified pair potentials, belongs to the realm of statistical physics. Liquids are particularly difficult to treat: simplifications as in solids due to symmetry or in gases due to dilution are not possible. Although the theory of liquids has advanced considerably and for simple atomic liquids with simple model potentials accurate results can be obtained, there is little hope that analytical theories will be successful for complex molecular liquids with arbitrary interactions.

The alternative to analytical theories are simulations: computer experiments that generate representative statistical samples of the system from which macroscopic properties can be derived. The theory used to generate simulations is much simpler than the statistical mechanical theories of liquids. Nevertheless the results can be far more accurate and less assumptions go into the derivation of the results. Computer simulations are thus often used to test the reliability of analytical theories, using model potentials. Using realistic potentials, computer simulations can be used to understand and predict properties of large systems of complex molecules. This is not restricted to simple molecular fluids but may extend to complicated macromolecules, liquid crystals, micels, surfaces and chemical reactions.

MC AND MD

There are two principal simulation methods: Monte Carlo (MC) and Molecular Dynamics (MD). MC methods generate a representative statistical equilibrium ensemble in which each configuration occurs with its Boltzmann probability, for a given temperature and volume or pressure. The result of the simulation is simply a tape with a large number (say, 10^6) of configurations. All static equilibrium properties can be obtained by computing that property for each configuration and averaging over the ensemble. MD, on the other hand, produces a trajectory of the system in phase space by solving the equations of motion of all particles in the system. If the trajectory is sufficiently long, it will also contain a representative ensemble of equilibrium configurations, as in the MC case. In addition the trajectory will contain dynamical information because the time ordering of events is conserved. Thus MD simulations produce everything MC can, but in addition contain dynamical information. Another advantage of MD above MC is that non-equilibrium properties can be more easily studied: the trajectories can be generated in the presence of constrained gradients that keep the system in non-equilibrium conditions. MC methods are easier to implement in a computer program: they require only the calculation of the total energy of each configuration, while MD requires the forces on each particle. The computer effort for a comparable statistical accuracy is of the same order of magnitude for MD and MC. Therefore MD will generally be the preferred method. We will concentrate on MD and only give a brief account of MC.

Monte Carlo simulations

The usual MC method is the Metropolis-method after Metropolis (1, 2). Given a starting configuration of a N-particle system with given interaction potential $V(\tilde{r}_1, .. \tilde{r}_N)$, a new configuration is generated by random displacement of one (or more) coordinates. The displacements should be such that in the limit of a large number of successive displacements the available cartesian space of all particles is homogeneously sampled. [This does not mean that sampling is necessary carried out in cartesian space, e.g. for a diatomic molecule the centre of mass can be sampled in cartesian space while the orientation is sampled in angle (Ω) space by homogeneous sampling of $\cos\theta$ in (-1, +1) and φ in ($-\pi$, $+\pi$) where θ and φ are the polar and azimuthal angle describing the orientation of the molecule. It is not allowed to sample θ homogeneously because the solid angle element is $2\pi \sin\theta d\theta d\varphi$]. The newly generated configuration is either accepted or rejected on the basis of an energy criterion involving the change ΔE of the potential energy with respect to the previous configuration. In the case of rejection, the previous configuration is counted again and used as a starting point for an other random displacement. The criterion is as follows: accept if $\Delta E \leq 0$ or, for $\Delta E > 0$, accept if $\exp(-\Delta E/kT) > R$, where R is a random number, homogeneously distributed over the interval (0, 1). Thus a Markoff chain of configurations is generated in which the probability of each configuration is proportional to $\exp(-E/kT)$ where E is the potential energy of that configuration. We can see the validity of this probability by considering detailed balance in the Markoff process: Consider two configurations ① and ②, with $E_1 < E_2$(fig. 1). Hence going from ② to ① the acceptance probability is 1; going from ① to ② the acceptance probability is $\exp\{-(E_2 - E_1)/kT\}$. In equilibrium the following relation must hold

$$p_2 . 1 = p_1 . \exp\{-(E_2 - E_1)/kT\}$$

or

$$\frac{p_1}{p_2} = \frac{\exp(-E_1/kT)}{\exp(-E_2/kT)}$$

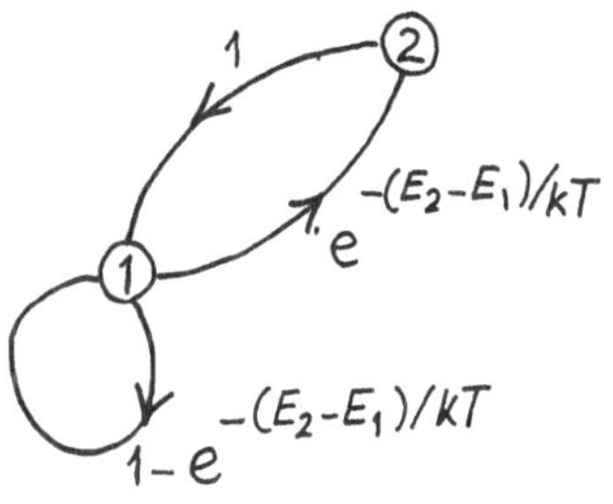

fig. 1 MC transition probabilities

Thus each configuration occurs with a probability proportional to its Boltzmann factor. The accepted samples are not randomly distributed but Boltzmann-distributed; often the term important sampling is used. The obvious advantage above random sampling is that most samples are relevant while with random sampling most computational effort would be spent on irrelevant high energy configurations. The best acceptance ratio for high computational efficiency is in the range of 50-70%. This ratio depends on the size of the random displacements: large displacements give low acceptance ratios but produce more rapid sampling of configuration space. If pair potentials are used the most efficient way to generate new configurations is to displace one particle at a time: in order to calculate ΔE only the neighbours of the displaced particle have to be considered. Other particles are then displaced in successive steps. As a rule of thumb one can estimate that for liquids each particle needs to be displaced about 1000 times for sufficient statistical accuracy. A 1000 particle system needs 10^6 MC steps.

Other methods have been designed to produce more efficient sampling in configuration space, by using force-bias displacements (3). Now in one move all particles are displaced, in a direction opposite tò the gradient of the potential energy. Although such methods can be somewhat more efficient than Metropolis sampling, the requirement that gradients have to be calculated make the computational effort comparable to that of dynamic simulations, and further reduces the relative advantage of MC versus MD.

Molecular Dynamic Simulations

MD computes a trajectory of a N-particle system by solving Newton's equations of motion. Let us first consider a system of N atoms in cartesian coordinates. This may be a fluid of atoms or a fluid of molecules made up of atoms with specified interactions between the atoms in a molecule. We will then describe methods of solution for these cartesian coordinates and velocities. Subsequently we will consider the case that molecules are treated as rigid bodies and cases in which molecules have internal degrees of freedom as well as internal constraints.

Consider a set of N particles with masses m_i at cartesian positions $\underset{\sim}{r}_1 \ldots \underset{\sim}{r}_N$. The particles are assumed to move according to the classical equations of motion

$$\frac{d^2 \underset{\sim}{r}_i}{dt^2} = \underset{\sim}{F}_i(\underset{\sim}{r}_1 \,..\, \underset{\sim}{r}_N)/m_i \qquad i = 1 \,..\, N \tag{1}$$

where

$$\underset{\sim}{F}_i = -\underset{\sim}{\nabla}_i \, V(\underset{\sim}{r}_1 \ldots \underset{\sim}{r}_N) \tag{2}$$

For simplicity, the force is assumed to be conservative, i.e., dependent on positions only. The first problem we encounter is the computation of the forces according to eq. (2). In the case of (effective) pair potentials we can write

$$V(\underset{\sim}{r}_1 \,..\, r_N) = \sum_{i<j} V_{ij}(\underset{\sim}{r}_{ij}) \tag{3}$$

where

$$\underset{\sim}{r}_{ij} \equiv \underset{\sim}{r}_i - \underset{\sim}{r}_j \tag{4}$$

$$\underset{\sim}{F}_i = \sum_j \underset{\sim}{F}_{ij} \tag{5}$$

$$\underset{\sim}{F}_{ij} = -\underset{\sim}{F}_{ji} = -\frac{\partial}{\partial \underset{\sim}{r}_{ij}} V_{ij}(\underset{\sim}{r}_{ij}) \tag{6}$$

In case the potential $V(\underset{\sim}{r}_{ij})$ is of spherical symmetry, which is often the case when atom-atom potentials are being used, $V(\underset{\sim}{r}_{ij})$ is a function of $r_{ij} = |\underset{\sim}{r}_{ij}|$ only, so that $\underset{\sim}{F}_{ij}$ becomes

$$\underset{\sim}{F}_{ij} = -\frac{dV_{ij}(r_{ij})}{dr_{ij}} \cdot \frac{\partial r_{ij}}{\partial \underset{\sim}{r}_{ij}} = -\frac{dV_{ij}(r_{ij})}{dr_{ij}} \cdot \frac{\underset{\sim}{r}_{ij}}{r_{ij}} \tag{7}$$

which expression is generally easy to evaluate. Summation must be carried out in principle over all $\frac{1}{2}N(N-1)$ pairs of particles; in practice over all neighbours of each particle within a certain cut-off range R_c.

At this point it is appropriate to define the boundary conditions of the simulated system. If we wish to simulate a bulk phase with a limited number of particles, the simulation of an isolated cluster or drop introduces unacceptable disturbances due to boundary effects. The usual method to avoid such effects is to use periodic boundary conditions: the simulated particles are put in a cubic (or triclinic) box that is repeated in a regular three-dimensional lattice (fig. 2). Thus in fact one simulates a crystal. Since the periodicity is an artefact of the

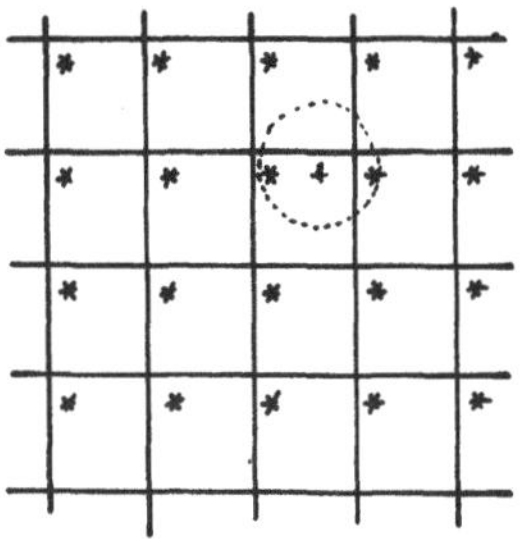

fig. 2 periodic box

computation, the effects of periodicity on the potential of interaction should not be significant. This means that a particle should not simultaneously feel the interaction with another particle and with its image. Consequently the box size R_{box} should exceed twice the cut-off radius:

$$R_{box} > 2R_c \tag{8}$$

For the force calculation only the nearest image is taken into account. Complications arise if the potential of interaction is of such long range that it is not feasible to use a box size satisfying eq. (8). These cases will be considered by Dr. Tildesley.

ALGORITHMS FOR MD

We now return to the solution of eqs. (1) and (2), assuming that the force calculation can be carried out for any configuration of the particles. Mathematically the problem constitutes an <u>initial value problem</u> (4), which can be simply formulated:

$$x'' = f(x) \tag{9}$$

$$x_0 = x(t_0);\ x_0' = x'(t_0) \tag{10}$$

where x, x', x'' are 3N-dimensional vectors and their derivatives with respect to t. The most appropriate approach to solving such a problem is the use of a <u>difference</u> or <u>step-by-step</u> method. The solution is approximated by its value at a sequence of discrete mesh points. Normally these points are assumed to be equally spaced:

$$t_n = nh,\ n = 0, 1, 2 \ldots \tag{11}$$

A step-by-step method provides a rule or algorithm for computing the approximation x_n at step t_n to $x(t_n)$ in terms of the values of x at t_{n-1} and possibly at preceding points. A k-value, l-step algorithm uses k preceding values of x or its subsequent derivatives up to and including t_{n-1}.

The existence and convergence of a solution depends on the properties of the function f and on the specified initial values x_0 and x_0'. In most MD applications the function f obeys the Lipschitz and differentiability conditions (4) for existence and convergence.

The choice of the available difference methods is determined by the fact that the force calculation (the evaluation of f) is the most time consuming part of the computation, involving a double summation over particles. This rules out any method of solution (such as Runge-Kutta and extrapolation methods) that require several function evaluations per step. General methods requiring only one force evaluation per step are multi-value predictor-corrector methods.

k-Value predictor-corrector methods first make a prediction y_{n+1} based on k previous values of x and its derivatives and then correct the prediction y_{n+1} to x_{n+1} using the available knowledge of the second derivative on the predicted point y_{n+1}. Such algorithms can be expressed in various representations (4). In the Nordsieck (5) or N-representation the values of x_n and (k-1) derivatives at t_n are used. We represent these by a vector $\underset{\sim}{x}_n$:

$$\underset{\sim}{x}_n^N = [x_n, hx_n', h^2x_n''/2, .. h^{k-1}x_n^{(k-1)}/(k-1)!]^T \tag{12}$$

The predictor step is given by a matrix $\underset{\approx}{A}$:

$$\underset{\sim}{y}_{n+1}^N = \underset{\approx}{A}\,\underset{\sim}{x}_n^N \tag{13}$$

The corrector step is expressed by a vector $\underset{\sim}{a}$

$$\underset{\sim}{x}_{n+1}^N = \underset{\sim}{y}_{n+1}^N + \underset{\sim}{a}\,\frac{h^2}{2}\,\{f(y_{n+1}) - y''_{n+1}\} \tag{14}$$

For $\underset{\approx}{A}$ usually a Taylor expansion from the previous values is used; the values of $\underset{\sim}{a}$ can be chosen for maximum stability and accuracy (4).

Elements of $\underset{\approx}{A}$ and $\underset{\sim}{a}$ can be found in (4) or (6). For k = 4, involving the derivatives of the force, the values are

$$\underset{\approx}{A} = \begin{pmatrix} 1 & 1 & 1 & 1 \\ 0 & 1 & 2 & 3 \\ 0 & 0 & 1 & 3 \\ 0 & 0 & 0 & 1 \end{pmatrix} \quad ; \underset{\sim}{a} = \begin{pmatrix} 1/6 \\ 5/6 \\ 1 \\ 1/3 \end{pmatrix} \tag{15}$$

One recognizes the binomial coefficients in the columns of A.

Since the second derivatives or forces are actually computed at every step, it is convenient to use a representation in which forces at previous steps are kept in stead of higher derivatives at the previous steps. We call this the force representation

$$\underset{\sim}{x}_n^F = [x_n, hx_n', \frac{h^2}{2} x_n'', \frac{h^2}{2} x''_{n-1}, \ldots \frac{h^2}{2} x''_{n-k+3}]^T \qquad (16)$$

Thus a k-value method retains forces at k-3 previous steps.

Now the predictor step is given by

$$\underset{\sim}{y}_{n+1}^F = \underset{\approx}{B}\, \underset{\sim}{x}_n^F \qquad (17)$$

and the corrector by

$$x_{n+1}^F = y_{n+1}^F + \underset{\sim}{b} \frac{h^2}{2} \{f(y_{n+1}) - y''_{n+1}\} \qquad (18)$$

The N- and F-representations are interchangeable by a matrix transformation

$$\underset{\sim}{x}_n^F = \underset{\approx}{T}\, x_n^N \qquad (19)$$

$$\underset{\approx}{B} = \underset{\approx}{T}\underset{\approx}{A}\underset{\approx}{T}^{-1} \qquad (20)$$

$$\underset{\sim}{b} = \underset{\approx}{T}\, \underset{\sim}{a} \qquad (21)$$

Elements of these conversion matrices are listed in ref. 6 for values of k up to 8. For k = 4 the F-representation matrix $\underset{\approx}{B}$ and vector $\underset{\sim}{b}$ of eqs. 17 and 18 are:

$$\underset{\approx}{B} = \begin{pmatrix} 1 & 1 & 4/3 & -1/3 \\ 0 & 1 & 3 & -1 \\ 0 & 0 & 2 & -1 \\ 0 & 0 & 1 & 0 \end{pmatrix} \quad ; \quad \underset{\sim}{b} = \begin{pmatrix} 1/6 \\ 5/6 \\ 1 \\ 0 \end{pmatrix} \qquad (22)$$

Values for other values of k up to 8 can be found in ref. 6. The Beeman algorithm (7) is an F-representation predictor-corrector method, although the coefficients of $\underset{\sim}{b}$ are not equivalent to those of Gear.

The use of higher-order algorithms may not always be profitable when the force has some unpredictable quality, for example as a result of cut-off errors, approximation, tabulation error or the explicit use of stochastic terms. In practice algorithms should at least be a 4-value type; i.e. they should contain the force derivative (or the force in the previous step) because the second derivative of the potential predominantly has a positive sign for favourably interacting particles and its omission would produce accumulating rather than random errors. The use of still higher derivatives, however, is of questionable practical interest.

A very successful algorithm is due to Verlet (8). It is a

4-value predictor algorithm in which the symmetry of forward- and backward prediction is explicitly incorporated into the algorithm, and the corrector step is omitted:

$$x_{n+1} = x_n + h\,\dot{x}_n + \frac{1}{2}h^2\,\ddot{x}_n + \frac{1}{6}h^3\,\dddot{x}_n$$
$$x_{n-1} = x_n - h\,\dot{x}_n + \frac{1}{2}h^2\,\ddot{x}_n + \frac{1}{6}h^3\,\dddot{x}_n$$
$$x_{n+1} = 2x_n - x_{n-1} + h^2\,\ddot{x}_n + o(\ddddot{x}_n) \quad (23)$$

This algorithm is extremely simple and it behaves very nicely as can be seen from a comparison with Gear-algorithm for a dynamics simulation of a macromolecule (ref. 6). Fig. 3 shows the rms error in the total energy of the system (which should be constant), compared with the rms fluctuation in kinetic

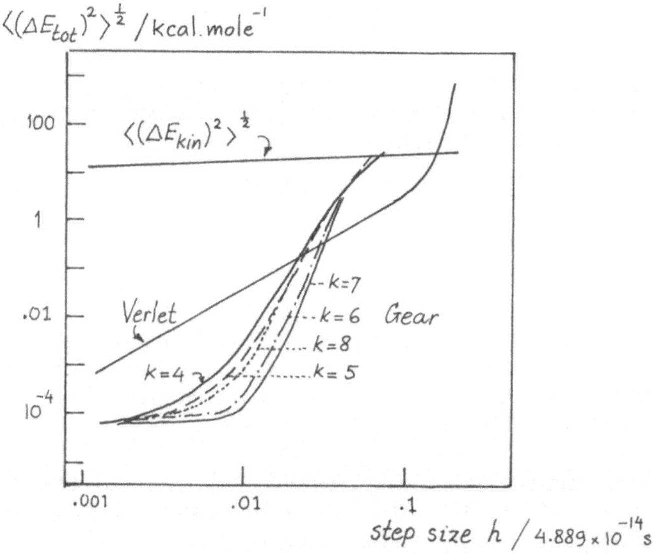

Fig. 3 R.m.s. fluctuation in total and kinetic energy in 100 steps of M.D. of the protein BPTI (ref.6), for the Verlet and Gear algorithms.

energy. A MD simulation can be considered reliable (statistically, that is) if $<\Delta E_{tot}^2>^{\frac{1}{2}}$ is less than 10% of $<\Delta E_{kin}^2>^{\frac{1}{2}}$. This criterion is somewhat arbitrary: some prefer a lower fluctuation, but it seems from experience that average quantities are still reliable at larger values of the fluctuation. The Gear algorithm gives better accuracy at low values for the step size than the Verlet algorithm does, but it more rapidly becomes inaccurate and unstable. It is interesting to note that k = 7 gives a better performance than k = 8. When the desired accuracy is not so high, the Verlet algorithm, being very simple to use, is to be preferred. Even for unacceptably large time steps from a point of view of accuracy (e.g. when $\langle \Delta E_{tot}^2 \rangle = \langle \Delta E_{kin} \rangle$), the Verlet algorithm remains stable, while the predictor-corrector algorithms are unstable.

The Verlet algorithm does not produce explicit velocities. If needed for analysis, velocities can be deduced from

$$v_n = \dot{x}_n = (x_{n+1} - x_{n-1})/2h, \tag{24}$$

which is accurate to first order only. Another way of deriving velocities is an alternative version of the Verlet algorithm, called the leap-frog scheme, which involves velocities at half-interval times:

Given $v_{n-\frac{1}{2}}$, x_n, the step involves

$$v_{n+\frac{1}{2}} = v_{n-\frac{1}{2}} + h\, f(x_n) \tag{25}$$

and

$$x_{n+1} = x_n + h\, v_{n+\frac{1}{2}} \tag{26}$$

Here x and v are 3N-dimensional vectors and $f(x_n)$ is the acceleration vector of force components divided by respective masses. Eqs. (25) and (26) are equivalent to eq. (23), as can be seen by inserting (25) into (26) and subsequently eliminating $v_{n-\frac{1}{2}}$ by using

$$v_{n-\frac{1}{2}} = (x_n - x_{n-1})/h, \tag{27}$$

which is the previous time step equivalent of eq. (26). As we shall see later, the leap frog scheme has advantages above the Verlet algorithm, whenever manipulations involving velocities are needed.

MONITORING DURING SIMULATIONS

The usual practice is to write all components of coordinates and velocities to tape at every time step, or - depending on requirements - at regular intervals of 5 or 10 time steps. Most structural and dynamic properties can be analyzed later from the recorded results and it is preferable to do so in order to keep the simulations clean. However, results that can only be obtained by summation over pairs should be monitored during the simulation. Two typical examples are radial distributions and pressure. The radial distribution $g(r)$ of atom pairs is monitored by collecting pair distances in distance intervals, thus constructing a histogram. Pressure is determined from the relation

$$P = \frac{2}{3V}(E_{kin} - \Xi) \tag{28}$$

where the virial Ξ is defined as

$$\Xi = -\tfrac{1}{2} \sum_{i<j} \mathbf{r}_{ij} \cdot \mathbf{F}_{ij} \tag{29}$$

Eq. (28) is strictly valid only for averages over an equilibrium ensemble. The virial involves a summation over pairs, also requiring forces, that should be evaluated when pair interactions are considered during simulation.

EQUATIONS OF MOTION IN GENERALIZED COORDINATES

Molecules are built of atoms that are bound together by strong covalent forces that constrain bond lengths and bond angles to a narrow range. If no "soft" internal modes are present, such as dihedral angles in e.g. hydrocarbons, the molecule can as a first approximation be considered as a rigid body. This approximation involves the assumption that intramolecular motional modes are decoupled from intermolecular modes, equivalent to a Born-Oppenheimer approximation in quantum mechanics. The assumption is justified only when the frequencies of intramolecular modes are well separated from those of intermolecular modes, which is true for fluids of weakly interacting molecules without soft internal degrees of freedom. HCl, CH_4, CH_2Cl_2, N_2 are examples where the rigid assumption is reliable; HF and H_2O with their strong hydrogen bonding intermolecular forces are border cases; butane and higher alkanes with soft dihedral modes cannot be treated as rigid bodies. In the latter case, however, bond lengths and bond angles can still be treated as internal constraints. Treating a molecule as a rigid body also means that internal constraints should be satisfied.

There are two different ways to obtain equations of motion in the presence of constraints:

a Equations of motion are set up in generalized coordinates involving all non-constrained degrees of freedom

b Equations of motion are still defined in cartesian coordinates, but modified so as to satisfy constraints.

In the first method, the equations of motion are derived from the definition of a Lagrangian in generalized coordinates, yielding generalized impulses and a generalized hamiltonian. For rigid bodies this procedure leads to the well-known Newton-Euler equations for rigid body translation and rotation (9). For molecules possessing internal degrees of freedom as well, the equations become rapidly unmanageable when the molecular complexity increases. Also the solution of rigid body Newton-Euler equations involves special care because division by the sine of an Euler angle occurs which requires transformation to a different set of angles for certain orientations (10). The second method does not have such disadvantages, is quite generally applicable even to complex macromolecules, and is computationally effective. Therefore we will concentrate entirely on the second method that has been described in detail before (11, 12). The consistent use of cartesian coordinates is particularly useful when the intermolecular potential is expressed as a sum of atom pair (or more generally site-site) interactions.

CONSTRAINT DYNAMICS

Algorithms

Molecular constraints belong to the class of scleronomous (i.e., time independent) holonomic constraints of the form

$$\sigma_k(\underset{\sim}{r}_1, \ldots \underset{\sim}{r}_n) = 0, \qquad k = 1 \ldots l \tag{30}$$

for the case of l internal constraints in a n-atom molecule. A completely rigid molecule of course has 3N-6 internal constraints. The constraint equations can be put in the form of distance constraints:

$$\underset{\sim}{r}^2_{ij} - d_{ij} = 0 \tag{31}$$

For example, in a triatomic molecule as H_2O three distance constraints (OH, OH, HH) determine the internal structure.

The problem now is to solve 3N Newton's equations of motion while satisfying l constraints. This is accomplished by applying Lagrange's method of undetermined multipliers (11)

$$m_i \ddot{\underset{\sim}{r}}_i = - \nabla_i (V + \sum_{k=1}^{l} \lambda_K \sigma_k) \tag{32}$$

This adds a zero term to the intermolecular potential energy V, while allowing to satisfy the constraints at all times by solving for λ_k. Thus λ_k are time-dependent multipliers.

The physical interpretation of eq. (32) becomes clear by rewriting it in terms of forces:

$$m_i \ddot{\mathbf{r}}_i = \mathbf{F}_i + \mathbf{G}_i \tag{33}$$

$$\mathbf{F}_i = -\nabla_i V \tag{34}$$

$$\mathbf{G}_i = -\sum_{k=1}^{l} \lambda_k \nabla_i \sigma_k \tag{35}$$

$\mathbf{F}$ is the total unconstrained force, derived from pair interactions as if no constraints were present; $\mathbf{G}$ is the constraint force, which compensates the components of $\mathbf{F}$ that act along the directions of the constraints.

In solving for λ_k one must be careful to select a method of solution that is of at least the same order in the time step h as the algorithm for solving Newton's equations, because otherwise the algorithm becomes unstable as errors accumulate. In connection with the Verlet or leap frog algorithm λ_k can be solved as outlined below. Modifications suitable for predictor-corrector algorithms can be found in ref. (6).

In the Verlet algorithm including constraint forces the following step is made

$$\mathbf{r}_i(t+h) = 2\mathbf{r}_i(t) - \mathbf{r}_i(t-h) + \frac{\mathbf{F}_i(t)}{m_i}h^2 + \frac{\mathbf{G}_i(t)}{m_i}h^2 \tag{36}$$

or $$\mathbf{r}_i(t+h) = \mathbf{r}_i' + \delta \mathbf{r}_i \tag{37}$$

where $$\delta \mathbf{r}_i = \frac{\mathbf{G}_i(t)}{m_i} h^2 \tag{38}$$

Here $\mathbf{r}_i'$ are the coordinates after a normal MD step disregarding all constraints,and $\delta \mathbf{r}_i$ are the corrections to be made as a result of the constraints. The constraint conditions are given by eq. (30), and eq. (35) yields the form of $\delta\mathbf{r}_i$:

$$\delta \mathbf{r}_i = \frac{h^2}{m_i} \sum_{k=1}^{l} \lambda_k \nabla_i \sigma_k (\mathbf{r}_1(t),..\mathbf{r}_n(t)) \tag{39}$$

If the constraints are defined in terms of distance constraints, as in eq. (31), $\delta\mathbf{r}_i$ becomes

$$\delta \mathbf{r}_i = \frac{2h^2}{m_i} \sum_k \lambda_k \mathbf{r}_{ij}(t), \tag{40}$$

where the sum extends over all constraints involving the i-th particle. Eq. (40) implies that corrections due to the constraint between particles i and j must be applied in the direction of vector $\mathbf{r}_{ij}(t)$. Corrections to $\mathbf{r}_i'$ and $\mathbf{r}_j'$ are in opposite direction and weighted by the inverse masses of i and j.

Consider for simplicity the two-particle case (fig. 4). There is one multiplier λ. The correction for particle 1 is

$$\delta \underset{\sim}{r}_1 = -\frac{2h^2}{m_1}\lambda \underset{\sim}{r}_{12} = \frac{g}{m_1}\underset{\sim}{r}_{12} \tag{41}$$

and for particle 2

$$\delta \underset{\sim}{r}_2 = \frac{2h^2}{m_2}\lambda \underset{\sim}{r}_{12} = -\frac{g}{m_2}\underset{\sim}{r}_{12} \tag{42}$$

Here g is an unknown parameter, to be solved from the constraint condition

$$\{\underset{\sim}{r}_1' + \frac{g}{m_1}\underset{\sim}{r}_{12} - (\underset{\sim}{r}_2' - \frac{g}{m_2}\underset{\sim}{r}_{12})\}^2 = d^2 \tag{43}$$

or

$$g^2(\frac{1}{m_1} + \frac{1}{m_2})^2 r_{12}^2 + 2g(\frac{1}{m_1} + \frac{1}{m_2})\underset{\sim}{r}_{12}'\cdot\underset{\sim}{r}_{12} + r_{12}'^2 - d^2 = 0 \tag{44}$$

In this case there is one equation of the second degree in g which can be solved exactly. In the general case there are l equations of the type (44), with l unknown parameters g_k, each equation involving several g_k's. Thus we have a system of l quadratic equations to solve. Two methods have been used (11). In the first the equations are linearized by neglecting the terms in g^2, after which a set of l linear equations is obtained that can be solved by matrix inversion. The first-order corrections thus obtained are applied and the procedure is iterated until all constraints are satisfied within a given tolerance.

The second method solves each constraint equation of the type of eq. (44) to first order, treating all constraints in succession, and iterating this procedure until all constraints are satisfied to within a specified tolerance. The procedure SHAKE ($\underset{\sim}{x}'\,|\,\underset{\sim}{x}$) - shake $\underset{\sim}{x}'$, given $\underset{\sim}{x}$ as reference - will make corrections to $\underset{\sim}{x}'$ in the direction of $\underset{\sim}{x}$ such that a specified list of constraints is satisfied by the new coordinates $\underset{\sim}{x}'$.

A complication arises when a constraint can not be satisfied to first order with length constraints, such as a linear triatomic molecule (e.g. CO_2) or a flat triangle or flat ring. Such cases have been considered by Ryckaert (13).

The following statements give a leap frog scheme with constraints:

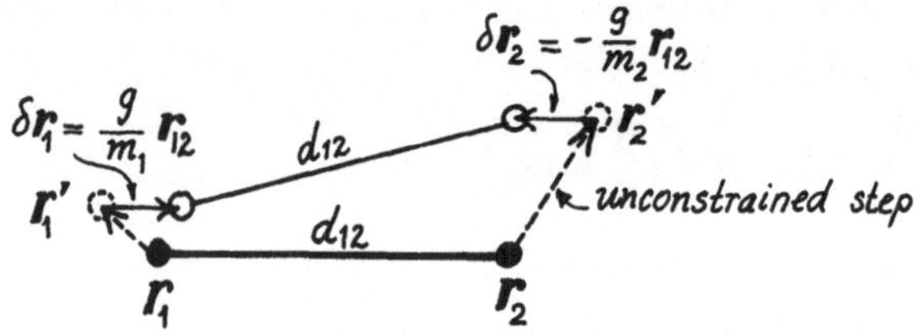

Fig. 4 Steps taken in correcting constraints

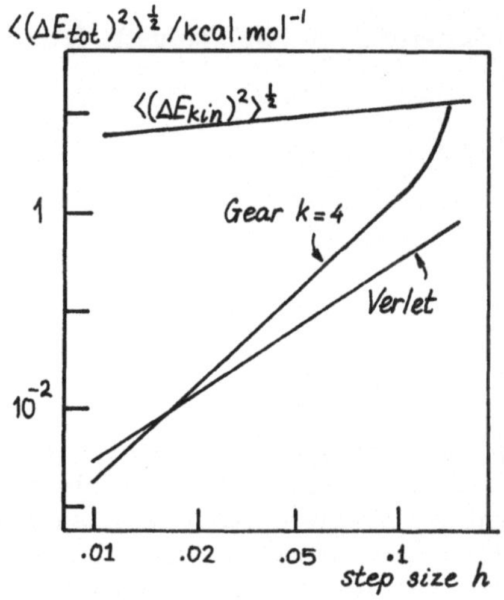

Fig. 5 Accuracy of Gear and Verlet algorithm in constraint dynamics

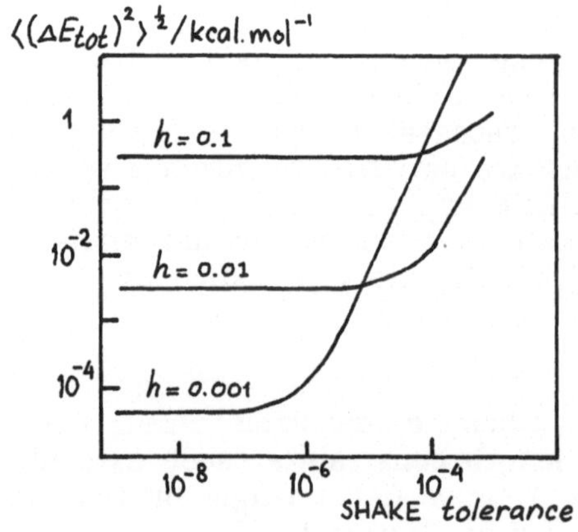

Fig. 6 Influence of relative tolerance in constraints on accuracy of Verlet algorithm

given $\underset{\sim}{v} = v(t - \tfrac{1}{2}h)$; $\underset{\sim}{x} = \underset{\sim}{x}(t)$

<u>repeat</u>:

1. compute $\underset{\sim}{a}$: $a_i = F_i(\underset{\sim}{x})/m_i$
2. $\underset{\sim}{v}' \leftarrow \underset{\sim}{v} + \underset{\sim}{a}h$ } unconstrained step
3. $\underset{\sim}{x}' \leftarrow \underset{\sim}{x} + \underset{\sim}{v}h$ }
4. SHAKE $(\underset{\sim}{x}'|\underset{\sim}{x})$ constrain positions
5. $\underset{\sim}{v} \leftarrow (\underset{\sim}{x}' - \underset{\sim}{x})/h$, constrain velocities
6. $\underset{\sim}{x} \leftarrow \underset{\sim}{x}'$; reset in box

Recovery of internal forces

As is immediately clear from eq. (38), the computation of δr_1 by SHAKE yields the internal constraint forces $\underset{\sim}{G}_1(t)$. They are easily recovered from the algorithm by saving the configuration $\underset{\sim}{x}'$ before SHAKE. In certain applications the measurement of $\underset{\sim}{G}(t)$ may be very useful, as $\underset{\sim}{G}(t)$ is the perturbing force on the intramolecular vibrational modes. Both its time dependence and statistical distribution is obtained from the simulation; these properties can be used in a perturbative treatment of the intramolecular vibrations. Thus the remarkable observation is made that details of intramolecular events can be derived from a rigid body simulation.

Constraints in higher order algorithms

Incorporations of constraints with the SHAKE method involves a correction to the forces (the second derivative of x) based on manipulation of coordinates. The Verlet algorithm has no corrector step and hence the effect of constraints need not be incorporated in x". Since the constraint forces can be calculated, corrections in x" are possible, but not in other derivatives. Therefore the N-representation is not useful and the F-representation should be chosen. The scheme how to incorporate SHAKE into predictor-corrector methods of any order has been described in ref. 6. In contrast to the results without constraints, it turns out that with bond length constraints in a macromolecule larger k-values than 4 in the Gear algorithm give no improvement. In fact the Verlet algorithm is to be preferred over the 4-value Gear algorithm unless very high accuracy is required. The constraints make it possible to increase the time step by a factor of 4 (12).

The better performance of low-order algorithms is undoubtedly a result of the less predictable character of the time dependent behaviour in the presence of constraints. The highly periodic nature of the bond length vibrations is now absent. Fig. 5 shows the relative accuracy (comparable to Fig. 3) in the constraint case. The influence of the relative tolerance used in SHAKE on the accuracy of the algorithm is shown in Fig. 6 (from ref. 6).

THE USE OF VIRTUAL SITES: FORCE REDISTRIBUTION

While retaining cartesian coordinates and internal constraints, forces should act on the atoms (masses) in the system. Often, potential interaction models use non-atomic sites as centres of interaction, because restriction to atomic sites may not give sufficiently accurate potential descriptions, especially for electrostatic terms. Examples (fig. 7) are a five-point charge model for a linear molecule (such as N_2) and the

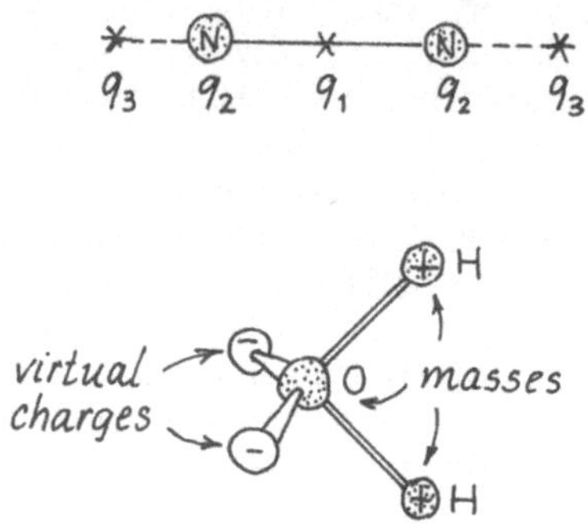

Fig. 7 Examples of virtual sites

ST2-model (14) for water. Such virtual sites are massless points, the positions of which are rigorously related to the position of the masses in the molecule. In the MD algorithm two additional steps are necessary:

a redistribution of forces on the virtual sites to atoms.

b reconstruction of the virtual sites after each step.

The latter is easily accomplished by the definition of the virtual sites. For example, for a virtual site on a linear molecule (fig. 8) the reconstruction is given by

$$\mathbf{r}_3 = -\frac{c}{b}\mathbf{r}_1 + \left(1 + \frac{c}{b}\right)\mathbf{r}_2 \qquad (45)$$

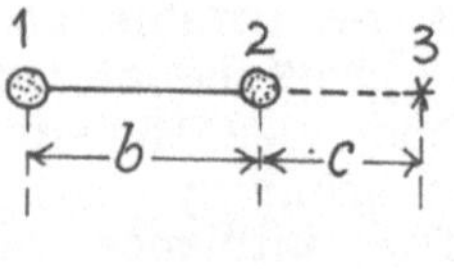

Fig. 8 Virtual site on diatomic molecule

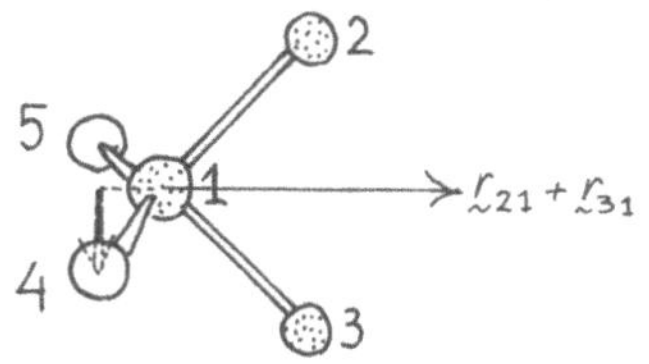

Fig. 9 ST 2 model

and for the virtual charges on the ST2 model (fig. 9)

$$r_4 = r_1 + \alpha(r_{21} + r_{31}) + \beta[r_{21} \times r_{31}] \tag{46}$$

where α and β are geometrical coefficients and $r_{ij} = r_i - r_j$. Thus a virtual position r is a function of the atomic positions $r_1 .. r_n$. The potential energy will have the form

$$V = V(r, r_1 .. r_n, ...)$$

which - because $r = f(r_1 .. r_n)$ - reduces to

$$V = V^*(r_1 .. r_n, ...)$$

The total force on the i-th mass is

$$F_i = -\frac{\partial V^*}{\partial r_i} = -\frac{\partial V}{\partial r_i} - \left\{\frac{\partial V}{\partial x}\frac{\partial x}{\partial r_i} + \frac{\partial V}{\partial y}\frac{\partial y}{\partial r_i} + \frac{\partial V}{\partial z}\frac{\partial z}{\partial r_i}\right\} \tag{47}$$

$$= F_i^{direct} + F_i^{distr}, \tag{48}$$

where

$$F_i^{distr} = F_x \frac{\partial x}{\partial r_i} + F_y \frac{\partial y}{\partial r_i} + F_z \frac{\partial z}{\partial r_i} \tag{49}$$

or, in matrix notation

$$F_i^{distr} = T F \tag{50}$$

with

$$\underset{\approx}{T} = \begin{pmatrix} \frac{\partial x}{\partial x_i} & \frac{\partial y}{\partial x_i} & \frac{\partial z}{\partial x_i} \\ \frac{\partial x}{\partial y_i} & \frac{\partial y}{\partial y_i} & \frac{\partial z}{\partial y_i} \\ \frac{\partial x}{\partial z_i} & \frac{\partial y}{\partial z_i} & \frac{\partial z}{\partial z_i} \end{pmatrix} \tag{51}$$

The distributed force on the i-th mass due to the force on the virtual particle (x, y, z) should thus be added to the direct force on the i-th mass. No transformation to a molecular coordinate system is needed.

Consider the example of a linear molecule (fig. 6, eq. 45). It is easily seen that

$$\underset{\sim}{F}_1^{distr} = F_x \begin{pmatrix} -c/b \\ 0 \\ 0 \end{pmatrix} + F_y \begin{pmatrix} 0 \\ -c/b \\ 0 \end{pmatrix} + F_z \begin{pmatrix} 0 \\ 0 \\ -c/b \end{pmatrix}$$

or

$$\underset{\sim}{F}_1^{distr} = -\frac{c}{b}\underset{\sim}{F} \tag{52}$$

Similarly

$$\underset{\sim}{F}_2^{distr} = (1 + \frac{c}{b})\underset{\sim}{F} \tag{53}$$

Thus the force is simply distributed over the masses in a ratio determined by the geometry. Note that the masses do not enter. This redistribution assures that both total force and total torque are conserved.

In the case of eq. (46) the relation between the redistribution of $\underset{\sim}{F}_4$ to e.g. particle 1 is given by

$$\underset{\sim}{F}_1^{distr} = \begin{pmatrix} 1-2\alpha & \beta z_{12} & \beta y_{13} \\ \beta z_{13} & 1-2\alpha & \beta x_{12} \\ \beta y_{12} & \beta x_{13} & 1-2\alpha \end{pmatrix} \underset{\sim}{F}_4 \tag{54}$$

METRIC TENSOR CONSIDERATIONS

Applying constraints implies that the particles move on a hypersurface in configuration space. The hamiltonian of the system is a function of non-constrained general coordinates and the constrained degrees of freedom and their conjugate momenta are removed from the hamiltonian. This is physically not equivalent to a system where constraints are actually "hard" degrees of freedom (e.g. harmonic oscillators with high force constant) and the behaviour of the remaining degrees of freedom is considered in the limit of infinite force constants for the hard degrees of freedom. This is a rather subtle problem, arising for molecules with internal degrees of freedom, that has been extensively discussed in the literature (12, 15-22). In the general case the consequence is that in constraint dynamics different regions of configuration space may get different weighting factors depending on the Jacobian of the transformation from cartesian to general coordinates. We will not repeat the arguments (see ref. 12), but only mention the results.

Consider for N particles a set of 3N generalized coordinates $\underset{\sim}{q}$ that can be divided into "soft" variables $\underset{\sim}{q}^{\alpha}$ and "hard" variables $\underset{\sim}{q}^{\beta}$. The potential energy of the system is composed of a soft potential V_s that is a function of the soft variables only and a harmonic hard potential V_h.

$$V(\underset{\sim}{q}) = V_s(\underset{\sim}{q}^{\alpha}) + V_h(\underset{\sim}{q}^{\beta}) \tag{55}$$

Under the assumption that the force constants of the hard modes are independent of $\underset{\sim}{q}^{\alpha}$, it can be shown that the expectation value of a variable A is given by

$$< A > = \frac{\int A \, |\underset{\approx}{G}|^{\frac{1}{2}} \exp\{-\beta \, V_s(\underset{\sim}{q}^{\alpha})\} \, d\underset{\sim}{q}^{\alpha}}{\int |\underset{\approx}{G}|^{\frac{1}{2}} \exp\{-\beta \, V_s(\underset{\sim}{q}^{\alpha})\} \, d\underset{\sim}{q}^{\alpha}} \tag{56}$$

Here $|\underset{\approx}{G}|$ is the determinant of the mass-metric tensor $\underset{\approx}{G}$:

$$G_{ij} = \sum_{k=1}^{3N} m_k \frac{\partial x_k}{\partial q_i} \frac{\partial x_k}{\partial q_j} \tag{57}$$

$|G|^{\frac{1}{2}}$ is a weighting factor in configuration space, involving the full dimensionality of q-space. Using constraints, rather than hard modes, the transformation is now limited to soft variables, yielding

$$< A >_{\alpha} = \frac{\int A \, |\underset{\approx}{G}^{\alpha}| \exp\{-\beta \, V_s(\underset{\sim}{q}^{\alpha})\} \, d\underset{\sim}{q}^{\alpha}}{\int |\underset{\approx}{G}^{\alpha}| \exp\{-\beta \, V_s(\underset{\sim}{q}^{\alpha})\} \, d\underset{\sim}{q}^{\alpha}}, \tag{58}$$

with

$$G_{ij}^{\alpha} = \sum_{k=1}^{3N} m_k \frac{\partial x_k}{\partial q_i^{\alpha}} \frac{\partial x_k}{\partial q_j^{\alpha}} , \tag{59}$$

a submatrix of G.
Thus, a constraint simulation does not yield the same physical equilibrium averages unless $|\underset{\approx}{G}| / |\underset{\approx}{G}^{\alpha}|$ is constant. One may, however, correct the constraint case by multiplying distribution functions by a factor

$$(|\underset{\approx}{G}| / |\underset{\approx}{G}^{\alpha}|)^{\frac{1}{2}} \tag{60}$$

Fixman (15) has derived a simple formula:

$$|\underset{\approx}{G}^{\alpha}| / |\underset{\approx}{G}| = |\underset{\approx}{H}| \tag{61}$$

where $|\underset{\approx}{H}|$ is the determinant of the tensor

$$H_{ij} = \sum_{k=1}^{3N} m_k \frac{\partial q_i^{\beta}}{\partial x_k} \frac{\partial q_j^{\beta}}{\partial x_k} \tag{62}$$

This determinant is much easier to evaluate than (60). The correction can be incorporated in the dynamics simulation by adding a potential:

$$V' = \tfrac{1}{2} kT \ln |\underset{\approx}{H}| \tag{63}$$

This potential can be viewed as the free energy due to the entropy $- \frac{1}{2} k \ln |H|$ associated with harmonic oscillators in the constrained degrees of freedom.

In what cases are the metric corrections necessary? If a molecule is completely constrained to a rigid body leaving no internal degrees of freedom, $|\underset{\approx}{H}|$ becomes a constant independent of $\underset{\sim}{q}^{\alpha}$ (which are now only translational and rotational degrees of freedom). So for rigid molecules there is no metric correction. If only bond lengths are constrained, the tensor $\underset{\approx}{H}$ will contain the angle between constrained bonds. For example, a triangular molecule (fig. 10) in which r_{13} and r_{23} are constrained, the matrix H becomes:

$$\underset{\approx}{H} = \begin{pmatrix} m_1 + m_3 & m_3 \cos\alpha \\ m_3 \cos\alpha & m_2 + m_3 \end{pmatrix} \tag{64}$$

and

$$|\underset{\approx}{H}| = (m_1 + m_3)(m_2 + m_3) - m_3^2 \cos^2\alpha \tag{65}$$

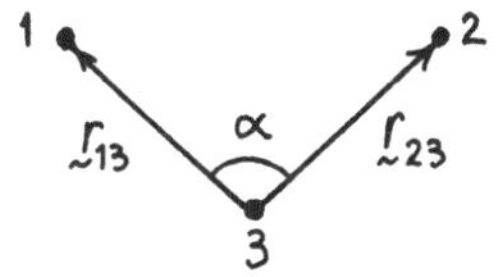

Fig. 10 triangular molecule

Hence there is an angle-dependent weighting factor in configurational space. However, if the angle α is restricted to a narrow range, $|\underset{\approx}{H}|$ may be safely assumed to be constant. Matters are different if rather soft internal degrees of freedom occur, such as the dihedral angles in alkanes. Constraining bond lengths only is still not significant because $|\underset{\approx}{H}|$ only contains the restricted bond angles. If, however, bond angles are constrained, then $|\underset{\approx}{H}|$ will depend on the dihedral angles as well and a correction as in eq. (63) should be made (17, 19).

As has been extensively tested for macromolecules (12, 23-25), the combined use of bond length and bond angle constraints in molecules with other internal degrees of freedom is not physically justified. There is too much coupling between bond angles and dihedral angles and not sufficient separation in frequency between bond angle vibrational modes and explicitly considered modes. Moreover, no computational gain is attained by constraining bond lengths in such cases. In contrast, constraining bond lengths only is both physically allowed and computationally efficient, leading to an increase of a factor 4 in the allowed time step.

The conclusions are straightforward: Complete constraints are allowed in rigid molecules; in flexible molecules only bond lengths should be considered. In neither case do complications arise because of metric tensor effects.

CONSTANT TEMPERATURE AND PRESSURE ALGORITHMS

Normally, MD is performed at constant volume and constant total energy, yielding a microcanonical ensemble. For various reasons this is not very convenient and many approaches have appeared in the literature to yield a type of dynamics in which temperature and pressure are independent variables rather than derived properties. When MD is performed in non-equilibrium situations (NEMD: non-equilibrium molecular dynamics) in order to study irreversible processes and transport properties, the need to impress external constraints or restraints is apparent. In such cases the temperature should be controlled as well in order to absorb the dissipative heat produced by the irreversible process. But also in equilibrium cases the automatic control of temperature and pressure as independent variables is very convenient. Thus slow temperature drifts that are the unavoidable result of truncation errors are corrected, while also rapid transitions to new desired conditions of temperature and pressure are more easily accomplished.

Several methods have been proposed, ranging from ad-hoc rescaling of velocities in order to adjust temperatures to consistent formulation in terms of modified Lagrangian equations of motion that ensure the dynamics to follow the desired constraints. We will not review these methods here but briefly describe a method that we have found to be very reliable and convenient, which is easy to implement and produces a stable algorithm even under adverse nonequilibrium conditions. The method achieves a coupling to an external bath with constant temperature and/or pressure with adjustable time constant of the coupling. This has the advantage that the coupling can be chosen sufficiently weak to avoid disturbance of the system and sufficiently strong to achieve the desired restraint. In contrast, methods based on modified Lagrangian equations of motion always contain a non-physical attribute to the potentials or forces that influences the physics of the system to a fixed non-adjustable extent. The full details of the method will be published elsewhere (26), where also a review of other methods is given.

Coupling to a P-T bath

The basic idea is to modify the equations of motions in such a way that the net result on the system is a first-order relaxation of temperature T and pressure P towards given reference values T_0 and P_0:

$$\left(\frac{dT}{dt}\right)_{bath} = \frac{T_0 - T}{\tau_T} \tag{66}$$

and

$$\left(\frac{dP}{dt}\right)_{bath} = \frac{P_0 - P}{\tau_P} \tag{67}$$

The modification of the equations of motion is such that local disturbances are minimized while the global effects of eqs. 66 and 67 are conserved. This effected (in the leap-frog scheme) by scaling velocities in every step with a factor λ:

$$v_i \leftarrow \lambda v_i \tag{68}$$

$$\lambda = \left[1 + \frac{\Delta t}{\tau_T}\left(\frac{T_0}{T} - 1\right)\right]^{\frac{1}{2}} \tag{69}$$

where Δt is the time step, and T is the actual temperature, and scaling coordinates and size l of the periodic box in every step with a factor μ:

$$x_i \leftarrow \mu x_i \; ; \; l \leftarrow \mu l \tag{70}$$

$$\mu = \left[1 - \frac{\Delta t}{\tau_P} \beta (P_0 - P)\right]^{1/3} \tag{71}$$

where β is the isothermal compressibility of the system and P is given by eqs. 28 and 29. For the calculation of the pressure internal forces within a molecule can be omitted (together with kinetic contributions of internal degrees of freedom). The equations can be modified to include non-isotopic systems. The coupling time constants τ_T and τ_P can be arbitrarily chosen, but should exceed 10 time steps to ensure stability of the algorithm. The value of the compressibility used in eq. 71 is not critical since it only influences the accuracy of the pressure coupling time constant τ_P.

ACKNOWLEDGEMENT

Several summer workshops of CECAM (Centre Européen de Calcul Atomique et Moléculaire) in Orsay have shaped our MD experience with stimulating contributions of dr. A. Rahman. Our collaborators J.P. Ryckaert, P. van der Ploeg and J.P.M. Postma have contributed to this work.

REFERENCES

1. Metropolis, N., Rosenbluth, A.W., Rosenbluth, M.N. Teller, A.H. and Teller, E. 1953, J. Chem. Phys. 21, 1087
2. Hansen, J.P. and McDonald, I.R., "Theory of simple liquids", Acad. Press 1976
3. Rossky, P.J., Doll, J.D. and Friedman, H.L. 1978, J. Chem. Phys.
4. Gear, C.W., "Numerical initial value problems in ordinary differential equations", Prentice-Hall 1971
5. Nordsieck, A. 1962, Math. Comput. 16, 22

6. Van Gunsteren, W.F. and Berendsen, H.J.C. 1977, Mol. Phys. 34, 1311
7. Beeman, D. 1976, J. Comput. Phys. 20, 130
8. Verlet, L. 1967, Phys. Rev. 159, 98
9. Goldstein, H. "Classical mechanics", Addison-Wesley, 1950
10. Rahman, A. and Stillinger, F.H. 1971, J. Chem. Phys. 55, 3336
11. Ryckaert, J-P., Ciccotti, G. and Berendsen, H.J.C. 1977, J. Comput. Phys. 23, 327
12. Berendsen, H.J.C. and Van Gunsteren, W.F., in "The physics of superionic conductors and electrode materials", J.W. Perram, ed. 1983, NATO ASI B92, 221, Plenum Press
13. Ciccotti, G., Ferrario, M. and Ryckaert, J.P. 1982, Mol. Phys. 11, 1
14. Stillinger, F.H. and Rahman, A. 1974, J. Chem. Phys. 60, 1545
15. Fixman, M. 1974, Proc. Natl. Acad. Sci. USA 71, 3050
16. Gō, N. and Scheraga, H.A. 1976, Macromol. 9, 535
17. Fixman, M., 1978, J. Chem. Phys. 69, 1527
18. Pear, M.R. and Weiner, J.H. 1979, J. Chem. Phys. 71, 212
19. Helfand, E. 1979, J. Chem. Phys. 71, 5000
20. Gottlieb, M. and Bird, R.B. 1976, J. Chem. Phys. 65, 2467
21. Chandler, D. and Berne, B.J. 1979, J. Chem. Phys. 71, 5386
22. Van Gunsteren, W.F. 1980, Mol. Phys. 40, 1015
23. Van Gunsteren, W.F. and Karplus, M. 1982, Macromol. 15, 1528
24. Van Gunsteren, W.F. and Karplus, M. 1980, J. Comput. Chem. 1, 266
25. Swaminathan, S., Ichiye, T., Van Gunsteren, W.F. and Karplus, M. 1982, Biochem. 21, 5230
26. Berendsen, H.J.C.,Postma, J.P.M., Van Gunsteren, W.F., DiNola, A. and Haak, J.R. 1984, to be published

COMPARISON OF THE RESULTS FROM SIMULATIONS WITH THE PREDICTIONS OF MODELS FOR MOLECULAR REORIENTATION.

R.M. Lynden-Bell

Department of Theoretical Chemistry, University of Cambridge.

Most simulations of liquids have angular momentum time correlation functions with a negative portion. This may be ascribed to a local cage. Reorientational correlation functions are related to angular momentum correlation functions by using a cumulant expansion. This is derived and used to compare the predictions from various models and approximations with the results from simulations, enabling one to guage the likely usefulness of these models and predictions in describing real liquids.

In the last ten years there have been many simulations of the molecular dynamics of simple liquids (1-12). It is now profitable to look at this range of simulations to see what can be learnt about the reorientation of small molecules in liquids and whether the various models and approximations that have been proposed are likely to be adequate.

1. ANGULAR MOMENTUM TIME CORRELATION FUNCTIONS

Let us begin by looking at angular momentum time correlation functions. These provide a rather direct probe of the causes of reorientation as the angular momentum, $\underline{J}$, is directly affected by the torque, $\underline{N}$, acting on the molecule by

$$\frac{d}{dt}\underline{J} = \underline{N} . \quad (1)$$

A. J. Barnes et al. (eds.), Molecular Liquids - Dynamics and Interactions, 501–518.

The correlation functions $C_{J\alpha}(t)$ and $C_J(t)$ are not experimentally accessible, but can be found from simulations. They are defined by

$$C_J(t) = \langle \underline{J}(t)\cdot\underline{J}(0)\rangle / \langle \underline{J}(0)\cdot\underline{J}(0)\rangle$$
$$C_{J\alpha}(t) = \langle \underline{J}_\alpha(t)\cdot\underline{J}_\alpha(0)\rangle / \langle \underline{J}_\alpha(0)\cdot J_\alpha(0)\rangle \qquad (2)$$

where $\underline{J}(t)$ is the total angular momentum at time t in a laboratory fixed axis system and $\underline{J}_\alpha(t)$ is the vector of the component of angular momentum along the molecular axis α at time t specified in a laboratory fixed frame. Angular brackets denote an ensemble average. There is no distinction between these two types of correlation function in spherical top and linear molecules, but symmetric and asymmetric tops often show considerable anisotropy in their reorientation.

Extremes of behaviour are illustrated by a freely rotating molecule and a harmonic solid. In the former case the angular momentum is constant, $C_J(t)$ is unity at all times and its time integral, τ_J, is infinite. In a harmonic solid the molecule librates about its mean orientation, so that $C_J(t)$ oscillates. Intermolecular forces will cause a damping of these oscillations so that τ_J tends to zero, although the "life time" of $C_J(t)$ during which it is significantly different from zero may be quite long. The behaviour of liquids lies between these two extremes. Examples of the range of types of angular momentum correlation functions are shown in figure 1.

The observed angular momentum correlation functions vary from gas-like simple decay (Figure 1.I) to solid-like oscillations (IV), but the typical behaviour of most simple liquids under conditions that are closer to the triple point than to the critical point is shown by III and II. Example IV is the angular momentum correlation function for the component of angular momentum parallel to the HH axis of water (1). It is not surprising that so structured a liquid shows distinct librations, and does not behave as a typical liquid. $C_J(t)$ functions of type I were found in several early simulations (N_2(2), HCl(3)), but such gas like behaviour is only found in low torque fluids. These are uncommon. Behaviour of types III and II have been found in many simulations including Cl_2, CO_2(4), CS_2(5), CBr_4(6,7) CF_4(8), CCl_4(9), CH_2Cl_2(10), $CHCl_3$(11); and this list is by no means exhaustive. An interesting example is a recent simulation of CH_3CN(12) in which the component of $\underline{J}$ parallel to the molecular axis shows type I behaviour, while the perpendicular component shows type III beharious. Simulations of Cl_2 at a range of state points (4), CX_4 at a range of state points (7) and a diatomic with a range of quadrupole moments (13) show that gas like behaviour (type I) only occurs in low torque fluids, which can be achieved by lowering the density and

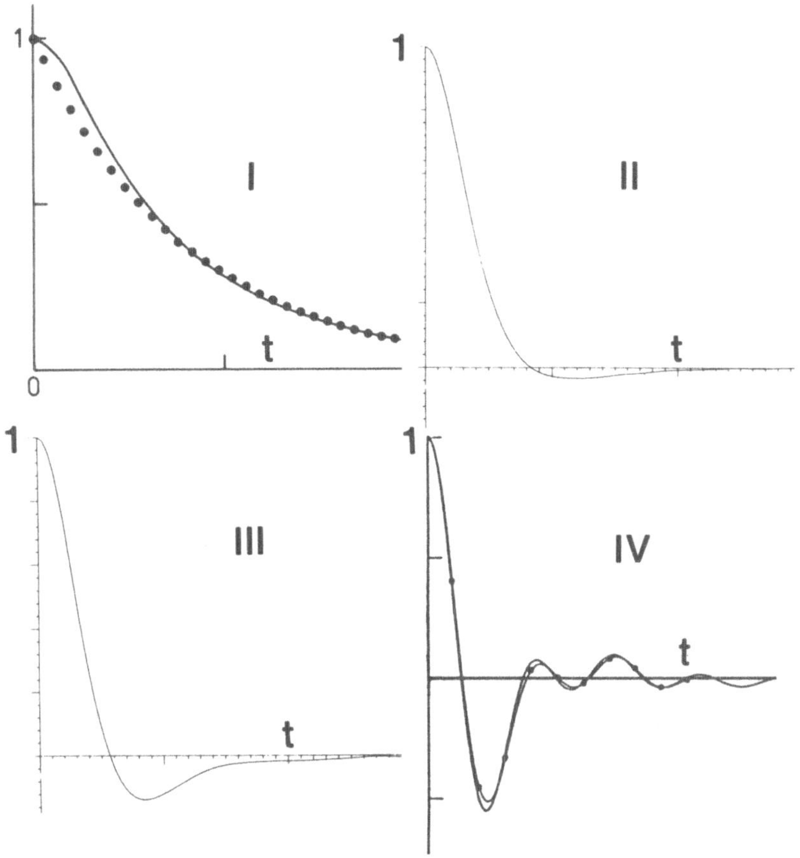

Figure 1. Types of angular momentum correlation functions found in simulations: I CBr_4 at low density and high temperature (6); II, CF_4 at 130K and atmospheric density (8); III CF_4 at 90K and atmospheric density (8); IV $C_{Jy}(t)$ for H_2O at 290K and atmospheric density (1).

increasing the temperature or by reducing the anisotropy of the potential.

The presence of the negative portion in the typical form III has implications for the correlation time τ_J and contains information about the reorientation process. The integral of $C_J(t)$, that is τ_J, is reduced by the negative part so that the correlation time defined as τ_J is less than the "life time" of the function during which it is appreciably different from zero. This latter time is not well defined but is a useful concept. The physical meaning of the negative part is that, on average, the angular momentum has a component in the reverse direction to its original value after a time lapse. Random uncorrelated collisions cannot cause such a reversal. Whether hard (as in the J diffusion model) or soft (as in a Fokker-Planck model) they lead to an exponentially decaying $C_J(t)$ of type I. But motion in a potential well does reverse angular momentum. We are led to attribute the negative portion to a cage effect, with the molecules moving in a temporary potential well caused by the local structure (the cage). These cages vary in size and shape and change continuously as the molecules move. The nature of the resulting motion is still rather different from that in a solid and it is misleading to refer to it as a libration.

The shape of $C_J(t)$ also has implications for the adequacy of models (14) used to describe reorientation. Models which are based on assumptions about the physical processes governing reorientation make predictions about $C_J(t)$ in addition to giving the reorientational functions themselves. Approximations, such as a truncated memory function expansion or Steele's torque approximation, do not give predictions about $C_J(t)$. Both Fokker-Planck and J diffusion models (14) assume uncorrelated collisions, which we have seen is an inappropriate assumption for most fluids. Some variant of the itinerant oscillator model may well prove to be more realistic, but it is easy to overestimate the librational character of the motion.

Both J diffusion and Fokker-Planck models have the merit of giving the correct short time free rotational behaviour for the reorientational correlation functions, and also give the expected exponential decay at long times, so one may ask whether they are "good enough". Does it matter if the $C_J(t)$ is not correct when it is the reorientation itself that concerns us? In order to answer this question we need to relate the angular momentum correlation functions to the reorientational correlation functions. A useful way of doing this is by means of a cumulant expansion which proves to provide insights into the connection between models, simulation results and real molecules.

2. MOLECULAR REORIENTATION

The instantaneous rate of reorientation of a molecule depends on its angular velocity, $\underline{\omega}$, which is related to its angular momentum by

$$\underline{\omega} = I^{-1}\underline{J} \tag{3}$$

where I is the moment of inertia tensor. For linear molecules and spherical tops $\underline{\omega}$ is proportional to $\underline{J}$ so that the angular velocity correlation function $C_\omega(t)$ is identical to $C_J(t)$. This is not true for less symmetrical molecules, but if α, β, γ are the principal axes of the moment of inertia tensor

$$C_{\omega\alpha}(t) = C_{J\alpha}(t) \tag{4}$$

where $C_{\omega\alpha}(t)$ is defined in an analogous way to that given in equation (2).

In order to describe the effects of the reorientation on a function $f(\Omega)$ of the orientation of a particular molecule, it is necessary to introduce the infinitesimal rotation operators $\hat{I}_x$, $\hat{I}_y$ and $\hat{I}_z$. These are identical to the familiar quantum mechanical angular momentum operators and obey the same commutation relations. The rate of change of f is

$$\frac{d}{dt} f(\Omega) = i\underline{\omega}(t)\cdot\underline{\hat{I}}\, f(\Omega) \tag{5}$$

which can be integrated to give

$$f(\Omega(t)) = \exp_o\left[i\int_o^t \underline{\omega}(t')\cdot\underline{\hat{I}}\,dt'\right] f(\Omega(0)) \tag{6}$$

Now consider an ensemble of molecules in an isotropic liquid. We want to know the average change in f due to reorientation in a time t. This can be found by taking an axis system that is different for each molecule (this is allowed as the system is isotropic) and which coincides with the molecular axis system at time zero (so that $\Omega(0) = (0, 0, 0)$). Then the average reorientation is described by the operator

$$\hat{C}(t) = \langle \exp_o i\int_o^t \underline{\omega}(t')\cdot\underline{\hat{I}}\,dt'\rangle \tag{7}$$

which can be called the reorientational propagator. The familiar orientational correlation functions are found by operating with $\hat{C}$ on spherical harmonics

$$\hat{C}_{L,mm'}(t) = \int Y_m^{L*}\hat{C}(t)\, Y_{m'}^L\, d\Omega \Big/ \int Y_m^{L*} Y_m^L\, d\Omega\ . \tag{8}$$

It is useful to think of the integral in equation (7) as the limit of a sum

$$\hat{C}(t) = \langle \exp_o i\sum_n \underline{\omega}(t_n)\cdot\underline{\hat{I}}\,\delta t\rangle\ . \tag{9}$$

As the components of $\hat{I}$ do not commute, each term in the sum must act in order, with early times acting first. Such an exponential operator is said to be time ordered; this is the meaning of the subscript "o".

3. CUMULANT EXPANSION OF $\hat{C}(t)$

The approximation to $\hat{C}(t)$ that we seek is found by using a cumulant expansion (15) of the ensemble average in equations (7) and (9). Cumulant expansions are used in probability theory to describe properties of random variables; here the values of $\underline{\omega}(t_n)$ for different molecules in the ensemble at different times can be considered to be random variables; $\underline{\omega}(t_i)$ and $\underline{\omega}(t_j)$ for the same molecule are not necessarily independent.

3.1. Definitions

Suppose $\{X_1,\ldots X_N\}$ are N random variables. The moment generating function (or characteristic function) is defined by

$$M(z_1,z_2\ldots z_N) = \langle \exp \sum_i z_i X_i \rangle \tag{10}$$

which can be expanded in terms of moments

$$M = \sum_{n_i,\ldots} \left\{ \prod_i \frac{z_i^{n_i}}{n_i!} \right\} \langle X_1^{n_1} X_2^{n_2} \ldots \rangle \quad . \tag{11}$$

The average in angular brackets is the $(n_1,n_2\ldots)$th moment. This is an obvious generalisation of the moment generating function and moments of a single variable

$$M(z) = \langle \exp zX \rangle = \sum_n \frac{z^n}{n!} \langle X^n \rangle . \tag{12}$$

For a single variable the cumulant generating function is defined as

$$K(z) = \ln M(z) \tag{13}$$

and the cumulants or cumulant averages, k_n, of the distribution are defined by its expansion

$$K(z) = \sum_n \frac{z^n}{n!} k_n \tag{14}$$

and may be related to the moments, m_i, by equating terms in the expansion of equation (14). The first few terms give

$$\begin{aligned} k_1 &= m_1 \\ k_2 &= m_2 - m_1^2 \\ k_3 &= m_3 - 3m_1m_2 + 2m_1^3 \\ k_4 &= m_4 - 3m_2^2 - 4m_1m_3 + 15m_1^2m_2 - 9m_1^4 \end{aligned} \tag{15}$$

Another notation used for the cumulants is

$$k_n = \langle X^n \rangle_C \; . \tag{16}$$

The usefulness of this expansion is illustrated by the fact that for a Gaussian (or Normal) distribution with mean m and variance σ^2 the only two non-zero cumulants are k_1 and k_2 with

$$\begin{aligned} k_1 &= m \\ k_2 &= \sigma^2 \; . \end{aligned} \tag{17}$$

The cumulant expansion can also be generalised to many variables by writing

$$K(z_1,z_2 \ldots) = \ln M(z_1,z_2 \ldots) \tag{18}$$

If one of the variables, X_1, is independent of the other variables the moment generating function becomes a product of terms

$$M(z_1,z_2 \ldots) = M(z_1)M(z_2,\ldots) \tag{19}$$

and the cumulant generating function becomes a sum of terms

$$K(z_1,z_2 \ldots) = K(z_1) + K(z_2,\ldots), \tag{20}$$

which shows that the cumulant average $\langle X_1^{n_1} X_2^{n_2} \ldots \rangle_C$ is zero if any one (or more) of the variables X_i is statistically independent of the others.

3.2. Application

The operator C(t) has the form of a moment generating function (compare equations (9) and (10)). Expanding it one obtains

$$\begin{aligned} \hat{C} = 1 - \sum_{i,j} \int_o^t \int_o^{t_1} \langle \omega_i(t_1)\omega_j(t_2) \rangle dt_2 dt_1 \hat{I}_i \hat{I}_j \\ + \sum_{i,j,k,\ell} \int_o^t \int_o^{t_1} \int_o^{t_2} \int_o^{t_3} \langle \omega_i(t_1)\omega_j(t_2)\omega_k(t_3)\omega_\ell(t_4) \rangle dt_4 \ldots dt_1 \\ \times \hat{I}_i \hat{I}_j \hat{I}_k \hat{I}_\ell - \ldots . \end{aligned} \tag{21}$$

where i, j, k, ℓ run over x, y, z in the t = 0 axis system. In deriving (21) we have used the fact that all odd moments are zero and the time ordering has been carefully preserved. This expansion does not converge rapidly as higher order moments exist for all even powers in the expansion. If we define the corresponding cumulant generating operator $\hat{K}$

$$\hat{C} = \exp \hat{K} \tag{22}$$

then the expansion of $\hat{K}$ can be found using a certain amount of algebra. It is

$$\hat{K} = -\sum_{i,j}\int_o^t\int_o^{t_1}\langle\omega_i(t_1)\omega_j(t_2)\rangle_C dt_2 dt_1 \hat{I}_i\hat{I}_j + \sum_{i,j,k,\ell}\int_o^t\int_o^{t_1}\int_o^{t_2}\int_o^{t_3}\langle\omega_i(t_1)\omega_j(t_2)\omega_k(t_3)\omega_\ell(t_4)\rangle_C dt_4\ldots dt_1 \times \hat{I}_i\hat{I}_j\hat{I}_k\hat{I}_\ell + \ldots \tag{23}$$

with integrands that are cumulant averages. As these are zero if the variables are independent this expansion converges more rapidly, especially in high torque fluids where the angular velocities are quickly randomised.

For a spherical top molecule the expression for $\hat{K}$ can be considerably simplified. The first term in (23), $\hat{K}_2$, is zero unless i = j, and the same for i = j = x or y or z. Hence

$$\hat{K}_2 = -\frac{1}{3}\int_o^t\int_o^{t_1}\langle\underline{\omega}(t_1)\cdot\underline{\omega}(t_2)\rangle_C dt_2 dt_1 \hat{\underline{I}}^2 . \tag{24}$$

The integrand depends only on the difference in times and is proportional to $C_\omega(t_1 - t_2)$. Equation (24) can be rewritten as

$$K_2 = -\int_o^t (t - \tau)C_\omega(\tau)d\tau\langle\omega_x^2\rangle \hat{\underline{I}}^2 \tag{25}$$

If we want to calculate $C_2(t)$, then using the fact that a spherical harmonic Y_M^L is an eigenfunction of $\hat{\underline{I}}^2$ with eigenvalue L(L + 1), the lowest order term gives

$$\ell n\, C_L(t) = -L(L + 1)\langle\omega_x^2\rangle \int_o^t (t - \tau)C_\omega(\tau)d\tau\ldots. \tag{26}$$

The same result is obtained for linear molecules. Equation (26) suggests that a useful way to compare correlation functions with different L values is to plot $[\ell n\, C_L(t)]/L(L + 1)$ versus t. More importantly it shows that the angular velocity correlation function determines the lowest order contribution to $C_L(t)$. Before we look at particular examples let us consider the general features of this approximation for $\ln C_L(t)/L(L + 1)$. At short times $C_\omega(t)$ is quadratic in time. At times longer than the "life time" of C_ω the integral gives

$$\int_o^t (t - \tau) C_\omega(\tau) d\tau = t\tau_\omega - A$$
$$\text{with } A = \int_o^\infty \tau C_\omega(\tau) d\tau \tag{27}$$

showing that the asymptotic behaviour is a linear decay with slope $(T_L)^{-1}$

$$T_L^{-1} = \langle \omega_x^2 \rangle \tau_\omega L(L + 1) \tag{28}$$

and an intercept $A\langle\omega_x^2\rangle L(L + 1)$ which is positive if C_ω is always positive (type I behaviour), but which may be negative for type III behaviour. The shape of the intermediate region depends on the detailed form of C_ω; if this becomes negative the K_2 contribution to $\ln C_L(t)$ has a point of inflexion.

Higher order terms contain integrals of four time, six time etc. correlation functions. The next term, which is shown in equation (23), contains integrands of the form

$$\langle \omega_i(t_1)\omega_j(t_2)\omega_k(t_3)\omega_\ell(t_4)\rangle_C \ . \tag{29}$$

Many of these are zero for spherical top molecules (e.g. x y y y) and others are related by symmetry. The ones that remain can be rearranged (7) to give

$$K_4(t) = k_1^{(4)} L(L + 1) + k_2^{(4)} L^2 (L + 1)^2 + k_{A_1}^{(4)} Q(\Gamma) \tag{30}$$

where the last term generates non-spherical effects, for example it would cause a difference between the E and T_2 Raman line shapes (16). We have found that it is negligible in all the spherical top molecules that we have investigated (6, 7, 8). The second term causes a deviation of the curves for different L values in the $[\ln C_L(t)]/L(L + 1)$ plot. As the higher order terms are negligible at short times, linear at long times and the "life time" of the integrand (29) is unlikely to be longer than that of $C_\omega(t)$, the main effect of these higher order times is to cause a fanning out of the $[\ln C_L(t)]/L(L + 1)$ curves due to changes in the asymptotic slopes. In most molecules it seems that the higher order terms cause the higher L curves to lie above those for lower Ls so that the slopes satisfy

$$L_1(L_1 + 1)/T_{L_1} < L_2(L_2 + 1)/T_{L_2} \quad \text{if} \quad L_1 > L_2 \ . \tag{31}$$

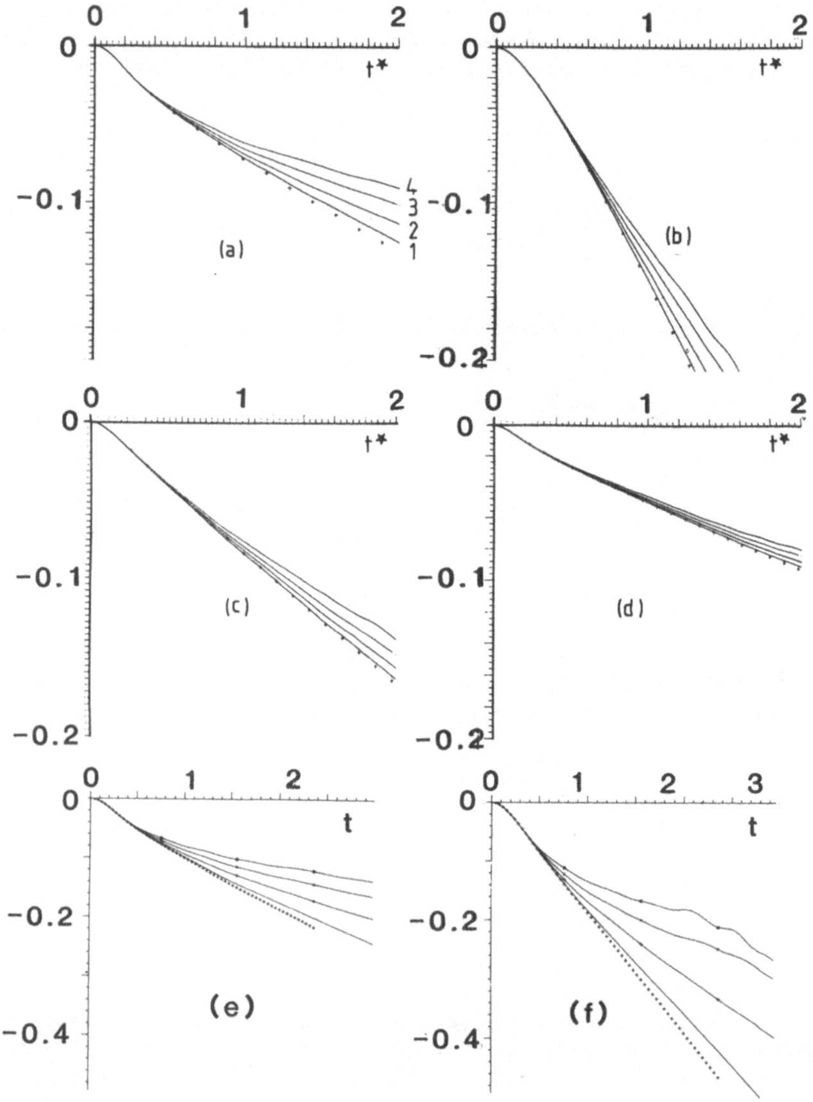

Figure 2. Curves showing $[\ell n\ C_L(t)]/L(L+1)$ for a range of simulations of tetrahedral molecules. (a) CBr_4 at 360 K; CX_4 at 345 K (b), at 330 K (c) and at 245 K (d); CF_4 at 90 K (e) and at 130 K (f). Time is in units of $(I/kT)^{\frac{1}{2}}$. (a)-(d) from (7); (e)-(f) from (8).

Some typical results from simulations are shown in figure 2, which contains results from simulations of CBr_4(6), CX_4(7) and CF_4(8). CX_4 is a modified potential for CBr_4 designed to make a more "knobbly" molecule. The range of observed behaviour was not found to differ much from CBr_4 or CF_4. The details of the potential do not appear to affect greatly the trends of behaviour with changing state conditions.

The curves in figure 2 show both the second order $K_2(t)$ computed from the simulation angular velocity correlation function (crosses) and the $\ell n C_L(t)$ functions for $L = 1$ to 4. The second cumulant provides a good approximation for $L = 1$, but higher order terms cause the higher L values to deviate, especially for low torque liquid states where the angular velocity correlation functions are longer lived. Curves (a) for CBr_4, (d) for CX_4 and (e) for CF_4, corresponding to angular momentum correlation functions of type III, show a negative intercept at $t = 0$ for the asymptotic straight line. Curves (c) for CX_4 and (f) for CF_4 have intercepts near zero; they correspond to $C_J(t)$ of type II. Curve (b) for CX_4 has a positive intercept and no point of inflexion as the $C_J(t)$ for this system is of type I.

3.3. Implications for correlation times

There are several different possible ways of defining correlation times for a time correlation function such as $C_L(t)$. Here we shall consider only two, the integral correlation time τ_L and the asymptotic slope correlation time T_L (equation (28)). Spectral line shapes are Lorentzian in the centre part, with a width determined by T_L, the slope correlation time; other relaxation measurements (notably NMR) are determined by the integral correlation times τ_L. It is important to realise that these quantities may differ. Figure 3 shows $\ell n C$ curves for type I and type II molecules together with the linear behaviour that would be shown by an exponentially decaying system with the same slope.

In I (gas like system) the actual $\ell n C$ curve is always higher than the exponential curve, so that its integral correlation time is greater than that for the exponential curve. However an exponential decay has equal slope and integral correlation times so that for systems with type I behaviour we have

$$\text{I} \qquad \tau_L > T_L \; . \tag{32}$$

The more common type III systems are also illustrated in figure 3. The reverse argument holds except at short times so that

$$\text{III} \qquad \tau_L < T_L \; . \tag{33}$$

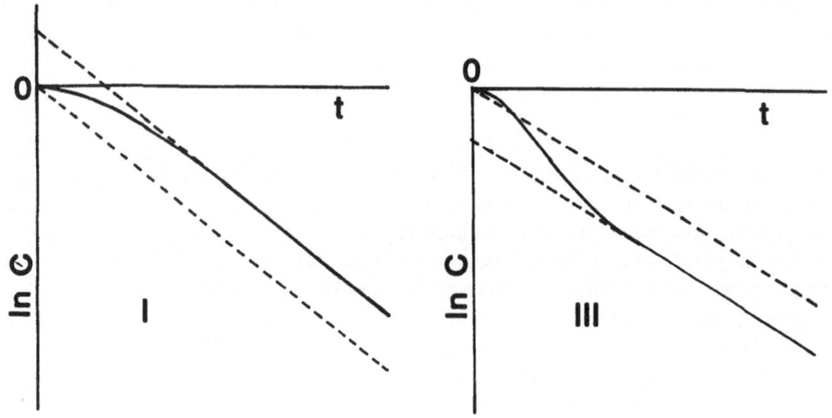

Figure 3. Comparison of diffusion predictions with observed results from type I and type III systems.

No conclusive argument can be given for systems of the intermediate type, except that the two correlation times cannot differ greatly for small L values.

4. COMPARISON WITH MODELS

4.1. Rotational diffusion

In this model the reorientation of any rigid molecule is described by a rotational diffusion tensor. Using the same infinitesimal rotation operators as in the discussion of cumulants, and choosing the principal axes of the diffusion tensor, the orientational propagator, $\hat{C}(t)$, for this model is

$$\hat{C}(t) = \exp-(D_{xx}\hat{I}_x^2 + D_{yy}\hat{I}_y^2 + D_{zz}\hat{I}_z^2)t. \tag{34}$$

This is reminiscent of equation (22) defining the cumulant generating operator, $\hat{K}$, so that for this model

$$\hat{K} = -(D_{xx}\hat{I}_x^2 + D_{yy}\hat{I}_y^2 + D_{zz}\hat{I}_z^2)\,t. \tag{35}$$

Comparing this with equation (23) we see that the diffusion model results if no higher order terms contribute to the cumulant expansion and if the constant D_{xx} is given by

$$D_{xx} = t^{-1}\int_0^t\int_0^{t_1} \langle\omega_x(t_1)\omega_x(t_2)\rangle \, dt_2 dt_1 \quad . \tag{36}$$

This expression is constant at times greater than the angular velocity correlation "life time". The diffusion model is therefore valid if $C_L(t)$ has not decayed significantly during the "life time" of $C_\omega(t)$. When this is true the shape of $C_\omega(t)$ is irrelevant and the components of the diffusion tensor $D_{\alpha\alpha}$ are given by the familiar expression

$$D_{\alpha\alpha} = \langle\omega_\alpha^2\rangle\tau_{\omega,\alpha} \qquad \alpha = x,y,z. \tag{37}$$

This model is good for large molecules. It can still be qualitatively useful for small molecules, but when the deviations are significant some other model is required.

4.2. Fokker-Planck and J diffusion models.

These two models (14) both vary between the limits of free rotation and diffusion as a single parameter, τ_ω, is changed. Both models assume successive uncorrelated instantaneous collisions, but differ in the strength of these collisions. In the Fokker-Planck model the angular impulse of each collision is so small that the angular momentum changes infinitesimally. Gordon's J diffusion model has strong collisions with large angular impulses that randomise the angular momentum at each step.

We have already seen that the assumption of uncorrelated collisions leads to an exponential decay of the angular momentum correlation function. This is rarely appropriate for the fluids that have been simulated. If $C_\omega(t)$ has the wrong shape the leading term in the cumulant expansion is incorrect. We deduce that these models are unlikely to be more useful than straight diffusion in most circumstances. However there are some "gas-like" low torque fluids and it is useful to be able to compare the two models.

Figure 4 shows the predicted curves from these models for an intermediate value of τ_ω, together with the second cumulant approximation to these models. The difference between them is striking, with no deviation between the different L values for the Fokker-Planck model, whose curves lie below the lowest cumulant approximation, and a distinct fanning out of the J diffusion curves. Comparing these with the results from a low torque CBr_4 simulation we found reasonable agreement with the J diffusion model (6).

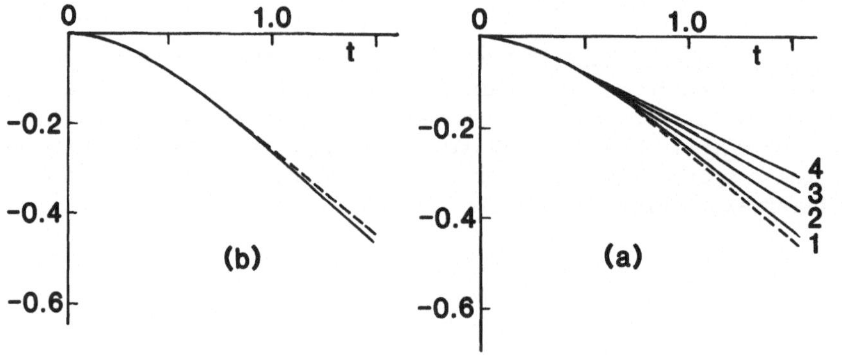

Figure 4. Curves of $[\ln C_L(t)]/L(L + 1)$ predicted by the J diffusion (a) and Fokker Planck (b) models for $\tau_\omega^{-1} = 2.5\ (kT/I)^{\frac{1}{2}}$. The dotted lines show the lowest cumulant approximation.

In spite of their popularity among spectroscopists it seems unlikely that the use of these models to fit spectral data is justified unless one has good reason to believe that the angular momentum correlation functions decay approximately exponentially with no cage effect.

4.3. Memory function approximations

The simplest type of memory function approximation is to truncate the memory function hierarchy by assuming that the n th order memory function decays exponentially. This leads to a truncated continued fraction for the Fourier transform of $C_L(t)$ with only one free parameter. Although the itinerant oscillator model in two dimensions is identical to a three term truncated continued fraction, in three dimensions there is no such simple relationship; this must be considered as an approximation with no prediction about $C_\omega(t)$.

This approximation proved to be unsuccessful in fitting the results of a high torque simulation of CBr_4. Figure 5 shows that if the limiting slopes are approximately correct oscillations are introduced into the intermediate part of the decay curve of $C_L(t)$;

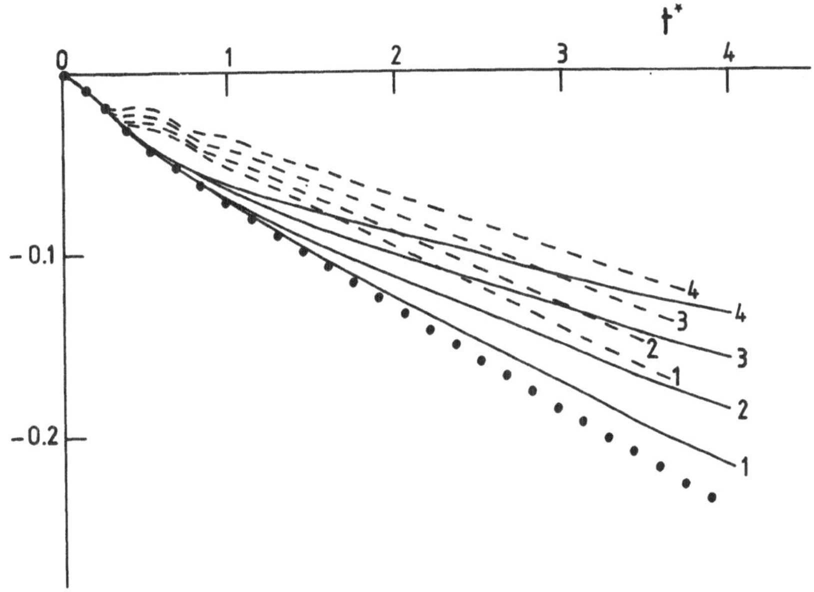

Figure 5. Comparison of best fit second order truncated memory function predictions (dashed lines) with simulation results for a high torque state of CBr_4 (6) and the lowest cumulant approximation (dotted line).

these do not match the observed simple inflexion due to the negative portion of $C_\omega(t)$.

A variant on this approach was proposed by Singer and his collaborators (17) who tried a range of functional forms for the time dependence of the second memory function. They obtained reasonable agreement for the L = 1 and L = 2 curves for a simulation of Cl_2 showing type III behaviour with a second memory function of the form

$$M_2(t) = M_2(0)\{\exp[-(t/t_1)^2] + ct^2\exp-(t/t_2)\} , \qquad (38)$$

where M_2 differs for different L values, but the three fitting parameters were taken to be the same. Although this type of model is instructive it contains too many parameters to be useful to spectroscopists. Their simpler approximations were less successful.

Another approach is due to Steele (18). He approximates the first memory function M_1 by the product of the free rotor memory function M_F and a gaussian decay. This may be compared with the memory function for the J diffusion model which is the product of the free rotor memory function and an exponential decay. The coefficient of t^2 in the gaussian decay term is chosen so that the coefficient of t^4 in $C_L(t)$ is correct. For a linear molecule this gives

$$M_1(t) = M_F(t) \exp - \left[\langle N^2 \rangle t^2 / 4 kT I \right] . \tag{39}$$

This "torque approximation" is most useful for low torque fluids where there is some evidence that it is better than the J diffusion model (18). It is not clear that it can reproduce the C_L curves of type III fluids. The published preductions for C_L do not show a point of inflexion.

4.4. Cumulant approximations

A simple approximation can be found by assuming that the only significant contribution to the correlation function comes from the second cumulant which is assumed to decay exponentially. The resulting form for C_L is

$$\ell n C_L / L(L + 1) = \tau_\omega^2 \langle \omega_x^2 \rangle \left[1 - t/\tau_\omega + \exp(-t/\tau_\omega) \right] . \tag{40}$$

This line shape (the Kubo or Rothschild line shape) has been used extensively for vibrational line shapes and less commonly for reorientation. Our previous discussion shows that it can only be adequate for gas like liquids (no point of inflexion). It is unlikely to be as good as J diffusion or Steele's torque approximation as the right hand side contains no L dependence.

4.5. Information theory approximations

Powles has introduced an approximation based on information theory (19). If one knows the second and fourth moments of a distribution then information theory can be used to find a "best" approximation to the moment generating function. Applying this to $C_L(t)$ is not completely straightforward, but Powles obtains an expression for the fourier transform of $C_L(t)$

$$\hat{C}_L(\omega) = A / \left[\exp(\beta + \lambda \omega^2) - 1 \right] \tag{41}$$

where the constants β and λ are determined by the values of the second and fourth moments of $C_L(t)$. These are given by St Pierre and Steele (20) and depend on the type of molecule (linear, symmetric or spherical top), the value of L and the mean square components of the torque. A is determined by the requirement that the integral of (41) is equal to 2π. Powles' published curves show a slightly better agreement with simulation results than Steeles torque model, but again points of inflexion are absent in the predicted $\ell n C_L$ curves.

5. CONCLUSIONS

To be useful, a model for molecular reorientation must be both simple and realistic. If spectroscopists are to use a model to describe results from a variety of techniques there must be a minimum of parameters to be fitted. In many cases the diffusion model introduced by Debye is both simple and adequate. However there are now an increasing number of examples where diffusion is not sufficiently realistic to be adequate and the short and intermediate time behaviour must be taken into account explicitly. Results from simulations show that the next step is not easy. Compressed gases and low torque liquids may reasonably be modelled by Steele's torque approximation, Powles' information theory expression or by the J diffusion models, although the evidence suggests that the Fokker-Planck model is not very successful. But commonly occurring high torque liquids cannot be successfully modelled by these techniques and there seems to be no simple alternative. It is likely that at least two parameters will be needed even for spherical and linear molecules. Further work to compare two parameter approximations with simulation results will show the most satisfactory next approximation.

I would like to thank Dr Nosé and Dr Klein for supplying a tape with two of their simulations of CF_4.

REFERENCES

(1) Impey, R.W., Madden, P.A. and McDonald, I.R. 1982, Molec. Phys. 46, pp. 513-539.
(2) Cheung, P.S.Y. and Powles, J.G. 1976, Molec. Phys. 32, pp. 1383-1405.
(3) Powles, J.G., McGrath, E. and Gubbins, K.E. 1980, Molec. Phys. 40, pp. 179-192.
(4) Singer, K., Singer, J.V.L. and Taylor, A.J. 1979, Molec. Phys. 37, pp. 1239-1262.
(5) Tildesley, D.J. and Madden P.A. 1983, Molec. Phys. 48, pp. 129-152.
(6) Lynden-Bell, R.M. and McDonald, I.R. 1982, Chem. Phys. Letts. 89, pp. 105-109.
(7) Lynden-Bell, R.M. and McDonald, I.R. 1981, Molec. Phys. 43, pp. 1429-1440.
(8) Nosé, S. and Klein, M.L. 1983, J. Chem. Phys. 78, pp. 6928-6938.
(9) Steinhauser, O. and Neumann, N. 1980, Molec. Phys. 40, pp. 115-128.
(10) Evans, M.W. and Ferrario, M. 1982, Adv. in molec. relax. and int. Processes 24, pp. 75-105.

(11) Evans, M.W. 1982, Adv. in molec. relax. and int. Processes 24, pp. 123-138.
(12) Bohm, H.J., Lynden-Bell, R.M., Madden, P.A. and McDonald, I.R. 1983,
(13) Steele, W.A. and Streett, W.B. 1980, Molec. Phys. 39, pp. 279-298.
(14) Steele, W.A. this volume.
(15) Kubo, R. 1967, J. Phys. Soc. Japan 17, pp. 1100-1139.
(16) Lynden-Bell, R.M. 1980, Chem. Phys. Letts. 70, pp. 477-480.
(17) Detyna, E., Singer, K., Singer, J.V.L. and Taylor, A.J. 1980, Molec. Phys. 41, pp. 31-54.
(18) Steele, W.A. 1981, Molec. Phys. 43, pp. 141-159.
(19) Powles, J.G. 1983, Molec. Phys. 48, pp. 1083-1092.
(20) St Pierre, A.G. and Steele, W.A. 1981, Molec. Phys. 43, pp. 123-140.

TOWARDS A MORE COMPLETE SIMULATION OF SMALL POLYATOMIC MOLECULES.

D.J. Tildesley

Department of Chemistry, Southampton University,
Southampton SO9 5NH, U.K.

This review contains an account of the models which have been used in the simulation of realistic molecular fluids. Methods for including the long-range dipole-dipole interaction and for calculating correlation functions and spectra are discussed. We review simulated dynamic properties with relation to experiment and describe the calculation of allowed and collision-induced infrared and light scattering spectra.

In the last decade the molecular dynamics technique has increased our understanding of a large number of condensed phase phenomena. We begin by guessing a simple potential model, the equations of motion of 256 molecules are solved forward in time for a hundred picoseconds and we hope to predict properties as diverse as the neutron diffraction pattern of water [1] and the librational frequencies of solid nitrogen [2]. At present the limiting factor in our understanding of condensed phases must be our crude knowledge of the interactions between molecules and we begin by discussing some of the common approximations used in building potential models for use in computer simulations.

I. Potential Models

I.1 Pairwise Additivity

In general the potential for a collection of N molecules can be built up from pair, triplet and higher order interactions.

$$U(\underset{\sim}{r}^{(N)},\underset{\sim}{\Omega}^{(N)})=\sum_{i>j}\sum u^{(2)}(r_{ij},\Omega_i\Omega_j)+\sum_{i>j>k}\sum\sum u^{(3)}(\underset{\sim}{r}_i\underset{\sim}{r}_j\underset{\sim}{r}_k,\Omega_i\Omega_j\Omega_k)+\cdots\cdots \quad (1)$$

A. J. Barnes et al. (eds.), Molecular Liquids - Dynamics and Interactions, 519–560.

where r_i describes the position of the centre of mass of the ith molecule and Ω_i the three Euler angles necessary to define its orientation. For an atomic fluid, such as argon, $u^{(2)}$ is well known and we have an approximate understanding of the three body contribution in terms of the Axilrod-Teller potential [3]. $u^{(2)}$ and $u^{(3)}$ have been used in a simulation study of liquid argon and give an adequate account of the thermodynamic properties of the fluid [4]. In contrast for a pair of N_2 molecules the exact two-body potential has not been established and it is not certain that the generalized triple-dipole potential is the dominant three body interaction in a molecular fluid. In these circumstances we are forced to consider an effective pair-potential

$$U(r^{(N)}\Omega^{(N)}) = \sum_{i>j} \sum u_{eff}^{(2)} (r_{ij} \, i \, j) \, . \qquad (2)$$

An example of one such effective pair interaction is the Lennard-Jones potential for argon (ε/k = 119.8K, σ =3.405Å) [5]. All reported simulation of molecular fluids and solids to date use this approximation.

The clear advantage in a pair potential is that we need only compute $N(N-1)/2$ independent interactions to move all molecules. An exact inclusion of the triplet term could increase the computing time by $N/3$ which would normally be prohibitively expensive. The disadvantage is that the effective potentials can be state dependent and since they are parameterised by fitting to a limited set of observable properties, they will often require readjustment when different properties are considered. In the case of N_2 the simple diatomic Lennard-Jones potential [6], (ε/k = 37.3K, σ = 3.31Å, $1/\sigma$ = 0.329), gives an excellent fit to the second virial coefficient [7]. An accurate effective liquid potential involves a small change in LJ parameters and the addition of a quadrupole [8], (ε/k = 35.3K, σ = 3.314Å, $1/\sigma$ = 0.332, Q = -4.93 x 10^{-40} Cm^2). A potential which describes the librational frequencies in the α-solid involves a further change in LJ parameters and a 10% reduction in the effective quadrupole moment [9] (ε/k = 36.4K, σ = 3.318Å, $1/\sigma$ = 0.331, Q = -3.91 x 10^{-40} Cm^2). With these limitations in mind we shall consider the pair potentials used in the simulation of simple molecular fluids.

I.2 Short-Range Interactions

The earliest models of molecular fluids were of the generalized Stockmayer form, comprising a spherical repulsive core and a long-range anisotropic interaction. These models have been particularly successful in describing molecular fluid mixtures and are reviewed in [10]. A class of models which attempt to include the geometry of the molecular core are the interaction site models (ISM). A molecule is constructed by fusing atoms into a rigid structure. Each atom acts as a force centre and interacts specifically with all the atoms in neighbouring molecules.

$$u(r_{ij}\Omega_i\Omega_j) = \sum_{\alpha=1}^{M_i} \sum_{\gamma=1}^{M_j} u_{\alpha\gamma}(r_{\alpha_i\gamma_j}) \tag{3}$$

There are M_i sites in molecule i and the potential $u_{\alpha\gamma}$ is radially symmetric. Two examples are shown in figure 1.

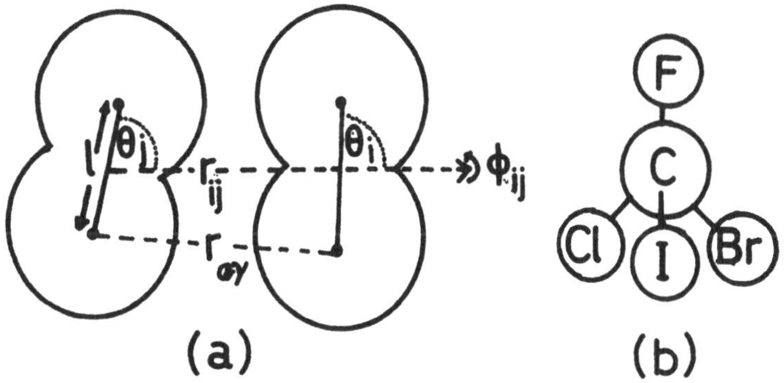

Figure 1: Interaction Site Models

In 1(a) the four site-site potentials between two chlorine molecules are equivalent and the four site-site distances, $\{r_{\alpha\gamma}\}$, and $\{r_{ij}\theta_i\theta_j\phi_{ij}\}$ are easily related. At the other extreme the potential for two molecules of the type shown in 1(b) consists of 25 interactions and 15 independent site-site potentials. A variety of ISM's have been studied by simulation: these include hard dumbells [11], soft-repulsive cores and LJ-sites [6, 12, 13] where

$$u_{\alpha\gamma}(r) = 4\varepsilon_{\alpha_i\gamma_j}\left((\sigma_{\alpha_i\gamma_j}/r)^{12} - (\sigma_{\alpha_i\gamma_j}/r)^{6}\right) . \tag{4}$$

Suitable energy and length parameters for LJ ISM's are given in table 1. $\sigma_{\alpha\gamma}$ decreases from left to right across a period and increases down the table. The well-depth, $\varepsilon_{\alpha\gamma}$, increases down the table, the larger dispersion energy corresponding to the large electron cloud. These parameters were not originally designed to be transferable, but they might constitute a good first guess in constructing a new core model. Interactions between unlike atoms are usually obtained from the Lorentz-Berthelot mixing rules.

$$\begin{aligned} \sigma_{\alpha_i\gamma_j} &= (\sigma_{\alpha_i\gamma_i} + \sigma_{\alpha_j\gamma_j})/2.0 \\ \varepsilon_{\alpha_i\gamma_j} &= (\varepsilon_{\alpha_i\gamma_i}\ \varepsilon_{\alpha_j\gamma_j})^{\frac{1}{2}} \end{aligned} \qquad (5)$$

There are a whole set of semi-empirical extensions of (5) which have yet to be tried in simulations [3].

			H[16]	He[3]
			8.6	10.2
			2.81	2.28
C[14]	N[7]	O[15]	F[13]	Ne[17]
51.2	37.3	61.6	52.8	47.0
3.35	3.31	2.95	2.83	2.72
		S[14]	Cl[13]	Ar[5]
		183.0	173.5	119.8
		3.52	3.35	3.41
			Br[13]	Kr[3]
			257.2	164.0
			3.54	3.83

Table 1. The upper number is $\varepsilon_{\alpha\gamma}/k$ in K, the lower is σ in Å. Several different parameter sets for the rare gases are available [3].

The interaction sites are often fixed at the position of the nuclei, but it is also possible to include additional sites in the middle of a bond to model the repulsion of the electron cloud [18]. As we tackle larger molecules it may be necessary to combine several atoms in an interaction group. In a recent simulation of a lipid bilayer [19], CH_3 and CH_2 units are represented as single sites and this idea has been applied to a simple six-site model of benzene [20]. There are complete sets of transferable potential parameters available for aromatic and aliphatic hydrocarbons [21] and for hydrogen-bonded fluids [22]. The William's potentials for hydrocarbons [21] have been used in a simulation of CH_4 [23], but the exponential repulsion is most profitably calculated using a table look-up.

ISM's lend themselves to the m.d. method because the total force and torque on a given molecule are readily calculated from the individual forces on the α sites of that molecule, $\mathbf{f}_{i\alpha}$

$$\mathbf{F}_i = \sum_{\alpha=1}^{M_i} \mathbf{f}_{i\alpha}$$
$$\boldsymbol{\Gamma}_i = \sum_{\alpha=1}^{M_i} \mathbf{h}_{i\alpha} \times \mathbf{f}_{i\alpha} \qquad (6)$$

where $\mathbf{h}_{i\alpha}$ is the bond vector from the centre of mass of i to the site α.

A working alternative to the ISM for the short-ranged interaction is the Gaussian overlap model [24]. The energy and length parameters are defined in terms of the overlap of Gaussian cores.

$$u(r_{ij}\Omega_i\Omega_j) = 4\varepsilon(\hat{\mathbf{h}}_i\hat{\mathbf{h}}_j)\left[\left(\frac{\sigma(\hat{\mathbf{h}}_i\hat{\mathbf{h}}_j,\hat{\mathbf{r}}_{ij})}{r_{ij}}\right)^{12} - \left(\frac{\sigma(\hat{\mathbf{h}}_i\hat{\mathbf{h}}_j,\hat{\mathbf{r}}_{ij})}{r_{ij}}\right)^{6}\right] \qquad (7)$$

where $\hat{\mathbf{h}}_i$ and $\hat{\mathbf{h}}_j$ are unit vectors along the molecular axes and

$$\varepsilon(\hat{\mathbf{h}}_i\hat{\mathbf{h}}_j) = \varepsilon_o[1 - \chi^2 (\hat{\mathbf{h}}_i\cdot\hat{\mathbf{h}}_j)^2]^{-\frac{1}{2}}$$

$$\sigma(\hat{\mathbf{h}}_i\hat{\mathbf{h}}_j\hat{\mathbf{r}}_{ij}) = \sigma_o\left[1 - \tfrac{1}{2}\chi\left\{\frac{(\hat{\mathbf{h}}_i\cdot\hat{\mathbf{r}}_{ij} + \hat{\mathbf{h}}_j\cdot\hat{\mathbf{r}}_{ij})^2}{1+\chi(\hat{\mathbf{h}}_i\cdot\hat{\mathbf{h}}_j)} + \frac{(\hat{\mathbf{h}}_i\cdot\hat{\mathbf{r}}_{ij} - \hat{\mathbf{h}}_j\cdot\hat{\mathbf{r}}_{ij})^2}{1-\chi(\hat{\mathbf{h}}_i\cdot\hat{\mathbf{h}}_j)}\right\}\right]^{-\frac{1}{2}} \qquad (8)$$

ε_o and σ_o are adjustable parameters and χ is a measure of the anisotropy, ($\chi > 0$: prolate). This model will handle large anisotropies without increasing the number of interaction sites. The original model has been extended to include an $\hat{\mathbf{r}}_{ij}$ dependent ε and can now be used to model a four-site linear chain [25]. The original model was used in the first molecular dynamics simulation of nematic ordering [26] and additional studies are in progress [27].

I.3 Point Multipoles

The interactions between molecules at large separations are conveniently represented by terms in the multipole expansion. For the interaction between two permanent dipoles we have

$$u(r_{ij}\Omega_i\Omega_j) = -\underset{\sim}{\mu}_i \cdot \underset{\approx}{T}^{(2)}(\underset{\sim}{r}_{ij}) \cdot \underset{\sim}{\mu}_j \tag{9}$$

where $\underset{\sim}{\mu}_i$ is the dipole moment of molecule i and T is the dipole interaction tensor

$$\underset{\approx}{T}^{(2)}(\underset{\sim}{r}_{ij}) = \frac{1}{4\pi\varepsilon_o r_{ij}^3}\left[\frac{3\underset{\sim}{r}_{ij}\underset{\sim}{r}_{ij}}{r_{ij}^2} - \underset{\approx}{I}\right] \tag{10}$$

The familiar expressions for other permanent electrostatic interactions in terms of T-tensors [28] and spherical harmonics can be found elsewhere [3]. Early simulation studies placed the permanent moments within a spherical LJ-core [29]. This generalized Stockmayer potential is now frequently used as a basic model to examine the effects of boundary conditions in simulations of polar fluids [30]. In the simulation of "realistic" liquids it is now more usual to graft the multipole moment onto an ISM core. The first simulations of this type of model involved a diatomic LJ-core and a permanent quadrupole to model N_2 and Br_2 [8, 12]. The point dipole interaction has also been used with a non-spherical core to model HF [31]. This particular simulation also included the quadrupole-dipole ($\theta\mu$) and quadrupole-quadrupole ($\theta\theta$) interactions. Point moments are readily included in an m.d. simulation. For linear molecules acting with a central anisotropic potential the force on molecule i is given by [32]

$$\underset{\sim}{F}_i = -\left\{\sum_j \hat{\underset{\sim}{r}}_{ij}\left[\frac{\partial u}{\partial r_{ij}} - \frac{\cos\theta_i}{r_{ij}}\frac{\partial u}{\partial\cos\theta_i} - \frac{\cos\theta_j}{r_{ij}}\frac{\partial u}{\partial\cos\theta_j}\right] + \left[\hat{\underset{\sim}{h}}_i\frac{\partial u}{\partial\cos\theta_i} + \hat{\underset{\sim}{h}}_j\frac{\partial u}{\partial\cos\theta_j}\right]\frac{1}{r_{ij}}\right\} \tag{11}$$

and the torque by,

$$\underset{\sim}{\Gamma}_i = -\hat{\underset{\sim}{h}}_i \times \left\{\sum_j + \hat{\underset{\sim}{r}}_{ij}\frac{\partial u}{\partial\cos\theta_i} + \hat{\underset{\sim}{h}}_j\frac{\partial u}{\partial\cos\psi}\right\}. \tag{12}$$

The appropriate angles are shown in fig. 1(a), $\cos\psi = \hat{h}_i . \hat{h}_j$ and the cap indicates a unit vector. The trigonometric functions required in the simulation are all calculated rapidly as vector products. In the case of non-linear molecules Stone [33] expands the potential in an orthonormal set of S-functions. General formulae are given for calculating the forces and torques obtained by differentiating these expansions.

In building simulation models it is natural to start by using the gas-phase values for the multipole moments. (The tables in Stogryn and Stogryn [34] are updated by an appendix in 'The Theory of Molecular Fluids' [35] which contains an extensive table of gas-phase dipoles, quadrupoles and polarizabilities gleaned from a variety of experimental and ab-initio techniques). Experience in simulating quadrupolar solids, N_2 [9], C_2H_2 [36], CO_2 [37] and alkali cyanide crystals [38] is that the best effective value of the quadrupole moment is 10-15% lower than the accepted gas-phase value if the higher order even multipoles are ignored. The addition of moments to ISM's often necessitates adjusting the original ε and σ values of the core. For N_2 these adjustments are small but in the case of CO_2 there is a substantial reduction in ε/k from 163.6K to 136.7K, as well as the 12% reduction in the size of the gas-phase quadrupole moment, (see model A1 of [37]). CO_2 is slightly unusual in that there is a balance between the LJ parameters and θ, and by lowering ε/k to 90.8K and raising σ to 3.529Å it is possible to use the gas-phase quadrupole, (model B of [37]). This is not true in the case of N_2 and the low effective quadrupole moment may be attributed to the neglect of higher order multipoles. We address this point in I.5.

I.4 Partial Charges

An alternative method of including electrostatic interactions is to distribute partial charges inside the core of a molecule. The hope is that by modelling the lowest non-zero moment correctly we may include some higher order moments. A simple partial charge model for N_2 is shown in figure 2(a).

The interaction between two N_2 molecules is

$$u(r_{ij}\Omega_i\Omega_j) = \sum_{\alpha=1}^{3} \sum_{\gamma=1}^{3} q_{i\alpha}\, q_{j\gamma}/4\pi\varepsilon_o\, r_{\alpha\gamma} \quad . \tag{13}$$

Since two of the charges are placed at the LJ centres, four of the nine site-site distances required in (13) are already evaluated for the underlying ISM potential. The ℓth moment of an N_2 molecule is given by

$$M_i^{\ell} = \sum_{\alpha=1}^{3} q_{i\alpha} (z_{i\alpha})^{\ell} \tag{14}$$

where $z_{i\alpha}$ is position of charge α along the bond. This simple model for N_2 predicts a hexadecapole which is six times too large and of the wrong sign [39]. Partial charges have been used to model the dipole in water [40]. The Rahman and Stillinger model shown in 2(b) predicts a dipole moment of 2.11D which is somewhat higher than the experimental value of 1.85D. A different tetrahedral arrangement of charges can be used to model the octopole moment of methane, see fig. 2(c) [41].

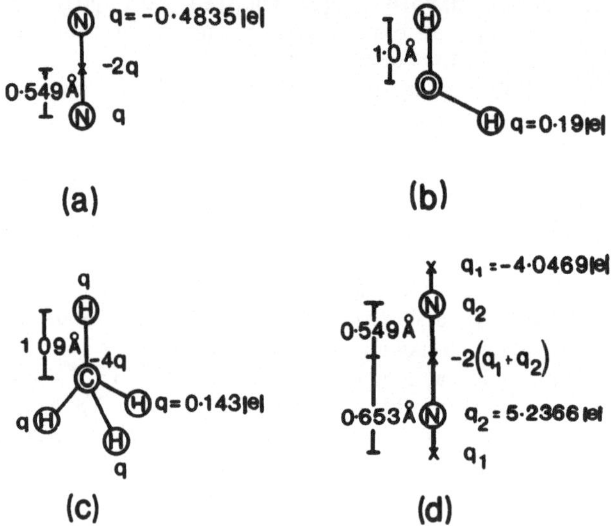

Figure 2(a) 3 charge model for N_2 ($\theta = -4.67\ 10^{-40}\ Cm^2$), b) The charge distribution for Rahman and Stillinger water ($\mu = 7.03 \times 16^{-30}\ Cm$), c) A five charge model for methane ($\Omega = 6.00 \times 10^{-50}\ Cm^3$). d) A five charge model for N_2 [48] ($\theta = -4.67 \times 10^{-40}\ Cm^2$, $\phi = -8.3 \times 10^{-60}\ Cm^4$). Note distances are in Å.

The charge distributions are normally fitted to experimental and ab-initio estimates of the lowest non-zero moment using generalizations of (14). Alternatively ab-initio results can be used to obtain direct estimates of the optimum charges on the atoms of a molecule. This method has been used recently in a simulation of CH_3CN [42]. However this method has the disadvantage that the position of the charges is completely fixed and cannot be varied by the simulator. Ab-initio studies which present tables of multipole moments are more flexible in designing models.

There are small but noticeable differences in potential curves from model 2(a) and a point quadrupole interaction [12]. Simulations of these hamiltonians produce small differences in liquid structure and thermodynamic results which can be rationalized in terms of the additional quadrupole-hexadecapole interaction for the charges [43]. Calculation of S(q, ω) for solid α-N_2 showed that the frequency of a particular librational phonon was insensitive to the method used to include the quadrupole-quadrupole interaction [45]. For the case where dipoles, quadrupoles and higher moments are all important comparison of the representations has not been made. A study of 2-d dipoles [44] reports different results for different representations at low temperatures.

Partial charges fit neatly into existing ISM codes, but the cpu times goes as α^2 where there are α sites. It pays to keep the charge distribution simple. Inclusion of a number of point multipoles is probably more cost-effective as α gets larger but this is a delicate balance since the algebraic expressions for higher point-moments are complicated.

I.5 Higher Moments

In small molecules such as HF and LiH higher order multipole moments are substantial [46]. This is important in simulation because the high moments certainly modify the liquid structure imposed by the lowest moments [47]. In a recent study of N_2 and CO_2 [48] models containing five partial charges have been used to mimic the higher order moments of the molecules accurately. (The 5q N_2 model is shown in 2(d)). This model predicts accurate lattice frequencies for α-N_2 solid and uses the gas-phase quadrupole moment. This supports the idea that the lower effective quadrupole moment in early models for the solid was a consequence of neglecting the higher moments. In CO_2 the higher moments change sign and tend to cancel, their inclusion is less important. In this case the reduction in θ can be balanced by changes in the core ε and σ [37]. This seems an important area for future study, which relies on accurate knowledge of higher moments. These will almost certainly come from SCF calculations with large basis sets and perturbation schemes for correlation. The 5q model for N_2 will take six-times longer to simulate than the basic N_2 ISM and the program modifications are minimal. It would be particularly interesting to see if the new model gives an improved agreement with the liquid state reorientational times [8]. Stone [49] has suggested modelling the higher moments by setting up charges, dipoles and quadrupoles at a number of points within the core. These distributed multipole expansions are more convergent than one-centre expansions, but the feasibility of including them in simulation has to be explored.

Apart from the slow convergence of the one-centre expansion there are regions of overlap where it is a poor representation of the interaction. Ng et al. discuss the effect of charge overlap on the quadrupole-quadrupole interaction [50]. Using their parameterization I have drawn the ISM + $\Theta\Theta$ interaction for two CO_2 molecules in the T-orientation, ($\theta_i = \pi/2$, $\theta_j = \phi_{ij} = 0$). Curve A represents,

$$u(r_{ij}\Omega_i\Omega_j) = \sum_{\alpha}\sum_{\gamma} u_{\alpha\gamma} + \sum_{m=-2}^{+2} \tilde{X}_{22}^{|m|} \left[\frac{4\pi}{5} Y_{2m}(\theta_i\phi_i)\, Y_{2-m}(\theta_j\phi_j)\right] \quad (15)$$

where

$$\tilde{X}_{22}^{|m|} = 4!\ \Theta^2 \Big/ \left[(2+|m|)!(2-|m|)!\ 4\pi\varepsilon_o r_{ij}^5\right] . \quad (16)$$

In curve B, $\tilde{X}_{22}^{|m|}$ is replaced by the appropriate coefficient including charge overlap. The vertical arrow represents the separation at which the ISM cores become repulsive. The difference in the potential curves A and B would be reflected in significant differences in local liquid structure of two fluids simulated with these potentials.

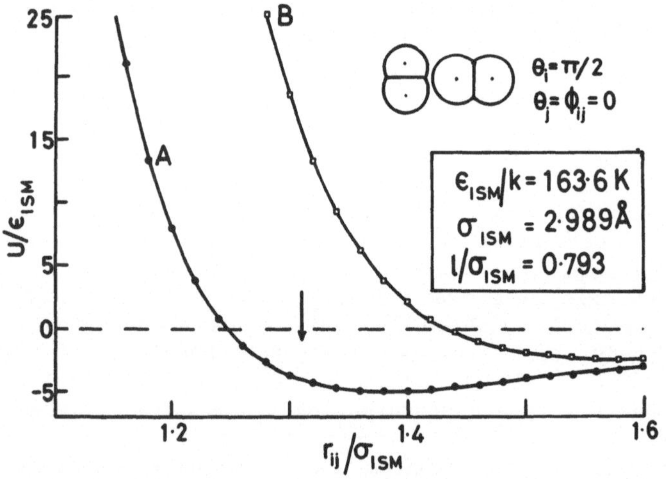

Figure 3. The effect of charge overlap on the $\Theta\Theta$ interaction for CO_2 in the T-orientation. The curves are calculated from [50].

The addition of multipoles to ISM cores affects liquid structure observed in simulations [8, 12, 42, 43, 44], because $X_{22}^{|m|}$ changes rapidly with distance in the overlap region. We should remain aware that the extrapolation of the multipole expansion to these regions is not well-founded.

I.6 Dispersion

The dispersion interaction in simulations of organic liquids is handled principally within the framework of the ISM core. The dispersion interaction for an ISM,

$$u_{dis}(r_{ij}\Omega_i\Omega_j) = \sum_{\alpha}\sum_{\gamma} - 4\varepsilon_{\alpha\gamma}(\sigma_{\alpha\gamma}/r)^6 , \qquad (17)$$

is anisotropic, has the correct r-dependence at long-range, but is not the exact result from quantum mechanical perturbation theory [51],

$$u_{dis}(r_{ij}\Omega_i\Omega_j) = -\frac{\overline{\Delta E}}{8} (\underline{\underline{\alpha}}_i.\underline{\underline{T}}_{ji}^{(2)}):(\underline{\underline{\alpha}}_j.\underline{\underline{T}}_{ji}^{(2)}) . \qquad (18)$$

$\overline{\Delta E}$ is the characteristic oscillator energy. This correct form of anisotropic dispersion has recently been used in a simulation of solid chlorine. An appropriate functional form for use in simulation studies is given in [52].

I.7 Non-Rigidity

In many systems of importance, hydrocarbon melts, polymers, lipid bilayers, internal rotation and vibration are important in determining equilibrium and dynamical properties. In the early simulation of N_2 and CO, atoms were attached by a stiff spring [53].

$$u(r) = \tfrac{1}{2}k\,(r-\bar{r})^2 \qquad (19)$$

where $\bar{r}$ is the equilibrium bond-length and r is the intra-molecular separation. Choice of a stiff spring constant appropriate to N_2 prevents energy transfer between the bond and the bath and the molecules are effectively rigid. The use of harmonic potentials in the simulation of sketal alkanes is common [54]. The spring constants are weaker and clear intramolecular distributions develop. The bond angles are controlled by potentials quadratic in the bond angle, with an equilibrium value of θ_0 = 113.3° and k = 1.3 x 10^5 J $mole^{-1}$. The simulation of these models is particularly simple. Atoms are treated independently and Newton's equations are solved using a leap-frog algorithm with a short time step, (Δt (H_2O) = 2.5 x 10^{-16}s [55], Δt (C_8H_{18}) = 2 x 10^{-15}s [54[).

In simulations of dihedral rotations, atoms are moved independently and bond-lengths and angles can be fixed at their equilibrium values by inverting the constraint matrix directly or by using the SHAKE algorithm. Chains adopt a particular dihedral angle according to the dihedral potential, $v(\phi)$. A typical potential for a hydrocarbon was first suggested by Ryckaert and Bellemans,

$$v(\phi) = \sum_{i=o}^{5} c_i (\cos\phi)^i \tag{20}$$

the coefficients are given in [56].

Momenta which are conjugate to constrained variables in a chain are zero and this introduces an extra determinental factor into the ensemble averages [57]. This factor can be included as an extra term in the hamiltonian appropriate to the unconstrained variables, but the correction is tedious to implement for long chains. Simulations without this correction can produce artificial structure in dihedral distributions because the coupling between bond-angle vibrations and dihedral rotations is suppressed. This artifact can be removed by relaxing the bond angle constraint using a quadratic bond - angle potential. Bond-length vibrations are at frequencies well separated from conformational rotations and they can be constrained without the metric correction.

Little attention has been given to including vibration directly in the simulation of organic liquids. It is essentially a quantum mechanical phenomenon which cannot be described using classical equations of motion. Recently Herman and Berne [58] have described a method of including quantum mechanical vibration in the adiabatic approximation.

II. Inclusion of Long-Range Forces

Once a suitable model is established, the simulation is performed and the valuable information is stored on tape. Most of the computer time in a simulation is spent evaluating the forces and torques acting on a given molecule. For LJ ISM's only interactions between near neighbours are considered; the force is cut-off in the range 2.5 - 3.0σ. This procedure is satisfactory for short range potentials, ($u \propto r^{-d}$, where d > the dimensionality), but is a serious approximation for the dipole-dipole interaction. For two dipoles the range of the interaction is much greater than the length of the simulation cell. There are a number of methods of including long-range interactions which we consider in this section.

II.1 Nearest Image Convention

The nearest image convention, (in which I include spherical

cut-off), amounts to ignoring the long-range part of the interaction. Only interactions inside the central cube are considered. This has been studied in the simulation of model dipolar fluids [30, 59] and realistic molecular fluids [31]. The consensus is that thermodynamic properties such as configurational energy and pressure are calculated accurately in this approximation. The single particle orientational correlation functions are not sensitive to the truncation.

A problem with the method is that the periodic boundary condition distorts the orientational space [30]. In fig. 4(a), molecule B' leaves the cell and enters as its image B". These two molecules have a different relative orientation to molecule A. Even for a moderate dipole (μ = 1.5D), this causes a rise in energy of 1.5% over 5000 timesteps. This phenomenon occurs for all anisotropic potentials but is most pronounced for the dipole-dipole

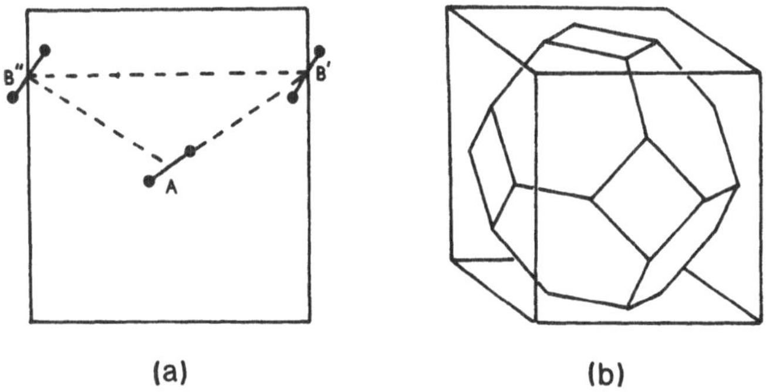

Figure 4(a) The nearest-image convention and relative orientations. (b) The truncated octahedron for simulating molecular fluids [62].

interaction. Adams et al. [30] suggest correcting for this by tapering the electric field of dipole, using a function f(r),

$$\underset{\sim}{E}^{mod}(\underset{\sim}{r}) = f(r)\, \underset{\sim}{E}(\underset{\sim}{r})$$

where

$$f(r) = \begin{cases} 1.0 & r < r_L \\ \exp(-a(r_L-r)^2) & r_L \leq r \leq r_T \\ 0.0 & r > r_T \end{cases} \qquad (21)$$

r_T is the truncation distance, r_L/r_T = 0.8 and $f(r_T)$ = 0.1. This slight modification is sufficient to defeat the energy pumping for moderate dipoles. Although nearest-image and spherical cutoff are suitable for simple properties they are not recommended for collective correlation functions, static dielectric properties and ℓ = 1 or 2 harmonic coefficients of $g(r_{ij}\Omega_i\Omega_j)$.

An interesting extension is in the use of non-cubic boxes [60, 61]. The truncated octahedron shown in 4(b) increases the distance for calculating distribution functions by 10% and provides a quasi-spherical environment for the isotropic fluid. The minimum image and periodic boundary conditions for this box are straightforward and readily vectorisable [62].

II.2 The Reaction Field

In this method the field on a dipole in the simulation consists of two parts: the first a short-range contribution from molecules situated within a truncation sphere; the second from molecules outside the sphere which are considered to form a dielectric continuum (ε') producing an Onsager reaction field at the centre of the cavity.

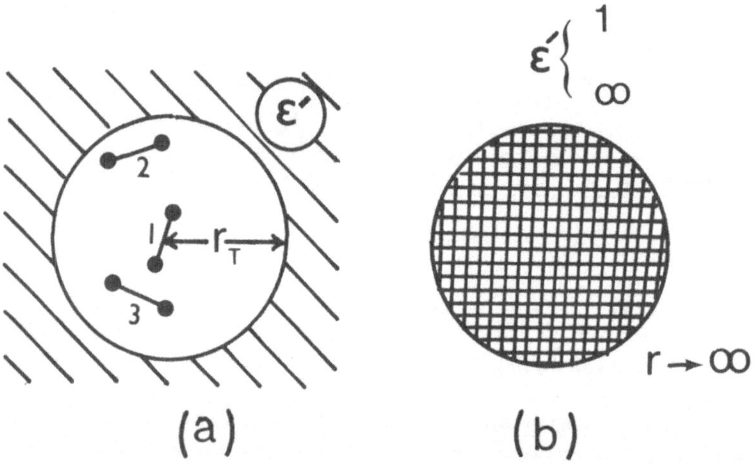

Figure 5(a) A cavity and reaction field. Molecule 2 and 3 interact directly with 1, the continuum produces a reaction field at 1. (b) The geometry of the Ewald sum.

The size of the reaction field is proportional to the moment of the cavity

$$\underset{\sim}{E}_R = \left(\frac{2(\varepsilon' - 1)}{2\varepsilon' + 1} \right) \frac{\underset{\sim}{M}}{r_T^{\,3}} \tag{22}$$

where $\underset{\sim}{M} = \sum_{i=1}^{N'} \underset{\sim}{\mu}$, ($N'$ molecules in the cavity), ε' is the dielectric constant of the continuum and r_T is the radius of the truncation sphere. Barker and Watts [63] first used this field in an M.C. study of water. It has not been used extensively in the simulation of realistic fluids because it assumes an a priori knowledge of ε'. The technique also produces more pronounced energy pumping than the nearest-image method. This can be cured by applying the normal tapering function to dipoles in the cavity and an additional tapering to the reaction field. This is achieved by modifying the total moment of the cavity [30].

$$\underset{\sim}{M}_{mod} = \sum_{i=1}^{N} \underset{\sim}{\mu}_i \, f(r_i) \tag{23}$$

where the sum is over all dipoles in the cube. The truncation radius used in (22) is also modified,

$$r_T^{\,3} = 3\int_0^\infty r^2 \, f(r) \, dr.$$

In an m.d. simulation, the reaction-field would not respond immediately to the reorientation of molecules in the central cavity. Efforts to implement a delayed reaction field have been made in a water simulation [64]. The method requires a complete knowledge of the dielectric response function, rather than a guess for ε', and there would be a dielectric loss as the reaction field does work on the central dipole [30].

The appropriate formula for calculating the dielectric constant for the R.F. is due to Kirkwood,

$$\frac{(2\varepsilon + 1)(\varepsilon - 1)}{9\varepsilon} = yg\,(\varepsilon') \tag{24}$$

where $y = 4\pi\mu^2\rho/9kT$ and

$$g(\varepsilon') = \frac{\langle |M|^2 \rangle - \langle M \rangle^2}{N\mu^2}$$

The g-factor depends on the dielectric of the continuum. When tapering is employed the moment is calculated in a sphere of radius r_L. We note there are problems in calculating correlation functions with an un-tapered reaction field, because of small discontinuities in these functions at the boundary. In addition collective correlation functions calculated using a reaction field differ from those evaluated using the full Ewald-Kornfeld sum.

In terms of implementation, the static reaction field is straightforward to include in a conventional program and involves only a modest increase in execution speed. The reaction field emphasises the continuum nature of fluids at long range, in contrast to the next technique which emphasises their lattice structure.

II.3 Ewald Sums

In this technique the nearest image convention is abandoned in favour of including all periodic images out to infinity. The simulation hamiltonian is,

$$H = \tfrac{1}{2} \sum_{n}{}' \sum_{i=1}^{N} \sum_{j=1}^{N} u\,(\underset{\sim}{r}_{ij} + \underset{\sim}{n},\ \Omega_i \Omega_j) \tag{26}$$

where the sum n is over all simple cubic lattice points with integer coordinates, $\underset{\sim}{n} = L\,(1,\ m,\ n)$. (The prime denotes terms i = j must be omitted for $\underset{\sim}{n} = o$). We are particularly concerned with dipole-dipole interactions where (26) is only conditionally convergent [65]. In this case the Ewald-Kornfeld representation of (26) gives

$$H = \tfrac{1}{2} \sum_{i=1}^{N} \sum_{j=1}^{N} \frac{1}{V} \left\{ \frac{4\pi}{3}\, \underset{\sim}{\mu}_i \cdot \underset{\sim}{\mu}_j - (\underset{\sim}{\mu}_j \cdot \underset{\sim}{\nabla})(\underset{\sim}{\mu}_j \cdot \underset{\sim}{\nabla})\ \Psi\,(r_{ij}/L;\ \tfrac{1}{2}) \right\} \tag{27}$$

where V is the volume of the cube, the gradients are with respect to $\underset{\sim}{r}_{ij}/L$ and Ψ (ρ, s) is an integral of a product of Jacobi theta polynomials, (for an explicit expression see (5) of [66]). Each theta function is evaluated using an infinite but rapidly convergent sum. The complicated expression for Ψ, has deterred all but the most determined from using the Ewald-Kornfeld representation [30, 65] and has prompted others to look for alternative methods for handling (26) [67].

If we choose to model the dipole moment by partial charges then we can use the more usual form of the Ewald sum, generally associated with ionic fluids. In the case where molecule i has n_{pi} partial charges, $q_{i\alpha}$ $(1 \leq \alpha \leq n_{pi})$, the potential at site r_i^α is given by

$$V_{i\alpha} = \sum_{j=1}^{N} \sum_{\gamma=1}^{n_{pj}} \sum_{\underset{\sim}{n}}{}' q_{j\gamma} \frac{\operatorname{erfc}(a|\underset{\sim}{r}_{ij}^{\alpha\gamma} + \underset{\sim}{n}|)}{|\underset{\sim}{r}_{ij}^{\alpha\gamma} + \underset{\sim}{n}|} +$$

$$+ \frac{1}{\pi L^3} \sum_{j=1}^{N} \sum_{\gamma=1}^{n_{pj}} \sum_{\underset{\sim}{h} \neq o} q_{j\gamma} \frac{\exp(-\pi^2 |\underset{\sim}{h}|^2 / a^2)}{|\underset{\sim}{h}|^2} \cos(2\pi \underset{\sim}{h} . \underset{\sim}{r}_{ij}^{\alpha\gamma}) - V'_{i\alpha} \quad (28)$$

where erfc(x) is the complementary error function, and h is the reciprocal lattice vector (1, m, n)/L. The first term in (28), the real space sum, is the potential due to the lattice of partial charges minus a diffuse Gaussian charge distribution of the same magnitude. The second term, the reciprocal lattice sum, adds a diffuse charge lattice of the opposite sign. Heyes [68] has recently pointed out that the third term $V'_{i\alpha}$ subtracts the normal Ewald self-term and an additional intramolecular contribution from sites in the same molecule;

$$V'_{i\alpha} = \frac{2aq_{i\alpha}}{\pi^{\frac{1}{2}}} + \sum_{\gamma \neq \alpha}^{n_{pi}} q_{i\gamma} \frac{\text{erf } (a|\underset{\sim}{r}_i^{\alpha\gamma}|)}{|\underset{\sim}{r}_i^{\alpha\gamma}|} \quad . \quad (29)$$

The parameter 'a' determines the width of the Gaussian distribution. If it is set to approximately 5/L, the complementary error function decays sufficiently rapidly to truncate the r-space sum at L/2. However the sharp Gaussian charge distribution can only be accurately represented by a larger number of reciprocal lattice vectors, (300 vectors in [65], but typically 150). The parameter 'a' may need to be adjusted for each simulation and is often chosen so that the thermodynamic properties are slowly varying with small changes in 'a'.

Recent theoretical studies [65, 69] have revealed the precise nature of the simulation when we employ the Ewald sum. Our small simulation box is reproduced throughout space to form an infinitely large sphere, surrounded by a dielectric continuum, (ε'), as shown schematically in 5(b). If $\varepsilon' = 1$, then the appropriate simulation hamiltonian is

$$H = \frac{1}{2} \sum_{i=1}^{N} \sum_{\alpha=1}^{n_{pi}} q_{i\alpha} V_{i\alpha} + \frac{2\pi}{3L^3} \left(\sum_{i=1}^{N} \sum_{\alpha=1}^{n_p} q_{i\alpha} \underset{\sim}{r}_{i\alpha} \right)^2 \quad (30)$$

The first term is the lattice sum effected by the Ewald formula (28), the second is a surface dipole term which depends on the symmetry of the infinite lattice and ε'. The last term in (30) can be set equal to zero if $\varepsilon' \to \infty$.

The appropriate formula for calculating the dielectric constant is given by [65]

$$\varepsilon = [2\varepsilon' + 1 + 6\varepsilon' yg(\varepsilon')]/[2\varepsilon' + 1 - 3yg(\varepsilon')] \tag{31}$$

where $g(\varepsilon')$ is calculated from the moment of the central cube. In summary the dielectric constant and thermodynamic properties do not depend on ε'. The g-factor and the simulation hamiltonian do. A sensible way to simulate ε is to use (30) without the surface term and put $\varepsilon' \to \infty$ in (31) to produce

$$\varepsilon = 1 + 3yg(\infty) \ . \tag{32}$$

There are clear advantages to the Ewald method. It is the technique which offers the most sensible route to the static and dynamic dielectric properties of polar fluids. It improves energy conservation, since there is no mechanism for energy pumping. There is a tendency to overemphasise the lattice nature of the liquid since the total dipole of the box is reproduced exactly throughout space. The method is expensive, using 100 wavevectors can increase the cpu time by 50%. Writing an efficient, vectorisable, Ewald code is not a task to be undertaken lightly. The CCP5 program library contains a version in the program MDMPOL [70].

III. Time Dependent Properties

III.1 Calculating Correlation Functions

Once the data from a simulation has been collected on tape, it can be analysed to determine the auto-correlation function (a.c.f), C(t), of some property $\underset{\sim}{A}_i$.

$$C(t) = \langle \underset{\sim}{A}_i(t) \cdot \underset{\sim}{A}_i(o) \rangle / \langle \underset{\sim}{A}_i(o)^2 \rangle \tag{32}$$

The subscript i refers to a particular particle and the brackets represent an ensemble average. To calculate many spectral properties in a fluid it is necessary to compute collective correlation functions,

$$C_c(t) = \left\langle \left(\sum_{i=1}^{N} \underset{\sim}{\mu}_i(t) \right) \cdot \left(\sum_{j=1}^{N} \underset{\sim}{\mu}_j(o) \right) \right\rangle \tag{33}$$

in this case $\underset{\sim}{\mu}_i$ represents the dipole moment of a molecule i.

The straightforward method of calculating an a.c.f. is shown in figure 6(a). The information at timestep 1 is read from the tape and stored. The information from timesteps 2 to 6 is read sequentially from the tape and correlated with 1, ($t_6 - t_1$ is the maximum time out to which we calculate C(t)). The tape is re-wound, we read in and store 2 and correlate this data with 3 through 7. In this way we move through the tape storing one step at a time. To calculate a velocity a.c.f. for N particles, we require only 3N words of memory and many a.c.f's can be computed simultaneously.

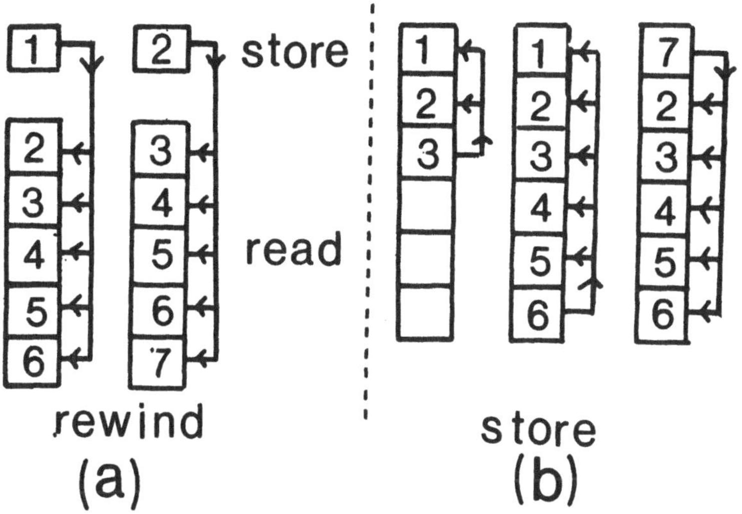

Figure 6. A schematic representation of the two methods of calculating time correlation functions.

The problem with method A is that the tape or disk rewinding is slow and this method of calculating correlation functions can take as long as the simulation. The second method involves reading t_{max} timesteps simultaneously into core, (t_{max} is the correlation length which is six in our example). As the data is read from tape to core, the last step is correlated with all the previously stored steps. When the core is full, the first step is overwritten with step $t_{max}+1$ and this is correlated with 2 through t_{max}. The overwriting continues until every step on the tape has been processed. There is no rewinding in this technique but storage requirements are high. Typically $t_{max}=150$ and the analysis of a single a.c.f. requires 450N words. This can be reduced by a factor ℓ, if we store every ℓ^{th} step as a time origin. Method (b) is the method of choice since the c.p.u. time for a single a.c.f. is under a minute. As storage requirements are high a.c.f.'s are calculated individually, but this restriction does not apply to a collective correlation function where only a single number is stored at each timestep.

III.2 Calculating Spectra

C(t) cannot be compared directly with experiment, but is normally

Fourier-transformed to produce a spectrum $S(\omega)$.

$$S(\omega) = (1/2\pi) \int_{-\infty}^{+\infty} C(t) \exp(i\omega t)\, dt \tag{34}$$

This transform is not straightforward, because the data, $C(t)$, can never be obtained over the whole range and it is not free from 'experimental' error. The noise in $S(\omega)$ can obscure interesting features present in $S_\infty(\omega)$. In particular the truncation in $C(t)$ causes spectral leakage which is often manifest as rapidly varying side lobes on a main peak and loss of resolution.

Windows are weighting functions applied to data to reduce the order of the discontinuity at the boundary of the periodic extension. Harris [71] has given a comprehensive review of the properties of over 30 windowing functions and we illustrate their properties with reference to a Blackman window, W_n.

$$W_n = 0.42 - 0.5 \cos \frac{2\pi(n + N - 2)}{2(N - 1)} + 0.08 \cos \frac{2\pi.2(n + N - 2)}{2(N - 1)} \tag{35}$$

$$n = 1, \ldots.N.$$

The N discrete values of the correlation function C_n are multiplied by W_n before taking the transform. Alternatively the Fourier transform of the windowing function, $\hat{W}$ is convoluted with $S(\omega)$ to produce the windowed spectrum $S_W(\omega)$

$$S_W(\omega) = \int_{-\infty}^{+\infty} S(\omega)\, \hat{W}\, (\omega-x)\, dx/2\pi \tag{36}$$

The coefficients in (35) are chosen so that $\hat{W}$ is sharply peaked, (leading to good resolution) and that the side lobes are reduced by a factor of 58 dB from those of a rectangular window. An example is shown in figure 7.

A model correlation function, C_n, is transformed

$$C_n = \exp(-0.1t) \cos 10t \qquad t = 0.1(n-1) \tag{37}$$

$$(n = 1, \ldots..35).$$

Application of the Blackman window removes the side lobes and gives a smooth spectrum with a peak at $10/2\pi$ Hz, but a lower intensity than obtained from the truncated data. Berens and Wilson [72] use a four-term Blackman-Harris window in computing the rotational spectra of CO and Ar from m.d. They note that by multiplying $S_W(\omega)$ by the inverse sum of the squares of the windowing function it is possible to correct the spectral band areas for the scaling effects of the windowing.

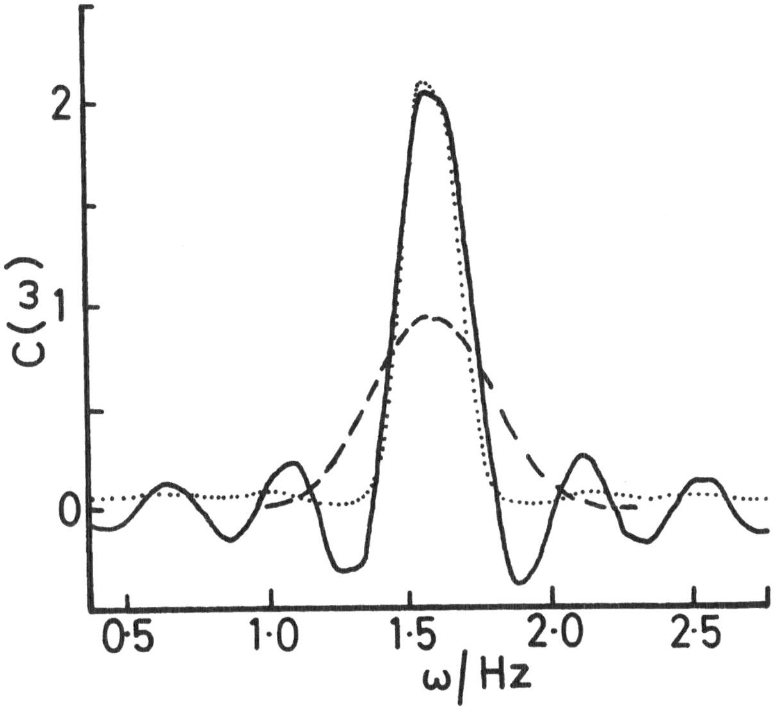

Figure 7. The spectrum obtained by Fourier transforming (37), – truncated data, – – – Blackman window, ... maximum entropy.

The maximum entropy method is a technique for computing the most uniform spectrum consistent with a set of data. We follow Gull and Daniells [73] and imagine a team of monkeys producing an enormous number of random spectra. If the transform of a particular spectrum is consistent with a measured C , it is sorted into a pile, different spectra in different piles. After every compatible spectrum has been sorted many times, the most likely $S(\omega)$ is found in the largest pile.

If we assume that C_n has Gaussian errors, with individual standard deviations σ_n, then the statistic $\sum_n |\hat{S}_n - C_n|^2/\sigma_n^2$ has a χ^2 distribution. $\hat{S}_n$ is the Fourier transform of the discrete trial spectrum S_j. The most probable fit to C_n is obtained by maximising

$$- \sum_j S_j \ln S_j - \frac{\lambda}{2} \sum_n | \hat{S}_n - C_n |^2 / \sigma_n^2 , \tag{38}$$

where λ is a Lagrange multiplier which constrains χ^2 to be equal the number of data points N. The first term in (38) is the entropy. We obtain

$$S_j = \exp \left[-1 + \lambda \sum_n \left\{ (C_n - \hat{S}_n)/\sigma_n{}^2 \right\} \exp(2\pi inj/N)\right] \tag{39}$$

which can be solved iteratively. For a particular λ, we begin with a uniform S_j transform to produce $\hat{S}_n$ and use (39) to recalculate S_j. The iterative sequence is completed for a number of λ's until we obtain a consistent S_j which gives $\chi^2 = N$. In practice the solution is unchanged for a large range in λ. The maximum entropy transform of (37) is shown in figure (7), where the method does an excellent job on smoothing the transform of the truncated data and maintains the peak height.

The difficulty in applying the maximum entropy method is in obtaining accurate values for σ_n. There is a small systematic error in computing C_n since there are less data points correlated to give a large value of n than for a small value of n. There is an error discussed by Zwanzig and Ailawadi due to replacing an ensemble average by a finite time average [74]. For a single particle, Gaussian a.c.f. calculated from N particle trajectories, the error Δ can be estimated as

$$\Delta = \pm \left(\frac{2\tau}{TN}\right)^{\frac{1}{2}} \tag{40}$$

where T is the total length of the run and τ can be thought of as the correlation time of the a.c.f. The factor $N^{-\frac{1}{2}}$ cannot be included for a collective correlation function [75] which may explain the substantial noise in those functions calculated by m.d.. Finally there is an error due to the cyclic boundary conditions. A density fluctuation moving at a speed of sound v across a box L will give the molecules an additional kick at a time L/v. This can be manifest as a sudden increase in the envelope of the correlation function and C_n values for larger times are meaningless. Presently, the only sure way to estimate the error in the correlation function is to repeat the simulation, (preferably with a different box-size). The expense of computing a reasonable σ detracts from the use of the M.E.M.

III.3 Diffusion Coefficient

The diffusion coefficient for a molecule, mass m, can be calculated from the area under the velocity a.c.f.

$$D = (k_B T/m) \int_0^\infty C_v(\tau)\, d\tau \qquad (41)$$

where C_v is defined in (32) with $\underset{\sim}{A}_i$ replaced by $\underset{\sim}{v}_i$, the c.o.m. velocity. Alternatively D can be calculated by monitoring the mean-squared displacement of molecules.

$$\lim_{t \to \infty} \left[\frac{\partial}{\partial t} \quad \Delta \; R_i^2 \; (t) \right] = 6D. \qquad (42)$$

Although the two routes are formally equivalent they do not appear to give the same results for molecular fluids [14, 76]. For CS_2 (41) gives values which are ~10% higher than (42). This can be attributed to a small finite negative tail in $C_v(t)$ which is lost in the noise at large t.

This discrepancy is illustrated most effectively by resolving D into its components parallel and perpendicular to the initial axis direction of a given molecule

$$C_v^{\parallel} = \frac{\langle(\underset{\sim}{v}_i(o) \cdot \hat{\underset{\sim}{h}}_i(o))(\underset{\sim}{v}_i(t) \cdot \hat{\underset{\sim}{h}}_i(o))\rangle}{\langle(\underset{\sim}{v}_i(o) \cdot \hat{\underset{\sim}{h}}_i(o))^2\rangle} \qquad (43)$$

and

$$C_v^{\perp} = \frac{\langle(\underset{\sim}{v}_i(o) \times \hat{\underset{\sim}{h}}_i(o)) \cdot (\underset{\sim}{v}_i(t) \times \hat{\underset{\sim}{h}}_i(o))\rangle}{\langle(\underset{\sim}{v}_i(o) \times \hat{\underset{\sim}{h}}_i(o))^2\rangle} \qquad (44)$$

these are shown for orthobaric CS_2 at 192K in figure 7.

By integrating under (43) and (44) we might conclude that $D_\parallel / D_\perp \sim 2$ which makes no sense for an isotropic fluid. In figure 8 we show the mean squared displacements parallel and perpendicular to the initial axis. The slopes become equal at surprisingly long times indicating that $D_\parallel = D_\perp$, $(D = 2D_\perp + D_\parallel)$. This anisotropy in the diffusion at short times must surely affect the spectroscopic measurement of D. It also warns us about using the integrated definition for the diffusion coefficient. Reasonable agreement between the experimental and simulated values of D have been obtained for N_2 [6, 8], the halogens [13], CS_2 [14] and CH_3CN [42]. Simulation studies of the halo-methanes have concentrated on reorientational motion and have not considered diffusion.

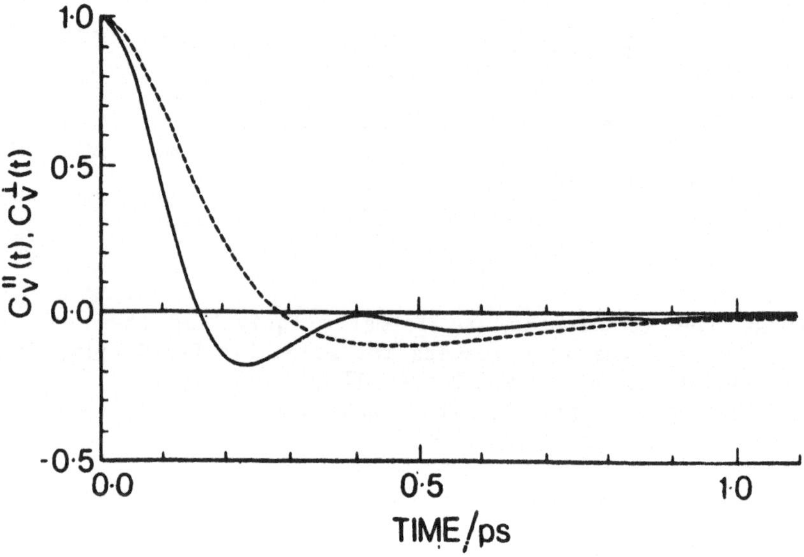

Figure 7. The velocity a.c.f. resolved parallel (....) and perpendicular (——) to the initial direction for liquid CS_2 at 192K.

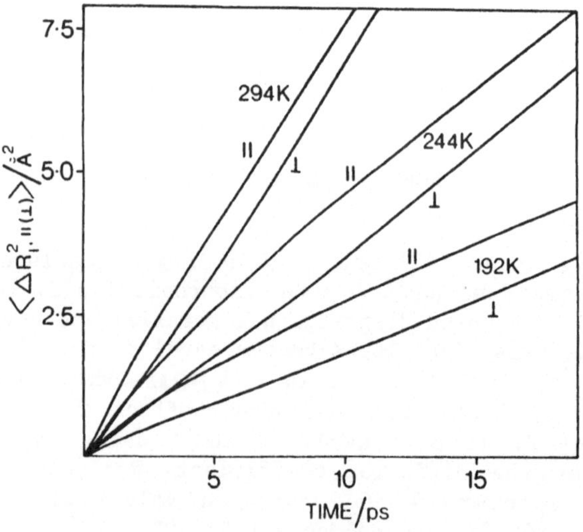

Figure 8. The mean-squared component of the molecular displacement with respect to the initial frame as a function of time [14]

III.4 Orientational Correlation Functions

The orientational correlation functions $C_\ell(t)$ are defined as

$$C_\ell(t) = \langle P_\ell\ (\hat{\underset{\sim}{h}}_i(t)\ .\ \hat{\underset{\sim}{h}}_i(o))\rangle \tag{45}$$

where P_ℓ is the ℓth Legendre polynomial and $\hat{\underset{\sim}{h}}_i$ is a unit vector fixed in the molecule. For a linear molecule, the unit vector is chosen to lie along the bond. For a non-linear molecule there are a number of $C_\ell(t)$'s corresponding to different vectors in the molecule, i.e. for CH_3Cl we might study the dipole vector and one orthogonal vector to comment on the anisotropy of rotational diffusion [77].

In figure 9 we have plotted $[\ell(\ell+1)]^{-1}\ln C_\ell(t)$ as a function of time for CS_2 at 192K. At short times the curves will be quadratic and coincident for all ℓ. If the Debye model for reorientation is obeyed the curves will be linear and coincident at long times. The curves are almost linear for t > 1.5ps but are not coincident for all ℓ. The curves contain a point of inflexion at the time when $C_\omega(t)$, (the angular velocity a.c.f.) passes through zero. The fact that $C_\ell(t)$ lies above its assymptotic exponential value means that the reorientational correlation times obtained by integrating under $C_\ell(t)$ are shorter than those obtained from the limiting slopes of the functions.

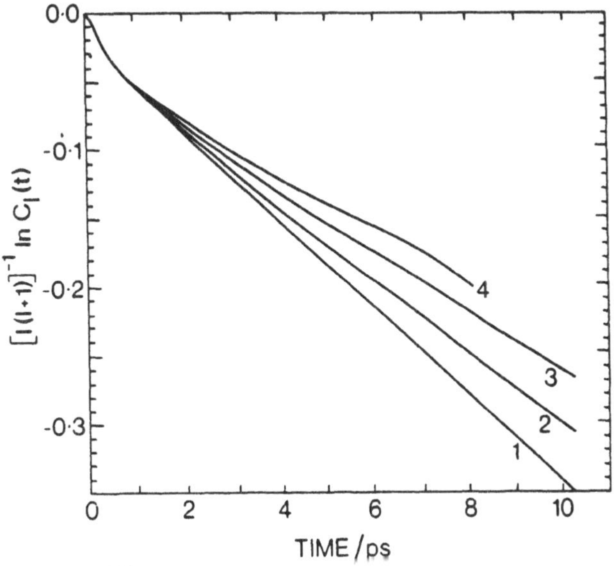

Figure 9. $[\ell(\ell+1)]^{-1}\ln C_\ell(t)$ as a function of time for CS_2 at 192K [14].

In CS_2 τ_2 can be measured by ^{13}C n.m.r. relaxation measurements and from the Raman line shape [78]. The n.m.r. measurements correspond to the value of τ_2 obtained by integration and the Raman measurements to the value of τ_2 from the slope of $C_2(t)$. As we expect the simulated and experimental values obtained from the slopes are ~ 10% larger than the values obtained by integrating under $C_2(t)$. For CS_2 the agreement between simulation and experiment is good at room temperature, but the simulated values are ~ 15% too high at the triple point.

In the same way that Raman and n.m.r. measurements give τ_2 for a molecule in a fluid, the infra-red rotational vibrational band-shape will give τ_1 for dipolar molecules. The broadening of the ν_1 (C-H stretch) in liquid chloroform leads to an orientational correlation function for the tumbling of the C_{3v} axis. Experimentally this correlation time is in the range 2.3 to 2.9 ps. The simulation of a five site model [77] gives τ_1^A = 3.6 ps. It is also possible to calculate τ_1^A for an axis perpendicular to the C-H bond, (this is a quantity which is not measured directly from experiment) and in the case of $CHCl_3$ the simulated τ_1^B = 3.9 ps. This indicates that the spinning around the C-H bond is slower than the tumbling of the C-H axis. The same anisotropy in the rotational diffusion is even more pronounced for bromoform [79]. The τ_1^A measured from the Raman C-H stretch in the range 2.0 ps to 5.1 ps, whereas τ_2^B from the Raman CBr_3 symmetric bend lies between 5.3 ps and 6.6 ps. The simulated values for τ_2^A and τ_2^B are 2.8 ps and 4.5 ps respectively. For CH_3CN [80] the correlation times are in the opposite sense. At 258K τ_2^A, from ^{14}N n.m.r. measurements, is 1.85 ps while the simulated value is 1.7 ps. The τ_2^B correlation time for the spinning of the hydrogens around the long-axis is twenty times faster and this reorientation is effectively two-dimensional.

An asymmetric top, such as acetone, has an independent reorientational time for each inertial axis. It has recently been modelled using a 4-site ISM with partial charges [81]. The simulated τ_1 times are A = 3.2 ps, B = 2.2 ps and C = 2.2 ps and the experimental values obtained from infra-red band shapes are 1.29, 1.01 and 1.11 ps (B is the dipolar axis and the infra-red bands are measured in dilute solution). This disagreement may be due to the ISM parameters, which were not adjusted to fit equilibrium properties. For such a strongly polar fluid the authors suggest that 108 particles may induce artificial ordering because of the small system. Finally we cannot be sure how modelling the CH_3 group as a single site will affect the dynamics. The accurate modelling of the a.c.f.'s of such a complicated organic liquid is at the frontier of simulation work and will probably require the use of 500 particle systems, an Ewald Sum, and preliminary fitting to equilibrium thermodynamic and structural data.

The fundamental link between a single particle reorientational time and the infra-red and Raman band shapes relies on a number of assumptions. Namely that there is no correlation between molecular orientation and vibrational normal coordinates. Coupling between different normal modes in the same molecule and normal modes on different molecules is neglected and there is no correlation between $\partial\mu_j/\partial q_\ell$ and q_ℓ [82]. These are all ideas which could be tested by simulation of the non-rigid models discussed in I.7.

Simulations of a number of molecular fluids have included studies of simple a.c.f.'s e.g. N_2 [6, 8], the halogens [86, 83], CO_2 [86], CS_2 [14], CH_3I [84], CH_2Cl_2 [85], $CHCl_3$ [77], $CHBr_3$ [78], CH_3CN [80] and CH_3CHO [81]. One of the limiting factors in this work is the accuracy of the experimentally determined τ_1 and τ_2 times. Since these times are simple to calculate in the simulation, they form a useful test of the modelling of the molecular dynamics. They are a sensible preliminary calculation to the computation of the more complicated spectral lineshapes. Where possible a variety of experimental techniques should be used to obtain these times with some estimate of the associated error.

III.5 Collective Correlation Functions

The far infra-red and Rayleigh spectra are related to the correlation function of the collective orientation density fluctuations

$$I_\ell(\omega) \propto \mathrm{Re} \int_0^\infty dt \exp(i\omega t) \lim_{\underset{\sim}{q}\to 0} G_{\ell,m}(q,t) \qquad (46)$$

where

$$G_{\ell,m}(q,t)=\left\langle \sum_i \sum_{j\neq i} [Y_{\ell,m}(\Omega_i)\exp(i\underset{\sim}{q}.\underset{\sim}{R}_i)](t)]Y_{\ell,m}(\Omega_j)\exp(-i\underset{\sim}{q}.\underset{\sim}{R}_j)^*(o)\right\rangle \qquad (47)$$

In this expression $Y_{\ell,m}(\Omega_i)$ is the ℓth order spherical harmonic of the orientation of molecule i. The vector $\underset{\sim}{q}$ is the scattering vector. For an isotropic fluid at low q the exponential factors are one and $G_{\ell,m}$ is independent of m. In computer simulations the periodic boundaries mean we study finite wave vectors of the form $\underset{\sim}{k} = (2\pi/L)(\ell,m,n)$ For a typical simulation box the smallest accessible wave vectors are three with magnitude $2\pi/L = k_1$, six of size $\sqrt{2}k_1 = k_2$ and four of size $\sqrt{3}k_1 = k_3$. The three values of $|\underset{\sim}{k}|$ are sufficiently low to compare with the limit as q tends to zero. In figure 10 we show $\bar{G}_2(\underset{\sim}{k}_i,t)$ for i = 1,2,3.

$$G_\ell(k_i,t) = [1/(N_i)(2\ell+1)] \sum_i \sum_m [G_{\ell,m}(\underset{\sim}{k}_i,t)/G_{\ell,m}(\underset{\sim}{k}_i,o)] \qquad (48)$$

where the sum over i runs over the N_i vectors of length k_i.

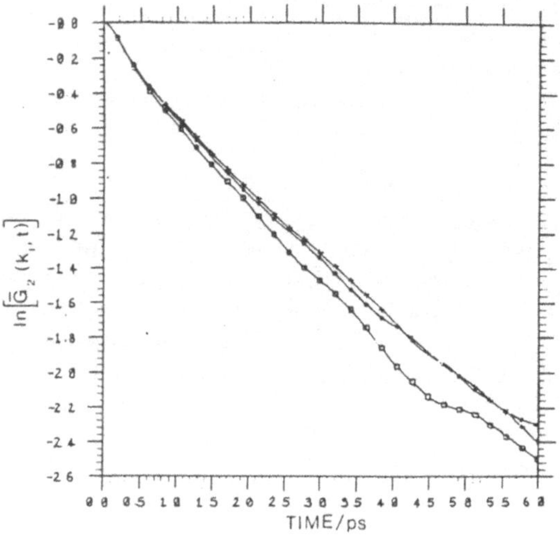

Figure 10. The collective orientational correlation at 244K calculated from (47). The functions are averaged over five values of m. The number indicates the index of $\underset{\sim}{k}_i$.
+++ i = 1, *** i = 2, □□□ i = 3.

The advantage of this technique is three-fold. First we perform a separate experiment for each value of m and k_i and the final averaging helps to compensate for the tremendous noise problem in these correlation functions [74, 75]. The correlation functions calculated for separate k_i and m values can be used to estimate the error which is useful in performing the transform using M.E.M.. Finally the exponential factors in $G_{\ell,m}$ reduce correlations at long range by a factor of $(\sin k_i R_{ij}/k_i R_{ij})$. Impey et al [87] have pointed out that for the static orientational correlation parameter, g_2, spurious correlations occur at the edge of the box because periodic images are perfectly correlated. This artifact will also affect the dynamical analogues of g_2 and g_1 given by (47). We note that the averaging over all m picks out the part of the correlation function which is symmetric with respect to the symmetry operations of the cube. This seems an appropriate quantity to compare with the isotropic fluid.

The collective correlation function ℓ=1 can be transformed to give the far infra red spectrum for a fluid of dipolar molecules [88]

$$A(\omega) = 2L\omega \tanh (\hbar\omega/2k_BT) D(\omega)/\varepsilon_o Vnc\hbar \quad (49)$$

where

$$D(\omega) = \frac{\mu^2}{3} \int_o^\infty \lim_{\underset{\sim}{k}_i \to o} \exp (i\omega t) \bar{G} (\underset{\sim}{k}_i, t) dt \quad (50)$$

n is the refractive index, V the volume of the fluid and L is a local field factor. We shall give an example of the use of (49) in the next section. Evans et al [77, 81, 85] have suggested computing the single particle part of $A(\omega)$ from the Fourier transform of $<\dot{\hat{\underset{\sim}{h}}}_A(t) \cdot \dot{\hat{\underset{\sim}{h}}}_A(o)>$ where $\dot{\underset{\sim}{h}}_A$ is the first derivative of the axial vector along the dipolar axis.

$$\dot{\hat{\underset{\sim}{h}}}_A = \underset{\sim}{\omega} \times \hat{\underset{\sim}{h}}_A \quad (51)$$

where $\underset{\sim}{\omega}$ is the angular velocity of a particular molecule. The spectrum obtained is shown for a five site model of CH_2Cl_2 in figure (11). The scatter in the points is due to difficulty in Fourier-transforming the correlation function due to a long-lived tail. When this is compared to experimental spectra obtained from a number of independent measurements the agreement is poor and the difference is almost certainly due to the cross-correlations which have been neglected.

Interestingly the simulated spectrum agrees well with an experimental spectrum of CH_2Cl_2 in CCl_4 (10%) indicating that in dilute solution the cross correlations may be unimportant.

The collective correlation functions used to obtain the simulated Rayleigh and far infra-red spectra require a substantial computational effort. A brute force attack on the problem requires runs of about 5×10^4 to 7.5×10^4 timesteps [86, 90]. We have described possible methods of improvement in this section.

III.6 Simulation of Collision Induced Effects

The first simulation work on the collision induced light scattering spectrum (CILS) of molecular fluids by Frenkel and McTague [90] calculated the effect of the DID on the Rayleigh and Raman spectra of diatomic molecules. A recent paper by Ladanyi [91] extends this work and we briefly outline the details of their calculations.

The polarizability of an assembly of N molecules is given by

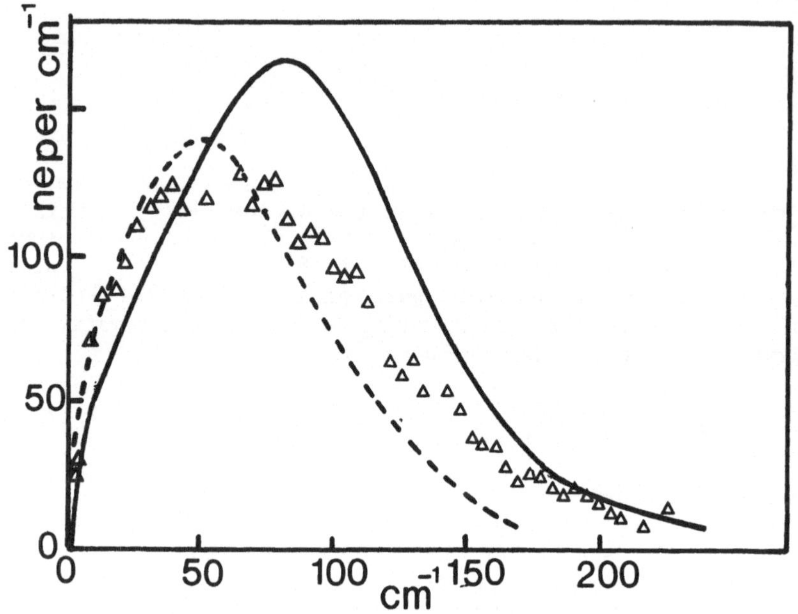

Figure 11. The far infra-red spectrum of CH_2Cl_2 [89] expt –; m.d. simulation single particle a.c.f. ΔΔΔ, expt spectrum 10% CH_2Cl_2 in CCl_4 (scaled) ----.

$$\underline{\underline{A}} = \sum_{i=1}^{N} \left(\underline{\underline{\alpha}}_i + \sum_{j\neq i}^{N} \underline{\underline{\alpha}}_i \cdot \underline{\underline{T}}^{(2)} \cdot \underline{\underline{\alpha}}_j \right) \tag{52}$$

where

$$\underline{\underline{T}}^n (\underset{\sim}{r}_{ij}) = (4\pi\varepsilon_o)^{-1} (\underset{\sim}{\nabla})^n (1/r_{ij}). \tag{53}$$

The total polarizability can be broken up into zero and second rank contributions

$$\begin{aligned} A_o &= \alpha N + \Delta A_o \\ \underline{\underline{A}}_2 &= \gamma \underline{\underline{Q}} + \Delta \underline{\underline{A}}_2 \end{aligned} \tag{54}$$

where α and γ are the trace and anisotropy of the polarizability tensor for a single molecule i and Q is a collective orientational variable

$$\underline{\underline{Q}} = \sum_{i=1}^{N} \frac{3}{2} \hat{h}_i \hat{h}_i - \frac{1}{2} \underline{\underline{I}} . \tag{55}$$

$$\Delta A_o = \alpha\gamma\phi_o^{(1)} + \gamma^2 \phi_o^{(2)} \tag{56a}$$

$$\Delta \underline{\underline{A}}_2 = \alpha^2 \phi_2^{(0)} + \alpha\gamma\phi_2^{(1)} + \gamma^2 \phi_2^{(2)} . \tag{56b}$$

Explicit expressions for $\phi_{0,2}^{(n)}$ are given as Cartesian tensors in [91] and in terms of Wigner rotational matrices in [90]. The former seem more appropriate for use in simulation. The isotropic (ℓ=o) CILS spectrum is the Fourier transform of the correlation function

$$C_{\ell=o}(t) = \langle \Delta A_o^{CI} (o) . \Delta A_o^{CI} (t) \rangle /N \tag{57}$$

where

$$\Delta A_o^{CI}(t) = \Delta A_o - \langle \Delta A_o \rangle . \tag{58}$$

ΔA_o is used to define an effective polarizability

$$\alpha_{eff} = \alpha + \langle \Delta A_o \rangle /N. \tag{59}$$

The anisotropic (ℓ=2) light scattering spectrum is slightly more complicated. The underlying correlation function contains a purely collective orientational part, a collision induced part, and a cross term.

$$C_{\ell=2}^{OR}(t) = (4\gamma_{eff}^2/9N) \left\langle Tr[\underline{\underline{Q}}(o) . \underline{\underline{Q}}(t)] \right\rangle$$

$$C_{\ell=2}^{CROSS}(t) = (2\gamma_{eff}/3N) \left\langle Tr[\underline{\underline{Q}}(o) . \Delta\underline{\underline{A}}_2^{CI}(t) + \underline{\underline{Q}}(t) . \Delta\underline{\underline{A}}_2^{CI}(o)] \right\rangle$$

$$C_{\ell=2}^{CI}(t) = (1/N) \left\langle Tr[\Delta A_2^{CI}(o) . \Delta A_2^{CI}(t)] \right\rangle \tag{60}$$

In (60)

$$\underline{\underline{A}}_2^{CI} = \Delta\underline{\underline{A}}_2 - \frac{2}{3} \Delta\gamma \underline{\underline{Q}} \tag{61}$$

when $\Delta\gamma$ has been chosen to project out of $\Delta\underline{\underline{A}}_2$ the part of the polarizability which depends on molecular orientation, i.e.

$$\frac{2}{3}\,\Delta\gamma = \frac{\langle \mathrm{Tr}\,(\underline{\underline{A}}_2\,(o)\,.\,\underline{\underline{Q}}(o))\rangle}{\langle \mathrm{Tr}\,(\underline{\underline{Q}}\,(o)\,.\,\underline{\underline{Q}}\,(o))\rangle} \quad . \tag{62}$$

$\Delta\gamma$ defines an effective polarizability anisotropy

$$\gamma_{eff} = \gamma + \Delta\gamma \quad . \tag{63}$$

In previous simulation studies of collision induced effects in rare gases, $\gamma=0$ and only the first term in (56b) contributes to $\Delta\underline{\underline{A}}_2$ [92]. The first and second rank collision induced correlation functions, (57) and (60), are notoriously difficult to calculate; they involve four point correlation functions, (i induced by j correlated with k induced by ℓ and the simulation runs in [90, 91, 92] are for tens of thousands of timesteps).

The main conclusions of these studies are:

a) α_{eff} and γ_{eff} increase with increasing temperature, decrease with increasing density and are more strongly density than temperature dependant. α_{eff} from the simulation agrees well with the experimental values from refractive index measurements,

$$(n^2 - 1)/(n^2 + 2) = 4\pi\rho\alpha_{eff}/3 \quad . \tag{64}$$

For CO_2 ($T^* = 1.61$, $\rho^* = 0.360$), $\alpha_{eff}/\alpha = 0.972$ from experiment, while simulation predicts 0.967 [89] and 0.973 [90].

b) The $2n^{th}$ spectral moment of depolarized scattering is given by

$$M_{2n} = (-1)^n \frac{\partial^{2n}}{\partial t^{2n}} \left[C^{OR}_{\ell=2}(t) + C^{CI}_{\ell=2}(t) + C^{CROSS}_{\ell=2}(t) \right]_{t=o} / C^{TOT}_{\ell=2}(o) \tag{65}$$

For $n \neq o$, the cross-correlation makes a substantial contribution to the spectral moment, and is in many cases the dominant collision induced effect.

c) For small molecules the timescale for the decay of the orientational and CI parts of the total correlation function are similar. For longer molecules $C^{OR}_{\ell=2}(t)$ is slower. The cross correlation function which starts at zero and is negative, shifts intensity into the wings of the spectrum. Indeed in this case the prospect of obtaining any clear separation of orientational and

CI effects in the Rayleigh and Raman spectra of small molecules is slim. For a molecule with a longer reorientational time such as CS_2, there is a clearer separation of time-scales. The three components for the $\ell=2$ Raman spectrum of the orthobaric fluid are shown in fig. 42. (The formalism for the DID Raman spectra is the same as for the Rayleigh spectrum (52), with one $\underline{\underline{\alpha}}$ replaced by $\partial\underline{\underline{\alpha}}/\partial q$).

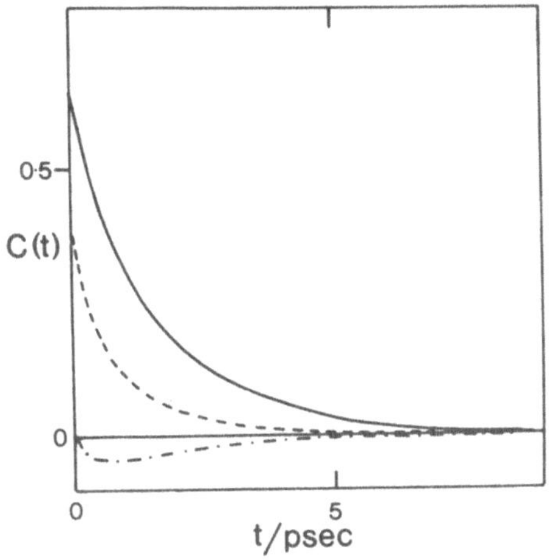

Figure 12. $C^{OR}_{\ell=2}(t)$ –, $C^{CROSS}_{\ell=2}(t)$ -.-.-., $C^{CI}_{\ell=2}(t)$ ... for a simulation of the second rank part of the Raman spectrum of CS_2. This simulation assumes a centre-centre $\underline{\underline{T}}^{(2)}_{ij}$.

Ladanyi has experimented with an alternative form of the collision induced polarizability. $\underline{\underline{T}}^{(2)}$ is replaced with $\underline{\underline{D}}^{(2)}$

$$\underline{\underline{D}}^{(2)}_{ij} = (1/4) \sum_{\alpha=1}^{2} \sum_{\gamma=1}^{2} \underline{\underline{T}}^{(2)} (r_{\alpha\gamma}) . \qquad (66)$$

(66) is suitable for a symmetric homonuclear diatomic. The most dramatic difference using the site model is that at high densities for the CO_2 model, $C^{CI}_{\ell=2}(t)$ and $C^{CROSS}_{\ell=2}(t)$ are much smaller relative to $C^{OR}_{\ell=2}(t)$ than for the corresponding centre-centre model. The site-site collision induced correlation functions decay more rapidly than the corresponding centre-centre functions because they have an additional rotational mechanism for relaxation. If (66) is closer to reality than (52) then for CO_2 the errors in measured relaxation times would be small. However it is worrying that the total line-shapes should depend so crucially on the detailed functional form of the DID

interaction.

The Rayleigh and Raman spectra discussed have backgrounds due to interaction induced effects. For some vibrations, (the u-symmetry modes in CS_2), the entire band is interaction induced. The DID mechanism cannot give rise to these Raman bands since $\partial\alpha/\partial q = 0$ (ν_2, ν_3). In a similar way the near infra-red spectrum of $\nu_1(\Sigma_g^+)$ band is interaction induced. We shall use this as an example of the calculation of the spectrum of a forbidden band. The induced dipole in i due to the quadrupole on j is given by

$$^{ij}m_\alpha = -(1/3)\,^{i}\alpha_{\alpha\beta}\, T^{(3)}_{\beta\gamma\delta}\,(\underset{\sim}{r}_{ij})\,^{j}\theta_{\gamma\delta} \tag{67}$$

where repeated Greek suffices are summed. We can break the polarizability and quadrupole into zero and second rank parts.

$$^{i}\alpha_{\alpha\beta} = \alpha\,^{i}S^{(o)}_{\alpha\beta} + (\gamma/3)\,S_{\alpha\beta}^{(2)} \tag{68}$$

$$^{i}\theta_{\alpha\beta} = (\theta/2)\,^{i}S^{(2)}_{\alpha\beta}$$

where

$$^{i}S^{(2)}_{\alpha\beta} = 3\hat{h}_{i\alpha}\hat{h}_{i\beta} - \delta_{\alpha\beta}$$

$$^{i}S^{(o)}_{\alpha\beta} = \delta_{\alpha\beta}\ . \tag{69}$$

Substituting (68) and (69) into (67) gives

$$^{ij}m_\alpha = (\alpha\theta/2)\,^{ij}\tilde{m}_\alpha^{(0,2)} + (\gamma\theta/6)\,^{ij}\tilde{m}_\alpha^{(2,2)} \tag{70}$$

where we use the short-hand

$$^{ij}\tilde{m}_\alpha^{(2,2)} = (-1/3)\,^{i}S_{\alpha\beta}^{(2)}\, T^{(3)}_{\beta\gamma\delta}\,(\underset{\sim}{r}_{ij})\,S_{\gamma\delta}^{(2)}.$$

We have now established our notation and can proceed to calculate infra-red spectrum associated with the vibrational coordinate q_1. The moment modulated by q_1 is given by

$$q_1{}^i \frac{\partial({}^{ij}m_\alpha + {}^{ji}m_\alpha)}{\partial q_1{}^i} = -(1/3)\left\{\left(\frac{\partial \underline{\underline{\alpha}}^i}{\partial q_1{}^i}\right)_{\alpha\beta} T_{\beta\gamma\delta}{}^{(3)}(\underset{\sim}{r}_{ij})\theta^j_{\gamma\delta}\right.$$

$$\left. + \alpha_{\alpha\beta}{}^j T_{\beta\gamma\delta}{}^{(3)}(\underset{\sim}{r}_{ji})\left(\frac{\partial \underline{\underline{\theta}}^i}{\partial q_1{}^i}\right)_{\gamma\delta}\right\} q_1{}^i. \quad (71)$$

We can sum (71) over all i and j≠i to give the total modulated moment and correlate this quantity at time t with the same value at time zero to obtain the Fourier transform of the ν_1 band. If we assume no coupling of ν_1 modes on neighbouring molecules and no vibrational dephasing we obtain,

$$B(\omega) = (2/3)L(\omega/\varepsilon_0\hbar nc)\tanh(\hbar\omega/2k_BT)\int_0^\infty dt \exp[i(\omega-\omega_1)t]$$

$$\times\left\{\sum_i\sum_{j,k}\left[\left(\frac{\alpha'\theta}{2}\right)^2\left\langle {}^{ij}_{\tilde{m}_\alpha}{}^{(0,2)}(t)\ {}^{ik}_{\tilde{m}_\alpha}{}^{(0,2)}\right\rangle\right.\right.$$

$$+\left(\frac{\gamma'\theta}{6}\right)^2\left\langle {}^{ij}_{\tilde{m}_\alpha}{}^{(2,2)}(t)\ {}^{ik}_{\tilde{m}_\alpha}{}^{(2,2)}\right\rangle$$

$$\left.+\left(\frac{\alpha'\gamma'\theta^2}{6}\right)\left\langle {}^{ij}_{\tilde{m}_\alpha}{}^{(0,2)}(t)\ {}^{ik}_{\tilde{m}_\alpha}{}^{(2,2)}\right\rangle\right]$$

$$+\left[\left(\frac{\alpha\theta'}{2}\right)^2\left\langle {}^{ji}_{\tilde{m}_\alpha}{}^{(0,2)}(t)\,{}^{ki}_{\tilde{m}_\alpha}{}^{(0,2)}\right\rangle\right.$$

$$+\left(\frac{\alpha\gamma(\theta')^2}{6}\right)\left\langle {}^{ji}_{\tilde{m}_\alpha}{}^{(0,2)}(t)\ {}^{ki}_{\tilde{m}_\alpha}{}^{(2,2)}\right\rangle$$

$$\left.+\left(\frac{\gamma\theta'}{6}\right)^2\left\langle {}^{ji}_{\tilde{m}_\alpha}{}^{(2,2)}(t)\ {}^{ki}_{\tilde{m}_\alpha}{}^{(2,2)}\right\rangle\right]$$

$$+\,2\left[\left(\frac{\alpha'\alpha\theta\theta'}{4}\right)\left\langle {}^{ij}_{\tilde{m}_\alpha}{}^{(0,2)}(t)\ {}^{ki}_{\tilde{m}_\alpha}{}^{(0,2)}\right\rangle\right. +$$

$$+ \frac{\alpha'\gamma\theta\theta'}{12} \left\langle {}^{ij}_{\tilde{m}_\alpha}{}^{(0,2)}(t)\ {}^{ki}_{\tilde{m}_\alpha}{}^{(2,2)} \right\rangle$$

$$+ \frac{\alpha\gamma'\theta\theta'}{12} \left\langle {}^{ij}_{\tilde{m}_\alpha}{}^{(2,2)}(t)\ {}^{ki}_{\tilde{m}_\alpha}{}^{(0,2)} \right\rangle$$

$$\left. \left. + \frac{\gamma\gamma'\theta\theta'}{36} \left\langle {}^{ij}_{\tilde{m}_\alpha}{}^{(2,2)}(t)\ {}^{ki}_{\tilde{m}_\alpha}{}^{(2,2)} \right\rangle \right] \right\} , \tag{72}$$

where α' is given by

$$\alpha' = (1/3) \left(\frac{\partial\alpha_{xx}}{\partial q_1^{\ i}} + \frac{\partial\alpha_{yy}}{\partial q_1^{\ i}} + \frac{\partial\alpha_{zz}}{\partial q_1^{\ i}} \right) \left\langle |q_1^{\ i}(0)|^2 \right\rangle^{1/2} ,$$

and γ' and θ' are similarly defined. We have calculated the ensemble averages in (72) to produce the spectrum shown in figure 13.

The amplitude of the simulated peaks was adjusted to match the experiment at the peak height. The experimental spectra are distorted on the low temperature side by hot bands and isotope features. All in all the agreement is excellent. We have also simulated the far infra-red and ν_3 Raman bands of CS_2. A detailed discussion of this work can be found in [93].

I should like to acknowledge useful discussions with Dr. David Adams, Dr. Paul Madden and Dr. S.F. O'Shea on various aspects of this work and to thank the authors of reference [48] for their preprint.

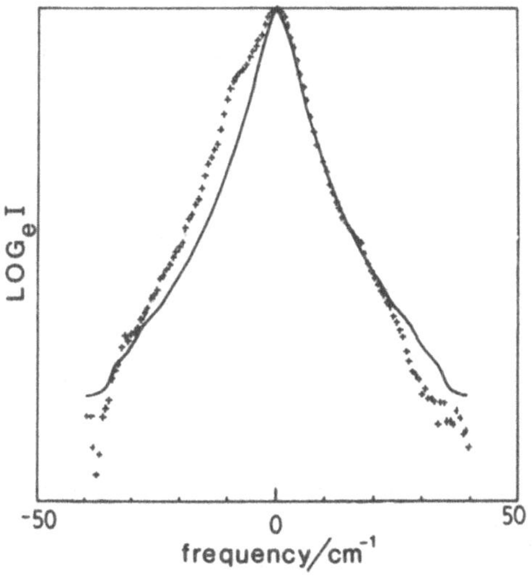

Figure 13. A comparison of simulated and experimental infrared line-shapes for orthobaric CS_2 at 297K. The crosses give the experimental points [1] and the full lines the results of the simulation.

References

[1] R.W. Impey, P.A. Madden and I.R. McDonald, Mol. Phys., 46 513, (1982).

[2] B. Quentrec, Phys. Rev. A12, 288, (1982).

[3] G.C. Maitland, M. Rigby, E.B. Smith and W.A. Wakeham, Intermolecular Forces, Oxford University Press, 1981.

[4] J.A. Barker, R.A. Fisher and R.O. Watts, Mol. Phys., 21, 657, (1971).

[5] L. Verlet, Phys. Rev., 159, 98, (1967).

[6] P.S.Y. Cheung and J.G. Powles, Mol. Phys., 30, 921, (1975).

[7] J.G. Powles and K.E. Gubbins, Chem. Phys. Lett., 38, 405, (1976).

[8] P.S.Y Cheung and J.G. Powles, Mol. Phys., 32, 1383, (1976).

[9] C.S. Murthy, K. Singer, M.L. Klein and I.R. McDonald, 41, 1387, (1980).

[10] C.G. Gray, Specialist Periodical Report: Statistical Mechanics, Volume 2, (Chem. Soc.). 300, (1975).

[11] W.B. Streett and D.J. Tildesley, Proc. Roy. Soc., A348, 485, (1976).

[12] W.B. Streett and D.J. Tildesley, Proc. Roy. Soc., A355, 239, (1977).

[13] K. Singer, A. Taylor and J.V.L. Singer, Mol. Phys., 33, 1757, (1977).

[14] D.J. Tildesley and P.A. Madden, Mol. Phys., 42, 1137, (1981).

[15] C.A. English and J.A. Venables, Proc. Roy. Soc., A340, 57, (1974).

[16] S. Murad and K.E. Gubbins, Computer Modelling of Matter, ed. P. Lykos, ACS Symposium Series, (1978).

[17] A.A. Clifford, P. Gray and N. Platts, J. Chem. Soc. Farad. Trans. 1, 73, 381, (1977).

[18] C.S. Hsu, D. Chandler and L.J. Lowden, Chem. Phys., 14, 213, (1976).

[19] P. van der Ploeg and H.J.C. Berendsen, J. Chem. Phys., 76, 3271, (1982).

[20] D.J. Evans and R.O. Watts, Mol. Phys., 32, 93, (1976).

[21] D.E. Williams, J. Chem. Phys., 47, 4680, (1967).

[22] W.L. Jorgensen, J. Amer. Chem. Soc., 103, 335, (1981).

[23] S. Murad, D.J. Evans, K.E. Gubbins, W.B. Streett and D.J. Tildesley, Mol. Phys., 37, 725, (1979).

[24] B.J. Berne and P. Perchukas, J. Chem. Phys., 56, 4213, (1972).

[25] J.G. Gay and B.J. Berne, J. Chem. Phys., 74, 3316, (1981).

[26] J. Kushick and B.J. Berne, J. Chem. Phys., 64, 1362, (1976).

[27] D.J. Adams, Southampton University U.K.; W.A. Steele, Pennsylvania State University, U.S.A.

[28] L. Jansen, Phys. Rev., 110, 661, (1958).

[29] S.S. Wang, C.G. Gray, P.A. Egelstaff and K.E. Gubbins, Chem. Phys. Lett., 21, 123, (1973).

[30] D.J. Adams, E.M. Adams and G.J. Hills, Mol. Phys., 38, 387, (1979).

[31] S. Murad, K.E. Gubbins and J.G. Powles, 40, 253, (1980).

[32] P.S.Y Cheung, Chem. Phys. Lett., 40, 19, (1976).

[33] A.J. Stone, Mol. Phys., 36, 241, (1978).

[34] D.E. Stogryn and A.P. Stogryn, Mol. Phys., 11, 371, (1966).

[35] C.G. Gray and K.E. Gubbins, Theory of Molecular Fluids, (1983), (OUP), in the press.

[36] Z. Gamba and H. Bonadeo, J. Chem. Phys., 76, 6215, (1982).

[37] C.S. Murthy, K. Singer and I.R. McDonald, Mol. Phys., 44, 135, (1981).

[38] M.L. Klein and I.R. McDonald, J. Chem. Phys., to appear.

[39] R.D. Amos, Mol. Phys., 39, 1, (1980).

[40] A. Rahman and F.H. Stillinger, J. Chem. Phys., 55, 3336, (1971).

[41] R. Righini, K. Maki and M.L. Klein, Chem. Phys. Lett., 80, 301, (1981).

[42] H.J. Böhm, I.R. McDonald and P.A. Madden, Mol. Phys., (in press), (1983).

[43] P.A. Monson, W.A. Steele and W.B. Streett, J. Chem. Phys., 78, 4126, (1983).

[44] R. Occelli, B. Quentrec and C. Brot, Mol. Phys., 36, 257, (1978).

[45] C.S. Murthy, K. Singer, M.L. Klein and I.R. McDonald, Mol. Phys., 40, 1517, (1980).

[46] J.T. Brobjer and J.N. Murrell, Chem. Phys. Lett., 77, 601, (1981).

[47] G.N. Patey and J.P. Valleau, J. Chem. Phys., 64, 170, (1976).

[48] C.S. Murthy, S.F. O'Shea and I.R. McDonald, (1983), Preprint.

[49] A.J. Stone, Chem. Phys. Letts. 83, 233, (1981).

[50] K.C. Ng, W.J. Meath and A.R. Allnatt, Mol. Phys., 33, 699, (1977).

[51] J. de Boer, Physica, 9, 363, (1942).

[52] E. Burgos, C.S. Murthy and R. Righini, Mol. Phys., 47, 1391, (1982).

[53] B.J. Berne and G.D. Harp, Adv. in Chem. Phys., 17, 63, (1979).

[54] T.A. Weber, J. Chem. Phys., 70, 4277, (1978) and references therein.

[55] F.H. Stillinger and A. Rahman, J. Chem. Phys., 68 666, (1978).

[56] J.P. Ryckaert and A. Bellemans, Faraday Disc., 66, 95, (1978).

[57] W.F. van Gunsteren, Mol. Phys., 40, 1015, (1980).

[58] M.F. Herman and B.J. Berne, J. Chem. Phys., 78, 4103, (1983).

[59] D. Levesque, G.N. Patey and J.J. Weiss, Mol. Phys., 34, 1077, (1977).

[60] S.S. Wang and J.A. Krumhansl, J. Chem. Phys., 56, 4287, (1977).

[61] D. Adams, Chem. Phys. Letts., 62, 329, (1979).

[62] D. Adams, NRCC Workshop, 9, The problem of long range forces in computer simulation of condensed media, (ed. D. Ceperley), 13, (1980).

[63] J.A. Barker and R.O. Watts, Mol. Phys., 26, 789, (1973).

[64] W.F. Van Gunsteren, H.J.C. Berendsen and J.A.C. Rullman, Faraday Disc., 66, 58, (1978).

[65] S.W. de Leeuw, J.W. Perram and E.R. Smith, Proc. Roy. Soc. A373, 27, (1980); ibid A373, 57, (1980).

[66] E.R. Smith and J.W. Perram, Mol. Phys., 30, 31, (1975).

[67] A.J.C. Ladd, Mol. Phys., 33, 1039, (1977).

[68] D. Heyes, CCP5 Newsletter, 8, 29, (1983); J. Chem. Phys., 74, 1924, (1981).

[69] B.U. Felderhof, Physica, 101A, 275, (1980).

[70] CCP5 Program Library, SERC, Darsbury Laboratory.

[71] F.J. Harris, Proc. I.E.E.E., 66, 51, (1978).

[72] P.H. Berens and K.R. Wilson, J. Chem. Phys., 74, 4872, (1981).

[73] S.F. Gull and G.J. Daniell, Nature, 272, 686, (1978).

[74] R. Zwanzig and N.K. Ailawadi, Phys. Rev., 182, 280, (1969).

[75] D. Frenkel, Intermolecular Spectroscopy and Dynamical

Properties of Condensed Systems, Proceedings of the International School of Physics, 'Enrico Fermi', (ed. J. van Kranendonk, North Holland), pp. 156-198, (1980).

[76] C.S. Murthy and K. Singer, private communication.

[77] M.W. Evans, Adv. in Mol. Relax, 24, 123, (1982).

[78] T.I. Cox, M.R. Battaglia and P.A. Madden, Mol. Phys., 38, 1539, (1979).

[79] V.K. Agarwal, G.J. Evans and M.W. Evans, J. Chem. Soc. Faraday Trans. 2, 79, 137, (1983).

[80] H.J. Böhm, I.R. McDonald and P.A. Madden, private communication.

[81] G.J. Evans and M.W. Evans, J. Chem. Soc. Faraday Trans. 2, 79, 153, (1983).

[82] W.A. Steele, Adv. Chem. Phys., 34, 1, (1976).

[83] W.A. Steele and W.B. Streett, Mol. Phys., 39, 279, (1980).

[84] M.W. Evans, private communication.

[85] M.W. Evans and M. Ferrario, Adv. in Mol. Relax, 24, 75, (1982).

[86] K. Singer, J.V.L. Singer and A.J. Taylor, Mol. Phys., 37, 1239, (1979).

[87] R.W. Impey, P.A. Madden and D.J. Tildesley, Mol. Phys., 44, 1319, (1981).

[88] C. Brot, Dielectric and Related Processes, 2, (A Specialist Periodical Report), ed. M. Davies, (Chem. Soc.), p.1., (1975).

[89] M. Ferrario and M.W. Evans, Adv. in Mol. Relax., 22, 245, (1982).

[90] D. Frenkel and J.P. McTague, J. Chem. Phys., 72, 2801, (1980).

[91] B.M. Ladanyi, J. Chem. Phys., 78 2189, (1983).

[92] B.J. Alder, J.C. Beers, H.L. Strauss and J.J. Weiss, J. Chem. Phys., 70, 4091, (1979).

[93] P.A. Madden and D.J. Tildesley, Mol. Phys., 49, 193, (1983).

SURVEY OF FUTURE DEVELOPMENTS IN MOLECULAR DYNAMICS

H.J.C. Berendsen

University of Groningen, the Netherlands

Molecular Dynamics will see important developments in the next decade in two directions: Methods and Instrumentation. The combination of improvements in both aspects will probably lead to applications to realistic complex molecular systems over time scales many orders of magnitude longer than are presently attainable. Accurate simulations of complex molecular liquids into the nanosecond range will become commonplace. Applications may include catalysis and behaviour of biological macromolecules, leading to catalyst and drug design.

Methods will improve for complex systems in the sense that not all details have to be simulated on a short time scale. Accurate potentials of mean force, in combination with stochastic methods, will enable the reliable neglect of unimportant degrees of freedom in a complex system. Thus the system size is reduced such that much longer time scales can be studied. In addition, new methods will be worked out for slow events that involve activated processes and barrier-crossing events.

Instrumentation for MD is the computer. The spectacular development of MD in less than 15 years - from non-existence to applications to complex molecular systems containing 10,000 degrees of freedom now - has been possible only because of the spectacular development of computer power. Expressed in floating point multiply operations per second, computer power has constantly increased by a factor of ten every 5 or 6 years since 1950 (see figure 1) and a levelling- off is not yet apparent. In figure 1 both the actual technology and the average university accessibility (with Groningen University as a typical example) are depicted. Universities lag some 5 years behind oil explorators, weather forecasters, and the military.

A. J. Barnes et al. (eds.), Molecular Liquids - Dynamics and Interactions, 561–564.

Projections for computer technology for the next decade are as follows (2):

Supercomputers:		
	CRAY 1	180 Mflop
	CRAY X-MP (1983)	360 Mflop
	CYBER 205 2-pipe	200 Mflop
	4-pipe	400 Mflop
	FUJITSU VP200 (1983)	500 Mflop
	HITACHI S810 (1983/84)	630 Mflop
	CYBER 2XX (1987)	8 Gflop
	NEC X2 (1984/85)	1,3 Gflop
	Japanese Project (1986/87)	10 Gflop

Hardware: Both Josephson-junction (liquid helium temperatures) and Gallium-arsenide technologies are being developed, the latter looking more promising. Josephson junction: less than 10 ps/gate, GaAs less than 30 ps/gate with densities around 3000 per chip. Widespread use expected around 1990. By then memories will be available with 10 ns access, 1 Gbyte capacity and 15 Gbyte/s transfer rate.

Architecture: There are three main types:
SISD: single instruction, single data stream
SIMD: single instruction, multiple data stream
MIMD: multiple instruction, multiple data stream.
Present day conventional computers are SISD, but the SISD concept is not extendable to very powerful systems because speed alone will ultimately be limited by signal propagation rates and packing density of gates. Future supercomputers will all, in one way or another, make use of parallelism. The SIMD computers (of which the ICL Distributed Array Processor is an example) are likely to provide high processing power for certain classes of problems that inherently possess a large potential for parallelism, such as image processing and possibly molecular dynamics. For general purpose machines the MIMD concept is more promising. For the nearest future it seems that the use of just a few complicated pipe-line processors is being preferred by the main supercomputer manufacturers (CRAY, CDC, FUJITSU). For the further future real MIMD computers using hundreds or thousands of processors can become increasingly promising, but only if concurrent software problems can be solved, dealing with subdivision of tasks, data stream organisation and synchronisation. The Denelco HEP computer is a start on the road of commercial MIMD machines.

Finally, for molecular dynamics production runs on complex systems, it seems already clear now that the most cost-effective machines will be dedicated computers with MIMD architecture, but built up from special purpose processors that are geared to specific tasks and constructed for optimal cost/performance ratio. The processors will be micro-programmable to retain as much flexibility as possible, but still the algorithms will be built into the machine. Such machines may become generally available when design and construction of algorithm-oriented processors will become fully automated. The Delft Molecular Dynamics Processor isthe first and at present only example of a special purpose MD machine (3,4). This machine handles a fluid of 16,000 particles with CRAY speed for a hardware cost of $40,000. The further future may see MD processors on VLSI chips.

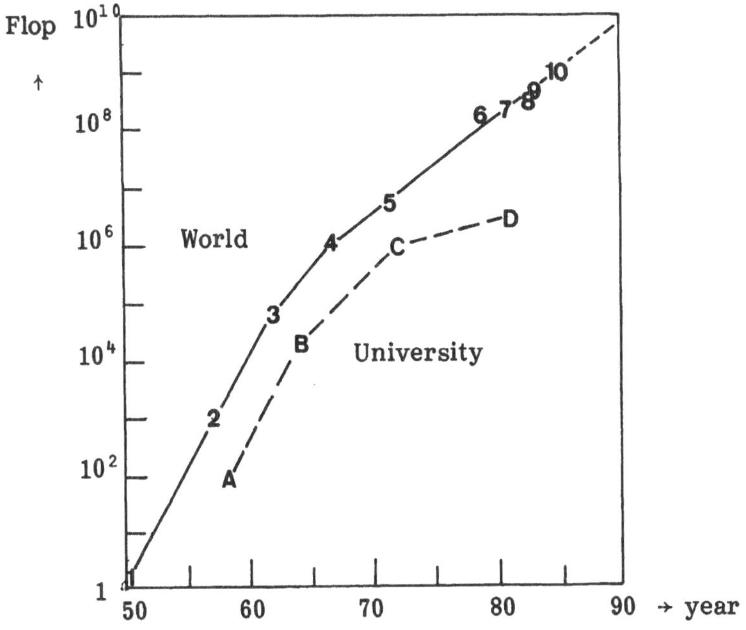

Fig. 1. Number of floating point multiply operations of new computers versus their first year of availability numbers, drawn line); same for computer available at the University of Groningen (letters, broken line) (ref. 1)

1: Frieden (electromechanical), 2: Deuce (valve), 3: IBM 7090 (transistor), 4: CDC 6600 (integrated circuit), 5: IBM 360/195; CDC 7600, 6: CRAY 1, 7: CYBER 205, 8: CRAY X-MP, 9: FUJITSU VP 200; HITACHI S810, 10: NEC X2 (projected), 11: CYBER 2XX (projected).

A: ZEBRA, B: Telefunken TR 4, C: CYBER 74, D: CYBER 170/760.

References

(1) Data partially from R.W. Hockney, Reading
(2) partly based on NATO Workshop on "High-Speed Computation", Jülich, Germany, June 1983 (to be published by Springer Verlag, J.S. Kowalski, editor)
(3) A.F. Bakker, C. Bruin, F. van Dieren and H.J. Hilhorst, Phys. Lett. 93A (1982) 67
(4) A.F. Bakker, Thesis, Technical University Delft, Sept. 1983.

LIST OF POSTERS PRESENTED AT THE NATO, ASI ON MOLECULAR LIQUIDS, FLORENCE, JUNE 26 - JULY 8th, 1983.

A. POSTERS ON 'SIMPLE' MOLECULAR LIQUIDS.

A1. Statistical-mechanical computations of the second virial coefficient and pair-correlation-coefficients of liquid N_2, O_2, F_2, CO_2 using a spherical-harmonic-expansion.
M. AssFalg and M. Zeidler, Institut Fur Physikalische Chemie, RWTH Aachen, Templergraben 59, 5100 Aachen, W. Germany.

A2. Saturation Properties: Model and Real Fluids.
J. Fischer and R. Lustig, Institut für Thermo-und Fluiddynamik, Ruhr-Universität Bochum, Postfach 10 21 48, D-4630 Bochum, W. Germany.

A3. The Thermodynamic Properties of Liquid Mixtures of Carbon Monoxide with Methane.
Jorge C.G. Calado, Henrique J.R. Guedes, Manuel Nunes Da Ponte and William B. Streett, Centro De Quimica Estrutural, Complexo I, I.S.T., Av. Rovisco Pais, 1096 Lisboa Codex, Portugal.

A4. Isotropic Raman Spectra of pH_2 and HD-Argon Mixtures in Liquid and Hypercritical Fluid.
M. Echargui, F. Marsault-Herail, G. Levi and J.P. Marsault, Laboratoire de Spectrochimie Moleculaire, Bat F, Universites (Pierre et Marie-Curie - Paris 7), 4 place Jussieu 75005, Paris, France, and
J. Bonamy and D. Robert, Laboratoire de physique Moleculaire - ERA - Faculte des Sciences, 25030 Besancon Cedex, France.

A5. Raman Spectrum of Rare Gas Bound Dimers and Collisional Pairs.
Y. Le Duff and R. Ouillon, Department des Recherches Physiques, Universite P. et M. Curie, Tour 22, 4 Place Jussieu, 75005 Paris, France, and
V. Chandrasekharan, Laboratoire des Interactions Moleculaires et des Hautes Pressions, Centre Universitaire Paris-Nord, Avenue Jean-Baptiste Clement, 93430 Ville-Aneuse, France.

A6. Infrared Study of N_2O mixed with Argon at Variable Temperature.
D. Balou, N. Brigot, J. Cartigny and C. DreyFus, Laboratoire de Recherches Physiques, Universite P. et M. Curie, 4 Place Jussieu 75230, Paris Cedex 05, France.

A7. Non-Exponential Decay of the Fluorescence from the $^1\Delta g$ State in Liquid Oxygen at High Excitation Intensities.
H. Klingshirn and M. Maier, NWF II - Physik, Universitat Regensburg, Postfach, D-8400 Regensburg, W. Germany.

A8. Dielectric Polarization of Non-polar Molecules at Low and High Densities.
W. Schroer, Dept. of Chemistry, University Bremen, Bremen, W. Germany.

A9. Collision Induced Light Scattering from the CH_4-CH_4, CH_4-Ar and CH_4-Xe pairs.
A.R. Penner, N. Meinander and G.C. Tabisz, University of Manitoba, Dept. of Physics, Winnipeg, Manitoba, Canada R3T 2NZ.

A10. The Inter-molecular Potential for SF_6.
J.G. Powles, J.C. Dore, M. Deraman and E.K. Osae, Physics Laboratory, University of Kent, Canterbury CT2 7NR, England.

A11. Rotation-Vibration Correlation in Infrared and Anisotropic Spectra of diluted Van der Waals Solutions. Effect of Intermolecular Forces.
G. Tarjus, S. Bratos, Laboratoire de Physique Theorique des Liquides, Universite Pierre et Marie Curie, 4 Place Jussieu, 75005, Paris, France.

A12. PVT and Dielectric Permittivity Data for Liquid Dichloromethane from 303 K to 423 K and up to 200 MPa.
R. Diguet, Universite de Nancy 1, Laboratoire de Chimie Theorique, B.P. 239 - 54506 Vandoeuvre-Nancy, Cedex, France, and
R. Deul and E.U. Franck, Institut für Physikalische Chemie und Elektrochemie, Lehrstuhl 1 der Universität Karlsruhe, Kaiserstrabe 12, 7500 Karlsruhe 1.

A13. Neutron Scattering Measurements on Methylene Chloride,
G. Jung and M.D. Zeidler, Inst. of Phys. Chemie, RWTH Aachen, Templergraben 59, D5100 Aachen, West Germany.

A14. The Structure of Liquid Cl_2 at 200°K and 290°K by Neutron Diffraction.
P. Bosi, F. Cilloco, F.P. Ricci and F. Succhetti, Istituto di Fisica G. Marconi, Universite di Rome - p.A. Moro 2 - 00185, Rome, Italy.

A15. Intermolecular Bonding or Lone Pairs in Solid Chlorine?
S.L. Price and A.J. Stone, University Chemical Laboratory, Lensfield Road, Cambridge, CB2 1EW, England.

A16. Intermolecular Potentials for CH_3F and CH_3Cl.
H.J. Böhm and R. Ahlrichs, Institut für Physikalische Chemie und Theoretische Chemie, Universität Karlsruhe, 75, Karlsruhe, West Germany.

A17. Thermodynamic Properties of Liquid Sulphur Dioxide Obtained by Molecular Dynamics Simulation.
Yves Guissani and Franjo Sokolic, Laboratoire de Physique Theorique des Liquides, Université Pierre et Marie Curie, Tour 16, 4, Place Jussieu, 75230 Paris, Cedex 05, France.

A18. The Memory-Function-Modeled Self-Diffusion Coefficient and Velocity Autocorrelation Function for Stockmayer Fluids of Linear Rotors.
Anthony G. St. Pierre, OSB, St. Vincent Archabbey, Latrobe, PA 15650, U.S.A., and
William A. Steele, Dept. of Chemistry, Pennsylvania State Univ., University Park, PA 16802, U.S.A.

A19. A Molecular Dynamics Study of Vibrational Dephasing in Liquid HCl.
S.F. O'Shea, Department of Chemistry, University of Lethbridge, Lethbridge, Alberta, Canada, T1K 3M4.

A20. Reorientation of Methyl Cyanide in a Computer Simulation.
H.J. Böhm, Institüt fur Physikalische Chemie und Elektrochemie, Universität Karlsruhe, 75 Karlsruhe, West Germany, and
R.M. Lynden-Bell, Department of Theoretical Chemistry, Lensfield Road, Cambridge, CB2 1EW, England, and
P.A. Madden, Royal Signals and Radar Establishment, Malvern, Worcs, WR14 3PS, England, and
I.R. McDonald, Department of Physical Chemistry, Lensfield Road, Cambridge, CB2 1EP.

A21. Far-infrared and Microwave Spectroscopic Studies of the Dynamics and Interactions of Acetonitrile in the Liquid and in Dilute Solution.
K.A. Arnold and J. Yarwood, Department of Chemistry, University of Durham, South Road, Durham, DH1 3LE, England, and
J.R. Birch, Division of Electrical Sciences, National Physical Laboratory, Teddington, Middx., TW11 0LW, England, and
A.H. Price, Edward Davies Chemical Laboratories, U.C.W., Aberystwyth, Dyfed, SY23 1NE, Wales.

A22. Contributions of Self and Distinct Pair Vibrational Correlation Function to Isotropic and Anisotropic Raman Spectra of the $\nu_2(A_1)$ Bands of Pure CD_3I and CH_3I.
J.P. Pinan, G. Tarjus, J. Loisel, Laboratoire de Recherche Physique, Université P. et M. Curie, Paris, France.

A23. Investigations of Local Concentrations in Binary Liquid Mixtures Containing CH_2Cl_2 and CH_3I by Analyzing Raman Band Shapes.
G. Döge, Institut für Physikalische Chemie, Hans-Sommer-Straße 10, D3300 Braunschweig, West Germany.

A24. Far-infrared, Raman and Microwave Spectroscopic Studies on the Molecular Dynamics of Methyl Iodide in the Liquid and in Solution in Hydrocarbon Solvents.
G.P. O'Neill and J. Yarwood, Department of Chemistry, University of Durham, South Road, Durham, DH1 3LE, England,
J.R. Birch, Division of Electrical Sciences, National Physical Laboratory, Teddington, Middx., TW11 OLW, England, and
A.H. Price, Edward Davies Chemical Laboratories, U.C.W., Aberystwyth, Dyfed, SY23 1NE, Wales.

A25. An Experimental Verification of Incoherent Light Scattering,
W. Härtl and H. Versmold, Physikalische Chemie, University of Dortmund, Postfach 500 500, D-4600 Dortmund, W. Germany.

A26. Molecular Orientation in Liquid Diphenyl Ether as Probed by Depolarised Light Scattering.
G. Fytas, D. Lilge and Th. Dorfmüller, Univ. of Bielefeld, Fakultat fur Chemie, Universitätstrasse, D4800 Bielefeld, 1, West Germany.

B. POSTERS ON 'MORE COMPLICATED' SYSTEMS

B1. Dielectric Relaxation in Phospholipid Vesicle Aqueous Solution.
V. Uhlendorf, K.D. Goepel, U. Kaatze, R. Pottel, Drittes Physikalisches Institut, Universität Gottingen, Buergerstr.42, D-3400 Göttingen, F.R. Germany.

B2. Structural Studies of Super-Cooled Water by Neutron Diffraction.
J.C. Dore and D.C. Steytler, Dept. of Physics, University of Kent, Canterbury, Kent, CT2 7NR, England, and
J. Teixeira and L. Bosio, ESPCI, Paris, France.

B3. A Small Angle Neutron Scattering Study of AOT Microemulsions.
J.C. Dore, B.H. Robinson, C. Toprakcioglu and A. Howe, Department of Physics, University of Kent, Canterbury, Kent, CT2 7NR, England.

B4. Hydrogen-Bonds in Associated Liquids - Conformational Change or Proton Delocalisation?
J.C. Dore, D.C. Steytler, Department of Physics, University of Kent, Canterbury, Kent, CT2 7NR, England, and
D.G. Montague, Department of Physics, Willamet University, Oregon, U.S.A.

B5. Collision Induced Contributions in the Isotropic and Anisotropic Rayleigh Spectra of Liquid H_2S.
A. De Santis and M. Sampoli, Instituto di Fisica, Universita di Venezia, 30123, Venice, Italy.

B6. Low Temperature Dielectric Properties of w/o Microemulsions.
D. Senatra, Physics Dept., University of Florence, Largo E. Fermi 2 (Ascetri) 50125, Florence, Italy.

B7. Angle-Dependent Rotational Function in Anisotropic Liquids.
G. Moro and P.L. Nordio, Institute of Physical Chemistry, University of Padua, Italy.

B8. Polarized-Rayleigh-Brilloin Spectra of Liquids Composed of Anisotropic Molecules: p-Anisaldehyde, Aniline.
B.L. O'Steen, Cal. tech, Pasadena, U.S.A.,
G. Fytas, Univ. Bielefeld, W. Germany, and
C.H. Wang, University of Utah, Salt Lake City, U.S.A.

B9. Dielectric Relaxation Measurements with Respect to Association of N,N-disubstituted Amides.
M. Stockhausen, M. Kessler, E. Seitz† and H. Utzel, Institut für Physikalische Chemie, Universität Munster and † Institut für Physik, Universität Mainz, F.R. Germany.

B10. Adsorption of Metal Ions from Water Solutions onto Surface Modified Silicas.
U. Köklü, Department of Chemistry, I.T.U.Fen-Edebiyat Fakultesi, Anorganik Kimya AnaBiliu Dali, Maslak, Istanbul, Turkey.

B11. Vibrational spectroscopic studies on the Hydrogen Bonding and Aggregation in Surfactant Solutions.
J. Yarwood and W.F. Pacynko, Department of Chemistry, University of Durham, Durham City, DH1 3LE, England, and
G.J.T. Tiddy, Unilever Ltd., Port Sunlight Research Laboratories, Bebington, Wirral, Merseyside, LG3 3JW.

B12. Molecular Dynamics Simulation of a Bilayer Membrane.
P. van der Ploeg and H.J.C. Berendsen, Laboratory of Physical Chemistry, University of Groningen, Nyenborgh 16, 9747 AG Groningen, The Netherlands.

B13. Computer Simulations involving a Polarisable Model for Water.
J.E. Quinn, Crystallography Department, Birkbeck College, Malet Street, London, WC1E 7HX.

B14. Non-linear Phenomena in Liquids Explored via a Theoretical Approach of Langevin Type.
P. Grigolini and F. Marchesoni, Dipartimento di Fisica, Piazza Torricelli 2, 56100, Pisa, Italy.

B15. Toward a Comprehensive Picture of the Molecular Dynamics of Water: Experimental Results.
D. Bertolini, M. Cassettari and P. Salvetti, Istituto di Fisica Atomica e Molecolare, Via del Giardino 6, 56100 Pisa, Italy.

B16. Solute-Solute Potential of Mean Force for Apolar Solutes in Water.
A. Tani, Istituto di Chimica Fisica, Universita di Pisa, 56100 Pisa, Italy.

B17. Non-Markoffian, non-linear Properties of a two dimensional Liquid System: Theory and Computer Simulation.
M. Ferrario, P. Grigolini and A. Tani,
Istituto di Fisica Atomica e Molecolare del CNR, via del Giardino 6, 56100 Pisa, Italy.

B18. Toward a Comprehensive Picture of the Molecular Dynamics of Water: Theoretical Results.
D. Bertolini, M. Cassettari, M. Ferrario, P. Grigolini and P. Salvetti, Istituto di Fisica Atomica e Molecolare del CNR, via del Giardino 6, 56100 Pisa, Italy.

INDEX